STATISTICS FOR
APPLIED ECONOMICS
AND BUSINESS

RICHARD L. MILLS

Whittemore School of Business and Economics
University of New Hampshire

STATISTICS FOR APPLIED ECONOMICS AND BUSINESS

McGRAW-HILL BOOK COMPANY

New York   St. Louis   San Francisco   Auckland   Bogotá   Düsseldorf
Johannesburg   London   Madrid   Mexico   Montreal   New Delhi
Panama   Paris   São Paulo   Singapore   Sydney   Tokyo   Toronto

STATISTICS FOR APPLIED ECONOMICS AND BUSINESS

34567890   DODO   78321098

This book was set in Times New Roman. The editors were J. S. Dietrich, Claudia A. Hepburn, and Annette Hall; the designer was Judith Michael; the production supervisor was Leroy A. Young. The drawings were done by Oxford Illustrators Limited.
R. R. Donnelley & Sons Company was printer and binder.

Library of Congress Cataloging in Publication Data

Mills, Richard L   date
    Statistics for applied economics and business.

    Bibliography: p.
    Includes index.
    1. Statistics. I. Title.
HA29.M635   519.5   76-20590
ISBN 0-07-042372-5

TO DOROTHEA F. AND
THE MEMORY OF LEWIS R.

CONTENTS

APPENDICES

PREFACE

What motivates an individual to introduce one more textbook in introductory statistics for business and economics? The potential increase in the author's income was a factor. But, let us analyze that potential. The movie rights are worthless. Textbook writing, like gold mining, has the characteristic: "You have a very small chance to strike it rich, and a very large chance that you cannot earn a subsistence wage in the endeavor." Thus, there must be some reason why I thought that the approach represented in this text would be a more effective learning device than the current multitude of competitors.

The author's income will directly depend upon that "reason" which is far more relevant to instructors and students anyway. My goals for this text are directly derived from my teaching environment and experience.

The author, as an instructor, is continually faced with a group of students (a) who have diverse educational backgrounds and (b) who have equally diverse future objectives. With respect to statistical methods, the common feature of those objectives is that each student could benefit from an *understanding* of the concept of statistical inference, a proposition that is the central theme of the first half of the text. The key problem is: "How do you develop an understanding of statistical inference when (a) the students are not all starting from the same place and (b) each student's interest in extending the concept of statistical inference may be different?"

Starting with the students' future objectives, I have used an imperfect classification scheme; I have attempted to distinguish between the student of business and the student of economics. Some subdisciplines within a business curriculum may have more in common with economics (e.g., finance) than others, but I would argue that the investigative tools of the economist (as a model builder and as a social investigator) have diverged from the tools required by a manager (especially a manager in the role of an action, on-line decision maker). Statistical inference will be critical to both interests; however, in introducing the concept of statistical inference, we should keep in mind the variety of ways in which we hope to incorporate that concept in developing more specialized statistical techniques. Thus, one purpose of this text is to develop the concept of statistical inference in such a manner that it can be extended to the interests of the economics student as well as the business student.

Next, consider the diverse educational background represented by potential readers of this text. The phrase "elementary algebra is all that is necessary" is suggestive but not very precise. If I further say that I place more emphasis on the word "elementary" than on the word "algebra," the required level of mathematical skill is still somewhat in doubt. In order to clarify the point, I assume that the reader understands such concepts and relations as: (1) the square root of a real number; (2) if $a + b = c$, then $a = -b + c$; (3) $(a - b)^2 = a^2 - 2ab + b^2$; and (4) if $a = b/c$, then $b = ac$. Essentially, all other concepts and relations are reviewed as we need them.

The preceding speaks to the minimum level of mathematical understanding, but what about the reader who exceeds that minimum? For those students whose investment in human capital has included more mathematics than the aforementioned minimum, the footnotes and an occasional "starred section" will give them a hint as to how calculus can assist in understanding the concepts presented. Some of the footnotes in Chapter 10 and some sections of Chapter 11 (all of which can be omitted) are enriched by knowledge of matrix algebra. The use of footnotes and starred sections is intended to provide an alternative approach for those equipped to use that information.

In summary, this text is intended to acknowledge the needs of two minority groups. First, neither economics nor business students, assuming their paths will eventually diverge, should be discriminated against. Second, with the exception of elementary algebra, no student should be prevented from understanding statistical inference because of a lack of mathematical background. Acknowledging the needs of minorities does not necessarily require ignoring the concerns of the majority. It is within that delicate balance that this text was attempted.

A cup of tea and a mixed drink have two things in common: (1) the final product is liquid and (2) there is an infinite array of proportions that their inputs can assume. Statistics books have similar properties, even within the goals I have outlined above. Although this text surveys several statistical techniques, it is not an encyclopedic presentation.

A few topics have been selected with the assumption that understanding those topics is more important than briefly treating a wide range of statistical techniques. I tried to implement that strategy with several stylistic techniques.

First, within the first several chapters, I have introduced each important concept in several different ways. The optimal path to St. Louis depends upon where your starting point is. Since each reader may start with a different background, my plan is that we all reach the middle of the text together. This repetitiveness may be bothersome if you understand the concept the *first* time I state it. While you should not ignore the cost of that repetitiveness, you should also recall the benefit of that prolixity when you did not initially understand the concept presented.

Second, this text frequently uses a "generalize the results of a specific experiment" approach to introduce a concept or statistical technique. Regardless of its value to statistical methods, this approach is a useful exercise for decision makers and social scientists.

Third, several statistical concepts can be presented (1) in words, (2) in mathematical expressions, and (3) in pictorial or graphic form. Frequently, it is very useful to be able to translate among the "equation description," the "word description," and the "pictorial description." In an attempt to expedite this translation, I have used a device which I call a "panel." A panel presents these "descriptions" of a concept side by side and usually on one page. In order to *understand* a statistical concept, the ability to form a "mental image" of the concept is assumed to be very important. However, in order to *apply* that concept, equations may be more efficient. To describe the results, the "word description" is important. Hence, the panel is also intended as an efficient reference, as the reader progresses through the text, to translate from one "description" of a concept to another.

Finally, the exercises at the end of each chapter were constructed for several purposes. The first several questions at the end of each chapter are designed to reinforce *individual* concepts and methods that are presented within the chapter. When number manipulation is required, these problems use "simple numbers" to minimize the arithmetic component of the exercise. For example, if a square root is required, the $\sqrt{4}$ is preferred to $\sqrt{3}$. The "higher-numbered" exercises tend to present situations that are more realistic. Further, they tend to (1) call for the use of more than one concept, (2) require an interpretation of what the numerical results indicate, and (3) question the assumptions used. Finally, the "case questions," which are numbered separately, refer to "real-world" data sets in the Appendices. Case questions can use computer assistance, but they do not require it. The student-reader can benefit from all these "types" of exercise. The "practice" exercises tend to reinforce concepts; the situational exercises help focus the strength and limitations of techniques; and the case questions offer exposure to the vulgarities of obtaining information from larger data sets.

To the student: In an introductory course in economics, finance, marketing, etc., you are subjected to certain hypotheses, models, and theories that highlight the discipline. To the extent that you are interested in those topics, you may pursue more advanced study. This introduction to statistics is a survey of a broader range of material.

The clause "it can be shown that" is frequently substituted for a mathematical proof. The emphasis is on understanding what a theorem, concept, or theory says. If that is accomplished, the importance of the mathematical proof, which can be found in more advanced treatments, is more appreciated.

Within this text each individual concept is relatively easy to understand. Semantics peculiar to statistics tend to cloud the issue. I suspect that you already understand the concept of an "average" of a set of numbers. Replacing the name tag "average" with the name tag "arithmetic mean" does not change the concept, and much of this book is concerned with the concept of an "average."

It is the synthesizing of a group of simple concepts into one statistical technique that requires some mental effort. Even that mental effort is more akin to reloading a stapler than it is to assembling an electronic computer.

One criticism of a statistics course is that it is boring. An important reason for that opinion is that the student loses interest before explicit treatment of statistical inference occurs. There is no easy solution to this problem if we assume, as this book does, that the reader has not yet been exposed to those concepts that are necessary for statistical inference. Few readers of this text will compare it to a novel; let us explore the analogy. One strategy for writing a novel is to simultaneously introduce several "characters" and let their interaction develop the plot. To get to statistical inference (the plot?), we have to introduce several "characters." To simultaneously introduce these "characters" would add confusion to the process; hence, the author has chosen to introduce these "characters" one at a time in the first several chapters. It is not until Chapters 5 and 6 that the plot, statistical inference, begins to take shape. As a concession to the student-reader, the number of characters (and, to some extent, how well we know them) has been kept to a minimum in the initial stages of the book. As the plot thickens, we have to introduce new characters (or concepts).

In summary, be patient. Statistical methods are understandable and powerful.

To the instructor: Consider the following hypothesis: Two decades ago, the optimal package of topics for the business student was equivalent to the optimal package of topics for the economics student. Within the last 20 years, the developments in operation research, marketing research, financial management, Bayesian methods, etc., have given a decision-making emphasis to the evolution of the optimal package of topics for the business student. At the same time, economics has placed relatively more emphasis on regression models for the social investigator. While the optimal set of topics for the two student groups may still overlap at the introductory level, the approaches and the goals have diverged.

On the assumption that this hypothesis has merit, I attempted to structure this text in a fashion that is amenable to both student groups. Although it covers Bayesian and nonparametric methods, the text is primarily an introduction to classical forms of statistical inference.

The first seven chapters are relatively standard in the sequence of topics (with the possibility of introducing descriptive statistics earlier according to pedagogical preference). Chapters 8 through 14 can be rearranged to fit the purpose of the course and the composition of the class. The *Instructor's Manual* offers more detail about the order of presentation.

No statistical concept herein discussed was originated by me. I have attempted only to disseminate knowledge, and I am grateful for those who originated and developed that knowledge. Finally, to the other members of our family, Kathy, Kylie, and Ricky, thank you for accepting the prolonged absence of the person who was in the next room.

Richard L. Mills

STATISTICS FOR
APPLIED ECONOMICS
AND BUSINESS

1

INTRODUCTION

Babe Ruth's lifetime batting average in professional baseball was 0.342.
There were 126 people on the airplane when it crashed.
Only 12,682 heads of beef were sold at the Kansas City stockyard this morning.

The police estimated that 60,000 attended the rally.
Twelve million people watched the live telecast.
One out of five American families changed residence during that 5-year period.

Tomorrow's high temperature will be between 75 and 78°F.
Real GNP is forecasted to increase between 1.5 and 2.0 percent during the next quarter.
About 6 million motorists will take to the highways during this holiday weekend.

The average time for a shipment to arrive from our Dallas office is between 16 and 20 hours.
The average life of these tires is 40,000 to 42,000 miles.
In our fair city, this year's median family income was between $10,600 and $10,900.

These twelve statements are paraphrased comments from the news media. A number is included in each of these statements. How are the uses of these numbers similar?

The first three statements imply that a complete enumeration was performed. All relevant items in the *population* of interest were counted. A *census* was conducted.

The second set of three statements implies that a *point estimate* was concocted. Obviously, a census or complete enumeration was not performed. Somehow, an estimate was derived, and that estimate is represented by a single number.

The third set of three statements refers to something that might happen in the future. It is not an *estimate*, because it has not yet happened. These three statements are *forecasts*, or *predictions*. All three may be the result of a very elaborate model of past behavior, but they use numbers in a context quite different from the previous six statements.

The fourth set of three statements implies some *interval estimate* in the use of numbers. They all present some range within which an average might fall. By presenting a range of possible values, it is doubtful that a complete enumeration was performed. But there is at least the implication that some data were used to present the interval estimate.

These twelve statements, of course, are not a complete representation of the statements, which include numbers, which we encounter daily. However, they do represent distinctions which we will refine throughout this book.

Among other things, we will be concerned about several questions: How do you present the results of a complete enumeration? How do you make an estimate, be it a point or an interval estimate? How do you interpret the estimates made by other people? How do you prepare forecasts? What do they mean?

Statistic and Statistics

The previous paragraphs never used the word "statistics"; certainly the twelve statements included statistics. The words "statistic" and "statistics" both have more than one meaning.

For example, *statistic* can mean "a single number or datum." If there are 500 pages in a book, then "500 pages" is a statistic. This is the most common interpretation of the word "statistic."

However, we will be concerned with a second definition for the word "statistic." A *statistic* is "some measure of the characteristics of a sample." Assume a class has 50 students; those 50 students comprise a population. The average weight of those 50 students is a parameter (i.e., a measure of the characteristics of a population). Assume that we randomly select 6 of those 50 students. The average weight of those 6 students is a statistic; it is a measure of the characteristics of a sample. What is important to recognize is that the second definition of "statistic" refers to the concept "average weight," and not to whatever numerical value is measured. The first definition allows the numerical value to be called "a statistic." For our purposes, a statistic refers to some measure of the characteristics of a sample.

As a plural noun, *statistics* refers to "a collection of numbers." It is the plural for the first definition of statistic.

The topic of this book refers to a second definition of statistics. *Statistics* refers to the science, or body of knowledge, which is concerned with the collection, presentation, analysis, and interpretation of numerical data. In this context, "statistics" is a singular noun.

Why Numerical Data?

Business can be categorized in several ways. Examples include profit versus nonprofit, wholesale versus retail, manufacturing versus service. Regardless of the category and regardless of the size of the organization, management requires information for decision making and control. In many situations, that information will take the form of numerical data. That simple logic offers a powerful argument for managers or potential managers to investigate statistical methods.

Economics, as a science, must find some relationship between empirical evidence and theory. Data can suggest a hypothesis or a theory. However we attempt to generalize from data, we will need some assumptions or theories. Empirical observation and theory are interwoven. Understanding statistical methods aids in understanding, extending, revising, and applying economic theory.

Numerical data are the result of measurement, and measurement is required for both business and economics.

1.1   MEASUREMENT SCALES

Each number is a measurement or designation of a particular characteristic or property of that which we are observing. For example, we use a variety of writing instruments in the period of a week. You may have a pencil, a fountain pen, a ball-point pen, and a felt-tip pen in your possession at this very moment. What measurements could we perform on these writing instruments? A partial list might include:

1.   Our writing instruments could be categorized as follows:

Each pencil is assigned the number 0.
Each fountain pen is assigned a 1.
Each ball-point pen is assigned a 2.
Each felt-tip pen is assigned a 3.

Then our collection of writing instruments could be described as 0, 0, 0, 0, 1, 2, 3, 3; that is, we have four pencils, one fountain pen, one ball-point pen, and two felt-tip pens.

2.   We could rank-order our preference for using each writing instrument:

Black felt-tip pen	1	→	Best
Blue ball-point pen	2	→	Next best
Longest no. 2 pencil	3	→	.
Next longest no. 2 pencil	4	→	.
Shortest no. 2 pencil	5	→	.
No. 3 pencil	6	→	.
Red felt-tip pen	7	→	.
Fountain pen	8	→	Least preferred

3,   We could measure the minimum temperature at which each writing instrument will function.

4.   We could weigh each writing instrument.

Each of these four examples represents the use of a different *measurement scale*. Obviously, they each measure a different characteristic of our collection of writing instruments. But more important, the *use* of numbers is different in each example.

In order to discern this difference in measurement scale, we ask three questions:

1. Does the sequence, or order, of the numbers have meaning?
2. Does the distance between two numbers have meaning?
3. Is the origin unique in interpretation?

The answers to these three questions determines what type of measurement scale is being used. Table 1.1 summarizes the possibilities.

The Nominal Scale

The nominal scale simply uses numbers as "name tags." In our first example, 0 was equivalent to "pencil." The Library of Congress catalog card number that appears in most books and Social Security numbers are uses of the nominal scale.

If one person has the Social Security number 314-42-6175 and another person is blessed with 314-42-6176, the second is not "greater" (by "1") than the first. The first person may not be younger or shorter, may not weigh less, or may even have received the number first. Considering all nine digits as "a number," the *order* of two numbers and the *distance* between two numbers do not have any special meaning. The origin is arbitrarily chosen.

When two things have two different numbers from a nominal scale, all we can say is: "Those two things are different." We shall make limited use of the nominal scale. In investigating the number of defects in a batch of parts, we may let a 1 represent a defective part and a 0 represent a nondefective part. A "subscript," which is discussed in the last section of this chapter, is another use of a nominal scale.

The advent of computers has increased the use of the nominal scale; names are frequently coded as Social Security numbers. Occasionally, this practice has been called dehumanizing. Yet some individuals, in the name of privacy, preferred the use of an "account number" rather than a "personal name" bank account long before the computer came onto the scene. The primary use of the nominal scale in this text is for the purpose of a shorthand notation.

TABLE 1.1 MEASUREMENT SCALES

Type of Scale	Does *Order* Have Meaning?	Does *Distance between* Numbers Have Meaning?	Does the *Origin* Have Unique Meaning?
Nominal	No	No	No
Ordinal	Yes	No	No
Interval	Yes	Yes	No
Ratio	Yes	Yes	Yes

The Ordinal Scale

By rank-ordering our preference for using each writing instrument, an ordinal scale was used to measure preference. The *sequence* in which the numbers appear has meaning. A lower number designates a more preferred writing instrument. We could have used $-10, -5, 0, +5, +10, +15, +20, +25$ to order preference. Obviously, the origin is arbitrarily chosen.

The *distance* between two numbers has no quantitative meaning with an ordinal scale. Although $3 - 2 = 1$ and $2 - 1 = 1$, that does not mean that the *difference* in satisfaction between the black felt-tip pen and the blue ball-point pen *is the same* as the *difference* in satisfaction between the blue ball-point pen and the longest no. 2 pencil. By using an ordinal scale, we can determine that one writing instrument is preferred to another, but we cannot determine by how much it is preferred.

Judges of beauty contests, spelling bees, and dog shows rank-order contestants. The "seeding" of tennis players and the grading of meat are measurements that use an ordinal scale.

For example, consider the annual sales of seven sales representatives shown in Table 1.2. Two salespersons are in the high-sales category; three, in the medium-sales range; two in the low-sales categories. The ordinal measurement scale W could be used to designate the three categories.

We know that $X_1 > Y_3$, but we do not know $X_1 - Y_3$. We know that the sales for $W = 1$ is greater than the sales for $W = 2$, but we do not know the difference in sales.

Most of the discussion of treating numbers from an ordinal scale, specifically rank-orderings, will occur in Chapter 9.

The Interval Scale

The interval scale (and the ratio scale) are characterized by a *unit of measurement*. The best-known interval scale is the one used to measure the degree of hotness or coldness: a temperature scale.

TABLE 1.2   RANK-ORDERING SALES
REPRESENTATIVES

Dollar Value of Sales		
High Sales $W = 1$ ($)	Medium Sales $W = 2$ ($)	Low Sales $W = 3$ ($)
X_1	Y_1	Z_1
X_2	Y_2	Z_2
	Y_3	

Consider the Fahrenheit temperature scale. Larger numbers mean a hotter temperature; the order of the numbers has specific meaning. Not only can we say "It is warmer today," but we can say "It is 10°F warmer today." The distance between numbers has specific meaning. The origin, however, is arbitrarily chosen. Under specific atmospheric conditions, water freezes at 32° on the Fahrenheit scale and at 0° on the Celsius scale. Gabriel D. Fahrenheit chose the origin of the temperature scale, 0°F, as that temperature produced by equal weights of snow and salt.

Another interesting feature of the interval scale is that the proportionality of differences is the same for various scales. For example, use both the Fahrenheit and Celsius scales to measure the following:

	°F	°C
Water freezes	32	0
Well water	50	10
Water boils	212	100

Using the Fahrenheit scale, the ratio of differences is

$$\frac{212 - 50}{212 - 32} = \frac{162}{180} = 0.90$$

and employing the Celsius scale,

$$\frac{100 - 10}{100 - 0} = 0.90$$

The ratio of differences is preserved regardless of scale because one scale is a linear transformation of the other. That is,

$$°F = 32 + (1.8)(°C) \qquad \text{or} \qquad °C = \frac{-160 + 5 \cdot °F}{9}$$

Note that because the origin is arbitrarily chosen, you cannot say that 40° is "twice as hot as" 20°.

Calendar time is another example. We are in the 1970s according to the Gregorian calendar; year "zero" is presumably the year after the birth of Jesus Christ. With reference to the Gregorian calendar, the Jewish calendar has its origin at 3761 B.C. (the creation), and the Muslim calendar has its origin at 622 A.D. (the year of Hegira). The length of a "year" is different for each calendar, and all three have a "leap year" concept. Converting from one calendar to another is not easy.

Test scores may be on an interval scale. A "zero" on an examination seldom means that the testee knows nothing about the subject. It is not meaningful to say that a person who scored 80 percent knows "twice as much" as the person who scored 40 percent.

The Ratio Scale

A ratio scale has both a unit of measurement and a nonarbitrary origin. Weight, height, liquid measure, money income, and direct costs are examples of measurement with a ratio scale.

An inch may be defined as the distance between two scratches on a platinum bar in Washington, D.C., and a meter may be the distance between two points on a platinum-iridium bar near Paris, France, but 0 inch and 0 meter both represent the same length. The origin is meaningful and has the same meaning for both scales.

A ratio scale is a ratio of the quantity under consideration to some standard unit of measurement. This "standard" may be arbitrarily established (i.e., the distance between two scratches on a bar), but the relative number of "standard units" is not arbitrarily determined. Given the two scratches which define 1 inch, the relative length of a pencil can be stated in inches. We can now say:

1. "A 6-inch (or 15.24-centimeter) pencil is longer than a 3-inch (or 7.62-centimeter) pencil." Order is meaningful.
2. "A 6-inch (or 15.24-centimeter) pencil is 3 inches (or 7.62 centimeters) *longer* than a 3-inch (or 7.62-centimeter) pencil." We can say how much greater one number is than another.
3. "A 6-inch (or 15.24-centimeter) pencil is twice as long as a 3-inch (or 7.62-centimeter) pencil." The origin is not arbitrary.

We can convert from one ratio scale to another by the multiplication of some constant. For example,

$$\text{Number of centimeters} = (2.54)(\text{number of inches})$$

Monetary units (that is, United States dollars, pound sterling, pesos) are ratio scales, and exchange rates are "conversion constants."

Summary

Most of the statistical methods developed in this book are meaningful *only* if the numbers are a result of measurement on an interval or a ratio scale. When ordinal measurements are also acceptable for a statistical method, it will be noted.

"Garbage in, garbage out" is a cliché frequently applied to computers. The formulas that we will occasionally use have the same property. The formula does not know whether the number fed it is from a nominal, ordinal, interval, or ratio scale. You, the investigator, have to determine when a formula is applicable and when it is not.

1.2 THE INTERMEDIATE GOAL: STATISTICAL INFERENCE

Your personal goal may be one of the following: (1) being able to improve your decision-making ability in uncertain situations; (2) being able to take empirical evidence and generalize how a household, a firm, a hospital, or an economy functions; (3) being able to take a model or a theory and test it in a particular situation. If your personal goal is one of those three, or if your personal goal is dependent upon one of those three, *statistical inference* will be beneficial.

In general, statistical inference is a logic which permits one to make statements about the characteristic of a population when only partial or sample data are available. The intermediate goal, our short-run objective, is developing a logic that will permit you to produce a statistical inference.

A model, or generalization, of a firm is represented in Figure 1.1a. An analogous model for "production of statistical inference" is presented in Figure 1.1b.

FIGURE 1.1 (a) Model of a firm; (b) analogous model for statistical inference.

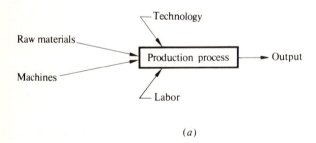

(a)

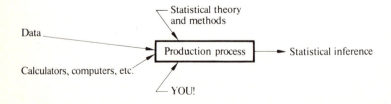

(b)

Our "raw material" is data, and we will have to know how to prepare it for use in the production process (e.g., Chapter 4). Our "capital equipment" consists of calculators, slide rules, computers, etc. Most of the time both the "production process" and the "raw materials" (i.e., numbers) will be kept simple enough that your labor can be substituted for the capital equipment. As in a firm, large-scale operations may call for the substitution of capital (e.g., a calculator) for labor if efficiency is a concern.

Our "technology," or "state-of-the-arts," component will generally be called "statistical theory and methods." The first several chapters are designed to develop the essential technology necessary to produce statistical inferences.

The last portion of the book is concerned with hybrid extensions of the general concept of statistical inference, and with some descriptive techniques that are rather unique to business and economics.

APPENDIX 1   SUMMATION NOTATION

What does an octagonal-shaped road sign mean to you? The octagon shape and the word "stop" are equivalent ways of *issuing an instruction*. Occasionally, you are provided with "additional information," such as the "four way" on the sign in Figure 1.2.

FIGURE 1.2   Stop sign.

We will make frequent use of a *symbol* to designate an *instruction*. The symbol is Σ, the uppercase Greek letter sigma. The instruction that sigma designates is "add up or sum that which follows." Occasionally, "$i = 1$" will appear below Σ and "n" will appear above Σ to provide additional information.

The *sum of the X values* from $i = 1$ to $i = n$ is defined as

$$\sum_{i=1}^{n} X_i = X_1 + X_2 + X_3 + \cdots + X_n \tag{1.1}$$

The *subscript i* can assume only sequential (positive) integer values: $i = 1, 2, 3, \ldots, n$. The subscript is a use of the nominal measurement scale. X_1 is only the first value of X that will be considered; X_1 is not necessarily either the smallest or the largest value of X.

For example, consider the five X values given here:

i		X_i
1	$(X_1 =)$	8
2	$(X_2 =)$	3
3	$(X_3 =)$	9
4	$(X_4 =)$	0
5	$(X_5 =)$	5

Using the definition of ΣX,

$$\sum_{i=1}^{5} X_i = X_1 + X_2 + X_3 + X_4 + X_5$$

$$= 8 + 3 + 9 + 0 + 5 = 25$$

and
$$\sum_{i=2}^{4} X_i = X_2 + X_3 + X_4 = 3 + 9 + 0 = 12$$

There are several properties of this "summation instruction" that will be useful throughout this book. First, note that if c is a constant, we have property 1.

Property 1 $$\sum_{i=1}^{n} c = nc$$ (1.2)

Why is this relationship valid? Start with the definition

$$\Sigma X_i = X_1 + X_2 + X_3 + \cdots + X_n$$

and let $X_i = c$ for all i. Then

$$\sum_{i=1}^{n} c = c_{(1)} + c_{(2)} + c_{(3)} + \cdots + c_{(n)} = nc$$

Adding c "n times" is the same as taking "n times c." For example, let $n = 5$ and $c = 2$:

$$\sum_{i=1}^{n} c = \sum_{i=1}^{5} (2) = 2 + 2 + 2 + 2 + 2 = 10$$

and
$$nc = (5)(2) = 10$$

If c is a constant, then we have property 2.

Property 2
$$\sum_{i=1}^{n} (cX_i) = c \sum_{i=1}^{n} X_i$$
(1.3)

Why? Let $Y_i = cX_i$. Then

$$\sum_{i=1}^{n} Y_i = Y_1 + Y_2 + Y_3 + \cdots + Y_n \qquad \text{by definition}$$

$$\sum_{i=1}^{n} (cX_i) = cX_1 + cX_2 + cX_3 + \cdots + cX_n \qquad \text{substitute } Y_i = cX_i$$

$$= c(X_1 + X_2 + X_3 + \cdots + X_n) \qquad \text{factor out } c$$

$$= c \sum_{i=1}^{n} X_i \qquad \text{by definition}$$

For example, let $c = 2$, $X_1 = 8$, $X_2 = 3$, and $X_3 = 9$.

$$\sum_{i=1}^{3} (2 \cdot X_i) = (2 \cdot 8) + (2 \cdot 3) + (2 \cdot 9)$$

$$= 16 + 6 + 18 = 40$$

and
$$2 \sum_{i=1}^{3} X_i = 2(8 + 3 + 9) = 2(20) = 40$$

Also, if c is a constant, then we have property 3.

Property 3
$$\sum_{i=1}^{n} (X_i + c) = \sum_{i=1}^{n} X_i + nc$$
(1.4)

Why? Because

$$\sum_{i=1}^{n} (X_i + c) = (X_1 + c) + (X_2 + c) + (X_3 + c) + \cdots + (X_n + c)$$

$$= (X_1 + X_2 + X_3 + \cdots + X_n) + (c_{(1)} + c_{(2)} + \cdots + c_{(n)})$$

$$= \Sigma X_i + nc$$

Or,
$$\Sigma(X_i + c) = \Sigma X_i + \Sigma c = \Sigma X_i + nc$$

For example, let $c = 2$, $X_1 = 8$, $X_2 = 3$, and $X_3 = 9$.

$$\sum_{i=1}^{3} (X_i + 2) = (8 + 2) + (3 + 2) + (9 + 2)$$

$$= 10 + 5 + 11 = 26$$

and

$$\sum_{i=1}^{3} X_i + 3(2) = (8 + 3 + 9) + 6 = 26$$

Note that if $c = -4$, then

$$\sum_{i=1}^{3} (X_i - 4) = (8 - 4) + (3 - 4) + (9 - 4) = 8$$

and

$$\sum_{i=1}^{3} X_i + 3(-4) = (8 + 3 + 9) - 12 = 8$$

It should be noted that, in general,

$$\Sigma X_i^{\,2} \neq (\Sigma X_i)^2$$

Why? Expand just the left side:

$$\Sigma X_i^{\,2} = X_1^{\,2} + X_2^{\,2} + X_3^{\,2} + \cdots + X_n^{\,2}$$

Now, consider the right side:

$$(\Sigma X_i)^2 = (X_1 + X_2 + X_3 + \cdots + X_n)^2$$

If it is not obvious that those two are not equal, then let $n = 2$:

$$\sum_{i=1}^{2} X_i^{\,2} = X_1^{\,2} + X_2^{\,2}$$

and

$$\left(\sum_{i=1}^{2} X_i \right)^2 = (X_1 + X_2)^2 = X_1^{\,2} + 2X_1 X_2 + X_2^{\,2}$$

Finally, if c is a constant, then we have property 4.

Property 4 $$\sum_{i=1}^{n} (X_i - c)^2 = \sum_{i=1}^{n} X_i^{\,2} - 2c \sum_{i=1}^{n} X_i + nc^2 \qquad \textbf{(1.5)}$$

Expanding $\Sigma(X_i - c)^2$ yields

$$(X_1 - c)^2 + (X_2 - c)^2 + \cdots + (X_n - c)^2$$
$$= (X_1{}^2 - 2cX_1 + c^2) + (X_2{}^2 - 2cX_2 + c^2) + \cdots + (X_n{}^2 - 2cX_n + c^2)$$
$$= (X_1{}^2 + X_2{}^2 + \cdots + X_n{}^2) - 2c(X_1 + X_2 + \cdots + X_n) + nc^2$$
$$= \Sigma X_i{}^2 - 2c\Sigma X_i + nc^2$$

For example, for $c = 2$, $X_1 = 8$, $X_2 = 3$, and $X_3 = 9$,

$$\sum_{i=1}^{3}(X_i - 2)^2 = (8 - 2)^2 + (3 - 2)^2 + (9 - 2)^2$$
$$= 36 + 1 + 49 = 86$$

and $\quad \displaystyle\sum_{i=1}^{3} X_i{}^2 - 2(2)\sum_{i=1}^{3} X_i + 3(2)^2 = (8^2 + 3^2 + 9^2) - 4(8 + 3 + 9) + 3(4)$

$$= (64 + 9 + 81) - 4(20) + 12$$
$$= 154 - 80 + 12 = 86$$

The reason for Equation 1.5 will become clear in Chapter 3. If we let $c = (\Sigma X)/n$, then $\Sigma(X_i - c) \equiv 0$ and $\Sigma(X_i - c)^2 = \Sigma X^2 - (\Sigma X)^2/n$.

EXERCISES

1. Which measurement scale is commonly used for each of the following?
a. Jersey numbers on a basketball team
b. Men's hat sizes
c. Water in a storage tank
d. Results of a horse race
e. Level of sound (a measure of noise pollution)
f. Elapsed time for the 100-meter foot race
g. Gasoline credit card number
h. Pencil lead hardness (i.e., no. 2, no. 3, etc.)

2. Which scale or scales could be used to measure:
a. Warehouse storage capacity
b. The "point difference" in football scores
c. Marital history as of a certain point in time (e.g., never married; divorced; single, mate deceased; married; etc.)
d. Market value of inventory
e. Time as measured before and after July 4, 1776
f. The difficulty of problems on an examination
g. Rail distances from a warehouse to retail outlets
h. The quality of 1-minute TV advertisements

3. For $i = 1, 2, 3$, expand (i.e., show operations without Σ) the following:
a. $a\Sigma(cX_i)$ where a and c are constants
b. $\Sigma(X_i + Y_i)$
c. $\Sigma(aX_i + cY_i)$ where a and c are constants
d. $\Sigma X_i - a$ where a is a constant

4. For $a = 2$, $c = 3$, $X_1 = 7$, $X_2 = 9$, and $X_3 = 8$, calculate the four expressions in exercise 3.

5. For $a = \frac{1}{2}$, $c = -2$, $X_1 = -1$, $X_2 = -2$, $X_3 = -3$, $Y_1 = 5$, $Y_2 = 6$, and $Y_3 = 7$, show that:
a. $a\Sigma(cX_i) = ac\Sigma X_i$
b. $\Sigma(X_i + Y_i) = \Sigma X_i + \Sigma Y_i$
c. $\Sigma(aX_i + cY_i) = a\Sigma X_i + c\Sigma Y_i$
d. $\Sigma(X_i + Y_i + c) = \Sigma X_i + \Sigma Y_i + nc$
e. $\Sigma(X_i - Y_i) = \Sigma X_i - \Sigma Y_i$
f. $\Sigma(aX_i)^2 = a^2\Sigma X_i^2$
g. $\Sigma(X_i - c)^2 = \Sigma X_i^2 - 2c\Sigma X_i + nc^2$

6. Show that $\Sigma i = \frac{1}{2}(n)(n + 1)$ for
a. $n = 3$
b. $n = 10$

Note: Selected solutions to these exercises appear in Appendix A.

CASE QUESTIONS

[At the end of the exercises for each chapter, you will find Case Questions. These questions refer to the data presented in Appendix B at the end of the book. For each state listed in Appendix B, there are many families, and for each family there is data on many family characteristics (e.g., size of the family, family income, occupation for the head of household, etc.). We shall call each characteristic a "variable." This data is real-world data collected during the 1970 U.S. Census of Housing and Population. By the time you have completed this book, you will have answered a large number of questions using this data. However, there are many other questions which may be of special interest to you. You are invited to glean whatever information you can, form whatever hypotheses occur to you, and reach whatever conclusions you develop from this data.]

1. For each variable (or family characteristic) listed in Appendix B, decide whether it is measured with a nominal, an ordinal, an interval, or a ratio scale. The "Key" in the Introduction to Appendix B will be very useful in this regard.

2

PROBABILITY AND SET THEORY: SOME BASIC ASPECTS

We are preparing to develop a theory of statistical inference. We would like to take a *sample* from a *population*, extract some information from that sample, and make statements about the population without observing every element within the population. Further, we would like to do this without developing extrasensory perception or becoming intimate with the proverbial crystal ball.

As you might suspect, statistical inference is constructed from a foundation of many facets. One portion of this foundation is a body of knowledge generally referred to as *probability theory*.

Many individuals would argue that statistical inference simply comes of an understanding of the intricacies of probability theory; the powers and limitations of statistical inference become obvious from a rather extensive knowledge of probability theory. There is considerable merit in that argument; however, it represents only one approach to statistical inference. Not everyone will be thrilled by statistical inference. For those who are thrilled, more probability theory is available in other sources.[1] For those who are not thrilled, and for the purpose of this text, we will develop those basic aspects of probability theory that will enable us to "taste" statistical inference.

Probabilities are numbers associated with the potential occurence of *events*. Set theory gives us a consistent way of keeping track of events; hence, set theory is a prelude to probability theory.

In summary, since our goal of statistical inference requires some exposure to probability theory, we will define several interpretations of probability, review set theory, and develop rules for calculating, manipulating, and interpreting probabilities.

2.1 DEFINITIONS OF PROBABILITIES, OR "DEALER'S CHOICE"

It should not be too surprising to find several definitions for the word "probability." Frequently we find both the common definition of a word, plus those technical definitions for particular disciplines.

---

[1] William Feller, *An Introduction to Probability Theory and Its Applications*, 3d ed., vol. 1, John Wiley & Sons, Inc., New York, 1968.

For example, the word "force" has a common interpretation plus a specific technical use in engineering and physics. The words "good," "growth," and "competition" have a common interpretation and a rather different meaning in economics. Thus, you may well ask "What particular definition of probability will we use?"

Surprisingly enough, we will use three different definitions of "probability." Before you conclude that this will just add confusion to confoundment, there are aspects of this polydefinitional situation which act as a saving grace.

First, the *rules of probability* are the same regardless of which definition is employed. Even if you do not yet know what those rules are, it should be a relief, if not "heartwarming," to know that you will not need a separate set of rules for each technical definition of probability.

Second, the appropriate definition of probability will depend upon the nature of the specific problem you are trying to solve. For many people, mathematics is synonymous with abstract. Therefore, it should be somewhat refreshing to discover that the definition of something as mathematical sounding as "probability" will depend upon *practical considerations!* Which definition you employ will depend upon which problem you consider. Which problem you choose is a matter of "dealer's choice."

2.1(a) "Objective" Definitions of Probability

It is always an extremely difficult proposition to identify the origin of an idea. The development of the mathematical theory of probability can be traced back to correspondence between two French mathematicians: Blaise Pascal and Pierre de Fermat.[2] It is curious that what was called "the mathematical theory of probability calculus" was developed and originated in letters between friends. Fermat seldom published anything; he just wrote to friends.

Around 1650, the Chevalier de Méré presented Pascal with several questions concerning games of chance. These questions, which are trivial by modern standards, caused Pascal to initiate the discussion of probability with Fermat. If you object to the use of gambling examples (i.e., dice throwing, etc.) in the following pages, please excuse three centuries of tradition.

The "classical" or a priori definition of probability   Consider some experiment or potential happening. As a result of the experiment or happening, there are several possible *events* that could occur. Choose one of the possible events and call it A; group all the other events and call them $A*$ (or "not-A").[3] The classical

---

[2] Pascal was variously known for his work in fluid mechanics and designated the inventor of "fine prose in France" by Saint-Beuve. Fermat was the father of the modern theory of numbers, conceived analytical geometry, and anticipated Newton's differential calculus.

[3] "$A*$ = the event not-A" is not standard notation. $\bar{A}$ is more common; however, we will use the notation $\bar{A}$ to indicate the arithmetic mean of the variable A. In order not to confuse the two concepts, we will use the asterisk.

definition of probability is based on the ratio of the number of relevant outcomes (i.e., when A occurs) to the number of possible outcomes.

More specifically,

1.   if there are m possible outcomes that would indicate the occurrence of event A,
2.   if there are k possible outcomes that would indicate the occurrence of event A^*, the nonoccurrence of A,
3.   if each possible outcome is equally likely, and
4.   if all outcomes are mutually exclusive, then the probability that event A will occur, $P(A)$, is defined as

$$P(A) = \frac{m}{m+k} = \frac{\text{number of outcomes that indicate } A}{\textit{total} \text{ number of possible outcomes}} \qquad (2.1)$$

Example 2.1   Single Toss of a Die

Assume our experiment is the single toss of a die.[4] This experiment would generate six possible outcomes; the top face of the die reads 1, 2, 3, 4, 5, or 6. Let us denote "the occurrence of an even number" as event A. (The nonoccurrence of A, or A^*, is "the occurrence of an odd number.")

In a single toss of this die, there are three possible outcomes (i.e., a 2, a 4, or a 6) that indicate the occurrence of event A; therefore, $m = 3$. There are three possible outcomes (i.e., a 1, a 3, or a 5) which would indicate the nonoccurrence of event A; thus, $k = 3$.

If we assume a "fair die," then each outcome is equally likely. The six possible outcomes are mutually exclusive because no two outcomes (e.g., a 1 and a 2, a 4 and a 3) could occur on a single toss of the die.

With all these conditions, the probability of getting an even number on a single toss of the die is

$$P(A) = \frac{m}{m+k} = \frac{3}{3+3} = \frac{3}{6} = \frac{1}{2}$$

Let us conduct a postmortem examination of example 2.1; reconsider every step in this example. "The single toss of a die" describes our *experiment*, or *hypothetical process*. The classical definition of probability requires an explicit experiment. The experiment generates possible *outcomes;* the hypothetical process tell us which specific outcomes can occur.

For example, what if we changed the question to: What is the probability of getting exactly one even number in two tosses of a die? Our experiment has changed;

---

[4] A *die* is a small solid cube marked on each side with from one to six dots. The dots are arranged so that opposite sides of the die always total seven. "Shooting dice" involves two such cubes which are thrown and allowed to come to rest in a random way on a flat surface.

each possible outcome is now a pair of numbers (e.g., a 2 on the first toss and a 3 on the second, a 2 on the first toss and a 4 on the second). *We must always know the explicit nature of the experiment!* Why? The denominator of Equation 2.1 is the total number of possible outcomes, which is determined from the explicit nature of the experiment.

Please note that we do not actually perform the experiment in order to list the possible outcomes. Instead, we hypothetically conduct the experiment to enumerate the possible outcomes. Since we can list the possible outcomes prior to physically performing the experiment, this definition of probability is sometimes called "a priori."

Let us reconsider the use of event *A* to denote the occurrence of an even number. An *event* is a collection of one or more (possible and/or impossible) outcomes. In example 2.1, event *A* was a collection of three possible outcomes, the occurrence of a 2, a 4, or a 6 as we stated it. What about the occurrence of an 8, a 10, a 12, and so on? Obviously they are impossible outcomes, given the nature of the experiment; however, they are included in our definition of event *A*.

From the single toss of a fair die there are only three *possible* outcomes that indicate event *A*. There are an infinite number of even integers which represent *impossible* outcomes from this experiment, but the classical definition of probability is only concerned with *possible* outcomes. Once again, it is the specific nature of the experiment that determines which outcomes are possible and which are impossible. For example, let event *B* be the occurrence of an 8. With the single toss of a fair coin, *m* is equal to 0 and *k* is equal to 6; hence, $P(B) = 0/(0 + 6) = 0$.

Now let us reconsider the enumeration of "possible outcomes" from the experiment. In example 2.1 did we list *all* possible outcomes? We said there were six possible outcomes, the six sides, from the single toss of a die. Have you ever tried to roll a die on a shag rug? How many times did it land on its edge?[5]

We could start by saying that we have not sufficiently specified the experiment. We could redefine the experiment to be a single toss of a die on a hard, level, flat, contained surface. That would take care of the shag rug, but is it not possible for the die to land on its edge on a hard, level, flat, contained surface? We hate to appear so "unscientific," but our intuition urges us to reply "Yes, it is possible, but it is very improbable." (You may want to dodge the word "improbable" and use "unlikely," but the effect is the same.) That is, in the actual performance of a physical experiment, you would say that it is possible but very unlikely that the die would land on its edge. [$P(\text{edge}) \simeq 0$, or $P(\text{edge})$ is essentially equal to 0, rather than $P(\text{edge}) = 0$.] In the hypothetical experiment of tossing a die, you simply define the occurrence of an edge

---

[5] On a cube, there are six "faces" (i.e., "surfaces" or "sides"), eight "corners," and twelve "edges." If the die lands on one face, then one face (as viewed from above) is showing, as are four corners and four edges. If the die lands on one of its edges, then only one edge (has the highest elevation and) is showing, as are two faces and two corners. If the die lands on a corner, then one corner, three faces, and three edges are showing

and the occurrence of a corner as impossible events. The hypothetical experiment is a *model* of the physical experiment, and in constructing the model we have made some simplifying assumptions.

These simplifying assumptions are not without some justification, if we are rolling the die on a flat, hard surface. First, we feel uncomfortable about increasing the denominator of Equation 2.1 to account for the edge and corner possibilities. Somehow, we feel the condition that "each possible outcome is equally likely" is violated when we include the edges and corners as possible outcomes. Finally, the "mutual exclusiveness" condition would be dubious; a 1 and a 2 could occur at the same time if the die landed on its edge.

A shag rug and our intuition have led to the distinction between (1) those outcomes which are impossible in both the physical and the hypothetical experiments (e.g., the occurrence of a 10), and (2) those outcomes which are remotely possible in the physical experiment but are assumed to be impossible in the hypothetical experiment (e.g., the occurrence of an edge). Our intuition has urged us to employ another more important concept.

You are willing, lacking other information, to assign an equal "weight" to the occurrence of each side of the die. The willingness to assign an equal weight to each of the six possible outcomes has been called the *principle of insufficient reason* for measuring a priori or classical probabilities. In general, this principle can be stated as follows:

Principle of insufficient reason   Given a finite, exhaustive list of the possible outcomes from a particular experiment, if there is no reason to believe that one outcome is any more likely than any other, then assume that all outcomes are equally likely.

In experiments such as tossing a die, this principle is a starting point. There is a circularity of reasoning here which has, at least, philosophical importance. We say that it is a "fair" die in order to say that each side is equally likely, but we normally define a "fair" die as one which offers an equal opportunity for each side to occur.

The relative-frequency definitions of probability   Consider the following questions:

1.   What is the probability that a particular die (thrown on a flat, hard surface) will land on its edge?
2.   What is the probability that half or more of your class will get a passing grade on the first statistics test?
3.   What is the probability that the Dow Jones Industrial Average will go up by 2 points or more on the next trading day?
4.   What is the probability that the automatic coffee-dispensing machine will overfill the next cup?

The hypothetical experiment, employed in the classical definition of probability, does not assist us in answering these questions. In order to answer any of these questions, we need more information.

From our previous discussion, enumerating possible outcomes of the single toss of a die will not really help us determine the probability that the die will land on its edge. Consider your statistics class; "half or more will receive a passing grade on the first examination" and "less than half will not" are two mutually exclusive events which exhaust the possibilities. But, are you comfortable in using the principle of insufficient reason to assign a probability of $\frac{1}{2}$ to each event? What happened to the Dow Jones Industrial Average on the last trading day? During the last week, or month? Do the answers to those questions have anything to do with the probability that the Dow Jones Industrial Average will go up 2 points or more? Will "overfill" and "not-overfill" suffice as the two possible outcomes? If so, can you obtain classical probabilities for those two events?

These four probability questions are unanswerable in the context of the classical definition of probability. Nevertheless, they are representative of questions a reasonable person could ask. The reason for another definition of probability should be clear.

One way to assess or estimate the probability that a particular die will land on its edge is to conduct a physical experiment. For example, actually toss the die 100,000 times.

Table 2.1 represents the results of 100,000 tosses of a single die.[6] The results provide a basis for making a statement about P(edge). We could use the proportion

---

[6] As noted before, there are eight different corners and twelve different edges. The occurrence of any edge was considered the occurrence of the event "an edge."

TABLE 2.1 100,000 TOSSES OF A DIE

Outcome	Number of Occurrences	Proportion of Occurrences	Classical Probabilities
Side 1	16,640	0.16640	0.16666...
Side 2	16,760	0.16760	0.16666...
Side 3	16,703	0.16703	0.16666...
Side 4	16,602	0.16602	0.16666...
Side 5	16,664	0.16664	0.16666...
Side 6	16,629	0.16629	0.16666...
Corner	0	0.00000	0.00000
Edge	2	0.00002	0.00000
Total	100,000	1.00000	1.00000

(or relative frequency) of occurrences of an edge as the probability of the occurrence of an edge on the single toss of this die, or $P(\text{edge}) = 0.00002$.

This is the concept used in the *relative-frequency definition of probability*:

Relative-frequency definition of probability   A well-defined experiment is conducted n times (which is frequently called n *trials*), the number of times that event A occurs is denoted as a, and the proportion (relative frequency) of times event A occurs is the probability that event A will occur, or

$$P(A) = \frac{a}{n} \tag{2.2}$$

In comparison with the classical definition, the relative-frequency definition of probability does *not* require (1) that the outcomes be mutually exclusive, (2) knowledge of all possible events, or (3) that each outcome be equally likely. It does require that a physical experiment be performed.

If we tossed the die ten more times than the 100,000 already tossed, unless all outcomes were corners or edges, the relative frequency (or proportion) of each outcome would not change much. Because of the magnitude of the numerator and denominator, the maximum possible change would be an alteration of the fourth decimal place in any given proportion from Table 2.1. If all 10 additional tosses produced an edge, the relative-frequency definition of probability would yield $P(\text{edge}) = 12/100,010 = 0.00012$.

If we rolled the die another 100,000 times, however, would we expect *exactly* the same results as those in Table 2.1? Of course not! Equation 2.2 just provides an estimate from 100,000 trials of a particular experiment. Another experiment, under the same conditions and with the same die, could yield slightly different results. This discrepancy in the results of the relative-frequency definition of probability has led to the *limit of the relative-frequency definition of probability*:

Limit of the relative-frequency definition of probability   If we conduct the experiment an infinite number of times (that is, $n \to \infty$), then the relative frequency of the occurrence of event A is the probability of event A, or

$$P(A) = \lim_{n \to \infty} \left(\frac{a}{n} \right) \tag{2.3}$$

Although Equation 2.3 has a certain mathematical eloquence, it is, in at least one respect, just a conceptual evading of the issue. The limit of the relative-frequency definition, per se, requires an infinite number of trials, a time-consuming process which restricts the functional aspects of this definition. The merit of this definition rests in the *empirical*, not necessarily the *logical*, consistency of its use. Relative frequencies may fluctuate a great deal for a small number of trials, but there is a tendency for relative frequencies to stabilize for a large number of trials. Additional trials produce very little fluctuation in the proportion of times a particular event occurs. (Recall the "let's throw the die 10 more times" comment above; the proportions could not change much.)

Again, our intuition may vote for Equation 2.3, but in practice we are forced to accept Equation 2.2—with a large number of trials—as our relative-frequency definition of probability.

The definition of subjective probability In the last three decades, there has been considerable attention drawn to what could be called unique experiments. Primarily these have been decision problems concerning situations that will happen only once.

Place yourself in the role of a decision maker for the Manhattan Project, the development of the first atomic bomb. A "well-defined experiment" does not, in this atom bomb case, suffice to provide a set of outcomes which are equally likely. Nor does the relative-frequency approach have much to offer. The premier scientists did not know exactly what would happen with that first bomb; however, they could agree on a subset of most likely outcomes. Each scientist may have had a different rank-ordering of which possible events would happen; that is, they each had an opinion of how likely each event was. Essentially, only a set of subjective probabilities was available to the process of making a decision to "go ahead" with the detonation of that first atomic bomb.

Less dramatic examples are the probability that the President of the United States will veto a particular bill, the probability that it will rain tomorrow, and the assessment of the demand for a new product. These examples rely on past data (similar to relative frequencies), experience, and attitudes (often in the form of nonquantifiable information or feelings) to make probability assessments.

Subjective probability is defined, or interpreted, as follows.

Subjective probability Given a well-defined experiment, situation, or happening, the probability of an event is the "degree of belief" assigned to the occurrence of this event by a particular individual; the only requirements are that: (1) $P(A) = 0$ represent certainty that event A will not occur, (2) $P(A) = 1$ represent certainty that event A will occur, and (3) $0 < P(A) < 1$ represents the degree of certainty that event A will occur.

That is, the subjective probability of the occurrence of event A is an index number, assigned by a given individual, which represents the composite of evidence and impressions available to this individual. Another person, with a different composite of experiences, may assign an entirely different index number, or probability, to the occurrence of event A.

Consider $P(A) = \frac{1}{4}$ where A is the event {it will rain next Thursday}. Of course, this statement could be made weekly, and $P(A) = \frac{1}{4}$ can have a relative-frequency interpretation. Also, experience may alter the value assigned to $P(A)$. But, the subjective *interpretation* of probability is different from that relative-frequency interpretation. Let event A refer to one and only one particular day. The value of $\frac{1}{4}$ reflects the decision maker's degree of certainty that A will occur; she or he assigns "odds" of 1-to-3 that it will rain next Thursday.

The definition of subjective probability represents a broader concept in the sense that either of the two "objective" definitions can be components in assigning a value to $P(A)$ by a particular individual. Further, each individual must be internally consistent in the sense that $P(A) + P(A*) = 1$, which forms the basis for a self-test of your degree of belief in the occurrence of event A.

Let event A be that the President will veto a given bill, to which you assign a probability of $\frac{1}{3}$. The probability that the President will not veto must be $P(A*) = 1 - P(A) = \frac{2}{3}$. Using these subjective probabilities, are you willing to take *both* sides of a bet? That is, are you willing to pay \$2 if the President vetos the bill in return for \$1 if he does not? And, are you equally willing to pay \$1 if he does not veto in return for \$2 if he does? If you cannot answer yes to both questions, then presumably your assessment of $P(A)$ is not $\frac{1}{3}$.

For the first-time-ever experiment, the subjective probability interpretation may be the only functional definition possible. For pedagogical reasons only, much of the discussion that follows will not employ the subjective probability concept.

2.2 INTRODUCTION TO SET THEORY: A CLASSIFICATION SCHEME

We shall employ set theory as a way of classifying and/or keeping track of events.

A *set* is any well-specified collection of identifiable objects. The objects are frequently called *members*, or *elements*, of the set. The adjective "identifiable" refers to the condition that we could either *list* all the distinct objects in the set (which is feasible when the number of objects is finite) or employ some *defining property*, or *characteristic*, of each element (which is useful when there is an infinite number of objects in the set). Also, note that the way you identify the elements dictates how you specify the set.

The adjective "well specified" simply refers to the requirement that for each object it must be perfectly clear whether that object belongs to (or is contained in) the set. Although "well specified" and "identifiable" seem to be—and in many cases will be—redundant, they need not be.[7] The following are examples of sets:

---

[7] A "sleight-of-hand" artist shows you a "queen of hearts"; the card is identifiable as a "queen of hearts," but did it come from the deck shuffled?

Set	Elements	Identifiable by
The students enrolled in a statistics class	Persons	Sight, or by name or social security number on a roster
The cards in a standard bridge (or poker) deck	Cards	Symbols on face of card and design on the back
Books in a library	Books	Title and author's name, or catalog index number
Passengers on an airplane	Persons	Name, seat number

Sets are frequently expressed in symbols as a shorthand notation. Sets are designated by capital letters (for example, A, B, S, Q); the "equal sign" denotes "is equivalent to" or "is defined as"; and the elements, separated by commas, are enclosed in braces. For example, the cards in a standard bridge deck could be defined in symbols as follows:

$C = \{$ace of spaces, ace of hearts, ace of diamonds, ace of clubs,
king of spades, king of hearts, king of diamonds, king of clubs,
queen of spades, queen of hearts, queen of diamonds, queen of clubs,
jack of spades, jack of hearts, jack of diamonds, jack of clubs,
10 of spades, 10 of hearts, 10 of diamonds, 10 of clubs,
9 of spades, 9 of hearts, 9 of diamonds, 9 of clubs,
8 of spades, 8 of hearts, 8 of diamonds, 8 of clubs,
7 of spades, 7 of hearts, 7 of diamonds, 7 of clubs,
6 of spades, 6 of hearts, 6 of diamonds, 6 of clubs,
5 of spades, 5 of hearts, 5 of diamonds, 5 of clubs,
4 of spades, 4 of hearts, 4 of diamonds, 4 of clubs,
3 of spades, 3 of hearts, 3 of diamonds, 3 of clubs,
2 of spades, 2 of hearts, 2 of diamonds, 2 of clubs$\}$

This represents the *list* of all elements and is obnoxiously long, even for just 52 elements. Thus, another shorthand notation is frequently used:

$$C = \{x \mid x \text{ is a card from a standard bridge deck}\}$$

We would read this notation as follows: "Set C is designated by x such that x represents each of the 52 symbols (face or number and suit) on the front of a playing card." This notation represents the *defining property* method of specifying the set. Please note that the words "standard bridge deck" imply the list of elements.

The *sample space*, or *universal set*, is a set whose objects are the possible outcomes from an experiment or a happening. The experiment may actually be performed, or it may be hypothetical. In example 2.1, on a single toss of a die, the sample space could be defined as

$$S = \{1, 2, 3, 4, 5, 6\}$$

The elements represent all possible outcomes from the hypothetical experiment.

A *subset* is another set whose elements are also elements of the *sample space*. The largest subset of S (i.e., the sample space) is S itself, where all the elements contained in S are contained in the subset. The smallest subset, by definition, is a set with no elements; the *empty*, or *null*, *set* contains no objects and is designated by the Greek letter phi, ϕ. Note: it is not safe to say $\phi = \{0\}$, because zero might be an element of the sample space; $\phi = \{$no elements$\}$ is correct.

An *event* is some subset of the sample space. An *elementary event* is a subset that contains only one element from the sample space. If we let

$E = \{$the occurrence of an even number in the single toss of a fair die$\}$

$T = \{$the occurrence of a 2 on the single toss of a fair coin$\}$

$V = \{$the occurrence of a 7 in the single toss of a fair coin$\}$

then E and T are "events" or subsets of S. Further, T is an elementary event. With reference to the sample space S, V contains no elements or outcomes in S; thus, V is contained in ϕ (the null set), which is a subset of S.

Set Operations

We are interested in sets because we are interested in probability. As a result, the words "set" and "event" will be used interchangeably. There are two important set concepts to be developed: union and intersection. Their importance rests in our ability to understand the rules of addition and multiplication for probability.

Panel 2.1   Union

Definition

The *union* ($\cup$) of two sets (or events) is that collection of elements (or outcomes) that belong to *either* set. $\{A \cup B\}$ is a "new" set; its elements consist of those that are contained in A or in B or are common to both A and B.

Diagrammatic Representation

$A \cup B$

S

A

B

There are a wide variety of ways to write this concept. The more common are $A \cup B =$ "A union B" = "A or B or both" = "A cup B" = "$A + B$."

Consider the set of all passengers on an airplane (S), and let $A =$ {female passengers} and $B =$ {passengers who smoke}. Then $A \cup B =$ {passengers who either smoke or are females}. We recognize that there may be some female smokers; there may be some elements of A that are contained in B.

Panel 2.2   Intersection

Diagrammatic Representation

Definition
The *intersection* ($\cap$) of two sets (or events) is that collection of elements (or outcomes) that belong simultaneously to both sets. $A \cap B$ is a "new" set; its elements are only those mutually contained in both set A and set B.

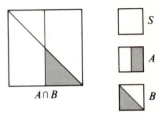

$A \cap B$

Again, the variety of notation for this concept includes: $A \cap B =$ "A intersection B" = "A and B" = "A cap B" = "$A \times B$" = "AB." Returning to our example of the airplane passengers, $A \cap B =$ {female smokers}. Those passengers who satisfy the joint classification of (1) being female and (2) being a smoker are those persons represented by $A \cap B$.

Additional Set Concepts

First, recall that we defined A^* as the event "not-A." The event "not-A" is called the *complement* of A. Also, note that $A \cup A^* = S$; "all the elements in A *or* all the elements not in A" designates all the elements in S. Further, note that $A \cap A^* = \phi$. By definition, A and A^* do not have any points in common.

Next, consider the possibility that events A and B do not have outcomes in common; there may be no female smokers on the airplane. In this case $A \cap B = \phi$, and we call A and B *mutually exclusive events*. If A and B are mutually exclusive events, then they simply cannot both occur at the same time.

2.3 ELEMENTARY PROBABILITY AXIOMS AND RULES

Regardless of how we define probability, there are some uniform characteristics which take the form of axioms.

Axiom 1 $0 \leq P(A) \leq 1$

The probability of event A occurring is represented by some number between—and inclusive of—0 and 1.

The two extreme values of the probability of event A, $P(A) = 0$ and $P(A) = 1$, represent certainties. To be more specific, $P(A) = 0$ means that there is no possible way you could observe event A; it cannot occur. In like manner, $P(A) = 1$ means that there is no possible way to avoid observing event A; it must occur.

For any particular value of $P(A)$ between 0 and 1, $0 < P(A) < 1$, event A may or may not occur. $P(A)$ is just the relative assessment that event A will occur *before you try* the experiment. After the fact, either event A occurred or it did not.

Axiom 2 $\qquad\qquad\qquad\qquad\qquad\qquad$ $P(S) = 1$

If S, the sample space, is the set of all possible outcomes, then the probability of getting event S in one trial of the experiment is 1, a certainty.[8]

Two Examples

To develop the rules of probability, we will employ two examples. One, the passengers on an airplane, will represent a sample space with a finite number of possible outcomes. The other, a square dart board, will represent a sample space with an infinite number of possible outcomes. The classical definition of probability will be our emphasis, but the rules developed will be applicable to all the definitions of probability.

Example 2.2 Passengers on an Airplane

Modern medical technology notwithstanding, we shall refer to the event "not-female" as the event "male." Of the 100 persons on this plane, 10 are female smokers; 20 are male nonsmokers; there are as many smokers as nonsmokers; and there are 40 female passengers. We will assume that each passenger has an equal chance to be selected (perhaps by a hijacker).

TABLE 2.2 PASSENGERS, BY SEX AND SMOKING HABIT*

	Female A	Not-Female $A*$	Total
Smokers (B)	10	40	50
Nonsmokers ($B*$)	30	20	50
Total	40	60	100

* If this "heading" seems awkward, consider the implications when the U.S. Bureau of the Census presents a table entitled "American Population Broken Down by Age and Sex."

---

[8] To mathematically develop the rules of probability a third axiom is required: If there are two mutually exclusive events, A and B, then $P(A \cup B)$ must equal $P(A) + P(B)$. We will interpret this axiom in the next section.

Example 2.3 A Square Dart Board

This square dart board (see Figure 2.1) measures 1 foot per side; thus, the total surface area is 1 square foot. Event C is landing a dart on the right half of this board; the area representing the occurrence of event C is $\frac{1}{2}$ square foot. The event D is landing a dart in the lower left-hand triangle of this board; the area of D is $\frac{1}{2}$ square foot. Please note that there are an infinite number of "points" or outcomes in S, C, and D.

We will consider any point on the board to be as likely as any other point.

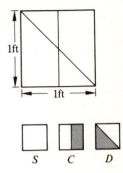

FIGURE 2.1 Square dart board.

Marginal Probabilities

For students who have had exposure to the discipline of economics, this is the most unfortunate of definitions. "Marginal," as it is used here, means only that the probabilities are calculated from numbers found in the "margins" of a table of data (e.g., in Table 2.2, the numbers 50, 50, 40, and 60). There is absolutely no correspondence between "marginal probabilities" and such concepts as marginal cost, marginal revenue, etc.

In example 2.2, $P(A)$, $P(A*)$, $P(B)$, and $P(B*)$ are all marginal probabilities. Let $\eta(A)$ be the number of outcomes that would indicate the occurrence of event A, and let $\eta(S)$ be the number of possible outcomes in a single selection. Since all passengers on the airplane are equally likely in the random selection of one passenger,

$$P(A) = P(\text{a female}) = \frac{\eta(A)}{\eta(S)} = \frac{40}{100} = 0.40$$

according to the classical definition of probability. Using the same procedure, it is a simple exercise to find that $P(A*) = 0.60$ and $P(B) = 0.50 = P(B*)$.

Now consider the marginal probabilities in example 2.3: $P(C)$, $P(C*)$, $P(D)$, and $P(D*)$. If each point on the dart board has an equal chance of receiving one randomly thrown dart,[9] then

$$P(C) = \frac{\eta(C)}{\eta(S)} = \frac{0.50 \text{ square foot}}{1.00 \text{ square foot}} = 0.50$$

---

[9] The use of a square, rather than a circular, dart board is intended to make the areas (which represent subsets) easier to calculate.

In this example, $\eta(C)$ is the surface area that represents event C. In like manner, show that $P(C^*) = P(D) = P(D^*) = 0.50$.

Joint Probabilities

A *joint probability* is the probability of the intersection of two events, the probability of an event which must simultaneously satisfy more than one classification. For example, in the selection of a single passenger, $P(A \cap B)$, $P(A^* \cap B)$, $P(A \cap B^*)$, and $P(A^* \cap B^*)$ are joint probabilities in example 2.2.

The probability of getting a female smoker is

$$P(A \cap B) = \frac{\eta(A \cap B)}{\eta(S)} = \frac{10}{100} = 0.10$$

There are 10 such outcomes out of 100 possible outcomes. Check to be sure that the probability of a female nonsmoker is 0.30, the probability of a male smoker is 0.40, and the probability of a male nonsmoker is 0.20.

In example 2.3, $P(C \cap D)$, $P(C \cap D^*)$, $P(C^* \cap D)$, and $P(C^* \cap D^*)$ are joint probabilities. The probability of a single dart landing in both C and D is

$$P(C \cap D) = \frac{\eta(C \cap D)}{\eta(S)} = \frac{\frac{1}{8}}{1} = 0.125$$

In like manner, the probability of a dart landing in C but not in D (or in C and not-D) is

$$P(C \cap D^*) = \frac{\eta(C \cap D^*)}{\eta(S)} = \frac{\frac{3}{8}}{1} = 0.375$$

Further, observe that

$$P(C^* \cap D) = \frac{3}{8} = 0.375$$

and

$$P(C^* \cap D^*) = \frac{1}{8} = 0.125$$

Conditional Probabilities

A *conditional probability* is the probability of getting an event when you are sampling from a *known subset* of the sample space. The conditional probability is written as $P(B \mid A)$ and is read as "the probability of event B occurring, *given* the fact that event A has already occurred."

If we randomly choose a female from our 100 airplane passengers, what is the probability that she is a smoker? Immediately we know we do not have to be concerned with the entire sample space; only that subset which represents females is relevant. Thus,

$$P(B \mid A) = \frac{\eta(A \text{ and } B)}{\eta(A)} = \frac{10}{40} = 0.25$$

Another way to visualize the conditional probability is: "What proportion of the female passengers smoke?" Please check the following $P(B^* \mid A) = {}^{30}\!/_{40} = 0.75$; "the probability of selecting a female given a smoker" $= P(A \mid B) = {}^{10}\!/_{50} = 0.20$; $P(A^* \mid B) = {}^{40}\!/_{50} = 0.80$; and, "the probability of getting a male given a nonsmoker" $= P(A^* \mid B^*) = {}^{20}\!/_{50} = 0.40$.

With our dart-dropping example, note that:

$$P(C \mid D) = \frac{\eta(C \text{ and } D)}{\eta(D)} = \frac{{}^{1}\!/_{8}}{{}^{4}\!/_{8}} = 0.25$$

$$P(C^* \mid D) = \frac{\eta(C^* \text{ and } D)}{\eta(D)} = \frac{{}^{3}\!/_{8}}{{}^{4}\!/_{8}} = 0.75$$

$$P(D \mid C) = \frac{\eta(D \text{ and } C)}{\eta(C)} = \frac{{}^{1}\!/_{8}}{{}^{4}\!/_{8}} = 0.25$$

and

$$P(D^* \mid C) = \frac{\eta(D^* \text{ and } C)}{\eta(C)} = \frac{{}^{3}\!/_{8}}{{}^{4}\!/_{8}} = 0.75$$

2.4 THE RULES OF ADDITION FOR PROBABILITIES

The General Rule of Addition for Two Events

This rule concerns the probability of observing either event A or event B or both events simultaneously:

$$P(A \cup B) = P(A) + P(B) - P(A \cap B) \tag{2.4}$$

This general rule is easily justified by considering the airplane passengers in example 2.2. We will select, at random, one passenger. What is the probability that this one passenger will be a female or a smoker? The number of females, $\eta(A)$, is 40; the number of smokers, $\eta(B)$, is 50. If we just add $\eta(A)$ and $\eta(B)$, then we have counted the female smokers *twice*. If each passenger is equally likely to be chosen (i.e., female smokers are not twice as likely as other passengers), then we must subtract from $\eta(A) + \eta(B)$ the 10 female smokers [or $\eta(A \cap B)$] in order to count each passenger only once. This gives

$$\eta(A \cup B) = \eta(A) + \eta(B) - \eta(A \cap B)$$

Dividing both sides by $\eta(S)$, the number of possible outcomes, yields

$$\frac{\eta(A \cup B)}{\eta(S)} = \frac{\eta(A)}{\eta(S)} + \frac{\eta(B)}{\eta(S)} - \frac{\eta(A \cap B)}{\eta(S)}$$

Using the classical definition of probability, this becomes Equation 2.4. Or, the probability of selecting a female or a smoker from the 100 passengers is

$$P(A \cup B) = {}^{40}\!/_{100} + {}^{50}\!/_{100} - {}^{10}\!/_{100} = 0.80$$

Please verify, from example 2.2, that $P(A \cup B^*) = 0.60$, $P(A^* \cup B) = 0.70$, and $P(A^* \cup B^*) = 0.90$.

From example 2.3, the addition rule allows us to state that "the probability of a single dart landing in area C or in area D or in both" is

$$P(C \cup D) = \frac{\eta(C)}{\eta(S)} + \frac{\eta(D)}{\eta(S)} - \frac{\eta(C \cap D)}{\eta(S)}$$

$$= \frac{{}^{1}\!/_{2}}{1} + \frac{{}^{1}\!/_{2}}{1} - \frac{{}^{1}\!/_{8}}{1} = {}^{7}\!/_{8} = 0.875$$

Again, we subtract the joint probability in Equation 2.4 to prevent the "double counting" of that area representing both C and D. You should find that $P(C \cup D^*) = {}^{5}\!/_{8}$, $P(C^* \cup D) = {}^{5}\!/_{8}$, and $P(C^* \cup D^*) = {}^{7}\!/_{8}$.

The Special Rule of Addition for Two Mutually Exclusive Events

If and only if two events, A and B, are mutually exclusive [i.e., the set $(A \cap B) = \phi$], then $P(A \cap B) = 0$ and

$$P(A \cup B) = P(A) + P(B) \tag{2.5}$$

Consider an experiment where a single card is drawn from a well-shuffled deck of cards. Let event $A = \{\text{the occurrence of an ace}\}$ and event $B = \{\text{the occurrence of a king}\}$. Events A and B are mutually exclusive because they could not both occur at the same time. Thus, $P(A \cup B) = {}^{4}\!/_{52} + {}^{4}\!/_{52} = {}^{2}\!/_{13} = 0.154$.

The General Rule of Addition for Three Events

Consider the probability of the occurrence of event A or event B or event C (which would include the simultaneous occurrence of any combination of the three events); the general rule for this case is given by

$$P(A \cup B \cup C) = P(A) + P(B) + P(C) - P(A \cap B) - P(A \cap C)$$
$$- P(B \cap C) + P(A \cap B \cap C) \quad (2.6)$$

In order to understand this rule, let us expand example 2.2 to include the marital status of the passengers.

First, let us look at Table 2.3 to determine $P(A \cup B \cup C)$ *without* using the general rule. Remember that we are selecting one of the 100 passengers at random and $P(A \cup B \cup C)$ asks "What is the probability of getting a female or a smoker or a married person?" Table 2.3a tells us that there are 60 married passengers. From Table 2.3b we observe that there are 32 passengers who either smoke (event B includes 20) or are female ($A \cap B^*$ accounts for 12) even though they are not married. Thus,

$$P(A \cup B \cup C) = \frac{\eta(A \cup B \cup C)}{\eta(S)} = \frac{60 + 32}{100} = 0.92$$

As a check, we note from Table 2.3b that there are only 8 unmarried males who do not smoke.

Now let us employ the rule given by Equation 2.6 to obtain the same result. From Table 2.2 we already know that $P(A) = {}^{40}/_{100} = 0.40$; $P(B) = {}^{50}/_{100} = 0.50$; and $P(A \cap B) = {}^{10}/_{100} = 0.10$. From Table 2.3a it is obvious that $P(C) = {}^{60}/_{100} = 0.60$.

Next, let us determine $P(A \cap C)$, or the probability of selecting a married female. From Table 2.3a we observe that there are 24 married females; therefore, $P(A \cap C) = {}^{24}/_{100} = 0.24$. In like manner, using Table 2.3a, we see that $P(B \cap C)$, the probability of selecting a married smoker, is ${}^{30}/_{100} = 0.30$. Furthermore, from

TABLE 2.3 PASSENGERS BY SEX, SMOKING HABIT, AND MARITAL STATUS

(a) Married: Event C				(b) Not Married: Event C^*			
	A	A^*	Total		A	A^*	Total
B	6	24	30	B	4	16	20
B^*	18	12	30	B^*	12	8	20
Total	24	36	60	Total	16	24	40

Table 2.3a, there are only 6 married females who smoke; thus, $P(A \cap B \cap C) = \frac{6}{100} = 0.06$. Finally, using Equation 2.6,

$$P(A \cup B \cup C) = P(A) + P(B) + P(C) - P(A \cap B) - P(A \cap C) - P(B \cap C)$$
$$+ P(A \cap B \cap C)$$
$$= 0.40 + 0.50 + 0.60 - 0.10 - 0.24 - 0.30 + 0.06$$
$$= 0.92$$

This airplane-passenger example demonstrates the calculational validity of Equation 2.6, but the *logical* validity of the general addition rule for three events may still be less than perfectly clear. To assist in clarifying the logical validity of Equation 2.6, we will expand example 2.3, a single toss (or drop) at the square dart board. Let event E be the lower half of the dart board, and the area which event E represents is $\frac{1}{2}$ square foot.

Consider the sets, or areas, given in Figure 2.2. We can see from Figure 2.2a that $P(C \cup D \cup E) = \frac{7}{8}$. Also, note that since the total area of the board is 1 square foot, then, as long as we measure all areas in square feet, the numerical value of the area

FIGURE 2.2   Subsets from example 2.3 used in Equation 2.6.

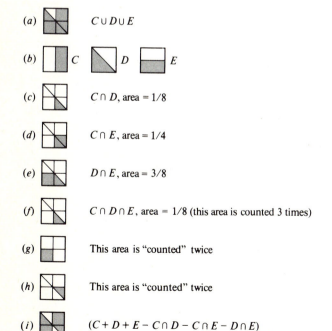

(a) $C \cup D \cup E$

(b) C D E

(c) $C \cap D$, area $= 1/8$

(d) $C \cap E$, area $= 1/4$

(e) $D \cap E$, area $= 3/8$

(f) $C \cap D \cap E$, area $= 1/8$ (this area is counted 3 times)

(g)   This area is "counted" twice

(h)   This area is "counted" twice

(i) $(C + D + E - C \cap D - C \cap E - D \cap E)$

representing an event is identical with the numerical value of the probability that that event will occur. That is, since $\eta(S) = 1$, then $P(C) = \eta(C) = \frac{1}{2}$.

First, consider the marginal probabilities on the right-hand side of Equation 2.6. From above, we can consider the sum of these three probabilities as the sum of three areas:

$$\eta(C) + \eta(D) + \eta(E)$$

This expression contains some areas of the dart board which are "counted" more than once. Those areas which are counted *only twice* are shown in Figure 2.2*g* and *h*. One area is counted *three* times, as represented in Figure 2.2*f*.

Next, consider the three negative terms in Equation 2.6. Carefully notice that if we subtract those areas represented by the three possible unions of two events (Figure 2.2*c, d,* and *e*), we count all relevant areas once *except* that we do not count the area representing $C \cap D \cap E$ at all! That is, Figure 2.2*i* represents

$$\eta(C) + \eta(D) + \eta(E) - \eta(C \cap D) - \eta(C \cap E) - \eta(D \cap E)$$

Thus, to count all the areas which represent $C \cup D \cup E$ only *once*, we must now add back in the area $\eta(C \cap D \cap E)$, to make

$$\eta(C) + \eta(D) + \eta(E) - \eta(C \cap D) - \eta(C \cap E) - \eta(D \cap E) + \eta(C \cap D \cap E)$$

Since these areas also represent the probability of each term, then

$$P(C \cup D \cup E) = \frac{1}{2} + \frac{1}{2} + \frac{1}{2} - \frac{1}{8} - \frac{1}{4} - \frac{3}{8} + \frac{1}{8}$$
$$= \frac{7}{8}$$

which conforms to Figure 2.2*a*.

In summary, the general addition rule for four events is given by

$$P(A \cup B \cup C \cup D) = P(A) + P(B) + P(C) + P(D) - P(A \cap B) - P(A \cap C)$$
$$- P(A \cap D) - P(B \cap C) - P(B \cap D) - P(C \cap D)$$
$$+ P(A \cap B \cap C) + P(A \cap B \cap D) + P(A \cap C \cap D)$$
$$+ P(B \cap C \cap D) - P(A \cap B \cap C \cap D) \qquad (2.7)$$

From this progression, you should be able to determine the general rule for *n* events if you should ever need it.

The Special Rule of Addition for *n* Mutually Exclusive Events

From the general rule for *n* events we can see if *all* the events are mutually exclusive, then *all* the joint probabilities will be zero. Thus,

$$P(A_1 \cup A_2 \cup A_3 \cup \cdots \cup A_n) = P(A_1) + P(A_2) + P(A_3) + \cdots + P(A_n) \quad (2.8)$$

2.5   THE RULES OF MULTIPLICATION FOR PROBABILITIES

The General Rule of Multiplication for Two Events

This rule concerns the probability of simultaneously observing two events in one outcome:

$$P(A \cap B) = P(A)P(B \mid A) = P(B)P(A \mid B) \tag{2.9}$$

which is read as "the probability of A and B both occurring is equal to the probability that event A will occur times the probability that event B will occur given that event A has occurred," which is equivalent to "the probability that event B will occur times the probability that A will occur given that B has occurred."

Again, we will justify this rule with those very busy airplane passengers in example 2.2. Recall that we defined a joint probability as $P(A \cap B)$, and Equation 2.9 could be viewed either as the relationship between a joint probability and a conditional probability or as a way of calculating the joint probability. Previously we found that the probability of selecting a female smoker, or $P(A \cap B)$, was 0.10; that result was obtained by observing that there are 10 female smokers among the 100 passengers. But what if we did not have the information provided by Table 2.2? Instead, we may know that "40 percent of the passengers are female" and that "25 percent of the female passengers smoke" (or that "50 percent of the passengers smoke" and "20 percent of the smokers are female"). Then, the (joint) probability of selecting a female smoker is represented by

$$\frac{\eta(A \cap B)}{\eta(S)} = \left[\frac{\eta(A)}{\eta(S)}\right]\left[\frac{\eta(A \cap B)}{\eta(A)}\right]$$

$$= \left(\frac{40}{100}\right)\left(\frac{10}{40}\right) = 0.10$$

$$= \left[\frac{\eta(B)}{\eta(S)}\right]\left[\frac{\eta(A \cap B)}{\eta(B)}\right]$$

$$= \left(\frac{50}{100}\right)\left(\frac{10}{50}\right) = 0.10$$

By saying the general rule of multiplication for two events concerns "the probability of simultaneously observing two events in one outcome" we do not intend to imply that there must be only one draw from the population. Let us define a new experiment where we select one passenger and then another. That is, we choose one passenger, and without returning that passenger to the population, we choose a second passenger. Suppose we are interested in the probability of selecting two consecutive female passengers, or $P(A_1 \cap A_2)$ where the subscripts refer to "first selection" and "second selection." From Equation 2.9,

$$P(A_1 \cap A_2) = P(A_1)P(A_2 \mid A_1)$$

the right-hand side of which can be interpreted as "the probability of choosing a female on the first selection" times "the probability of choosing a female on the second selection given a female was chosen on the first selection." Or,

$$P(A_1)\,P(A_2\,|\,A_1) = \left[\frac{\eta(A_1)}{\eta(S_1)}\right]\left[\frac{\eta(A_2)}{\eta(S_2)}\right]$$

$$= \left(\frac{40}{100}\right)\left(\frac{39}{99}\right) = 0.158$$

Thus, the general rule of multiplication can be employed for the probability of observing two events in two draws. Note that $\{A_1 \cap A_2\}$ is still considered as *one outcome* from the experiment.

The Special Rule of Multiplication for Two Independent Events

Reconsider our square dart board as represented in Figure 2.2. From Figure 2.2*d* we can see that $P(C \cap E) = \frac{1}{4}$, and from other parts of Figure 2.2 we see that $P(C) = P(E\,|\,C) = P(E) = P(C\,|\,E) = \frac{1}{2}$.
    By using Equation 2.9, we find that

$$P(C \cap E) = P(C)\,P(E\,|\,C) = (\tfrac{1}{2})(\tfrac{1}{2}) = \tfrac{1}{4}$$
and $$P(C \cap E) = P(E)\,P(C\,|\,E) = (\tfrac{1}{2})(\tfrac{1}{2}) = \tfrac{1}{4}$$

It is important to note that $P(C) = P(C\,|\,E)$ and that $P(E) = P(E\,|\,C)$. That is, the knowledge that event E has occurred does not affect the probability that event C will occur (and vice versa).
    In general, *when $P(A) = P(A\,|\,B)$ [or $P(B) = P(B\,|\,A)$]*, then A and B are said to be *independent* events. Thus, the special rule of multiplication for two independent events is

$$P(A \cap B) = P(A)\,P(B) \qquad\qquad (2.10)$$

For example, in our experiment where two successive passengers are selected, if we *replaced* the first passenger before we randomly selected the second passenger, then the probability of choosing a female on the second selection would not be affected by the result of the first selection; hence, $P(A_2) = P(A_2\,|\,A_1) = P(A)$, or events A_1 and A_2 are independent. With the replacement of the first person before we choose the second, the probability of observing two successive females is given by Equation 2.10: $P(A_1 \cap A_2) = (^{40}/_{100})(^{40}/_{100}) = 0.160$. If two events A and B are independent, then whether B occurs will have no influence on the probability that A will occur; $P(A) = P(A\,|\,B)$.

To check your understanding of independence, from Table 2.2 show that events A and B are *not* independent [that is, $P(A) \neq P(A|B)$ and $P(B) \neq P(B|A)$] and from Tables 2.2 and 2.3 show that events A and C are independent (as are events B and C).

The General Rule of Multiplication for Three Events

The probability of the occurrence of an outcome which simultaneously satisfies three different classifications is given by

$$
\begin{aligned}
P(A \cap B \cap C) &= P(A)\,P(B|A)\,P(C|A \cap B) \\
&= P(A)\,P(C|A)\,P(B|A \cap C) \\
&= P(B)\,P(A|B)\,P(C|A \cap B) \\
&= P(B)\,P(C|B)\,P(A|B \cap C) \\
&= P(C)\,P(A|C)\,P(B|A \cap C) \\
&= P(C)\,P(B|C)\,P(A|B \cap C) \qquad \textbf{(2.11)}
\end{aligned}
$$

In the set $\{A \cap B \cap C\}$, events A, B, and C can be written in any order without changing the outcome represented by $\{A \cap B \cap C\}$. Thus, the last five expressions in Equation 2.11 are redundant. From Tables 2.2 and 2.3, we can find that the probability of randomly selecting a married, female smoker from the 100 passengers is

$$
\begin{aligned}
P(A \cap B \cap C) &= P(A)\,P(B|A)\,P(C|A \cap B) \\
&= (^{40}\!/_{100})(^{10}\!/_{40})(^{6}\!/_{10}) = 0.06
\end{aligned}
$$

Given the general rule of multiplication for four events to be

$$
P(A \cap B \cap C \cap D) = P(A)\,P(B|A)\,P(C|A \cap B)\,P(D|A \cap B \cap C)
$$

the progression developing should indicate the nature of the general rule for n events.

The Special Rule of Multiplication for Three Independent Events

If and only if events A and B are independent, events A and C are independent, and events B and C are independent, then the special rule is

$$
P(A \cap B \cap C) = P(A)P(B)P(C) \qquad \textbf{(2.12)}
$$

It is important to remember that Equation 2.12 *is only valid when all possible pairs of events are independent.* In the airplane-passenger example, we have already seen that events A and B are not independent even though events "A and C" and "B and C" are independent.

Consider three tosses of a fair die. Let $A_1 = \{$a 6 on the first toss$\}$, $A_2 = \{$a 6 on the second toss$\}$, and $A_3 = \{$a 6 on the third toss$\}$. What is the probability that a 6 will appear on all three tosses? The outcome on the first toss does not affect the probability of getting a 6 when you pick up the die and toss it again. All possible pairs (that is, A_1 and A_2; A_1 and A_3; and A_2 and A_3) are independent. Thus,

$$P(A_1 \cap A_2 \cap A_3) = P(A_1)P(A_2)P(A_3) = (\tfrac{1}{6})(\tfrac{1}{6})(\tfrac{1}{6}) = \tfrac{1}{216} = 0.0046$$

It should not be surprising to find that if (and only if) all possible pairs of events are independent, then

$$P(A_1 \cap A_2 \cap A_3 \cap \cdots \cap A_n) = P(A_1)P(A_2)P(A_3) \cdots P(A_n)$$

Bayes' Rule

Will Durant has described the period 1715–1789 as follows:

> Science too was offering a new revelation. . . . Two priesthoods came into conflict: the one devoted to the molding of character through religion, the other to the education of the intellect through science.[10]

The Reverend Thomas Bayes straddled the fence in that conflict. As a minister and as a mathematician, he was concerned with cause-effect relationships. The use of both the theorem which bears his name and the concept of subjective probability has produced a revolution in our own time.

Bayes' rule is a technique for calculating conditional probabilities which was not then, and is not now, very exciting. As a probability rule there is no debate as to its validity. The "debate" or "revolution of thought" concerns the use of subjective probabilities. Bayes' rule is the facilitator in the use of subjective probabilities to treat decision problems in the face of uncertainty. The process or body of knowledge developed has been generally titled *Bayesian decision theory*. It is in this context that Bayes will probably exceed in notoriety, in modern Western civilization, the premier mathematicians (and possibly the theologians) of his time.

Bayes' interest was in developing a method of finding the probability of a specific cause when a particular effect was observed. That is, event B has occurred, what is the probability that it was generated by event A_1 (a possible cause)? Or by event A_2 (another possible cause)?

We start with Equation 2.9,

$$P(A_1)P(B \mid A_1) = P(B)P(A_1 \mid B)$$

---

[10] Will and Ariel Durant, *The Age of Voltaire*, Simon & Schuster, Inc., New York, 1965, p. 507.

and rewrite it in the form of

$$P(A_1|B) = \frac{P(A_1)P(B|A_1)}{P(B)} \tag{2.13}$$

Assume that there are n mutually exclusive events (that is, $A_1, A_2, A_3, \ldots, A_n$) which could cause event B. Event B (the effect) had to come from one of those causes; thus, the probability of event B occurring can be represented by

$$P(B) = P[(A_1 \cap B) \cup (A_2 \cap B) \cup (A_3 \cap B) \cup \cdots \cup (A_n \cap B)]$$

Since the A's are mutually exclusive, then $(A_i \cap B)$ and $(A_j \cap B)$ must be mutually exclusive for all $i \neq j$. Therefore, employing the special rule of addition, we obtain

$$P(B) = P(A_1 \cap B) + P(A_2 \cap B) + P(A_3 \cap B) + \cdots + P(A_n \cap B)$$

Recalling the general rule of multiplication for two events, we find

$$P(B) = P(A_1)P(B|A_1) + P(A_2)P(B|A_2) + \cdots + P(A_n)P(B|A_n)$$

Substituting this result into Equation 2.13 yields:

Bayes' rule

$$P(A_1|B) = \frac{P(A_1)P(B|A_1)}{P(A_1)P(B|A_1) + P(A_2)P(B|A_2) + \cdots + P(A_n)P(B|A_n)} \tag{2.14}$$

Let us review what these terms mean:

$P(A_1|B)$ = given the event (or "effect") B occurred, what is the probability that event A_1 caused it?

$P(A_i)$ = the probability of event (or "cause") A_i occurring.

$P(B|A_i)$ = given the event (or "cause") A_i occurs, what is the probability that it will generate effect B?

Bayes' rule is, in a sense, what we hope a medical doctor is doing in diagnosing a patient. The doctor knows the symptoms associated with each disease, $P(B|A_i)$, in a long list of diseases, as well as the relative frequency of each disease, $P(A_i)$. What is observed with a specific patient is a symptom (or effect), B, and the doctor must assess the probability that this patient has a particular disease given that symptom, $P(A_i|B)$. For less drama but with an equal understanding of the mechanics of Bayes' rule, consider the following example.

Example 2.4 Bookstore Orders

Three women write all special orders at the university bookstore. From past records we know that Ms. W writes 50 percent of the special orders; Ms. X writes 20 percent; and Ms. Y writes 30 percent. We also know that W makes a mistake on 1 out of 100 orders; X makes a mistake 4 percent of the time; and Y makes a mistake 2 percent of the time. A student receives the wrong book and correctly maintains that there was a mistake in taking the order. What is the probability that Ms. W wrote the order? Let $M = \{mistake\}$.

$$P(W|M) = \frac{P(W)P(M|W)}{P(W)P(M|W) + P(X)P(M|X) + P(Y)P(M|Y)}$$

$$= \frac{(0.50)(0.01)}{(0.50)(0.01) + (0.20)(0.04) + (0.30)(0.02)}$$

$$= \frac{0.005}{0.005 + 0.008 + 0.006} = \frac{0.005}{0.019} = 0.263$$

In like manner, $P(X|M) = 0.008/0.019 = 0.421$ and $P(Y|M) = 0.316$. Thus, even though Ms. W fills most of the special orders, she is least likely to have filled this one.

The results of using Bayes' rule are graphically shown in Figure 2.3. A mistake has occurred; we are dealing *only* with the shaded area. The probability that Ms. W made the mistake is the ratio of the area represented by $P(W \cap M)$ to the *total shaded* area:

$$P(W|M) = \frac{P(W \cap M)}{P(M)}$$

Bayes' rule will be encountered again in Chapter 15, where we take a closer look at modern decision theory. In Equation 2.14, the various values of $P(A_i)$ are called *a priori probabilities; $P(A_i)$* represents the probability assessment that event A_i will occur *before* further sample information is obtained. Event B is interpreted as a particular result from further sample information. By employing Bayes' rule, we can find $P(A_i|B)$ for each i. The probability that event A_i will occur may be altered by the occurrence of event B from the sample information, and Bayes' rule provides a consistent mechanism for altering the value of $P(A_i)$. The values of $P(A_i|B)$ are called *posterior probabilities;* they represent the revised probability that event A_i will occur *after* the sample result B has been observed.

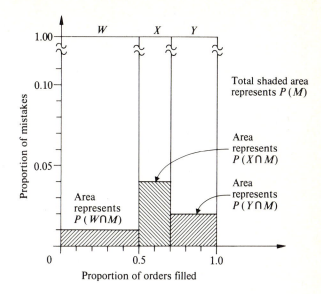

FIGURE 2.3 Graphical representation of Bayes' rule.

SUMMARY

This chapter has presented various concepts of sets, sample spaces, events, and set relationships such as the union and intersection of sets. The three definitions or interpretations of probability were followed by the rules of addition and multiplication for probabilities. Such concepts as collectively exhaustive events, mutually exclusive events, and independent events influence how these rules of addition and multiplication can be employed and interpreted. Finally, Bayes' rule was formulated using these rules and concepts.

This chapter is introductory in nature and is not intended as a complete and thorough treatment of either set theory or probability theory. Probabilities are used to qualify statistical inferences. We have simply taken the first step in developing techniques for statistical inference.

APPENDIX 2   COUNTING THE COUNTABLE

The a priori definition of probability required the experiment to be explicit enough to list all possible outcomes. Even if we could list all possible outcomes, it may be an extremely time-consuming process.

Consider a deck of 52 playing cards. If we shuffled the deck, randomly picked one card, recorded its "face value" (e.g., jack, queen, etc.) and its "suit" (e.g., hearts, spades, etc.), and repeated the process 4 more times, we would have a "five-card list." We are sampling, replacing the selected card after each draw, from a population of 52 cards. How many different "five-card lists" could we generate from this experiment? Well, there are 52 different cards that could appear as the first item on one list of five items. Since we replace the card each time, then on the second (third, fourth, and fifth) draw there are 52 possible cards that could appear as the second (third, fourth, or fifth) item on our list. Thus, there are

$$(52)(52)(52)(52)(52) = 380,204,032$$

different "five-card lists" from this experiment.

If each of the N elements in a population is equally likely to occur, and if we are taking a sample of K elements, with replacement, from this population, then there are N^K different samples that could be selected.

Now assume we perform a similar experiment. We take the first five cards off the top of a well-shuffled deck and record them *in the order they are selected*. By sampling without replacement, we have a smaller number of "five-card lists." Previously, we could have had five "ace of spades" entries for one list; now if we select the "ace of spades" on the first card, there is no way to see that card again in the next four. The top card on the deck could be any one of the 52, but once it is drawn, the second card can be any one of 51 cards, and so forth. Hence, there are only

$$(52)(51)(50)(49)(48) = 311,875,200$$

different "five-card lists" from this experiment.

To generalize this result, we need to introduce a new definition. The term "n factorial" is written as $n!$ and it represents the *operation* of multiplying together all the integer values from 1 to n. That is,

$$n! = (n)(n-1)(n-2) \cdots (3)(2)(1)$$

For example, "5 factorial" is $5! = (5)(4)(3)(2)(1) = 120$, "6 factorial" is $6! = (6)5! = (6)(5)(4)(3)(2)(1) = 720$, and "7 factorial" is $7! = (7)6! = (7)(6)5! = (7)(6)(5)(4)(3)(2)(1) = 5,040$. Check your knowledge of this concept by showing that "10 factorial" is 3,628,800. As a matter of definition, $1!$ and $0!$ are both equal to 1.

With this definition of factorials, the *permutation* of n things taken k (where $k < n$) at a time is calculated from

$$\frac{n!}{(n-k)!}$$

Note that in our "five-card lists" where we sampled without replacement, the total number of possible outcomes could be considered as "the permutation of 52 things taken 5 at a time" or

$$\frac{n!}{(n-k)!} = \frac{52!}{(52-5)!} = \frac{52!}{47!}$$

Fortunately, you do not have to calculate either 52! or 47! to get an answer. Please note that

$$\frac{52!}{47!} = \frac{(52)(51)(50)(49)(48)47!}{47!} = (52)(51)(50)(49)(48)$$

which (we already discovered) is equal to 311,875,200.

The permutation considers the ordering of the five cards to be important. That is, it considers the following two 5-card lists to be different.

First card:	Ace of Spades	King of Spades
Second card:	King of Spades	Ace of Spades
Third card:	Queen of Spades	Queen of Spades
Fourth card:	Jack of Spades	Jack of Spades
Fifth card:	10 of Spades	10 of Spades

Why? Because the order in which the "ace" and "king" appeared is different. What if we did not care about the order in which the five cards appeared? As long as the same five cards were in two samples, and the samples differed only by the order in which the five cards appeared, we want that to be called "one possible outcome." That is, in order for outcome A to be "different" from outcome B, at least one card in B has to be different from any of the five cards in outcome A.

When the order of the elements within a sample is not important, the *combination* of n things taken k at a time is calculated from

$$\frac{n!}{k!\,(n-k)!}$$

For example, the number of different five-card hands that could occur by dealing off the top five cards from a deck of 52 playing cards is

$$\frac{n!}{k!\,(n-k)!} = \frac{52!}{5!\,(52-5)!} = \frac{52!}{5!\,47!}$$

$$= \frac{(52)(51)(50)(49)(48)47!}{(5)(4)(3)(2)(1)47!} = 2,598,960$$

Example 2.5   The Ice Cream Store

Dippy Whip has five flavors of ice cream: *A*lmond, *B*lueberry, *C*hocolate, *S*trawberry, and *V*anilla.

(a) *The Triple-Decker Cone.* The "kiddie favorite" is three dips where each dip may or may not be the same flavor as any other dip. How many different triple-decker cones can be made with the five flavors?

First, we have to decide what "different" means. Some children want the top two dips to be chocolate and the bottom dip to be vanilla; others want the top dip to be vanilla and the bottom two dips to be chocolate. Hence, the following are examples of different flavor arrangements:

C	V	C	C	S	C	S	V	V	V
C	C	V	V	C	S	V	S	C	V ⋯
V	C	C	S	V	V	C	C	S	V

For the bottom dip, there are five possible flavors. Since the next dip can repeat the flavor of the first, then there are five possible flavors for the middle dip. Likewise, there are five possible flavors for the top dip. Thus, there are

$$N^k = (5)^3 = (5)(5)(5) = 125$$

different triple-decker cones that could be made with five flavors.

(*b*) *The Skyscraper-Rainbow.* The "teenie bopper" favorite is a four-dip cone where each dip is a different flavor. Individuality is not lost here; the order in which those four flavors are stacked on the cone is important to these customers. How many different skyscraper-rainbow cones can be constructed from five flavors?

In this case, all the following represent "different cones":

A	B	A	B	A	A	A	A
B	A	B	A	C	C	S	S ⋯
C	C	S	S	B	S	B	C
S	S	C	C	S	B	C	B

Since the order of the elements in a sample is important, then this is a problem of the permutation of n things taken k at a time.

$$\frac{n!}{(n-k)!} = \frac{5!}{(5-4)!} = \frac{5!}{1!} = \frac{(5)(4)(3)(2)(1)}{(1)} = 120$$

different skyscraper-rainbow cones that can be made from five flavors.

(*c*) *The Bananaless Split.* The diet-conscious adult ice cream addicts are fond of a sundae which consists of three different flavors in a plastic bowl. How many bananaless splits can be concocted from five flavors?

Since the three different flavors are placed in a bowl, the customer can choose the order in which each flavor is devoured. A *A-B-C* bananaless split is no different than a *B-A-C* split. Since the order of flavors is not important, this is a problem of the combination of n objects taken k at a time. There are

$$\frac{n!}{k!\,(n-k)!} = \frac{5!}{3!\,(5-3)!} = \frac{(5)(4)(3)2!}{(3)(2)(1)2!} = 10$$

different bananaless splits that can be constructed from five flavors.

EXERCISES

1. Given the following sets:
$A = \{1, 2, 3, 4, 5\}$ $C = \{1, 3, 5, 7, 9\}$
$B = \{1, 4, 9, 16, 25\}$ $D = \{0, 2, 4, 6, 8\}$
what are the elements contained in the sets

a.	$A \cap B$	**g.**	$C \cup D$
b.	$A \cup B$	**h.**	$C \cap D$
c.	$A \cap C$	**i.**	$A \cup B \cup C$
d.	$A \cup C$	**j.**	$A \cap B \cap C$
e.	$A^* \cap C$	**k.**	$(B \cap C) \cup D$
f.	$A^* \cap B$	**l.**	$B \cap (C \cup D)$

2. Given the following sets:
$A = \{6, 7, 8, 9\}$ $C = \{1, 3, 5, 9\}$
$B = \{0, 5, 10\}$ $D = \{0, 4, 9\}$
what elements are contained in

a.	$A \cup B$	**e.**	$A \cap (B \cup C)$
b.	$A \cap B$	**f.**	$B \cap D$
c.	$A \cap C$	**g.**	$(A \cup B) \cap D$
d.	$(A \cap B) \cup C$	**h.**	$B \cup C \cup D$

3. The unemployment rate for the next quarter is forecasted by an economic model. The model's forecast could be described as one of five events:
$A_1 = \{$Unemployment will be 10 percent or more$\}$
$A_2 = \{$Unemployment will be 8 percent or more, but below 10 percent$\}$
$A_3 = \{$Unemployment will be 6 percent or more, but below 8 percent$\}$
$A_4 = \{$Unemployment will be 4 percent or more, but below 6 percent$\}$
$A_5 = \{$Unemployment will be below 4 percent$\}$
Let B_i represent actual unemployment according to the same five classifications (for example, $B_1 = \{$Unemployment was 10 percent or more$\}$).
a. Are events A_1 through A_5 mutually exclusive? Are they collectively exhaustive?
b. In words, what do the following events indicate?
$A_2 \cap B_3$
$A_3 \cup A_4$
$A_i \cap B_i$
$A_i \cap B_j$ where $i > j$

4. A stock portfolio contains four common stocks. On a particular trading day, let
$A = \{$more than half of the stocks will go up in price$\}$
$B = \{$more than half of the stocks will go down in price$\}$
$C = \{$more than half of the stocks will not change in price$\}$
a. What does the event $\{A \cup C\}$ indicate?
b. What does the event $\{A \cap B\}$ indicate?
c. Are events A and B mutually exclusive? What about A and C? B and C?
d. Are A, B, and C collectively exhaustive events?
e. In defining events A, B, and C, if "more than half" were changed to "half or more," would A and B be mutually exclusive? What about A and C? B and C? Are A, B, and C collectively exhaustive?

5. A fair coin is flipped twice.
a. List all possible outcomes.
b. What probability is associated with each outcome?

c. What is the probability of obtaining *exactly* one head?
d. What is the probability of obtaining *at least* one head?

6. Two dice, one red and one green, are tossed.
a. List all possible outcomes.
b. What is the probability that both show a 1.

7. The computer programmers in a department have the following characteristics:

A_1	29 program in (at least) FORTRAN
A_2	21 program in (at least) BASIC
A_3	21 program in (at least) COBOL
$A_1 \cap A_2$	12 program in both FORTRAN and BASIC
$A_1 \cap A_3$	15 program in both FORTRAN and COBOL
$A_2 \cap A_3$	13 program in both BASIC and COBOL
$A_1 \cap A_2 \cap A_3$	9 program in all three

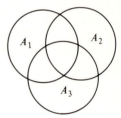

a. Use a *Venn diagram*, the three overlapping circles at the right, to determine how many programmers are in the department.
b. Find $P(A_1 \cup A_2)$.
c. Find $P(A_1 \cap A_2^*)$.
d. Find $P(A_1 \cup A_2 \cup A_3^*)$.

8. Consider the following place-of-work versus place-of-residence data:

Zone of Work (W)	Number of Workers			
	Zone of Residence (R)			Total Workers
	1	2	3	
1	80	40	10	130
2	70	60	20	150
3	50	50	20	120
4	200	150	50	400
Total Residents	400	300	100	800

For example, of the 300 people who live in zone 2, 40 work in zone 1, 60 in zone 2, 50 in zone 3, and 150 in zone 4. Land use in zone 4 is for commercial-industrial purposes only. There are 800 workers who live on this island; no one leaves the island for work; no one comes onto the island for work.

Let $R_i = \{$a worker who resides in zone $i\}$ and $W_i = \{$a worker who works in zone $i\}$.

Experiment A. A randomly selected worker is telephoned at home. (The selection process was performed in complete ignorance of zone of residence or zone of employment.)
a. What is the probability that the worker resides in zone 2?
b. Find $P(R_1)$.
c. What is the probability that the worker works in zone 2?

d. Find $P(W_1)$.

e. Are your answers to parts *a* to *d* marginal, joint, or conditional probabilities?

f. What is the probability that the worker both resides and works in zone 1?

g. What is $P(R_2 \cap W_1)$?

h. What is $P(W_2 \cap R_1)$?

i. Are your answers to parts *f* to *h* marginal, joint, or conditional probabilities?

j. Suppose the worker tells you that he/she lives in zone 2.

    (1) What is the probability that he/she works in zone 1?

    (2) Use your last answer, the answer to part *a*, and the multiplication rule to find $P(W_1 \cap R_2)$. Why should your answer be identical to part *g* above?

    (3) In part (2), why is $P(W_1 \cap R_2) \neq P(W_1) \cdot P(R_2)$?

k. Suppose the worker tells you that he/she works in zone 1.

    (1) What is the probability that the worker lives in zone 3?

    (2) What is $P(R_2 | W_1)$?

    (3) Use your answers from part (2) and part *d* to find $P(W_1 \cap R_2)$. Note that this is the same as the answer to part *j*(2).

l. What is the probability that the worker either resides or works (or both) in zone 1?

    (1) Use just the table above to answer.

    (2) Use the general rule of addition to answer.

    (3) Why not use the special rule of addition?

m. Which of the following are mutually exclusive events?

    (1) R_1 and W_2 (4) R_1 and R_2

    (2) R_4 and W_4 (5) W_2 and R_2

    (3) W_4 and R_2 (6) W_1 and W_4

Experiment B. Aggregate the zone of work classifications into two groups: industrial area versus residential area. That is, let $X_1 = \{W_1 \cup W_2 \cup W_3\} = \{$a worker who works in an area which is zoned for residential use$\}$, and let $X_2 = W_4 = \{$a worker who works in the area zoned only for commercial-industrial use$\}$. Randomly select one worker.

n. Find $P(X_1)$ and $P(X_2)$.

o. Find $P(X_1 | R_1)$, $P(X_1 | R_2)$, and $P(X_1 | R_3)$.

p. Are area of work place (that is, X_1 or X_2) and zone of residence independent? Why or why not?

Experiment C. Aggregate data according to "works in zone of residence" versus "works in zone outside of zone of residence." Let $Y_1 = \{$a person who resides and works in the same zone$\} = \{(R_1 \cap W_1) \cup (R_2 \cap W_2) \cup (R_3 \cap W_3)\}$, and $Y_2 = \{$a person who works in one zone and resides in another$\}$. Randomly select one worker.

q. Find $P(Y_1)$ and $P(Y_2)$.

r. Find $P(Y_2 | R_1)$, $P(Y_2 | R_2)$, and $P(Y_2 | R_3)$.

s. Are the Y's and R's independent?

t. Express your answer to part *s* in words.

u. Without calculating anything else, does $P(R_1) = P(R_1 | Y_1) = P(R_1 | Y_2)$?

Experiment D. You go to zone 4 during work hours and randomly select one and then another (but different) worker.

v. What is the probability that they both reside in zone 2?

w. For part *v*, why is the special rule for multiplication inappropriate?

x. In terms of the worker's zone of residence, list all possible results to this experiment. (*Hint:* R_3, $R_2 = $ the first person selected resides in zone 3 and the second person resides in zone 2.)

y. Find P (*at least* one selected worker resides in zone 1).

9. Which of the following represent three events which are both (1) collectively exhaustive and (2) pairwise mutually exclusive (that is, $\{A \cap B\} = \{A \cap C\} = \{B \cap C\} = \{\phi\}$).
a. $P(A) = 0.6$, $P(B) = 0.2$, $P(C) = 0.1$, and $P(A \cap B) = 0$
b. $P(A) = 0.1$, $P(B) = 0.4$, $P(C) = 0.5$, $P(A \cup B) = P(C)$, $P(A \cup C) = 0.6$, and $P(B \cap C) = 0$
c. $P(A) = P(B) = 0.2$, $P(C) = 0.6$, $P(A \cap B) = 0$, and $P(A \cup C) = P(B \cup C) = 0.8$
d. $P(A) = P(B) = P(C) = 0.35$ and $P(A \cap B) = P(A \cap C) = 0$

10. A theater has only one movie projector. The bulb in the projector is working; the probability that it will burn out before the movie ends is 0.40. Of the 20 spare bulbs, one is unnoticeably defective. Of the remaining spare bulbs, the probability is 0.20 that they will burn out before the movie ends.
a. What is the probability that the current bulb will burn out and, randomly selecting a spare, the defective bulb will be chosen? Assume independence.
b. What is the probability that the current bulb will burn out, a nondefective bulb will be randomly chosen to replace it, and that bulb will burn out before the movie ends?

11. You have been appointed as project director for an advisory group in a foreign country. En route you are studying some background information concerning your new staff, which you have not met. You find:

Staff Classification	Job Classification	Number of Persons
(A_1) Professional Staff	(B_1) Economist	10
	(B_2) Chemist	6
	(B_3) Engineer	14
(A_2) Support Staff	(B_4) Secretary	20
	(B_5) Runner (Go-for)	5

Everyone has just one job classification (i.e., no one is both a chemist and an engineer). Half of the secretaries, chemists, and economists are female, but only four of the engineers and none of the runners are of that gender. Everyone speaks English except fifteen secretaries and all the runners. Everyone speaks French except one male and one female chemist, six of the male engineers, and two of the female economists. Let M = males, M^* = females, F = French-speaking, and E = English-speaking.

Experiment A. One person is selected at random from these 55.
a. Which of the following pairs of events are mutually exclusive?
  (1) Runner and English-speaking
  (2) B_4 and F^*
  (3) F^* and M^*
b. Are sex and job classification independent?
c. Are sex and staff classification independent?

Experiment B. The professional staff draws straws to see who will meet you at the airport.
d. What is the probability that you will be met by a French-speaking female? Or, $P(\{F \cap M^*\}|A_1) = ?$
e. What is the probability that you will be met by a chemist or an economist? Or, $P(\{B_1 \cup B_2\}|A_1) = ?$

Experiment C. Not to be outdone, the·support staff draw straws to send one of their members to the airport.
f. Find $P(\{F \cap M^*\}|A_2)$.
g. Find $P(E)$.

Experiment D. The day after you arrive, the entire professional staff is scheduled to visit Big Joe's Plantation. Two runners are with you and Big Joe; the rest of the professional staff is scattered all over the plantation. Big Joe asks a question which requires consultation with either an engineer or a chemist. You tell a runner to summon either an engineer or a chemist; he replies "Yes, boss!" and takes off.

There are a few things you do not know. First, the runner does not know an economist from a chemist or an engineer. But, you are a new boss, and he does not want to ask questions. Second, in order to get right back for the next message, the runner will run near the first professional staff member he sees (a random selection), blurt out the message in French, and return too quickly for anyone else to give him a message. Finally, the runner is never sure who speaks French, and he assumes that everyone does.

h. What is the probability that the message is delivered to a French-speaking professional staff member? Or, $P(F|A_1) = ?$

i. What is the probability that the message is delivered to a chemist or an engineer who speaks French? Or, $P[\{(F \cap B_2) \cup (F \cap B_3)\}|A_1] = ?$

Experiment E. Two neighboring plantation owners arrive. They both have economic-type questions, but they are not anxious to have each other included in the discussion. Before the first runner returns, you send the second runner (whose behavior is identical to the first) in search of two economists.

j. What is the probability that two economists will show up? Or, $P[(\{B_1 \cap F\}|A_1)_{\text{first}} \cap (\{B_1 \cap F\}|A_1)_{\text{second}}] = ?$

k. As the discussion continues, you find that the two visiting plantation owners hate intelligent women. What is the probability that two women will show up? $P(\{M^* \cap F|A_1\}_{\text{first}} \cap \{M^* \cap F|A_1\}_{\text{second}}) = ?$

12. A T-group (i.e., a sensitivity training group) of sixteen persons is meeting for the first time. The group consists of 6 males, two of them have previous T-group experience, three are black, and three of the white males have not had previous T-group experience. Of the females, three are black, eight have not had previous T-group experience, and only one white female has had previous T-group experience.

a. Complete the following table.

	Male (F*)			Female (F)		
	Black	White	Total	Black	White	Total
Previous T-group Experience (E)						
No previous T-group Experience (E*)						
Total						

Experiment. Only one person is selected at random.

What is the probability of selecting:

b. a female? $[P(F) = ?]$

c. a white male? $[P(W \cap F^*) = ?]$

d. a white female with previous T-group experience? $[P(W \cap F \cap E) = ?]$

e. a person with previous T-group experience? $[P(E) = ?]$

f. a male or a black person? $[P(F^* \cup B) = ?]$

g. a male given the person has no T-group experience? $[P(F^*|E^*) = ?]$

h. a person with T-group experience given a white person? $[P(E|W) = ?]$

i. Consider *only* those persons without previous T-group experience. From this "subpopulation," are "race" and "sex" independent? [*Hint:* Is $P(B|F^*) = P(B|F) = P(B)$? Or, is $P(F|B) = P(F|W)$ $= P(F)$?]

j. Consider *only* those persons with previous T-group experience. From this subpopulation are "race" and "sex" independent?

13. The pointer on the car radio's tuning knob is broken; you can change stations, but you do not know which station you have. From past experience you know the percentage of programming time, for the only four stations in town, is:

	Radio Station			
	ROCK	**JAZZ**	**DANC**	**WILD**
News and Weather	2%	10%	10%	7%
Community Affairs	8	15	5	30
Drama Series	0	15	30	8
Music	80	50	55	30
Athletic Events	10	10	0	25
	100%	100%	100%	100%

a. If you have station WILD, what is the probability that either drama or music will be on?

b. An auto mechanic was listening to the radio while working on your car, and you assess the probability of the radio being tuned to a particular station to be $P(ROCK) = 0.5$, $P(WILD) = 0.3$, and $P(JAZZ) = P(DANC) = 0.1$. You pick up the car, turn on the radio, and the broadcast is a community affairs program. Use Bayes' rule to find $P(ROCK|CA)$.

c. Does the assumption of independence make any sense here? That is, if there is a news broadcast on one station, does that affect the probability that there will be a news broadcast on some other station?

CASE QUESTIONS

1. Consider the 100 households in the Illinois sample as a population. Find the number of households in each cell of the following two tables.

Table 1 Households broken down by (*a*) gas-heated (i.e., V22 < 2), (*b*) number of persons in household (V3), and (*c*) number of bedrooms (V23). (Use the following form.)

V23 Number of Bedrooms	V3 Number of Persons 1 2 3 4 5 or more	Total
0 1 $\vdots$ 5+ 6(NA)		
Total		

Table 2 Households broken down by (a) not heated by gas (i.e., V22 ≥ 2), (b) number of persons, and (c) number of bedrooms.

V23 Number of Bedrooms	V3 Number of Persons					Total
	1	2	3	4	5 or more	
0						
1						
⋮						
5+						
6(NA)						
Total						

[*Hint:*   Table 2 is the same as Table 1 except that V22 ≥ 2 in Table 2. Go through the 100 values for V22 and "check off" those values greater than 1. Table 1 is for those households *not* checked, and Table 2 is for checked households. To complete Table 1, go back through the unchecked housing units and tally (mark the appropriate cell) each housing unit according to the value of V3 and V23. To complete Table 2, repeat the process for checked housing units.]

Experiment.   One housing unit is randomly selected from the Illinois sample.
a.   Write, in your own words, the meaning of the following sets: $(\{V22 < 2\} \cap \{V3 = 2\})$, $\{V3 > V23\}$, $(\{V23 = 0\} \cup \{V22 \geq 2\})$, and $(\{V23 \leq 1\} \cap \{V3 \geq 4\})$.
b.   Find: $P(V22 < 2)$, P(housing unit is not heated by gas), $P(V23 = 1)$, P(housing unit has no bedroom), $P(V23 \leq 1)$, P(housing unit has more than one bedroom), $P(\{V22 < 2\} \cap \{V23 \leq 1\})$, P(housing unit is occupied by three persons and is heated by gas), $P(\{V22 < 2\} \cap \{V23 = 2\} \cap \{V3 = 3\})$, P(the housing unit is either not heated by gas or it has five or more persons living there), $P(V3 > V23)$, and $P(\{V22 \geq 2\} \cup \{V23 \leq 2\} \cup \{V3 \leq 2\})$.
c.   "Four persons per family is about average" is frequently contended. For this data, find $P(V3 = 4)$, $P(V3 > 4)$, $P(V3 < 4)$, and $P(V3 \neq 4)$.
d.   For these 100 housing units, would you say that gas is a popular source of home heat?
e.   Which of the following pairs of events are mutually exclusive? $\{V22 < 2\}$ and $\{V23 = 4\}$, $\{V22 \geq 2\}$ and $\{V23 = 4\}$, and $\{V3 = 1\}$ and $\{V23 \geq 5\}$.

2.   Consider the Illinois sample as a population.
Experiment A.   One family is selected at random. Find the probability that:
a.   The family income will be $10,000 or more; find $P(V32 \geq 100)$.
b.   The head of household is female; find $P(V34 = 1)$.
c.   The head of household attended college; find $P(V36 \geq 15)$.
d.   The head of household worked less than 40 hours last week; find $P(V38 \leq 3)$.
e.   The family income is below the poverty cutoff; find $P(V42 \leq 2)$.

Experiment B.   Two families are randomly selected with replacement. Find the probability that:
f.   Both families will have an income of $10,000 or more.
g.   The head of household will be male for both families.
h.   Only one of the two heads of household has attended college.
i.   At least one of the heads of household worked less than 40 hours last week.
j.   At most one of the families is below the poverty cutoff.

Experiment C.   Two families are randomly selected without replacement. Repeat parts *f* through *j* from experiment B above.

3

RANDOM VARIABLES
AND PROBABILITY DISTRIBUTIONS

INTRODUCTION

In Chapter 2, we discussed the probability of particular events. We recognize that an experiment might result in several different outcomes (or events). With the concepts from Chapter 2, we should be able to list all possible outcomes and find (or assess) the probability associated with each outcome. However, there are some practical problems which accompany that procedure. For example, if there are an infinite number of possible outcomes, then we can conceptually discuss the list of all possible outcomes, but to actually construct such a list is impractical.

It would be convenient to have some relation between each outcome and its associated probability of occurrence. Then we could discuss and investigate that relation, as a single entity, rather than a list of outcomes and probabilities. Thus, we need to introduce the concepts of a *random variable* and a *probability distribution*.

Different probability distributions will have different characteristics, and we will attempt to describe those characteristics with such measures as the *expected value*, the *variance*, and *skewness*. In the next chapter, we will investigate actual or empirical distributions and similar measures for their characteristics. Our final task in this chapter will be the definition of two specific theoretical distributions.

We are still in the process of building the foundations for statistical inference.

3.1 RANDOM VARIABLES—SOME CONCEPTUAL DIFFERENCES

Consider the following experiments:

Experiment A   We take a part from an assembly line and classify it as defective or nondefective.

Experiment B   We have a new coffee blend. We ask a potential consumer to try a cup and classify the taste as "very bitter," "slightly bitter," or "mild."

Experiment C   Out of the country's 200 largest cities, we randomly select one city and find the total tax revenue collected in property tax during 1975.

Experiment D   From a set of used pencils, we are going to randomly select one and measure its length.

In experiment A there are only two outcomes: defective and nondefective. By carefully defining the conditions which would cause the part to be called "defective" and by sufficient inspection of the part, these two outcomes are collectively exhaustive (no other outcome could occur) and mutually exclusive (a part could not be simultaneously defective and nondefective).

The two outcomes are denoted by words. The words "defective" and "nondefective" are quite clear in suggesting all possible outcomes, but they are cumbersome in the attempt to describe all outcomes with one variable. We could use a nominal scale and assign one number to the occurrence of a defect and another number to the occurrence of a nondefective part. For example, let 1 denote the occurrence of a defect and 0 denote the occurrence of a nondefect, or

---

Rule A $$W = \begin{cases} 0 & \text{if part is nondefective} \\ 1 & \text{if part is defective} \end{cases}$$

We could now call W a *discrete random variable*. W is a variable because it can assume more than one value; it is a random variable because the value $W = 1$ will occur by chance. The adjective "discrete" will be reconsidered after we look at all four experiments. For the moment note that W can assume only a finite number of values: 0 and 1.

The value $W = 1$ is not greater than the value $W = 0$; the 0 and 1 just denote different outcomes. Hence, W is measured on a nominal scale. We could have used any two numbers to denote the pair of possible outcomes; 0 and 1 are commonly used because of their algebraic convenience. Finally note that the 0 and 1 only have significance insofar as they are defined by rule A. Rule A is a well-specified mechanism for assigning a number to each possible outcome of experiment A.

In experiment B we have a slightly different situation. We have three different outcomes expressed in words. The three different outcomes represent one person's opinion (or rank ordering) about the relative bitterness of the coffee. In assigning a number to each outcome, we would probably want to preserve the order that the possible outcomes represent; we would prefer the use of an ordinal scale to that of a nominal scale. For example, we might create a bitterness scale as follows:

Rule B $$B = \begin{cases} 0 & \text{if the response is "very bitter"} \\ 1 & \text{if the response is "slightly bitter"} \\ 2 & \text{if the response is "mild"} \end{cases}$$

Of course, we could reverse the order of these three numbers, if we wish to define B in that fashion. Or, as long as we specify what the three numbers indicate (in a manner similar to rule B), we could use any three different numbers to denote the three possible outcomes. Once again, we could call B a *discrete random variable*.

In experiment C, we have an easier task when we attempt to assign numbers to each possible outcome, but the number of possible outcomes is considerably larger. The variable of interest, the city's property tax revenue for 1 year, is measured on a ratio scale. For example, let

---

Rule C $T =$ the city's 1975 property tax revenue, rounded to the nearest dollar

---

We could also call T a *discrete random variable*. Obviously, T can assume a multitude of values, and it is random because the value of T will depend upon the chance selection of a city. But, why is it discrete?

Consider a finite but relevant range of possible values for T; say, $\$0 \leq T \leq \$300$ billion. Within that rather sizeable interval, according to rule C, T can assume only a finite number of values. Specifically, T can have the values 0, 1, 2, ..., 300 billion, which is 300,000,000,001 possible values. Now, that may be large, but it is finite. Within a finite interval, $\$0 \leq T \leq \$300$ billion, T can assume only a finite number of values; therefore, we say T is a *discrete* random variable.

Recall that in experiments A and B the *values* of the discrete random variables W and B had meaning only if you knew rules A and B, respectively. While that may seem less important in experiment C, it is also true that the meaning of each value of T is uniquely dependent upon rule C. Conceptually, we could define T as "*twice* the property tax revenue" or "the property tax revenue *squared*," and T would still be a discrete random variable. Such definitions of T would make the value of T much more difficult to interpret, but that may be your purpose. Confidential financial data may be coded by some such weird rule, and only if you knew the rule or decoding formula would the recorded number have any meaning.

In experiment D, we have our last case before we attempt to generalize the results. The length of the pencil is measured on a ratio scale, and we could let

---

Rule D $L =$ length of the pencil using a metric scale

We would call L a *continuous random variable*. It is a variable because it could assume more than one value. It is a random variable because it is by chance that any given pencil is selected. But, why is L called a *continuous* random variable?

Consider some finite range of possible values for L, say $0 \leq L \leq 10$ centimeters. There are an infinite number of possible values for L within that interval. Our measuring device may limit us to the nearest millimeter or the nearest angstrom (one hundred-millionth of a centimeter), but the *exact* length of that pencil can be any one of an infinite number of values between 0 and 10 centimeters in length.[1] Thus, we call *L* a *continuous* random variable.

We can now generalize the results of investigating these four experiments. First, we start with an experiment. Second, we consider all possible outcomes. Finally, we define a *random variable* as some well-specified rule or mechanism by which we can assign a numerical value to each outcome. The outcomes must represent mutually exclusive events. Whatever well-specified rule we choose to use for a particular experiment, *each outcome must receive one and only one particular value.* However, two outcomes may be assigned the same value. From rule C, for example, *T* is assigned the dollar value of the property tax revenue for the particular city that we select; for a given outcome there is only one value that can be assigned to *T*. This does not prevent two cities from having *exactly* the same property tax revenue; both cities would be assigned the same value of *T*.

Finally, random variables come in two flavors: discrete and continuous. To distinguish between them, we have suggested the following test: (1) Define the random variable; specify some rule for assigning a value of X to each possible outcome. (2) Consider some finite, nonzero, and relevant range of possible X values; say, $-N \leq X \leq +N$, where N is finite. (3) If the *number* of possible X values within that range is finite (i.e., there are a limited number), then X is a *discrete* random variable; if the *number* of possible X values within that range is infinite (or unlimited), then X is a *continuous* random variable.

3.2 DISCRETE RANDOM VARIABLES
AND PROBABILITY DISTRIBUTIONS

In many practical business problems, a discrete random variable is used for the enumeration of how many items fall within a particular category. For a sample of size 10, the number of defective flash bulbs in a production batch, the number of accounts overdue in the records of a retail store, the number of families with more than one automobile in a particular city, and the number of checking accounts in a bank with a minimum balance of $10 or less are all examples using a discrete random variable.

Random variables are used to assign numbers to outcomes. If we can find the probability of occurrence for each outcome, then we can associate a probability with each value of the random variable. This process will lead us to the concept of a *probability distribution.*

---

[1] The scale represents a continuum of numbers. In mathematical terms, the interval contains all the *real* numbers between 0 and 10. It contains all the *rational* numbers: the integers (0, 1, 2, etc.) and the ratio of integers ($1/4$, $3/16$, $1/9$, etc.). For example, the fraction $1/9 = 0.1111 \ldots$ is an endless decimal, but the digits occur in a repeating pattern. Also, it includes irrational numbers, such as $\sqrt{19} = 4.359\ldots$, which are endless, nonrepeating decimals.

To review the concept of a random variable, to demonstrate the use of a discrete random variable for enumeration of items in particular categories, and to demonstrate how we might associate probabilities with values of a discrete random variable, we will start with a simple coin-flipping example.

Example 3.1   Three Tosses of a Fair Coin

The experiment is three tosses of a fair coin. What are the possible outcomes of this experiment?

HHH	HHT	HTT	TTT
	HTH	THT	
	THH	TTH	

Now, there are several random variables, or rules to assign a numerical value to each outcome, which can be used for this experiment. First, consider Q = "the number of heads" in the three tosses. The possible values of Q are 0, 1, 2, 3. How you define the rule which assigns a number to each outcome depends upon what aspect of each outcome interests you. Presumably, to use Q we are interested in the number of heads in the three tosses.

If we are interested in the number of tails in three tosses, we simply define R = "the number of tails," and R can assume values of 0, 1, 2, and 3 as shown in Table 3.1. Imagine we are interested in the number of consecutive pairs of heads in three tosses, and call that random variable S. Then S can have values of 0, 1, and 2. Also, we could consider the number of consecutive pairs, be they heads or tails. Call that random variable V which may have values of 0, 1, and 2. Or we could conceive a situation where you received $10 for each head and you paid $5 for each tail. Let W = net gain from three tosses; thus W may have values of $-15, 0, 15$, and 30 only. Finally, let X = the square root of the number of heads; therefore, X can be equal to only 0, 1, 1.414, and 1.732.

The choice of which rule is arbitrary; however, the rule itself is not arbitrary. Instead, the rule for generating values of the random variable should reflect that information (i.e., the number of heads, or the number of consecutive pairs) which is most relevant to the problem you wish to solve.

TABLE 3.1   VALUES OF SEVERAL RANDOM VARIABLES TO REPRESENT EACH OUTCOME

Outcome	Q	R	S	V	W	X	Probability
HHH	3	0	2	2	30	1.732	$1/8$
HHT	2	1	1	1	15	1.414	$1/8$
HTH	2	1	0	0	15	1.414	$1/8$
THH	2	1	1	1	15	1.414	$1/8$
HTT	1	2	0	1	0	1.000	$1/8$
THT	1	2	0	0	0	1.000	$1/8$
TTH	1	2	0	1	0	1.000	$1/8$
TTT	0	3	0	2	-15	0	$1/8$

Is the random variable Q discrete or continuous? Choose some finite range of possible values, say between -10 and $+10$. Since there are only a finite number (i.e., four) of possible values of Q in that range, Q is a discrete random variable.

Note that we have listed the eight possible outcomes for the three tosses, but we have not assigned a probability of occurrence to each outcome. Recognize that this is the type of experiment where we may employ the classical or a priori definition of probability.

First, is each possible outcome equally likely? Assuming they are, there are eight mutually exclusive outcomes; we could assign a probability of $\frac{1}{8}$ to each outcome. Each outcome is equally likely, but we will want to demonstrate the proposition.

By the term "fair coin" we mean that on a single toss a head and a tail are equally likely. If we assume that there are only two possible mutually exclusive outcomes [i.e., assume $P(\text{edge}) = 0$], then $P(H) = P(T)$ and $P(H) + P(T) = 1$. This implies that $P(H) = P(T) = \frac{1}{2}$. Also, we can assume that the outcome of one toss will not influence the probability of a head (or a tail) on the next toss; the three tosses are independent events. Thus,

$$P(H \cap H \cap H) = P(H)P(H)P(H) = (\tfrac{1}{2})(\tfrac{1}{2})(\tfrac{1}{2}) = \tfrac{1}{8}$$

$$P(H \cap H \cap T) = P(H)P(H)P(T) = (\tfrac{1}{2})(\tfrac{1}{2})(\tfrac{1}{2}) = \tfrac{1}{8}$$

$$P(H \cap T \cap H) = P(H)P(T)P(H) = (\tfrac{1}{2})(\tfrac{1}{2})(\tfrac{1}{2}) = \tfrac{1}{8}$$

and so forth.

The Concept of the Discrete Probability Distribution

In example 3.1 we specified several random variables for one experiment: the three tosses of the coin. It was almost as an afterthought that we calculated the probability for each of the eight mutually exclusive possible outcomes. A *probability distribution* is a paired list of all possible values of a random variable and the probability associated with each value of that random variable.

Let us investigate the definition of probability distribution step by step. (1) We must have an exhaustive list of the possible *values* of the random variable. Remember that one value of the random variable may represent more than one possible outcome from the experiment (e.g., "one head" or $Q = 1$ in example 3.1, could be generated by three outcomes: HTT, THT, TTH). (2) We need the *probability* of each value of the random variable occurring. If each value of the random variable is mutually exclusive of all other possible values and if we have an exhaustive list of possible (random variable) values, then the sum of the probabilities will equal 1.

Example 3.2   Probability Distribution for the "Number of Heads"

Consider the random variable Q, the number of heads in three tosses of a fair coin, from example 3.1. The four possible values for Q are 0, 1, 2, and 3. The probability that Q will equal 0 is the probability of getting only tails in the three tosses: $P(Q = 0) = P(T \cap T \cap T) = \frac{1}{8}$. What is the probability that Q will equal 1? What is the probability,

TABLE 3.2 PROBABILITY DISTRIBUTION FOR Q

Possible Values for Q	Probability of Each Q Value
0	$\frac{1}{8}$
1	$\frac{3}{8}$
2	$\frac{3}{8}$
3	$\frac{1}{8}$
	$\frac{8}{8}$

in three tosses of a fair coin, of getting only one head? $P(Q = 1) = P[(\text{HTT}) \cup (\text{THT}) \cup (\text{TTH})] = ?$ Since each of the three possible outcomes where only one head occurs is mutually exclusive of the other two, then $P(Q = 1) = P(\text{HTT}) + P(\text{THT}) + P(\text{TTH}) = \frac{1}{8} + \frac{1}{8} + \frac{1}{8} = \frac{3}{8}$. In like manner, $P(Q = 2) = \frac{3}{8}$ and $P(Q = 3) = \frac{1}{8}$. The *probability distribution* for Q is given in Table 3.2.

Note that since Q is a discrete random variable, the resulting distribution is called a *discrete probability distribution*.

The discrete probability distribution is frequently represented in both tabular and diagrammatic form. The probability distributions for the six random variables in example 3.1 are presented in Table 3.3. Be sure to check each probability distribution with the information in Table 3.1.

3.3 CONTINUOUS RANDOM VARIABLES AND PROBABILITY DISTRIBUTIONS

If the experiment of interest generates a finite number of possible outcomes, then the random variable will have only a finite number of values which yields a discrete random variable. If the experiment of interest could generate an infinite number of possible outcomes, then the random variable may be either discrete or continuous. The test remains: in a finite range of values, can the random variable assume a finite or an infinite number of values? If it can assume an infinite number of values, then we must consider a continuous random variable.

For example, consider the circle in Figure 3.1 which has a pointer pivoted at the center. A clock face, which would be numbered from 0 to 12, or a compass, which would be numbered 0 to 360, would also serve; however, we will use Figure 3.1 because it's easier to compare with example 3.1. Note that the circumference of this circle is measured from 0 to 8. If this circular device were lying level and we spun the pointer, the pointer could come to rest at any point on the circumference. Since there are an infinite number of points on the circumference of a circle, there are an infinite number of possible outcomes from this experiment.

TABLE 3.3 PROBABILITY DISTRIBUTIONS FROM EXAMPLE 3.1

Q	$P(Q)$
0	$\frac{1}{8}$
1	$\frac{3}{8}$
2	$\frac{3}{8}$
3	$\frac{1}{8}$
	$\frac{8}{8}$

Q = number of heads

R	$P(R)$
0	$\frac{1}{8}$
1	$\frac{3}{8}$
2	$\frac{3}{8}$
3	$\frac{1}{8}$
	$\frac{8}{8}$

R = number of tails

S	$P(S)$
0	$\frac{5}{8}$
1	$\frac{2}{8}$
2	$\frac{1}{8}$
	$\frac{8}{8}$

S = number of consecutive heads

V	$P(V)$
0	$\frac{2}{8}$
1	$\frac{4}{8}$
2	$\frac{2}{8}$
	$\frac{8}{8}$

V = number of consecutive heads or tails

W	$P(W)$
-15	$\frac{1}{8}$
0	$\frac{3}{8}$
$+15$	$\frac{3}{8}$
$+30$	$\frac{1}{8}$
	$\frac{8}{8}$

W = net gain

X	$P(X)$
0.000	$\frac{1}{8}$
1.000	$\frac{3}{8}$
1.414	$\frac{3}{8}$
1.732	$\frac{1}{8}$
	$\frac{8}{8}$

X = the square root of the number of heads

FIGURE 3.1 Spinning pointer.

Let us now consider the random variable R, which is concerned with whether the pointer lands in the top half or the bottom half of the circle. That is, let $R = 0$ if the pointer comes to rest above the "6–2 line" and let $R = 1$ if the pointer stops on or below the "6–2 line." Within a finite range of values for R, say between -10 and $+10$, there are only a finite number of values that R may assume (that is, 0 and 1). R is a discrete random variable.

Not many economic problems are more clearly understood by spinning a pointer (although several national policies could not be damaged by such a procedure); therefore, this circle must have some relevance besides being another example of a discrete random variable. Indeed, consider the circumference to be 8 units in length, and let X be a random variable which represents the distance (measured in a clockwise direction) from point 0 to where the pointer stops. Within any finite range (e.g., between 2 and 3), there are an infinite number of possible X values. X is a *continuous* random variable.

What is so special about a continuous random variable? Nothing. The special concerns arise from trying to assign probabilities to particular values of the continuous random variable. Since the variable has an infinite number of values, the "list" will be quite long. But, this is not much of a problem; we can use functional notation. We do, nevertheless, have some problems.

First, what is the probability that X will exactly equal 2? We will define the probability that X is equal to any specific value as being equal to 0†, or $P(X = k) = 0$. Therefore, we will refer to the probability for a specific *range* of X values. For example $P(2 \leq X \leq 3)$ equals the probability that the pointer will stop on or between 2 and 3. Using the classical, or a priori, definition of probability and noting that the total circumference is 8 units (e.g., inches, millimeters, etc.), we can denote that infinite number of points between 2 and 3 as $(3 - 2 =)$ 1 unit. Representing the infinite number of points on the entire circumference as 8 units implies that

---

† Consider the fraction $1/n$. As n becomes larger, the fraction $1/n$ becomes smaller (that is, $\frac{1}{2} = 0.5$, $\frac{1}{3} = 0.33$, $\frac{1}{4} = 0.25$, etc.). The fraction $1/\infty$ is infinitesimal (or incalculably small) and is approximated by calling it equal to 0. Since there are an infinite number of points on the circumference of the circle, and with the classical definition of probability, the probability of stopping on one point is $1/\infty$, or 0.

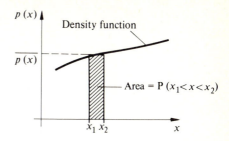

FIGURE 3.2 Hypothetical probability density function.

$P(2 \le X \le 3) = (1 \text{ unit})/(8 \text{ units}) = \frac{1}{8}$. In like manner, $P(2 \le X \le 5) = \frac{3}{8} = 0.375$ and $P(2 \le X \le 2.1) = 0.1/8 = 0.0125$. In general, if $x_1 \le x_2 \le 8$, then $P(x_1 \le X \le x_2) = (x_2 - x_1)/8$, or the circumferential difference between x_2 and x_1 divided by the total circumference.

Second, if the $P(X = x_1) = 0$ for a continuous random variable, how do we generate a probability distribution for X? In a sense, we don't! Instead, we use a new concept, the *probability density function*. The probability density function, $p(x)$, is defined as that function where, between x_1 and x_2, the area under $p(x)$ is equal to $P(x_1 \le X \le x_2)$. In addition, the total area under the probability density function must equal 1; the pointer must come to rest somewhere on the circumference. (See Figure 3.2.)

The spinning pointer in Figure 3.1 and the random variable X (i.e., that distance from 0 to some resting point) have a unique property that permits us to generate the probability density function without much formal mathematics. If we had a rectangle, then the base times the altitude would give us the area of the rectangle. Since we know x_1 and x_2 ($x_2 - x_1 =$ base) and the probability that X will be in that range $[P(x_1 \le X \le x_2) = \text{area}]$, we should be able to find the average value of the density function, $\overline{p(x)}$, for each range of X values.

With the a priori definition of probability we previously reasoned that $P(2 \le X \le 3) = \frac{1}{8}$. Thus, in the range from 2 to 3 the average value of the probability density function is $\overline{p(x)} = [P(x_1 \le X \le x_2)]/(x_2 - x_1) = (\frac{1}{8})/(3 - 2) = \frac{1}{8}$. From what we have previously calculated, consider the average values for the probability density function given in Table 3.4.

What does Table 3.4 suggest? The average value of the probability density function is equal to $\frac{1}{8}$ regardless of the width of the X interval. One possibility, which is the case, is that $p(x) = \frac{1}{8}$ for $0 \le X \le 8$ and $p(x) = 0$ for $X < 0$ or $X > 8$. What does this probability density function of X look like? Observe Figure 3.3.

First notice that the total area under the probability density function $[(\frac{1}{8}) \cdot (8)]$ is equal to 1. Again this represents the assumption that if we spin the pointer, it will come to rest at some spot on the circumference of the circle. Next, the shaded area

TABLE 3.4 SAMPLE CALCULATIONS FOR AVERAGE VALUES OF THE PROBABILITY DENSITY FUNCTION FOR X

Range of X Values $x_2 - x_1$ [Base of Rectangle]	Probability $P(x_1 \leq X \leq x_2)$ [Area of Rectangle]	Average Value of Density Function $\overline{p(x)}$ [Altitude = Area/Base]
$3 - 2 = 1$	$\frac{1}{8}$	$(\frac{1}{8})/1 = \frac{1}{8}$
$5 - 2 = 3$	$\frac{3}{8}$	$(\frac{3}{8})/3 = \frac{1}{8}$
$2.1 - 2.0 = 0.1$	$0.1/8$	$(0.1/8)/0.1 = \frac{1}{8}$

should represent $P(2 \leq X \leq 3)$. Since $(3 - 2) \cdot (\frac{1}{8}) = \frac{1}{8}$, we have some confidence that $p(x) = \frac{1}{8}$ conforms to our definition of a probability density function. This is an example of a *uniform* probability density function.

Comparison of a Uniform Continuous and a Uniform Discrete Distribution

Referring to example 3.1, the three tosses of a fair coin, let us define the random variable T as follows:

T	Outcome	Probability
0.5	HHH	$\frac{1}{8}$
1.5	HHT	$\frac{1}{8}$
2.5	HTH	$\frac{1}{8}$
3.5	THH	$\frac{1}{8}$
4.5	HTT	$\frac{1}{8}$
5.5	THT	$\frac{1}{8}$
6.5	TTH	$\frac{1}{8}$
7.5	TTT	$\frac{1}{8}$

FIGURE 3.3 Probability density function for continuous random variable x.

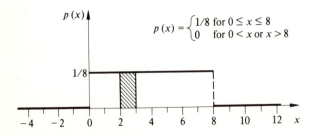

$$p(x) = \begin{cases} 1/8 & \text{for } 0 \leq x \leq 8 \\ 0 & \text{for } 0 < x \text{ or } x > 8 \end{cases}$$

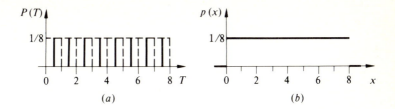

FIGURE 3.4 Comparison of a discrete and continuous distribution. (*a*) Three tosses of a coin; random variable *T*; (*b*) one spin of the pointer in Figure 3.1; random variable *X*.

The random variable T is being used simply as a coding device; it does not refer to the number of heads, etc. The discrete random variable T has a probability distribution as shown by the heavy dark lines in Figure 3.4*a*. Even though T is discrete, we could ask the question: What is the probability that T will be between 0 and 1? Or, $P(0 \le T \le 1) = ?$ Between 0 and 1, T can assume only one value $(T = 0.5)$; however, $P(0 \le T \le 1)$ is a valid statement. Thus, $P(0 \le T \le 1) = P(T = \frac{1}{2}) = \frac{1}{8}$. The same is true for $P(1 \le T \le 2), P(2 \le T \le 3), \ldots, P(7 \le T \le 8)$. The dashed lines in Figure 3.4*a* diagrammatically present the same concept.

Random variables T and X refer to entirely different experiments; T is discrete and X is continuous. Yet the dashed-line version of the probability distribution for T is strikingly similar to the probability density function for X.

Nonuniform Probability Density Functions

The uniform probability density function of the type shown in Figure 3.3 is perhaps the easiest to understand, but unfortunately it is seldom useful for problems in business and economics. Even a casual view of Table 3.3 demonstrates several nonuniform probability distributions. Discrete random variables Q, R, V, and W appear to have inverted U shapes in Table 3.3. Variable S has a reversed J-shaped probability distribution, and variable X has a longer tail to the left than to the right.

In section 3.6 we will consider some frequently used theoretical density functions. In order to visualize an experiment whose density function is similar but intuitive and easy to calculate, let us consider one more example.

Example 3.3 "Stop-Action Picture" of a Pendulum

In Figure 3.5*a* we have a iron arrowhead attached to an arm to form a pendulum. The pendulum is fastened to a semicircular board whose circumference is numbered from 0 to 4, as shown. Assuming some air friction, if we hold the pendulum at 0 and let go, the pendulum will swing back and forth but not quite reach the altitude it attained on the last swing.

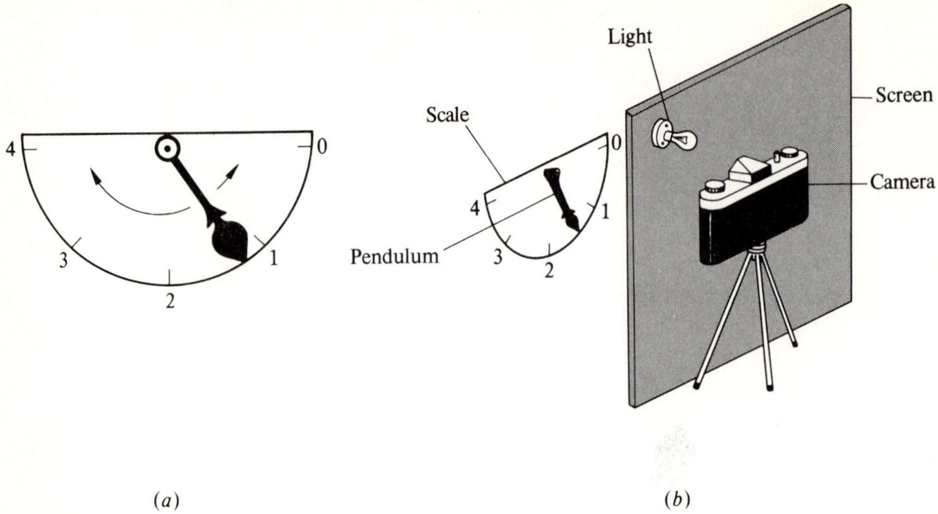

FIGURE 3.5 (a) Pendulum on a scale; (b) experimental setup.

Further, suppose we set up a screen or blind in front of the pendulum scale. The screen has a small hole in it through which we can focus a camera that can take stop-action-type photographs. We place nontransparent tape over the viewfinder on the back of the camera and find a person who has absolutely no idea what the experiment is about. Assume it takes the pendulum 8 minutes to come to a complete stop. Also assume that a light at the top of the screen goes on the moment the pendulum is released from the 0 position on the scale.

We tell our unsuspecting friend that after the light goes on, she or he has 8 minutes in which the camera may be triggered only once.[2] Now let us define the random variable Z as that point on the scale where the pendulum is "caught" by the camera. Z could assume an infinite number of values between 0 and 4; however, unlike the case of the pointer in Figure 3.1, these values of Z are not all equally likely. At least intuitively, there is a much higher probability that the picture would reveal the pendulum between 1.8 and 2.2 than between 0 and 0.4 or between 3.6 and 4.0.

So what does the probability density function for this continuous random variable look like? A rough approximation is presented in Figure 3.6a. For different possible intervals of Z, the area under the probability density function should be the probability that the camera will catch the pendulum within that interval. The area of a right triangle is $\frac{1}{2} \cdot$ (altitude) $\cdot$ (base). First consider $P(3.6 \leq Z \leq 4.0)$. The altitude is given by the density function; $p(z = 3.6) = 1 - (3.6)/4 = 1 - 0.9 = 0.1$. The base is $4.0 - 3.6 = 0.4$; therefore, the area, or $P(3.6 \leq Z \leq 4.0)$, equals $(\frac{1}{2})(0.1)(0.4) = 0.02$.

---

[2] We could use our "spinning pointer" in Figure 3.1 to randomly select the time when the picture is taken.

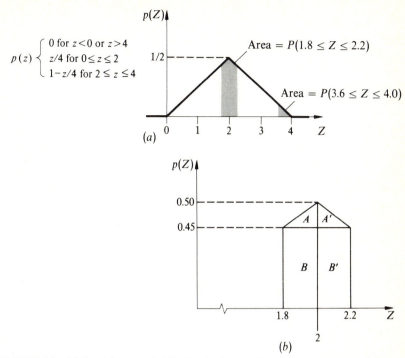

FIGURE 3.6 (*a*) Approximate probability density function for random variable Z; (*b*) Four sections of $P(1.8 \leq Z \leq 2.2)$.

In order to find $P(1.8 \leq Z \leq 2.2)$, there are several areas that we must add together. In Figure 3.6*b* we have designated the four areas as A, A', B, and B'. The area of A is equal to the area of A', as is true for B and B'. In order to find the altitude of area B, we must evaluate the probability density function at $Z = 1.8$; or $p(Z = 1.8) = (1.8)/4 = 0.45$. [Note that $p(Z = 2.2) = 1 - (2.2)/4 = 1 - 0.55 = 0.45$.] Thus, area B (and B') is equal to $(2 - 1.8)(0.45) = 0.09$.

The altitude of area A is $0.50 - 0.45 = 0.05$, and its base is $2 - 1.8 = 0.2$; therefore the area of triangle A is $(0.5) \cdot (0.2) \cdot (0.05) = 0.005$, which is identical to the area of A'. Hence, $P(1.8 \leq Z \leq 2.2) =$ area $A +$ area $A' +$ area $B +$ area $B' = 0.005 + 0.005 + 0.090 + 0.090 = 0.190$.

Finally, notice that the probability density function for Z is symmetric: the right-hand side is a mirror image of the left-hand side. The camera is just as likely to catch the pendulum on the right side of the scale as on the left side, or $P(0 \leq Z \leq 2) = P(2 \leq Z \leq 4) = 0.5$. Also since the area under the density function is equal to 1, then $P(0 \leq Z \leq 3.6) = P(0 \leq Z \leq 4.0) - P(3.6 \leq Z \leq 4.0) = 1 - 0.02 = 0.98$.

You will no doubt be happy to learn that, excluding the problems at the end of this chapter and the appendices, we will not have to calculate probabilities *directly* from density functions in the rest of this book. Although we will consider several probability density functions, the "areas" under these functions (i.e., the probabilities) have been calculated and placed in tables at the end of the text. The importance of understanding the concept of the probability density function is that it reduces the mystery of where those tables came from and why you have to add or subtract "areas."

* Postscript on Formalizing the Probability Density Function[3]

Although these comments are not necessary to understand the concept of the density function, one should not discriminate against those students who have studied the calculus. From the definition of the probability density function, the calculus student should recognize that an alternative definition could be written as

$$P(x_1 \le X \le x_2) = \int_{x_1}^{x_2} p(x)\, dx$$

This simply states the probability that the continuous random variable X will lie between x_1 and x_2 is the area under the density function between x_1 and x_2. From example 3.3, we could find that

$$P(1.8 \le z \le 2.2) = \int_{1.8}^{2.0} \frac{z}{4}\, dz + \int_{2.0}^{2.2} \left(1 - \frac{z}{4}\right) dz$$

$$= \frac{z^2}{8}\bigg|_{1.8}^{2.0} + \left[z - \frac{z^2}{8}\right]\bigg|_{2.0}^{2.2}$$

$$= (0.500 - 0.405) + (1.595 - 1.500)$$

$$= 0.095 + 0.095 = 0.19$$

3.4 MATHEMATICAL EXPECTATIONS AND VARIANCE

At this point, two characteristics of a probability distribution, and their relation to a probability density function, should be obvious. Table 3.5 summarizes two characteristics that are inherent in the definitions of probability.

These two characteristics will assist us in determining whether a probability distribution has been properly constructed; however, these two characteristics are

---

[3] Starred sections, such as this one, can be omitted without loss of continuity.

TABLE 3.5   TWO CHARACTERISTICS

(a) Probability Distribution (Discrete Random Variable)	(b) Density Function (Continuous Random Variable)
1. $P(X) \geq 0$ for all possible values of X	$p(x) \geq 0$ for all real values of X
2. $\sum_{i=1}^{N} P(X_i) = 1$	$\int_{-\infty}^{+\infty} p(x)\, dx = 1$; the area under the density function is equal to 1

true for all probability distributions. Hence, they do not help us "visualize" a particular probability distribution.

If we return to Table 3.3, we see that the probability distributions for random variables Q and R are identical. In that same table, using Q as a reference, we note that the probability distribution of S has a "tail" extending to the right; the probability distribution of V is symmetrical but different; W's is symmetrical but more "spread out"; and X's has a longer "tail" to the left than to the right. Either the tabular or the diagrammatic presentation in Table 3.3 perfectly describes these distributions and their differences. But what if the experiment had been "in 20 tosses of a fair coin"? Each of the random variables in Table 3.3 would have had many more possible values (for example, Q could assume values of 0, 1, ..., 20). The tabular values would be harder to decipher, and the diagrams would be more intricate.

It would be convenient to have just a few pieces of information, or measured characteristics, about a distribution that would permit us to visualize how two distributions generally differ from each other. We might be willing to sacrifice the complete details about two distributions in order to know in general how the two distributions differ or are similar. For that objective, we could consider the following:

1. *Measures of Location.* Given that some values of the random variable may have higher probabilities of occurring than others, we want some sort of "average" value for the random variable—a single number that would indicate the location of the distribution along the horizontal axis.

2. *Measures of Dispersion or Variation.* Is the distribution relatively tightly packed or very spread out? (Compare Q and W in Table 3.3.) In one number we would like to have some impression of the compactness of the distribution.

3. *Measure of Skewness.* Is the distribution symmetric? If a distribution is asymmetric, or lopsided, we say it is "skewed." If the distribution has a longer tail to the right than to the left (and if the horizontal axis has

algebraically larger numbers as we go to the right), we say the distribution is skewed to the right, or *positively* skewed. Conversely, if the left tail is longer, we say the distribution is skewed to the left, or *negatively* skewed. Although these definitions are susceptible to Freudian slip, they are among the more lucid in statistical nomenclature.

In Chapter 4 we will need more than one measure for each of these three properties of an empirical (i.e., generated from "real world" data) distribution. For theoretical distributions, however, we will generally need to consider only one measure of location and one measure of variation. We do not need a measure of skewness for theoretical distributions because we usually have sufficient information, without calculating anything, to know whether the theoretical distribution is symmetric, skewed to the right, or skewed to the left. Our measure of location is called the *mathematical expectation*, or the *expected value*, of the random variable; our measure of dispersion is the *variance* of the random variable.

Mathematical Expectation

Suppose you are responsible for creating and running a game booth at the annual SHS (Society for Humanizing Statisticians) Fund Raising Fair which is traditionally held on February 30. From past experience you are aware that the SHS Fair is usually attended by only those who are hooked on fairs. Your game must be very simple and, since you are running it, efficient. You decide upon a game and call it "Hat Money." You procure a top hat and place 5 one-dollar bills, 4 five-dollar bills, and 1 ten-dollar bill in the hat. Each contestant is permitted to reach into the hat and pull out only one bill (if more than one bill is selected, the contestant gets nothing—a sufficient incentive to exclude this as a possibility) which the contestant may keep. Now you must decide what to charge a contestant for playing this game.

Suppose that this game will be played many times during the day and that you would like to average $1 per person in net revenue, or profit. That is, (price to play) − (average "winnings" per play) = $1. In order to decide how much you will charge to play Hat Money, we will need to calculate the "average winnings per play," which we shall call the expected value of the discrete random variable W.

Let W equal the amount of money a contestant could win in a single draw from the hat; W can assume values of 1, 5, and 10 only. Let us assume that each bill, regardless of denomination, has an equal chance of being selected; therefore, using the a priori definition of probability, the probability distribution for W is given in Table 3.6.

For the sake of argument, suppose the game were played ($N =$) 100 times during the day. Hence, we would expect that $[P(W_1)N = (0.5)(100) =]$ 50 times the contestants would draw a one-dollar bill for an SHS loss of $50; 40 times the contestants would draw a five-dollar bill for an SHS loss of $200; and 10 times the contestants would draw a ten-dollar bill for an SHS loss of $100. Since our total

TABLE 3.6 PROBABILITY DISTRIBUTION AND EXPECTED VALUE FOR THE DISCRETE RANDOM VARIABLE W

Probability Distribution (a)		Calculations for $E(W)$ (b)	Diagrammatic Representation of Distribution and $E(W)$ (c)
W	$P(W)$	$W \cdot P(W)$	
1	0.50	0.50	
5	0.40	2.00	
10	0.10	1.00	
	1.00	$E(W) = 3.50$	

SHS loss (or contestant's "total winnings") is $50 + $200 + $100 = $350 for 100 contestants, our average loss per contestant is $350/100 = $3.50. In order to average $1 in profit per customer, we should charge $4.50 for the privilege of drawing a bill from the hat.

Another way of considering the average SHS loss per contestant is:

$$\frac{P(W_1) \cdot N \cdot W_1 + P(W_2) \cdot N \cdot W_2 + P(W_3) \cdot N \cdot W_3}{N}$$

$$= P(W_1) \cdot W_1 + P(W_2) \cdot W_2 + P(W_3) \cdot W_3$$
$$= (0.5)(1) + (0.4)(5) + (0.1)(10)$$
$$= 3.5$$

Note that N drops out of the calculation. We may now generalize these results.

The *expected value*[4] (or mean or mathematical expectation) of the discrete random variable X is defined as

$$E(X) = \sum_{i=1}^{N} [P(X_i) \cdot X_i] = P(X_1) \cdot X_1 + P(X_2) \cdot X_2 + \cdots + P(X_N) \cdot X_N \qquad \textbf{(3.1)}$$

---

[4] If X is a continuous random variable with a probability density function given by $p(x)$, then

$$E(X) = \int_{-\infty}^{\infty} p(x) \cdot x \cdot dx$$

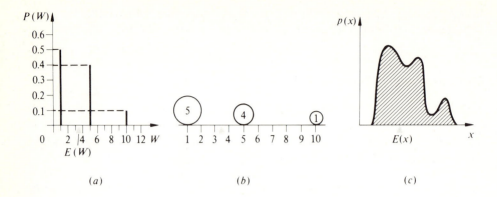

FIGURE 3.7 The "balancing" characteristic of the expected value. (*a*) Random variable *W*; (*b*) "weights" analogy; (*c*) hypothetical density function.

We started by looking for a measure of location; so, how is the expected value such a measure? First, note that *expected value* of a random variable represents an *average value* of that random variable. Each possible value of the random variable is included in the calculation of the expected value. Further, since each value of X may not be of equal importance, each value of X is "weighted" by the probability of that value occurring [that is, $P(X_1) \cdot X_1$].

The expected value is an average value of the random variable if the experiment is conducted many times. Note that the expected value may not be a value that the random variable can assume in one try of the experiment [for example, $E(W) = 3.5$ even though the possible values of W are 1, 5, and 10].

In a more mechanical sense, the expected value is the point along the horizontal axis that will "balance" the distribution. Consider the analogy provided in Figure 3.7.

Figure 3.7*a* is simply a slightly different version of Table 3.6*c*. In the best tradition of classical physics, assume a weightless board 9 feet long. At the left end place a 5-pound weight and at the right end a 1-pound weight; 4 feet from the left end place a 4-pound weight. Where would we place a fulcrum in order to balance this system of board and weights? By placing the fulcrum $2\frac{1}{2}$ feet from the left end of the board we can balance this system.

In like manner, the "balancing point" of a probability density function is given by the expected value of the continuous random variable. Figure 3.7*c* depicts this relationship for a hypothetical random variable X.

To check your understanding of the expected value, return to Table 3.3, calculate the expected value for each of the discrete random variables, and draw in the fulcrum on each diagram in Table 3.5.[5] Compare your values for the expected value with the general location of the distribution on the horizontal axis.

The Variance of a Random Variable

Now that we have a way of measuring the location of the probability distribution, the question now becomes: how do we represent, preferably in one number, whether the distribution is spread out or close together? If we intend to use the expected value in conjunction with a measure of dispersion to describe a distribution, then the distribution about the expected value is a relevant (although not unique) method for considering the dispersion of the possible values of a random variable. We shall use the expected value as a reference point.

First, we should note that this concept of dispersion requires some measure of the *distance* between a particular value of the random variable and the expected value [that is, $X_1 - E(X)$]. This distance would be all that is necessary if all values of a discrete random variable had equal importance (or equal probability of occurring). Frequently this is not the case; more often than not, some values of the random variable will have a higher probability of occurrence than others. Thus, we need some way of weighing each distance to reflect their differences in relative importance.

This line of reasoning would tend to indicate that all we need to do is measure the distance between each value of the discrete random variable [that is, $X_i - E(X)$] and weigh this distance by the probability of that value occurring (that is, $[X_i - E(X)]P(X_i)$). By adding all these weighted distances, $\Sigma[\{X_i - E(X)\}P(X_i)]$, we would have our measure of dispersion. Our logic is fine, but this measure will not permit us to distinguish one probability distribution from another.

---

[5] Your answers should be $E(Q) = 1.5$, $E(R) = 1.5$, $E(S) = 0.5$, $E(V) = 1$, $E(W) = 7.5$, and $E(X) = 1.12175$. Further, for those who wish to mess with the calculus, from Figure 3.3,

$$E(X) = \int_{-\infty}^{+\infty} p(x)\, dx = \int_{-\infty}^{0} (0)x\, dx + \int_{0}^{8} \tfrac{1}{8}x\, dx + \int_{8}^{\infty} (0)x\, dx = \frac{x^2}{16}\Big|_0^8 = 4.$$

From Figure 3.6,

$$E(z) = \int_{-\infty}^{0} (0)z\, dz + \int_{0}^{2} \left(\frac{z}{4}\right)z\, dz + \int_{2}^{4} \left(1 - \frac{z}{4}\right)z\, dz + \int_{4}^{\infty} (0)z\, dz = 0 + \frac{z^3}{12}\Big|_0^2 + \frac{z^2}{2}\Big|_2^4 - \frac{z^3}{12}\Big|_2^4 + 0$$

$$= (\tfrac{2}{3} - 0) + (8 - 2) - (\tfrac{16}{3} - \tfrac{2}{3})$$

$$= 2$$

Recall that the two characteristics presented in Table 3.5 did not permit us to distinguish one probability distribution from another because they are true for all probability distributions. Unfortunately,

$$\Sigma\{[X_i - E(X)]P(X_i)\} = 0$$

for all probability distributions[6]; therefore, it is not useful as a measure of dispersion. It yields the same numerical value (that is, 0) regardless of which distribution is being considered.

There are several ways of avoiding this difficulty and maintaining our "weighted distance" idea for measuring dispersion. By using the difference of a particular value from the expected value (that is, $[X_i - E(X)]$), our measure of "distance" was sometimes positive and sometimes negative. By squaring that difference (that is, $[X_i - E(X)]^2$) we still have a measure of distance, but the numerical value is always positive. Then we can weight our "new" measure of distance by the probability of that value of X occurring, and we have a measure of dispersion. Mathematicians have traditionally used this measure, which they call the *variance* of the random variable.

The *variance* of a discrete[7] random variable X is defined as

$$
\begin{aligned}
\text{Var}(X) = \sigma_X^2 &= \sum_{i=1}^{N} \{[X_i - E(X)]^2 P(X_i)\} \\
&= [X_1 - E(X)]^2 P(X_1) + [X_2 - E(X)]^2 P(X_2) + \cdots \\
&\quad + [X_N - E(X)]^2 P(X_N)
\end{aligned}
\tag{3.3}
$$

---

[6] Proof:

$$
\begin{aligned}
\Sigma\{[X_i - E(X)] \cdot P(X_i)\} &= [X_1 - E(X)] \cdot P(X_1) + [X_2 - E(X)] \cdot P(X_2) + \cdots \\
&\quad + [X_N - E(X)] \cdot P(X_N) \\
&= [X_1 P(X_1) + X_2 P(X_2) + \cdots + X_N P(X_N)] \\
&\quad - E(X)[P(X_1) + P(X_2) + \cdots + P(X_N)] \\
&= E(X) - E(X)(1) = 0
\end{aligned}
$$

[7] If X is a continuous random variable and $p(x)$ is its probability density function, then

$$\text{Var}(X) = \int_{-\infty}^{+\infty} [x - E(x)]^2 p(x) \, dx$$

For example, from Figure 3.3, $\text{Var}(X) = 5.33$; and from Figure 3.6, $\text{Var}(Z) = \frac{2}{3}$.

There are several ways of representing the variance of X. $\mathrm{Var}(X)$, while descriptive, is sometimes cumbersome; σ_X^2 or σ^2 is more frequently used.[8] Now let us consider how the numerical value of σ^2 reflects the dispersion about the expected value.

Consider the six probability distributions in Panel 3.1. Each distribution has the same expected value. In Panel 3.1a the discrete random variable G can assume only the value 2; therefore, there is no dispersion about the expected value, and the variance is 0. The distribution of G is uninteresting except to demonstrate the minimum possible value of the variance.

Note that as we go from the probability distribution for random variable G to the one for M, the expected value or measure of location is the same, but the variance is larger [that is, $\mathrm{Var}(G) < \mathrm{Var}(H) < \mathrm{Var}(J) < \mathrm{Var}(K) < \mathrm{Var}(L) < \mathrm{Var}(M)$]. Random variables H, J, and K can assume the values 1, 2, and 3; however, the probabilities of them assuming those values differ. The fact that $\mathrm{Var}(K) > \mathrm{Var}(J) > \mathrm{Var}(H)$ reflects only the importance of the "weights" or probabilities in our measure of dispersion. By comparing the probability distributions of H and L, we can observe the influence of "distance" on the value of the variance of H and L.

From Panel 3.1 we should be able to infer that *the smaller the value of the variance, the more compact or tightly packed is the distribution about its expected value.* Or, the larger the value of the variance, the more spread out is the distribution from its expected value.

It should be clearer why a business executive, in considering an investment project, would use the variance as a crude measure of the risk associated with that investment. That is, if our random variable X is the net return in a particular year from an investment project, then the investor assigns (subjective) probabilities to various possible net returns occurring that year. The investor might want to maximize the expected return, $E(X)$, and minimize the risk, $\mathrm{Var}(X)$. Compare random variables G and M in Panel 3.1. The situation denoted by G is "a certainty of earning 2" (whatever 2 means); M represents a "4 or nothing" gamble which has more risk associated "with earning something." Although $E(G) = E(M)$, $\mathrm{Var}(M) > \mathrm{Var}(G)$.

Finally, note that if our random variable were measured in dollars, then the units on the expected value would be dollars; however, the units associated with the variance would be "dollars squared." Inches are units of measurement for length, and "inches squared" may represent area, but it is difficult to associate "dollars squared" with a similar concept. "Dollars squared" simply does not mean much to anyone. In order to keep the units on our measure of dispersion meaningful, we frequently use the square root of the variance, which is called the *standard deviation*, whose symbolic representation is, quite obviously, σ. Further, we always use the

---

[8] σ is the lowercase Greek letter sigma, while Σ is the uppercase representation of the same Greek letter. σ_X^2 is read as "sigma sub-X squared" and is just a symbol to represent "the variance of X." While σ^2 denotes some numerical value, Σ is a symbol for an *operator*. That is, Σ by itself does not have a numerical value; instead Σ is an "instruction" to *sum* the terms which follow it.

Panel 3.1 Probability Distributions with the Same Expected Values and Different Variances

(a) $E(G) = (1)(2) = 2$

$\text{Var}(G) = \sigma_G{}^2 = \Sigma\{[G_i - E(G)]^2 P(G_i)\}$

$\quad = (2 - 2)^2(1) = 0.0$

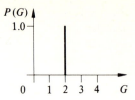

(b) $E(H) = (0.2)(1) + (0.6)(2) + (0.2)(3) = 2$

$\text{Var}(H) = \sigma_H{}^2 = \Sigma\{[H_i - E(H)]^2 P(H_i)\}$

$\quad = (1 - 2)^2(0.2) + (2 - 2)^2(0.6)$

$\quad + (3 - 2)^2(0.2)$

$\quad = 0.40$

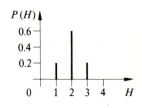

(c) $E(J) = (\tfrac{1}{3}) \cdot 1 + (\tfrac{1}{3}) \cdot 2 + (\tfrac{1}{3}) \cdot 3 = 2$

$\text{Var}(J) = \sigma_J{}^2 = \Sigma\{[J_i - E(J)]^2 P(J_i)\}$

$\quad = (1 - 2)^2(\tfrac{1}{3}) + (2 - 2)^2(\tfrac{1}{3})$

$\quad + (3 - 2)^2(\tfrac{1}{3})$

$\quad = \tfrac{2}{3} = 0.67$

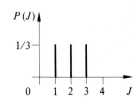

(d) $E(K) = (0.4) \cdot 1 + (0.2) \cdot 2 + (0.4) \cdot 3 = 2$

$\text{Var}(K) = \Sigma\{[K_i - E(K)]^2 P(K_i)\}$

$\quad = (1 - 2)^2(0.4) + (2 - 2)^2(0.2)$

$\quad + (3 - 2)^2(0.4)$

$\quad = 0.80$

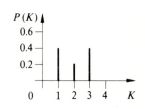

(e) $E(L) = (0.2) \cdot 0 + (0.6) \cdot 2 + (0.2) \cdot 4 = 2$

$\sigma_L{}^2 = (0 - 2)^2(0.2) + (2 - 2)^2(0.6)$

$\quad + (4 - 2)^2(0.2)$

$\quad = 1.60$

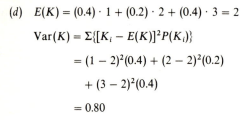
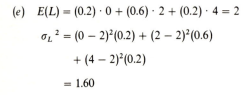

(f) $E(M) = (0.5) \cdot 0 + (0.5) \cdot 4 = 2$

$\sigma_M{}^2 = (0 - 2)^2(0.5) + (4 - 2)^2(0.5)$

$\quad = 2 + 2 = 4.00$

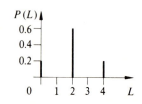

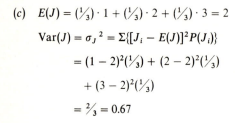
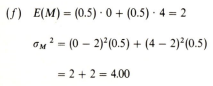

positive root of the variance; thus, the relationship between relative magnitudes of standard deviations is the same as relationships between the values of variances. For example, from Panel 3.1 we can see that $\sigma_H = \sqrt{0.4} = 0.632$ is less than $\sigma_J = \frac{2}{3} = 0.818$.

3.5 A DISCRETE THEORETICAL PROBABILITY DISTRIBUTION: THE BINOMIAL

The appropriateness of a theoretical probability distribution obviously depends upon the problem and/or experiment of interest to the investigator. The discrete theoretical distributions appropriate for many business and economics problems are the binomial, the multinomial, the Poisson, and the hypergeometric.

In order to pull together the several concepts presented thus far, we will consider the binomial distribution in this section and the normal, or Gaussian, distribution in the next. We will introduce other theoretical distributions as we need them.

The binomial distribution was presumably conceived by Jacobus (or Jacques I or Jahob) Bernoulli (1654–1705), and is occasionally called the Bernoulli distribution. Since Bernoulli was the first of at least ten notable mathematicians from the same lineage, Will Durant suggests that it was "... a family as remarkable as the Bachs, the Brueghels, and the Couperins for the social heredity of genius."[9] The problem of assigning a probability to "heredity of genius" could, in a crude form, be approached with a Bernoulli distribution. With no historial disrespect intended, we shall define a Bernoulli experiment and call the resulting discrete probability distribution a binomial distribution.

The Bernoulli Experiment

The Bernoulli experiment consists of n trials. The n trials must have the following properties:

1. On each trial, the *outcome* must be due to *chance*.
2. On each trial, all possible outcomes must be classified into *two mutually exlusive categories*. (Although they do not have to signify two degrees of "goodness," these two mutually exclusive categories are frequently called "success" and "failure.")
3. Each trial is *independent* of all the others. That is, the probability of a success on one trial is not influenced by the outcome of any other trial.
4. The *probability* of observing each category is *constant from one trial to the next*. (If π = probability of observing a success, then π has the same numerical value for all trials within the experiment.)

---

[9] Will and Ariel Durant, *The Age of Louis XIV*, Simon & Schuster, Inc., New York, 1963, p. 501.

This list is concise, but not too limiting. For example, consider two tosses of a fair coin. Since there are two tosses, we would say there are ($n = 2$) two trials. Each toss results in a chance outcome. If we let "success" = (heads) and "failure" = (tails or edge), then all possible outcomes on each trial have been classified into two mutually exclusive categories. Since we are using the same coin for both tosses, it is reasonable to assume that the probability of getting a head on the first toss is exactly the same as that of getting a head on the second toss; therefore, π is a constant from one trial to the next. Further, there is no reason to believe that the outcome on the first toss will influence the probability of getting a head on the second toss; the trials are independent events.

Consider these three experiments:

1. Randomly select three individuals as they enter the Empire State Building (as opposed to the men's gym or the women's room of a hotel) and record their sex.
2. Randomly select 10 automobiles as they pass a particular point on a limited-access highway and record them as American-made or foreign-made. (You would, of course, have to decide how to classify Capris, Opels, etc.)
3. Record the number of faulty labels on the four 1-gallon cans in a randomly selected case of house paint.

The number of trials is 3, 10, and 4 respectively. The outcome of each trial is determined by chance. There are only two outcomes: male or female, American or foreign, faulty or not faulty. The probability of a male, from experiment 1, or an American-made car, from experiment 2, is the same from one trial to the next; and the trials are independent. For the paint cans, however, we would want more knowledge of the sequence of labeling and packing paint cans before we would be willing to say that, in a given case, a faulty label on one can will not influence the probability of a faulty label on the next can. We would be suspicious of the independence of the four trials associated with the 1-gallon paint cans.

Example 3.4   Four Bernoulli Experiments

Consider two hypothetical societies: Balsex and Purdued. In Balsex, there is an equal probability of having a male or a female child. In Purdued, the probability of having a female child is 0.6; the probability of a male child is 0.4. We will assume that the sex of a particular child is independent of the sex of previous children in that family. We are interested in the number of females in a family of a particular size (e.g., two children).

Part A   We randomly select a two-child family from each society. What is the probability that there will be no female children in a family? One female child? Two female children?

How does this situation conform to the conditions for a Bernoulli experiment? The experiment is the personal experience of each family. Within a given family, each "live birth" is a Bernoulli "trial." The "outcome" of each trial is the child's sex, which is assumed to be a chance event. "Male" and "female" are the two, and only two, mutually exclusive categories for each outcome. We have assumed independence; thus, the probability of the child's sex being female (or male) is the same for all trials.

By listing first the sex of the first-born, the four possible sex sequences in a two-child family are (F, F), (F, M), (M, F), and (M, M). Because each trial is independent, the probability of each sex sequence is given by the following:

Sample Space	Number of Females	Probability	Balsex $P(F) = 0.5$	Purdued $P(F) = 0.6$
F, F	2	$P(F)P(F)$	$(0.5)(0.5) = 0.25$	$(0.6)(0.6) = 0.36$
F, M	1	$P(F)P(M)$	$(0.5)(0.5) = 0.25$	$(0.6)(0.4) = 0.24$
M, F	1	$P(M)P(F)$	$(0.5)(0.5) = 0.25$	$(0.4)(0.6) = 0.24$
M, M	0	$P(M)P(M)$	$(0.5)(0.5) = 0.25$	$(0.4)(0.4) = 0.16$
			1.00	1.00

If we let X = the number of females (or number of "successes"), then the *probability distributions* for X are:

Balsex		Purdued	
X	$P(X)$	X	$P(X)$
0	0.25	0	0.16
1	0.50	1	0.48
2	0.25	2	0.36
	1.00		1.00

Note that there are two mutually exclusive ways of obtaining exactly one female child: (F, M) and (M, F). Thus,

$$P(X = 1) = P[(F \cap M) \cup (M \cap F)] = P(F)P(M) + P(M)P(F)$$

Also, note that $P(F)P(M)$ will always equal $P(M)P(F)$ under the conditions of a Bernoulli trial; therefore,

$$P(X = 1) = 2P(M)P(F)$$

These two probability distributions are both called *binomial* distributions for ($n = 2$) two trials. There is a different probability associated with each value of X *not* because the points in the sample space are any different, but because the probability of a success [that is, $P(F)$ in this case] in one society is different from the other. Let us see what we can learn from increasing the number of trials.

Part B We randomly select a four-child family from each society. What is the probability that there will be X female children in a family? The discrete random variable X can now assume values of 0, 1, 2, 3, and 4. The sixteen possible sex sequences, and their respective probabilities of occurrence, are:

Sample Space	Number of Females	Balsex $P(F) = 0.5$	Purdued $P(F) = 0.6$
F, F, F, F	4	$(0.5)(0.5)(0.5)(0.5) = 0.0625$	$(0.6)(0.6)(0.6)(0.6) = 0.1296$
F, F, F, M	3	0.0625	$(0.6)(0.6)(0.6)(0.4) = 0.0864$
F, F, M, F	3	0.0625	$(0.6)(0.6)(0.4)(0.6) = 0.0864$
F, M, F, F	3	0.0625	$(0.6)(0.4)(0.6)(0.6) = 0.0864$
M, F, F, F	3	0.0625	$(0.4)(0.6)(0.6)(0.6) = 0.0864$
F, F, M, M	2	0.0625	$(0.6)(0.6)(0.4)(0.4) = 0.0576$
F, M, F, M	2	0.0625	$(0.6)(0.4)(0.6)(0.4) = 0.0576$
F, M, M, F	2	0.0625	$(0.6)(0.4)(0.4)(0.6) = 0.0576$
M, F, F, M	2	0.0625	$(0.4)(0.6)(0.6)(0.4) = 0.0576$
M, F, M, F	2	0.0625	$(0.4)(0.6)(0.4)(0.6) = 0.0576$
M, M, F, F	2	0.0625	$(0.4)(0.4)(0.6)(0.6) = 0.0576$
F, M, M, M	1	0.0625	$(0.6)(0.4)(0.4)(0.4) = 0.0384$
M, F, M, M	1	0.0625	$(0.4)(0.6)(0.4)(0.4) = 0.0384$
M, M, F, M	1	0.0625	$(0.4)(0.4)(0.6)(0.4) = 0.0384$
M, M, M, F	1	0.0625	$(0.4)(0.4)(0.4)(0.6) = 0.0384$
M, M, M, M	0	0.0625	$(0.4)(0.4)(0.4)(0.4) = 0.0256$
		1.0000	1.0000

Binomial Probability Distributions				
Balsex			**Purdued**	
X	$P(X)$	X		$P(X)$
0	$(1)(0.0625) = 0.0625$	0		$(1)(0.0256) = 0.0256$
1	$(4)(0.0625) = 0.2500$	1		$(4)(0.0384) = 0.1536$
2	$(6)(0.0625) = 0.3750$	2		$(6)(0.0576) = 0.3456$
3	$(4)(0.0625) = 0.2500$	3		$(4)(0.0864) = 0.3456$
4	$(1)(0.0625) = 0.0625$	4		$(1)(0.1296) = 0.1296$
	1.0000			1.0000

Generating the binomial probability distributions in this fashion takes a while, but the procedure is the same as the two-child-family case. Figure 3.8 shows our results in graphical form.

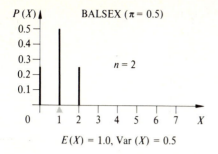

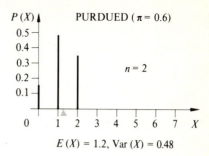

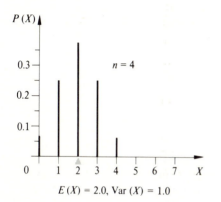

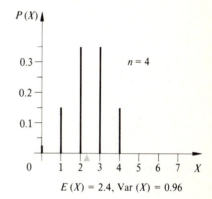

FIGURE 3.8 Binomial probability distributions.

What we have done is take the probability rules from Chapter 2 and apply them to a Bernoulli experiment to generate a binomial distribution. What if we ask: When we select a 10-child family, what is the probability that it will have two females? That is, for $n = 10$, what is $P(X = 2)$? In this situation, there are $2^{10} = 1,024$ sex sequences that we could observe. We could list them all and calculate the resulting distribution for X in the manner we just suffered through. There should be an easier way. There is.

The General Form of the Binomial Distribution

In order to understand the easier way of obtaining binomial probabilities, we have to carefully observe the calculations performed in example 3.4. In part B of example 3.4, note that the *number* of sex sequences that included exactly two females is the same regardless of the probability of a live female being born on any given trial. (Compare Balsex with Purdued.) We could generalize this by saying that we need a technique for calculating the number of ways we can get k successes from n trials.

Next, we should recognize that

$$P(F)P(F)P(F)P(M) = P(F)P(F)P(M)P(F)$$

$$= P(F)P(M)P(F)P(F) = P(M)P(F)P(F)P(F)$$

This is a direct result of the independence of trials and the condition that the probability of success [that is, $P(F)$] does not change from one trial to another. Since they are all equal, we need to choose only one of them for calculation. By convention, we choose the one that lists all the "successes" first and the "failures" last [i.e., in this case, $P(F)P(F)P(F)P(M)$].

Summarizing our two observations, we could say that a binomial probability is given by

$P(X = k) =$ (the number of elementary events which yield k successes from n trials) times (the probability of just one of those events occurring).

Indeed, this is the general form of a binomial probability.

Given the number of independent trials (n) and the probability of observing a "success" in one trial (π), then the probability of observing exactly k successes in n trials, from a Bernoulli experiment, is given by

$$P(X = k \mid n, \pi) = \left[\frac{n!}{k!\,(n-k)!} \right] [\pi^k (1 - \pi)^{n-k}] \qquad \textbf{(3.5)}$$

The first term, $n!/[k!\,(n-k)!]$, is the number of different ways k successes can occur in n trials. In part B of example 3.4, the number of ways we could get one success out of four trials is the number of different ways we could observe one female and three male children. The female could have been the first born, the second born, the third born, or the last born.

We could always (1) carefully list all possible outcomes, (2) group those outcomes that contain an equal number of successes (or females, in this case), and (3) count the number of outcomes that have exactly k successes. This is what we did in example 3.4. The *combination* formula, $n!/[k!\,(n-k)!]$, simply provides an alternative way of counting the number of ways k females may occur in a family of n children.

For one female ($k = 1$) out of four children ($n = 4$), there are

$$\frac{n!}{k!\,(n-k)!} = \frac{4!}{1!\,(4-1)!} = \frac{(4)(3!)}{(1)(3!)} = 4$$

different ways we could observe that result. There is only one way of getting zero successes from four trials, the event {M, M, M, M}. Using the first term of the binomial expression in Equation 3.5,

$$\frac{n!}{k!\,(n-k)!} = \frac{4!}{0!\,(4-0)!} = \frac{4!}{(1)(4!)} = 1$$

we easily verify that result. You may want to return to Appendix 2 for more details on using factorial notation.

The second term, $\pi^k(1-\pi)^{n-k}$, is just a shorthand expression for that particular case where all the k successes are first and the $(n-k)$ failures are listed last. That is,

$$\pi^k(1-\pi)^{n-k} = [(\pi)(\pi)(\pi)\cdots\;(\pi)\;(\pi)][(1-\pi)(1-\pi)\cdots(1-\pi)(1-\pi)]$$

$$\begin{array}{ccccccc} \uparrow\ \uparrow\ \uparrow & & \uparrow & \uparrow & \uparrow & \uparrow & \uparrow & \uparrow \\ 1,\ 2,\ 3,\ \ldots, & k-1, & k & k+1, & k+2, & \ldots, & n-1, & n \end{array}$$

$$= \underbrace{[P(F)P(F)\cdots P(F)P(F)]}_{k\ \text{of these}}\ \underbrace{[P(M)P(M)\cdots P(M)P(M)]}_{n-k\ \text{of these}}$$

in example 3.4.

Consider selecting a four-child family from Purdued in example 3.4. What is the probability of finding exactly three females in that family? Since $n = 4$, $\pi = 0.6$, and $k = 3$, then $n - k = 1$ and $(1 - \pi) = 1 - 0.6 = 0.4$. Thus,

$$P(X = 3) = \{4!/[(3!)(1!)]\}[(0.6)^3(0.4)^1] = (4)(0.6)(0.6)(0.6)(0.4) = (4)(0.0864) = 0.3456$$

which checks with our previous calculations.

Although it is important for you to know how to calculate binomial probabilities, it is not important that you become proficient in these calculations. Once you understand how binomial probabilities are calculated, you will be able to use a computer or the tables at the end of the book.

It is more important to recognize when you have a problem that can use the binomial probability distribution. The problem *must* satisfy the conditions of a Bernoulli experiment before the binomial probability distribution is a correct representation.

The binomial is the simplest discrete probability distribution, and its business applications are numerous. We will reconsider the binomial distribution after Chapter 4.

3.6 A CONTINUOUS THEORETICAL PROBABILITY DISTRIBUTION: THE NORMAL DISTRIBUTION

Generally regarded as the most productive mathematician of his time, Karl Frederick Gauss was also a German astronomer. In the calculation of the paths of stellar bodies, Gauss hypothesized that accidental errors of measurement must occur. If

Gauss could figure out how these measurement errors were distributed, he could more accurately determine stellar movements. It was in this context that Gauss developed the use of the "normal curve of error," or what is sometimes called *Gauss' curve.*

As we shall consider later, the binomial distribution approaches a normal curve as the number of trials becomes large. While Abraham DeMoivre published this result (in 1733) prior to Gauss' work, Gauss popularized the use of the normal curve. Further, since the normal curve of error does not always represent the distribution of measurement errors, we will simply call this distribution "the normal curve."

Finally, the normal curve, or normal distribution, is the best known and most frequently used theoretical distribution. There are so many physical phenomena well suited to using the normal distribution as a descriptive model of their behavior that we sometimes forget that the normal distribution is only a mathematical model, not a law of nature.

For example, if we correctly aimed a rifle at a bull's-eye on a target and secured the rifle in a vise, we would find that not all of the many shots fired would exactly hit the bull's-eye. If we measured the distance from the center of each bullet hole to the bull's-eye, those distances, when plotted against the frequency with which each occurs, would approximately map out a normal curve.

The amount of cola resulting from an automatic dispensing machine will not always be the same. Infrequently your cup may "runneth over," or your coin may dissappear with no tangible results. But even when you receive a full cup, you will not get exactly 8 ounces. If we precisely measured the volume of cola dispensed by a particular machine on many occasions, we might find that the volume of liquid dispensed is approximately normally distributed.

The examples of both the rifle in a vise and the automatic dispensing machine refer to experiments which generate a normally distributed *population.* Not all populations are normally distributed. Thus, why is the normal distribution so important? The answer is found in the fact that many sampling distributions, regardless of the shape of the population sampled, are approximately a normal distribution for large samples. This rather amazing situation is discussed further in Chapter 5. For now, please accept the proposition that the normal curve is the eminent distribution in statistical methods.

The general form of the *normal distribution,* or the normal probability density function, is given in Panel 3.2.

Panel 3.2 The Normal Distribution

Verbal Description
The normal distribution is a bell-shaped, smooth curve that is completely specified by the mean and standard deviation of the continuous random variable *x*. It is completely symmetric about its mean; the height and spread are given by the standard deviation.

Graphic Representation

Var (B) = Var (B')

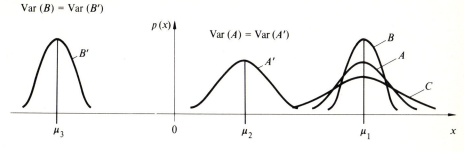

Var (C) > Var (A) > Var (B)

FIGURE 3.9   Graphic representation of the normal distribution.

Mathematical Description

$$p(x) = \frac{1}{\sigma\sqrt{2\pi}} e^{-1/2[(x-\mu)/\sigma]^2} \qquad \text{for } -\infty \leq x \leq +\infty \qquad (3.6)$$

where x = a value of the continuous random variable

   μ = the mean or expected value of x

   σ = the standard deviation of x, or the positive square root of the variance

   π = the constant, 3.1416... (π is used for the population proportion in all other sections of this text)

   e = the base of the Naperian, or natural, logarithms, 2.718...

Characteristics of the Normal Curve

If we can understand what Equation 3.6 says, we can confidently use the results without even directly manipulating it. To visualize Equation 3.6, we will break it into two parts. Call $1/(\sigma\sqrt{2\pi})$ the "coefficient component" and everything else the "exponential component."

1. Equation 3.6 was defined as a probability density function. Thus, by definition, when we measure in the same units as x is measured, *the total area under the normal curve must equal 1.* For example, x could be length measured in millimeters, inches, feet, or miles. (If x, μ, and σ are all in inches, then the definite integral of Equation 3.6 from $-\infty$ to $+\infty$ would equal 1.)
2. The "exponential component" provides the general shape to the normal curve. As we can see in Figure 3.9, the normal curve is *bell-shaped.*

3. Recall from algebra that any (nonzero real) number raised to power 0 is defined to be equal to 1; that is, $Y^0 = 1$. Now consider the exponent, $-\frac{1}{2}[(x - \mu)/\sigma]^2$. When $x = \mu$, the exponent is equal to 0 and the "exponential component" is equal to 1. For any value of x not equal to μ, the exponential component is less than 1. Let $x = \mu + k$, then $x - \mu = k$ and the exponent is $-\frac{1}{2}(k^2/\sigma^2)$. But, if $x = \mu - k$ (because we square the difference from the mean), we get the same exponent. Thus, the normal curve is *symmetric* about the mean (μ) of X. Consider the normal curve marked A in Figure 3.9. The mean of X is μ. The right half of the curve is the mirror image of the left half.

4. A normal curve is *completely defined by* three constants (that is, π, e, and 2) and two parameters, the *mean* and *standard deviation* of X. Recalling that the exponential component is equal to 1 when $x = \mu$, then the highest point on the normal curve must be given by the coefficient component, or $1/(\sigma\sqrt{2\pi})$. Note that the standard deviation of X (that is, σ) is in the denominator of the coefficient component. Thus, when σ is small, the peak is higher; when σ is large, the peak is lower. But, remember that the area under the curve must always equal 1 (x unit). When measuring two normal distributions, in the same x units, the one with the highest peak must have the tail come down to the x axis much more rapidly than the distribution with the lower peak. In summary, the mean (μ) locates the peak on the x axis, and the standard deviation (σ) determines the height of the peak and the "spread" of the distribution. Consider normal curves A and B in Figure 3.9. They have the same mean, μ_1, but Var(B) is smaller than Var(A); thus, σ_B is smaller than σ_A. Curve B is more tightly packed about μ_1 than curve A is. In like manner, curve C is more spread out than curve A even though they both have the same mean. Curves A and A' have the same variance but different means. Curves B and B' have the same variance, but curve B' has a negative mean (perhaps X is temperature) whereas curve B has a positive mean.

5. Since X is a *continuous* random variable, it may assume any real value between $-\infty$ and $+\infty$. Note in Figure 3.9 that the normal curve *does not touch the x axis*. Instead, as x increases (or decreases) away from the mean, the curve is *asymptotic* to the x axis. Indeed, when x gets more than 3σ units away from the mean, the vertical distance between the curve and the x axis is extremely small; however, for finite values of x, the curve is always above the x axis.

The Standard Normal Distribution

Recall from section 3.3 that we must find the area under the density function in order to find the probability of X occurring. This would indicate that to find $P(x_1 \leq X \leq x_2)$, we would have to integrate Equation 3.6 with respect to x from x_1 to x_2. That does not sound like much fun, especially since the normal curve is useful in many real-world problems that have different values for the mean and the standard deviation. Fortunately, we can avoid integrating anything.

If we recall that (1) a random variable is any arbitrarily chosen rule that consistently assigns numbers to events, and (2) that for any *given* distribution of X, the mean and standard deviation are constants, then we can *transform* any given normal distribution of X into a new random variable z by letting

$$z = \frac{X - \mu_X}{\sigma_X} \tag{3.7}$$

We can always retrieve our x values by letting $x = \mu_X + z\sigma_X$. Equation 3.7 is so important that it is given a special name; z is called the *standard normal deviate*. With this procedure we will "standardize" a normal distribution. The term "deviate" has no social implications; instead it is derived from the fact that $(x - \mu_X)$ are deviations from the mean.

How does this clever algebraic trick help us avoid integrating Equation 3.6? Note that we have not specified the values of μ_X and σ_X; we have only insisted that they be known. Well, as long as μ_X and σ_X are finite and σ_X is greater than 0 (not much of a restriction), then the mean and standard deviation of z are *constant*. That is, $\mu_z = E(z) \equiv 0$ and $\text{Var}(z) = \sigma_z^2 \equiv 1$ or $\sigma_z \equiv 1$.

Now, if we go back to Equation 3.6 and plug in $\mu = 0$, $\sigma = 1$, and $z = x$, we could integrate (or allow someone else the pleasure of integrating) $f(z)\,dz$ between

FIGURE 3.10   Transformation of X to z.

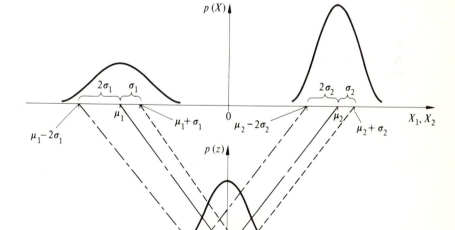

various values of z and create a table of probabilities. Such a table is found at the end of the book. (See Appendix Table IV.)

With the assistance of Figure 3.10 let us review what has happened. We can transform any normal distribution of X, where μ_X is finite and $0 < \sigma_X < \infty$, to a standard normal distribution by letting $z = (x - \mu_X)/\sigma_X$, and the mean of z will always be 0 and the standard deviation is always equal to 1.

The most important point is that the areas under X and z are comparable. Since the areas are probabilities, then $P(x \geq x_1) = P[z \geq (x_1 - \mu_X)/\sigma_X]$ and $P(x \leq x_2) = P[z \leq (x_2 - \mu_X)/\sigma_X]$. Thus, given a table of probabilities for various ranges of z values, we can find the probabilities of the comparable range of x values for any normally distributed random variable X.

Use of the Table of Values of z

On pages 598–9 you will find Appendix Tables IV(A) and IV(B). The total area under the *standard normal distribution,* or z curve, is equal to 1. Since the curve is symmetric, one half the area must lie to the right of $z = 0$ and one half to the left. Without consulting either z table, we already know that $P(z \geq 0) = P(z \leq 0) = 0.5000$. Also, remember that we are dealing with a continuous probability distribution; therefore, $P(z = 0)$ or $P(z = k)$ is always equal to 0. We can consider only the probability of z occurring in a particular *range* of values.

In both z tables given in the Appendix, the left-hand border gives the *value of z* to the first decimal place, and the top border gives the second decimal place. In Table IV(A), the body of the table (i.e., "the entries") presents the *probability* that z will occur between 0 and the value of z given by left-hand and top borders. The body of Table IV(B) presents the *probability* that z will be greater than the z value given by the two borders.

Example 3.5   An Automatic Cola-dispensing Machine

When activated, usually by the appropriate number of coins, an automatic dispensing machine releases a given amount of liquid which we shall call a "fill." The fills are normally distributed with mean of 8.1 ounces and a standard deviation of 0.2 ounce.

(a) If cups are manufactured to hold exactly 8.5 ounces, what is the probability that any given cup will overfill?

(b) What is the probability that an individual would receive less than 8 ounces?

(c) What is the probability that an individual would receive less than 7.7 ounces?

(d) What is the probability that a given cup will receive at least 8 ounces but will not overfill?

(a) The question asks:

$$P(X > 8.5) = P(8.50 < X < +\infty) = ?$$

Since $P(X > X_1) = P[z > (X_1 - \mu)/\sigma]$, then $P(X > 8.5) = P[z > (8.5 - 8.1)/0.2]$ = $P(z > 0.4/0.2) = P(z > 2)$. From Table IV(B), we find that $P(X > 8.5) = P(z > 2) = 0.0228$. That is, a little more than 2 percent of the time the machine will overfill the cup.

(b) The question asks for $P(X < 8)$, or $P(-\infty < X < 8)$. If we let $z_2 = (X_2 - \mu)/\sigma$ = $(8.0 - 8.1)/0.2 = -0.1/0.2 = -0.5$, then $P(X < 8) = P(z < -0.5)$. But there are no negative values for z in either Table IV(A) or IV(B). Recall that the z distribution is symmetric; thus, the area to the left of $z = -0.5$ is equal to the area to the right of $z = +0.5$. Or $P(z < -0.5)$ = $P(z > +0.5)$, and from Table IV(B) we find $P(z > 0.50) = 0.3085$. Hence, $P(X < 8\text{ ounces}) = 0.3085$, or about 31 percent of the time an individual would receive less than 8 ounces.

(c) The question asks for

$$P(X < 7.7) = P(-\infty \leq X < 7.7).$$

Let $z_3 = (X_3 - \mu)/\sigma = (7.7 - 8.1)/0.2 = -2.00$. Then $P(X < 7.7) = P(z < -2.00)$, which is equal to $P(z > +2.00)$, and in part a we already found that to be 0.0228. Therefore, $P(X < 7.7) = 0.0228$.

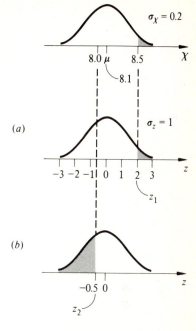

(a)

(b)

(c)

(d)

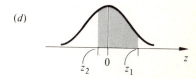

(d) Now the question asks for $P(8.0 \leq X \leq 8.5)$. Carefully note in the above diagram that we want $P(z_2 \leq z \leq z_1)$. Also, note that Appendix Table IV(A) provides only for $P(0 \leq z \leq z_i)$; therefore, to work this problem, we must add the areas. First, the critical values of z are $z_1 = (8.5 - 8.1)/0.2 = +2.00$, and $z_2 = (8.0 - 8.1)/0.2 = -0.50$.

$$
\begin{aligned}
P(8.0 \leq X \leq 8.5) &= P(-0.50 \leq z \leq +2.00) \\
&= P(-0.50 \leq z \leq 0) + P(0 \leq z \leq 2.0) \\
&= P(0 < z < +0.50) + P(0 \leq z \leq 2.0) \\
&= 0.1915 + 0.4772 = 0.6687
\end{aligned}
$$

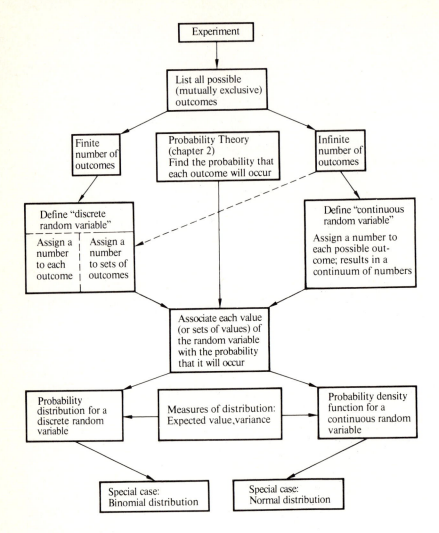

FIGURE 3.11 Summary.

Also, note that

$$P(8.0 \leq X \leq 8.5) = 1 - P(X < 8) - P(X > 8.5)$$
$$= 1 - P(z < -0.5) - P(z > 2)$$
$$= 1 - 0.3085 - 0.0228 = 0.6687$$

Please note that the adding and subtracting of areas in example 3.5 is the same procedure as adding and subtracting areas (i.e., triangles and rectangles) in example 3.3. Indeed, conceptually we are doing the same thing in both examples.

SUMMARY

This concludes our initial development of theoretical probability distributions. The concepts and techniques presented in Chapters 2 and 3 provide a theoretical model. We now need to investigate techniques for submitting real-world data to the theoretical model. Then, the combination of the theory and the data will permit us to state propositions that we were unable to state before.

One word of caution. This treatment of probability, random variables, and probability distributions is, at best, a skeletal presentation. In the next several chapters we will not need more than is presented here, but we will need every bit that is presented. As we need a more extensive treatment in later chapters, we will develop it.

Figure 3.11 is a summary of the process that took us to this point.

EXERCISES

1. Consider two roles of a single die. Let $X = \{$number of dots showing on the first roll$\}$ and $Y = \{$number of dots on the second roll$\}$.
a. List all possible outcomes.
b. What is $P(X = 1)$, $P(X = 2)$, etc.?
c. What is $P(Y = 1)$, $P(Y = 2)$, etc.?
d. Let $Z = X + Y$.
   (1) What are the possible values of Z?
   (2) What is $P(Z = 1)$, $P(Z = 2)$, etc.?
e. Is Z a random variable? Is Z discrete or continuous?
f. Does your answer to part d constitute a probability distribution?
   (*Hint*: Are $P(Z = i) \geq 0$ and $\sum_i P(Z = i) = 1$ for all values of i?)
g. Let $D = X - Y$.
   (1) What are all possible values of D?
   (2) What is $P(D)$ for each value of D?
h. Does your answer to part g constitute a probability distribution?

i. Find $E(X)$, $E(Y)$, $E(Z)$, and $E(D)$.

j. Find $\text{Var}(X)$, $\text{Var}(Y)$, $\text{Var}(Z)$, and $\text{Var}(D)$.

k. If X and Y are independent random variables, then $E(X + Y) = E(X) + E(Y)$ and $E(X - Y) = E(X) - E(Y)$. Use these two relationships to check your answers in part i.

l. If X and Y are independent random variables, then $\text{Var}(X + Y) = \text{Var}(X) + \text{Var}(Y)$, and $\text{Var}(X - Y) = \text{Var}(X) + \text{Var}(Y)$. Does your answer to part j demonstrate those two relationships? Does $\text{Var}(Z) = \text{Var}(D)$?

2. Consider one roll of two dice (e.g., maybe one die is red and the other is green). Are the results to any part of Exercise 1 any different? Why or why not? In the game of "craps," if the first roll of two dice results in a total of 7 or 11, you have an instant winner. Is the probability of an instant winner equal to $P(\{Z = 7\} \cup \{Z = 11\})$ from part d of Exercise 1?

3. A used-car sales representative has the opportunity to work for either dealer A or dealer B. The sales person assesses the sales prospects at each dealership as follows:

Dealer A		Dealer B	
A = number of cars/week that could be sold	$P(A)$	B = number of car/week that could be sold	$P(B)$
0	0.4	0	0.2
1	0.3	1	0.6
2	0.2	2	0.2
3	0.1	3	0.0
	1.0		1.0

a. Graph both probability distributions.

b. Find $E(A)$ and $\text{Var}(A)$.

c. Find $E(B)$ and $\text{Var}(B)$.

d. In the long run, would the salesperson expect to sell any more cars with dealer A than with dealer B?

e. Would the "weekly sales record" be any more consistent with dealer A than with dealer B?

f. Dealer A offers the prospective salesperson a "commission" of $100 per car sold. What is the expected value of the salesperson's weekly earnings with this offer? What is the variance in weekly earnings with this deal? [*Hint:* Let $X = 100A$. Since the events do not change, only the random variable which denotes each outcome changes, and $P(A_i) = P(X_i)$.]

g. Dealer B is quite anxious to acquire the services of this salesperson. Dealer B offers $200 per car as a "commission." If we let $Y = 200B$, find $E(Y)$ and $\text{Var}(Y)$.

h. Let b be some constant. Then, $E(bX) = bE(X)$ and $\text{Var}(bX) = b^2 \text{Var}(X)$. Check your answers to parts f and g using these relationships.

i. In terms of weekly earnings, which dealer offers the highest expected earnings? Which offer yields the most variation in weekly earnings? (*Suggestion:* Graph X and Y to interpret your numerical results.)

j. Dealer A comes up with a counteroffer; the sales representative will receive $100 per week plus $100 per car sold. Let $W = 100 + X$ (or $100 + 100A$). What are $E(W)$ and $\text{Var}(W)$?

k. How do the expected weekly earnings compare now? With which dealer would the salesperson expect the least variation in weekly earnings? Is $\text{Var}(W)$ any different from $\text{Var}(X)$ in part f?

l. If c is a constant, $\text{Var}(c + X) = \text{Var}(X)$. Use this relationship to check your answer to part j.

4. You are the only economist on the Isle of Sow Fee. The monetary unit on this isle is the "pound copper" or C. Your forecast for next year's Net National Income (Y) is given by:

(Millions of C) Y_i	$P(Y_i)$
3	0.3
4	0.5
5	0.1
6	0.1
	1.0

a. Find $E(Y)$ and $\text{Var}(Y)$.

b. Oil is discovered off the coast of Sow Fee. The foreign oil company which made the discovery offers the government 1 million C per year for drilling rights. (There are royalties involved, but they will not occur for several years.) How does this affect your forecast? (*Note:* The fee is exogenous to the Sow Fee economy; let $X = Y + 1$.)

c. Find $E(X)$ and $\text{Var}(X)$ from the probability distribution for X. Is $E(X) = E(Y + 1) = 1 + E(Y)$? What is the relation between $\text{Var}(X)$ and $\text{Var}(Y)$?

d. Suppose the exchange rate with U.S. dollars is \$2 per C. (Let $W = 2X$.) How would you report your estimate of Net National Income to the United States ambassador? (What is the probability distribution for W?)

e. Find $E(W)$ and $\text{Var}(W)$ from part d.

f. Is $E(W) = E(2X) = 2E(X)$? Is $\text{Var}(W) = \text{Var}(2X) = 2^2 \,\text{Var}(X)$?

g. Suppose the consumption function for the Isle of Sow Fee is $C = 2 + (0.8)X$. What is the probability distribution for C (in millions of C)?

h. Is $E(C) = 2 + 0.8E(X)$? Is $\text{Var}(C) = 0.64 \,\text{Var}(X)$?

5. Consider the following probability density function for x.

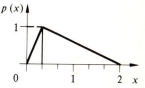

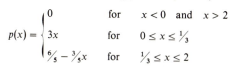

$$p(x) = \begin{cases} 0 & \text{for} \quad x < 0 \quad \text{and} \quad x > 2 \\ 3x & \text{for} \quad 0 \le x \le \tfrac{1}{3} \\ \tfrac{6}{5} - \tfrac{3}{5}x & \text{for} \quad \tfrac{1}{3} \le x \le 2 \end{cases}$$

a. By using the area of a right triangle, show that $P(0 \le x \le \tfrac{1}{3}) = \tfrac{1}{6}$.

b. Find $P(\tfrac{1}{3} \le x \le 2)$.

c. Find $P(x > 1)$.

d. Find $P(0 \le x \le \tfrac{2}{3})$.

6. Consider the uniform probability density function:

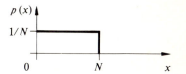

$$p(x) = \begin{cases} 0 & \text{for } x < 0, \ x > N \\ 1/N & \text{for } 0 \le x \le N \end{cases}$$

It should not be a surprise to find that $E(x) = N/2$ (i.e., consider the weight analogy in Figure 3.7). It is less obvious that $\text{Var}(x) = N^2/12$ or $\sigma_x = N/\sqrt{12} = N/3.4641$.

a. For $N = 1$, find $P[E(x) - \sigma_x \le x \le E(x) + \sigma_x]$.

b. For $N = 2$, find $P[E(x) - \frac{1}{2}\sigma_x \leq x \leq E(x) + \frac{1}{2}\sigma_x]$.

c. For $N = \frac{1}{2}$, find $P[E(x) - \frac{1}{2}\sigma_x \leq x \leq E(x) + \frac{1}{2}\sigma_x]$.

d. For $N = 2$, find $P[E(x) - 2\sigma_x \leq x \leq E(x) + 2\sigma_x]$.

e. How does your answer to part *b* relate to your answer to part *c*?

7. Assume that the probability that an adult resident will be home between 6 P.M. and 7 P.M. is 0.90. A door-to-door salesperson plans to make seven calls between 6 P.M. and 7 P.M.

a. Let r = number of homes with an adult present when the salesperson arrives. Assume independence and find the probability distribution for r.

b. What is the probability that an adult resident will be home for all seven calls?

c. What is the probability that the salesperson will find five or more homes with an adult resident present?

8. A fair die is rolled 10 times. A 1 or a 2 is considered a success. What is the probability that:

a. *Exactly* five successes will occur?

b. *At least* five successes will occur?

c. *At most* five successes will occur?

d. Five successes *or more* will occur?

e. *No more than* five successes will occur?

f. *Fewer than* five successes will occur?

g. *More than* five successes will occur?

9. In Mud Lake 60 percent of the fish are under the legal size limit. Assume that all the fish are equally gullible for a fisherman's bait and that the fisherman's trials are independent. If a fisherman catches five fish, what is the probability that:

a. All five are above the legal size limit?

b. None are above the legal size limit?

c. At least one is above the legal size limit?

d. At least two are below the legal size limit?

e. At least three are below the legal size limit?

10. Given a normal distribution with $\mu_x = 13$ and $\sigma_x = 2$, find:

a. $P(X \geq \mu)$

b. $P(X \leq 14)$

c. $P(12 \leq X \leq 14)$

d. $P(X > 14)$

e. $P(X = 14)$

f. $P(\mu + \sigma \geq X \geq \mu - \sigma)$

g. $P(\mu - 2\sigma \leq X \leq \mu + 2\sigma)$

h. $P(\mu - 3\sigma \leq X \leq \mu + 3\sigma)$

i. $P(\mu \leq X \leq \mu + 1.96\sigma)$

j $P(\mu - 1.96\sigma \leq X \leq \mu)$

k. $P(\mu - 1.96\sigma \leq X \leq \mu + 1.96\sigma)$

11. Given a normal distribution with $\mu = 103$ and $\sigma = 4$, find:

a. $P(X \geq 100)$

b. $P(X < 100)$

c. $P(X > 104)$

d. $P(X < 104)$

e. $P(100 \leq X \leq 104)$

f. $P[(X < 100) \cup (X > 104)]$

g. $P(96.42 \leq X \leq 109.58)$

h. $P[(X < 96.42) \cup (X > 109.58)]$

12. Given a normal distribution with $\mu = 100$ and $\sigma = 10$, find:

a. X_1 such that $P(X \leq X_1) = 0.5000$

b. X_2 such that $P(X \leq X_2) = 0.1000$

c. X_3 such that $P(X \geq X_3) = 0.1000$

d. $P(X_2 \leq X \leq X_3)$

e. X_4 such that $P(X \leq X_4) = 0.0250$

13. You are given a normal distribution with $\mu = 200$ and $\sigma = 5$.

a. Find $P(0 \leq X \leq 205)$.

b. Find $P(0 \leq X \leq 210)$.

c. Use your answers to parts a and b to show that $2[P(0 \leq X \leq \mu + \sigma)] \neq P(0 \leq X \leq \mu + 2\sigma)$.

d. Find $P(X > 210)$ using only the answer to part b and then using Appendix Table IV.

e. Find $P(X > 215)$.

f. Find $P(X > 210) - P(X > 215)$ from parts d and e. Draw a normal curve and show that $P(210 \leq X \leq 215) = P(X > 210) - P(X > 215)$.

g. Find $P(X < 190)$. Why is $P(X < 190) = P(X > 210)$?

h. Find $P(189 \leq X \leq 211)$.

14. A producer of instant coffee has a bottling machine which dispenses coffee into jars in such a manner that the weight of coffee in each jar is normally distributed with mean equal to 80.217 ounces and standard deviation of 0.100 ounce. What is the probability that one randomly selected jar filled by this machine will contain:

a. less than 80.117 ounces?

b. more than 80.117 ounces?

c. less than the 80 ounces (5 pounds) advertised on the jar?

d. between 80.021 and 80.413 ounces?

e. 80.089 ounces or more?

15. Assume that on Mondays, between 7 A.M. and 8 A.M., the water demanded by a municipal water system is normally distributed with mean of 200 million gallons and standard deviation of 10 million gallons. The reservoir system has a capacity of 4,240 million gallons. Whenever the reservoir falls 10 percent below capacity, a "fire warning" is issued.

Experiment A. Assume that on one randomly selected Monday morning, between 7 A.M. and 8 A.M., the reservoir is 5 percent below capacity, and there is no inflow of water to the reservoir. What is the probability that:

a. 200 million gallons or more will be demanded?

b. Less than 1 percent of the reservoir's capacity (not its content) will be demanded?

c. A fire warning will have to be issued at 8 A.M.?

d. Between 180 and 210 million gallons will be demanded?

Experiment B. Assume that the reservoir is 5 percent below capacity and is resupplied at a rate which is normally distributed with mean of 100 million gallons per hour and standard deviation of 30 million gallons per hour. It is still 7 A.M. on one randomly chosen Monday morning. What is the probability that:

e. Less than 10 million gallons will be supplied to the reservoir by 8 A.M.?

f. Less than 10 million gallons will be supplied to the reservoir *and* 222 million gallons or more will be demanded between 7 A.M. and 8 A.M.? This situation would require issuing a fire warning. (Assume that supply to the reservoir and demand from it are independent.)

16. The checkout counter at a pharmacy has space for only one weekly publication, and the owner has narrowed the field to *TV Weekly* (*TVW*) and *This Week in Tennis* (*TWIT*). *TVW* sells for 35¢ per copy and costs 10¢ per copy; the unsold copies are discarded. *TWIT* sells for 50¢ and costs 25¢; the unsold copies will receive a refund of 5¢ per copy. The demand schedules are given below.

Possible Demand (Copies per Week) d	Probability *TVW* $P(A = d)$	*TWIT* $P(B = d)$
0	0.02	0.10
1	0.03	0.10
2	0.10	0.10
3	0.20	0.10
4	0.30	0.20
5	0.20	0.10
6	0.10	0.10
7	0.03	0.10
8	0.02	0.10
	1.00	1.00

a. Calculate $E(A)$, $Var(A)$, $E(B)$, and $Var(B)$ and compare the results.

b. Note that:

$$\text{Weekly profit} = (\text{price})(d) - (\text{cost/copy})(\text{no. stocked}) + (\text{refund/copy})(\text{no. stocked} - d)$$

or

$$W = P \cdot d - C \cdot N + R \cdot (N - d) \qquad \text{for } N \geq d$$

(1) Use $E(A) = d = N$ and calculate the weekly profit for *TVW*.
Use $E(B) = d = N$ and calculate the weekly profit for *TWIT*.
(2) For both *TVW* and *TWIT*, assume eight issues are stocked each week and derive $W = F(d)$.

c. Which weekly publication should the owner place at the checkout counter and why?

CASE QUESTIONS

1. In Chapter 2, case question 2A*a*, you were asked to find $P(V32 \geq 100)$. Basically, your answer is the *proportion* of families, within the Illinois sample, which had incomes of $10,000 or more in 1969. Your answer should have been $P(V32 \geq 100) = 0.61$.

Experiment A—Data set represents the population of interest ; n = 4; sampling with replacement.

We assume that we will select four families from the Illinois sample in a manner that will conform to a Bernoulli experiment. For each family, let the event $\{V32 \geq 100\}$ be defined as a success; the probability of a success is given by $\pi = 0.61$. Let $X = $ the number of successes and $p = X/n = $ the proportion of successes.

a. Find the probability distribution for X.

Note: You should calculate each probability using $\pi = 0.61$; however, you can check the reasonableness of your answers by using Appendix Table III for $n = 4$, $\pi = 0.60$, and $\pi = 0.625$. For example,

$$P(X = 0 \mid \pi = 0.625) < P(X = 0 \mid \pi = 0.610) < P(X = 0 \mid \pi = 0.600)$$

or

$$0.0198 < P(X = 0 \mid \pi = 0.61) < 0.0256$$

b. Find the probability distribution for p.
c. Let $\{V32 < 100\}$ be a success. Find the probability distributions for X and p.
d. From part *a* above, find $E(X)$ and $\text{Var}(X)$.
e. From part *b* above, find $E(p)$ and $\text{Var}(p)$.

Experiment B—Data set represents the population of interest; n = 4; sampling without replacement.

For this experiment, the probability of a success changes slightly with each "draw" of a family from the population. Define the event $\{V32 \geq 100\}$ as a success.

a. Find the probability distribution for X.

Hint: Set up a table with the following headings:

Result of Draw 1 2 3 4	Number of S's	Probability of Result 1 2 3 4 Total

In the four left-hand columns, list all possible sample outcomes by filling in each cell with an S ($=$ success) or F. Next find the probability of each cell result occurring; remember that the probability of getting a success on the third draw depends upon how many successes and failures have been removed from the population on the first two draws. Next, for each line, total the number of S's in the first four columns to get the X value for that outcome. Then, for each sample or line, multiply the four probabilities and enter their product under "total." This has the effect of

$$P(A \cap B \cap C \cap D) = P(A)P(B \mid A)P(C \mid A \cap B)P(D \mid A \cap B \cap C)$$

Finally, the probability distribution of X results from grouping together all like values of X and their associated probabilities.

b. Find the probability distribution of p.
c. Calculate $E(X)$ and $\text{Var}(X)$.
d. Compare your answer to part *d* from experiment A with your answer to part *c* above.

4

EMPIRICAL DISTRIBUTIONS

We now turn from probability theory and theoretical distributions to experience, observation, or data and empirical frequency distributions. The topics in this chapter are often called the methods of *descriptive statistics*. Basically, we want to take a large set of data and describe it without listing every observation. We wish to summarize or consolidate information contained in the data.

These descriptive techniques are so common in everyday human affairs that we can easily hypothesize that these techniques will be of life-long benefit. The construction of a frequency distribution, for example, often requires judgment. Because of this "judgment" component, such quotes as "You can prove anything with statistics" and "Statistics lie" have stereotyped descriptive statistics. In the last section we will discuss some of the methods of misinformation that plague these otherwise useful techniques.

4.1 DATA AND INFORMATION: THE COMPROMISE

Let us start with a set of data and consider how we might logically proceed to describe these data without listing every observation. Perhaps for the purpose of estimating water usage, garbage pick-up, or traffic flow, let us assume that we are interested in the number of persons per household in a particular suburban-residential city block. There are 30 housing units (i.e., one half of a duplex is a "housing unit"), and we interview each household to find out how many individuals reside there as of a particular date.

Table 4.1 presents the data as it is received at our project headquarters. Although it is listed by address to make sure we have not missed any housing unit, in this form it is not very useful. If the even-numbered addresses are on one side of the street and the odd-numbered addresses are on the other, it does not even tell us the allocation of persons along the block.

We might take the number of persons at each address and order them in an ascending (or a descending) manner. Such an arrangement is called an *array* of the number of persons per household and is presented in Table 4.2.

Note that this is slightly more information about the distribution of household size along 2200 Main Street. Further, since an address can be associated with each number in Table 4.2, we have exactly the same data as contained in Table 4.1.

TABLE 4.1 NUMBER OF PERSONS RESIDING AT EACH ADDRESS; 2200 MAIN STREET

Address	Number	Address	Number	Address	Number
2201	2	2211	3	2221	4
2202	4	2212	1	2222	0
2203	5	2213	4	2223	4
2204	1	2214	5	2224	5
2205	4	2215	2	2225	2
2206	6	2216	7	2226	6
2207	2	2217	4	2227	3
2208	4	2218	1	2228	1
2209	3	2219	7	2229	5
2210	4	2220	3	2230	3

If we are interested in the distribution of household size, then we could *tally* the number of households of each size, as shown in Table 4.3. Note that while this yields a better visual picture of number of households in each size class, we do lose the information concerning address. Also note that the tally could be generated from either Table 4.1 or Table 4.2 since both reflect the original data.

From this tally, we can construct what is called an *empirical frequency distribution*, or what we will call simply a frequency distribution. Recall from Chapter 3 that the number of persons per household is a discrete variable. Thus, the "number of households" tells us how frequently that value of the discrete variable occurred. The frequency distribution can be represented in either tabular or graphical form. See Table 4.4 and Figure 4.1.

In many respects, this problem is trivial. Given the assignment of presenting the data in pictorial form, and enough time, you probably would arrive at something similar to Figure 4.1. However, it is not always preferable to use each possible value of a discrete variable in a frequency distribution.

For example, in the process of ascertaining the number of persons in each housing unit, suppose we also found the household income for the previous tax year.

TABLE 4.2 ARRAY OF THE NUMBER OF PERSONS PER HOUSEHOLD; 2200 MAIN STREET

0	2	3	4	5
1	2	3	4	5
1	2	4	4	6
1	3	4	4	6
1	3	4	5	7
2	3	4	5	7

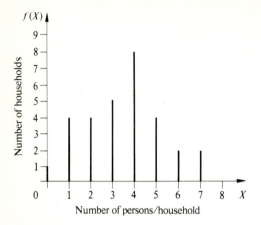

FIGURE 4.1 Frequency distribution: graphical form.

Table 4.5 presents such data, and Table 4.6 places it in an array. Let the discrete random variable Y be the income per household. We could graph the array; the horizontal axis would allow for the 14,500 values of Y, and the vertical axis (numbered 0, 1, 2, in this case) would show how frequently each value occurred. But that would not be a very effective way to summarize the income data.

Instead, let us discuss ranges or intervals of income. In Table 4.7 we have *chosen* $2,000 income intervals and found how many households fall within each income *class interval*. Thus, the tabular form of the income frequency distribution is presented in Table 4.8. It is important to note that with a quick glance at the frequency distribution presented in Table 4.8 we get a reasonably clear picture of the income distribution, whereas it takes some study of the array in Table 4.6 to obtain the same picture. Also, from Table 4.8 we could not tell the exact income of any one house-

TABLE 4.3 TALLY OF NUMBER OF HOUSEHOLDS BY NUMBER OF PERSONS PER HOUSEHOLD

Number of Persons per Household	Tally	Number of Households
0	/	1
1	////	4
2	////	4
3	//////	5
4	////// ///	8
5	////	4
6	//	2
7	//	2
		30

TABLE 4.4 FREQUENCY
DISTRIBUTION: TABULAR FORM

Number of Persons per Household (Value of the Discrete Variable) X	Number of Households (Frequency of Occurrence) $f(X)$
0	1
1	4
2	4
3	5
4	8
5	4
6	2
7	2
	30

hold. To paraphrase, Table 4.6 gives us more detail about every tree in the forest; Table 4.8 doesn't give as much information about each tree, but it does give us a better view of the forest.

Much of the published data in business and economics is presented in a manner similar to Table 4.8. Of the billions of bits of data gathered for the 1970 U.S. Census

Panel 4.1   Income Data by Household

TABLE 4.5   RAW DATA (BY ADDRESS)

(2201)	$ 6,025	$8,045	$ 9,200	$ 5,012	$ 7,800
(2202)	8,000	7,029	10,800	9,500	9,826
.	10,000	4,629	10,955	9,750	12,000
.	10,222	8,500	9,500	0	13,750
.	3,058	8,726	11,628	14,500	11,750
	6,466	9,000	7,500	11,630	11,989

TABLE 4.6   ARRAY

$ 0	$7,029	$8,726	$ 9,826	$11,630
3,058	7,500	9,000	10,000	11,750
4,629	7,800	9,200	10,222	11,989
5,012	8,000	9,500	10,800	12,000
6,025	8,045	9,500	10,955	13,750
6,466	8,500	9,750	11,628	14,500

TABLE 4.7 TALLY

Income Class	Tally	Number of Households
0– 1,999	/	1
2,000– 3,999	/	1
4,000– 5,999	//	2
6,000– 7,999	/////	5
8,000– 9,999	///// /////	10
10,000–11,999	///// ///	8
12,000–13,999	//	2
14,000–15,999	/	1
		——
		30

TABLE 4.8 FREQUENCY DISTRIBUTION: TABULAR FORM

Income Class (Range of Values for Discrete Variable) Y	Number of Households (Frequency of Occurrence) $f(Y)$
0– 1,999	1
2,000– 3,999	1
4,000– 5,999	2
6,000– 7,999	5
8,000– 9,999	10
10,000–11,999	8
12,000–13,999	2
14,000–15,999	1
	——
	30

of Population, most of it is published in the tabular frequency-distribution framework.

By the same logic, our graphical representation of the frequency distribution cannot be the line graph similar to Figure 4.1. True, the household income variable Y is a discrete variable. However, from Table 4.8 only (and that is the way we most often receive our data), we know only how many households fall in each income classification (for example, $2,000 to $3,999) and *not* the income of each household. There are two commonly used graphical representations: the *histogram* and the *polygon*.

As shown in Figure 4.2, the histogram is a sequence of rectangles where the width of the rectangle represents the class interval (for example, 0 to 1,999; 2,000 to 3,999; etc.), and the height represents the frequency with which observations fall

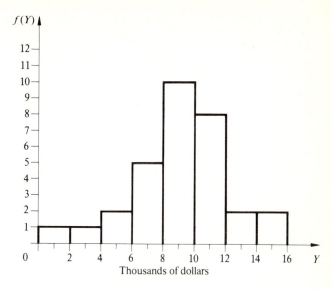

FIGURE 4.2   Frequency distribution: histogram.

within that interval. The polygon shown in Figure 4.3 is constructed by plotting a point, whose vertical distance is the frequency, at the *midpoint* of each class interval. Since the class interval to the left of the first class interval (i.e., the class for −$2,000 to −$1) and the class interval to the right of the last interval (that is, $16,000 to

FIGURE 4.3   Frequency distribution: polygon.

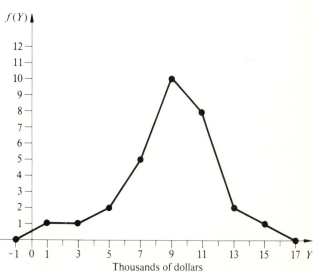

$17,999) have a frequency of 0, then using their midpoints (that is, −$1,000 and $17,000), we connect all the "points" to form the polygon.[1]

The crux of constructing and graphing a frequency distribution is in understanding the following concepts: the limits of a class interval, the width of the class interval, and the midpoints of a class interval. The specification of the "class limits" are both arbitrary and a function of the manner in which the data was collected.

First, arbitrary situations in scientific disciplines usually call for the evolution of criteria within which decisions should be made. The concept of "class limits" is no different. The following criteria enjoy acceptance among statisticians:

1. *The number of class intervals.* Primarily because of the limited ability of humans to grasp the totality of a table of numbers, the recommended number of classes is between 5 and 15. The exact number will depend upon your purpose in using the data, the nature of the variable (i.e., for income data, intervals of $1,000 are more readable than intervals of $1,968), and the range of the data.

2. *The width of the class interval should be convenient for the reader.* Again, for income data, intervals of 0 to 999, 1,000 to 1,999, etc., are somehow more readable than intervals of 0 to 1,968, 1,969 to 3,937, etc.

3. *The class intervals must be mutually exclusive, and contiguous and include all the observed values.* "Mutually exclusive" simply means that any given observation can be classified in only one class interval; class intervals of $1,000 to $1,999, $1,500 to $2,499, $2,000 to $2,999, etc., would not be permissible for a given frequency distribution. "Contiguous" simply means that it should not be possible for an observation to fall between one class and the next class interval. Since the frequency distribution is constructed from already collected data, there is no reason why the class intervals should not include all the existing observations.

4. *Within a single frequency distribution, all class intervals should be of equal width.* Whenever feasible, the class intervals should all be of the same width. In Table 4.8 all class intervals had a width of $2,000. Unequal class widths tend to deceive when the graphical representation of the frequency distribution is employed. Also, "open-ended" class intervals (e.g., for income, "$40,000 or above") prohibit certain calculations without access to the raw data.

To consider the influence of how the data was collected, reconsider Panel 4.1. In Table 4.5, the data is given in dollars, not cents. Was the household income rounded

---

[1] The "next lower" and "next higher," class intervals are used, even though their frequencies are 0, to make sure the area under the histogram is equal to the area under the polygon. Recall that the area under the probability density function (in Chapter 3) was equal to the probability of the random variable falling within that range. When we get to the point of merging the theoretical distributions with the practical techniques of this chapter, those areas will be very important.

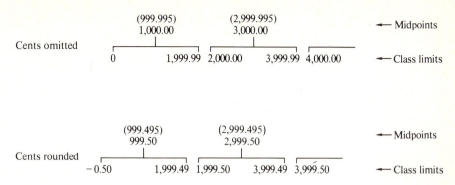

FIGURE 4.4 Influence of data collection rules on class limits.

to the nearest dollar? Were the cents simply omitted? The answer will influence the values of the class limits as shown in Figure 4.4. If the cents were ignored or omitted, then incomes from $1,999.00 to $1,999.99 would be recorded as $1,999; if the income were rounded off to the nearest dollar, then incomes from $1,998.50 to $1,999.49 would be recorded as $1,999. The 50-cent difference in the class limits is reflected in the midpoints as well.

Both cases shown in Figure 4.4 demonstrate the same class interval: $2,000. Recalling that there are 10 integers when counting from 0 to 9, then from 0 to 199,999 pennies represents 200,000 different "penny values," or a $2,000 interval. The same is true for the interval −$0.50 to $1,999.49.

As another example, consider the ages of humans. When asked our age, we normally reply with our age as of our last birthday. When your friendly insurance salesperson figures your life insurance premium, he or she will most likely use your "age" as of your closest birthday.

Panel 4.1 presents the basic steps in constructing an empirical frequency distribution. Starting with the raw data, we find that the construction of the array is a rather mechanical process. Before the tally stage, however, you must make several subjective decisions. How the data was collected and the four criteria listed above will assist in using *correct* procedures for deciding upon the number of classes, the width of the class interval, and the class limits (which will yield the midpoints). From any given set of raw data, there may be several possible frequency distributions that can be constructed by following correct procedures. Regardless of what judgments are made within the confines of the "correct procedures," the measures of the characteristics of the frequency distribution will be similar if not exactly the same.

That is, although the frequency distribution (e.g., Table 4.8 and Figures 4.2 and 4.3) may give us an improved overall view of this income data, we hope to be able to develop the talent of gaining the same "mental image" from the following information:

1. The arithmetic mean income is $8,893.00.
2. The median income is $9,350.
3. The standard deviation is $3,052.48.

It is to that purpose that we devote the remaining portion of this chapter.

In this section we have investigated a few techniques for summarizing a large set of numerical data; we have constructed a frequency distribution from the collected data. In the next two sections, we will investigate techniques for measuring the characteristics, properties, or attributes of a frequency distribution.

In section 3.4 we briefly discussed the measures of location, dispersion, and skewness for a theoretical distribution. Since empirical distributions may not have all the convenient mathematical properties of a theoretical distribution, we will need several measures in each of those three classifications.

Measures of location, dispersion, and skewness are necessary only because there is variability in the data. In Panel 4.1, if all 30 households had the same annual income, say $9,000, there would be no problem in describing that data. Most economic data has variability, and we must be able to deal with variability as it arises.

4.2 MEASURES OF LOCATION

A *measure of location* is semantically descriptive of the concept. What are we trying to locate? The frequency distribution. When we locate a city, we use longitude and latitude; to locate a house we might use a street address. Relative to what conceptual device (i.e., longitude, latitude, or address) do we locate a frequency distribution? The units of the (discrete or continuous) variable of interest! That is, from Table 4.1 we used the number of persons per household and from Table 4.5 the income per household. A frequency distribution is composed of (1) several classification intervals of some variable and (2) the number of observations that fall within each classification. We wish to "locate" the observed data within all possible values for the variable of interest. Specifically, we want some value of the variable of interest that will represent the center or middle of the observed data.

4.2(a) The Arithmetic Mean

Any student who gets to college can calculate the average of a set of numbers. The arithmetic mean is no more and no less than that familiar concept of the "average." There are several other ways, probably less familiar, to average a set of numbers. The arithmetic mean is just the technical name given to the average in order to distinguish it from those other measures which might also be called "averages."

When the original, or raw, data is used, the *population mean* is defined as the sum of all the observations divided by the number of observations, or

$$\mu = \frac{\sum_{i=1}^{N} X_i}{N} \qquad (4.1)$$

where μ = the population mean

X_i = the *i*th observation of X

N = the number of observations in the population

The Greek letter (mu) μ is exclusively used to denote the arithmetic mean of the population. How did "population" creep into the discussion? In Chapter 1 we made the distinction between a sample and a census: a sample is a subset of a census. We are often interested in the mean of a population (which would require a census), but all we have (or can get) is a sample from that population. That is, we usually want to know the population mean μ, but all we have is the arithmetic mean of a sample.

As a measure of location, the concept is the same for both the population and the sample. We use different notations as follows:

When original (or raw) data is used, the *sample mean* is defined as

$$\bar{X} = \frac{\sum_{i=1}^{n} X_i}{n} \qquad (4.2)$$

where $\bar{X}$ = the sample mean (and is read as "X bar")

X_i = the *i*th observation of X

n = the number of observations in the sample

Example 4.1   Arithmetic Mean of Persons per Household

Consider the number of persons per household as given in Table 4.1. Assume that the 2200 block of Main Street is our population of interest (rather than a sample of all residential blocks on Main Street). Then the (arithmetic) mean number of persons per household is

$$\mu = \frac{1}{N} \sum_{i=1}^{N} X_i = \frac{1}{30} \sum_{i=1}^{30} X_i$$

$$= (\tfrac{1}{30})(2 + 4 + 5 + 1 + \cdots + 1 + 5 + 3)$$

$$= (\tfrac{1}{30})(105)$$

$$= 3.5 \text{ persons per household}$$

That is, what you have commonly thought of as "the average number of persons per household" is given by the arithmetic mean.

If you have the raw data, then Equation 4.1 (or 4.2) will be all that is required to calculate the arithmetic mean. Even when the number of observations is quite large (i.e., several hundred), a computer or modern calculator renders the calculations relatively painless. However, what happens when the data is available only in the tabular form of a frequency distribution? We do not know each value of X!

An approximation of the arithmetic mean for grouped data When the data is available only in a frequency distribution, we say we have *grouped data* rather than raw or original data. Return to Table 4.4 where each class interval contains one and only one possible value of the variable X, the number of persons per household. In this special case, we know all the original values of X from the frequency distribution. The frequency for the first class, $f(X = 0)$, is 1; therefore, there was only one unoccupied house. Since $f(X = 1) = 4$, there were four houses with only one individual living there. And so forth.

It is calculationally quicker and algebraically the same thing if we take the value of X times its frequency:

$$(0)(1) + (1)(4) + (2)(4) + (3)(5) + (4)(8) + (5)(4) + (6)(2) + (7)(2)$$

$$= \sum_{j=1}^{K} [X_j \, f(X_j)] = X_1 f(X_1) + X_2 \, f(X_2) + \cdots + X_8 \, f(X_8) = 105 \quad \textbf{(4.3)}$$

rather than add up all the original values (i.e., from the array in Table 4.2):

$$\sum_{i=1}^{30} X_i = (0) + (1 + 1 + 1 + 1) + (2 + 2 + 2 + 2)$$

$$+ \cdots + (7 + 7) = 105$$

For this *very* special case, the arithmetic mean could be calculated by $\mu = (1/N)/\sum_{j=1}^{K} [X_j f(X_j)]$ where K is the number of values of the discrete variable X for which the frequency is recorded.

It is this same concept which we will employ to approximate the arithmetic mean from grouped data. Further, we need one assumption. We assume that each value within a class interval can be represented by the midpoint of that class interval.[2] Thus, in Equation 4.3 we replace X_j with the midpoint M_j. Since the frequency for a particular class interval may represent more than one value of X [thus, $f(X_j)$ is not appropriate notation], we replace $f(X_j)$ with f_j, which represents the frequency of the jth class regardless of how many different values of X could fall in that class interval.

Our *approximation of the arithmetic mean* is given by

$$\mu = \frac{1}{N} \sum_{j=1}^{K} (M_j \, f_j) \qquad (4.4)$$

where the frequency distribution has K *class intervals*, $M_j =$ the *midpoint* of the jth class, and $f_j =$ the *frequency* of the jth class. Of course, the sample mean is approximated by

$$\bar{X} = \frac{1}{n} \sum_{j=1}^{K} (M_j \, f_j) \qquad (4.5)$$

Example 4.2   Approximate Arithmetic Mean of Income per Household

Using the income data from households on the 2200 block of Main Street, assume that we have only the frequency distribution given in Table 4.8. The necessary calculations for the approximate mean, Equation 4.4, are given in Table 4.9a.

The approximate mean is

$$\mu = \frac{1}{N} \sum (M_j \, f_j) = \frac{1}{30} (268{,}000) = \$8{,}933.33$$

How much error is involved in this approximation? In order to answer that question, let us return to the original data in an array; see Table 4.6. In the frequency distribution, the $0 to $1,999 class included one observation which Table 4.9a estimates as the midpoint (or $1,000). From Table 4.6 we can see that that observation was $0; therefore, we made a $1,000 error in our approximation for that class.

In the next class interval, we have one observation and a midpoint of $3,000. Since the actual income for the household in that interval was $3,058, we made a $58 error in

---

[2] This assumption can be stated several ways. Assume there are m values of X within one class interval. What we would like to know is $\sum_{i=1}^{m} X_i$, but we do not. If we let M_j be the midpoint of this class interval and f_j the frequency, then we assume that $\sum X_i = M_j f_j$. Or, we could assume that the mean of only those values within the class interval, $(\sum X_i)/f_j$, is equal to the midpoint M_j. All three statements result in the same error.

TABLE 4.9 WORKSHEET FOR APPROXIMATING THE MEAN

(a) Approximate the Mean			(b) Actual Data ΣX_i for Each Class	(c) Error $\Sigma X_i - M_j f_j$
Midpoints M_j	Frequency f_j	$M_j f_j$		
1,000	1	1,000	0	−1,000
3,000	1	3,000	3,058	58
5,000	2	10,000	9,641	−359
7,000	5	35,000	34,820	−180
9,000	10	90,000	90,047	47
11,000	8	88,000	88,974	974
13,000	2	26,000	25,750	−250
15,000	1	15,000	14,500	−500
Total	30	268,000	266,790	−1,210

this class. In the $4,000 to $5,999 class, the two observations actually totaled $9,641, and we approximated that total at $10,000 by multiplying the midpoint times the frequency; thus, we got a $359 error.

Table 4.9b uses the actual data. From Equation 4.1, the actual arithmetic mean is

$$\mu = \frac{1}{N} \Sigma X_i = \frac{1}{30}(266,790) = \$8,893.00 \text{ per household}$$

Table 4.9c shows the error for each class interval. The error in calculating the mean is

$$\text{Error} = \text{actual mean} - \text{approximated mean}$$

$$= \$8,893.00 - \$8,933.33 = -\$40.33(= -1,210/30)$$

Is this $40.33 error in the arithmetic mean serious? If we have the original data, this error of approximation should certainly be avoided. But if we do not have the original data, please note that $40.33 relative to $8,893.00 represents less than 0.5 percent [that is, $(40.33/8,893)(100) = 0.45$ percent]. Thus, even if we are stuck with only grouped data, our approximation technique introduces a very small error.

Advantages and disadvantages of the arithmetic mean The arithmetic mean has the distinct advantage of being understood by virtually everyone. You may wish to describe it as "the average" (although "an average" would be more appropriate if no more precise), but the public has some concept of what the arithmetic mean represents.

In addition (no pun intended), the mean is relatively easy to calculate and (in Equation 4.1 at least) includes the influence of all observations in the data set. Less obvious but important nonetheless, the arithmetic mean is amenable to further calculation. The relevance of this fact will be demonstrated throughout this text.

The principal disadvantage of the mean is that it is not always the "most" representative measure of location. If a frequency distribution is very skewed, the mean is not necessarily the "best" locational measure. Also, the arithmetic mean is severely influenced by extreme values. For example, in Panel 4.1, we considered the income of the unoccupied house to be $0. For some objectives, the mean income of occupied households would be more relevant. By omitting this one "extreme value" the arithmetic mean of occupied households is [$266,790/29 =] $9,199.66, which represents a $306.66 increase over the previously calculated mean.

Finally, a word of caution. Essentially, everyone understands what the arithmetic mean is; however, you can reduce confusion in interpreting the mean if you always include the units (e.g., income per household in dollars) when stating the mean.

4.2(b)   The Median

The second most important measure of location for economic data is the *median*. Unlike the arithmetic mean, the value of which is algebraically determined, the value of the median is determined by its position within an array.

First, we must deal with the obvious question: Why do we need another measure of location? The answer is subtle but understandable. When a frequency distribution is skewed, either to the right or to the left, the median is more representative of a "central value" than the arithmetic mean.

For example, consider a very large statistics course in a college. Assume that we take three samples of 5 from the students and determine their earned income during the previous calendar year. The data is as follows (in thousands of dollars):

Sample 1	1	2	2	2	3	$\bar{X}_1 = {}^{10}\!/_5 = 2$
Sample 2	2	2	2	2	12	$\bar{X}_2 = {}^{20}\!/_5 = 4$
Sample 3	0	2	2	2	2	$\bar{X}_3 = {}^{8}\!/_5 = 1.6$

In sample 1, the data is symmetric about the mean, and the arithmetic mean is sufficient as a central value. In sample 2, however, we may be somewhat uncomfortable about using $4,000 as the "central value" of this distribution, which is skewed to the right. After all, $4,000 is twice the income of 80 percent of the persons in sample 2. In sample 3 the distribution is skewed to the left. Now $1,600 may "feel OK" as a representation of the central value, but we still have 80 percent of the sample with income above the mean value. The extreme values (i.e., the 12 in sample 2 and the 0 in sample 3) are included in calculating the mean value.

Note that all three samples are presented in an array. In all three samples, the value of the middle observation in the array is 2, or $2,000. It is this *middle value in an array* which we shall call the *median*. Half the incomes are above and half are below the median of $2,000. It is in this "central-value sense" that the median is commonly used for economic data, particularly for income data which is frequently skewed to the right.

Raw data: "even or odd?" in determining the median In order to determine (rather than calculate) the value of the median for a set of raw or original data, the following steps are necessary:

1. Place the data in an *array*.
2. *a.* If there are an *odd* number of observations in the array, then the *median* is the value of the $(n + 1)/2$ observation.
   b. If there are an *even* number of observations in the array, then the *median* is the *mean* of the $n/2$ and the $(n + 2)/2$ observation.

Example 4.3

Find the median for (a) 2, 2, 2, 2, 12 and (b) 0, 12, 2, 1.

Answer
(a) The data is an array. There are *five* (an odd number of) observations. The third [that is, $(5 + 1)/2 = 3$] observation has a value of 2; therefore, the median equals 2.

(b) An array for the data is 0, 1, 2, 12. There are *four* (an even number of) observations. The second (that is, $^4/_2 = 2$) observation has a value of 1, and the third [that is, $(4 + 2)/2 = 3$] observation has a value of 2; therefore, the median equals [$(1 + 2)/2 =$] 1.5. Since there is no single middle value in an even number of observations, we take the mean of the middle two observations.

To check your understanding of the median, verify that the median value in Table 4.2 is 4 persons per household. In Table 4.6 is the median $9,350?

Grouped data: an approximation for the median Since the median is the value of the middle observation in an array and since we do not know the values of each observation in grouped data (e.g., see Table 4.8), we must approximate the value of the median.

For the grouped data, *after* the median class has been located, the value of the *median* is approximated by

$$\text{Median} = L + \left(\frac{n/2 - F}{f_{\text{med}}}\right) c \tag{4.6}$$

where L = the *lower* class limit of the *median class*
n = the number of observations in the data set
F = the *sum* of the frequencies up to, but not including, the median class
f_{med} = the frequency of only the median class
c = the *width* of the median class interval

Do not be deceived by the apparent complexity of Equation 4.6; it is a shorthand notation for a straightforward and logical approximation.

In a frequency distribution, we do not know the value of each observation, but the class intervals are in an array. Our first step is to *find, by inspection, the class interval that contains the middle value*, and then use Equation 4.6 to determine how far into that class interval we will go to approximate the median.

Example 4.4   Median of Income from the 2200 Block of Main Street

Find the median of the income data presented in Table 4.8. In order to make the class limits more specific, let us rewrite Table 4.8 as follows:

Income Class	Frequency	Cumulative Frequency
0.00– 1,999.99	1	1
2,000.00– 3,999.99	1	2
4,000.00– 5,999.99	2	4
6,000.00– 7,999.99	5	9
8,000.00– 9,999.99	10	19 Median class
10,000.00–11,999.99	8	27
12,000.00–13,999.99	2	29
14,000.00–15,999.99	1	30
	30	

First, we must locate the median class. Since our value of the median is an approximation anyway, we will always look for the value of the $(n + 1)/2$ observation, regardless of whether the number of observations is odd or even.

Since there are $(n =)$ 30 observations, we are in pursuit of the $[(30 + 1)/2 =]$ 15.5 observation (or "halfway between the fifteenth and sixteenth observation"). The Cumulative Frequency heading shows that there are nine household incomes below $8,000 and ten observations within the $8,000 to $9,999.99 range. Hence, the fifteenth and sixteenth observations are somewhere in the $8,000 to $9,999.99 class interval, which is why we shall call it the "median class." The width of that class interval, c, is $2,000.

Starting at the lower class limit of the median class (that is, $L =$ $8,000.00), how far into that interval should we venture (or how much of the $2,000 interval width should we use) to be approximately between the fifteenth and sixteenth observations? Equation 4.6 assumes that the observations within the median class are equally distributed. In this case, the 10 incomes are equally distributed in the $2,000 interval.

The proportion[3] of the $2,000 that we want to add to the lower class limit of $8,000

---

[3] From the "equally spaced" assumption it follows that the 10 values are assumed to be $8,100, $8,300, $8,500, ..., $9,900. If we are looking for the $(n + 1)/2$ item, then the proportional distance into the median class is $[(n/2) - F]/f_{med}$ and *not* $[(n + 1)/2 - F]/f_{med}$.

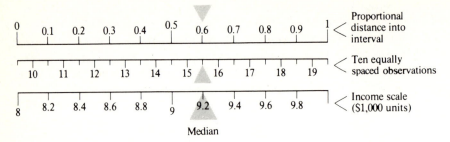

FIGURE 4.5   Approximation within the median class interval.

is given by

$$\left(\frac{n/2 - F}{f_{med}}\right)c = \left(\frac{30/2 - 9}{10}\right)(\$2,000) = \frac{6}{10}(\$2,000) = \$1,200$$

The nine observations that occur in lower class intervals are represented by F. The term $(n/2 - F)$ represents the distance, in "units of frequency," that we want to go into the median class; the frequency of that class, f_{med}, is the number of "units of frequency" available; therefore $[(n/2 - F)/f_{med}]$ represents a proportion of the interval that we approximate as relevant to the value of the median.

The center scale in Figure 4.5 denotes the 10 equally spaced observations. Since the first nine incomes were lower than $8,000, these 10 observations would be the tenth through the nineteenth in an array of the original data. We want to be halfway between the fifteenth and sixteenth observations, which the top scale in Figure 4.5 shows to be 0.6, or 60 percent, of the interval. Translating that into dollars yields $9,200 on the bottom scale of Figure 4.5.

Drawing the three scales similar to Figure 4.5 every time we want to approximate the median is a pain. Hence, we use our algebraic equivalent in the form of Equation 4.6.

$$\text{Median} = L + \left(\frac{n/2 - F}{f_{med}}\right)c = \$8,000.00 + \left(\frac{30/2 - 9}{10}\right)(\$2,000)$$

$$= \$8,000.00 + (0.6)(\$2,000) = \$9,200 \text{ per household}$$

Note that from Table 4.6 we found the actual median to be $9,350. As a percentage of $9,350, the $150 error is quite small; the error was due to the fact that the 10 observations in the "median class," as we can see from Table 4.6, are not equally distributed within the interval. Without the luxury of the raw data, our approximation was close to the value of the median.

Advantages and disadvantages of the median   As a positional, rather than an arithmetic, measure of location, the median is not influenced by extreme values. The median is easy to understand; half the data are below it, and half the data

distributions are almost always skewed (and usually to the right[4]). Official United States government publications include the medians with their frequency distributions because of these advantages.

Among the disadvantages, the most important is that the median is not amenable to further calculation. The median does not use much information from a set of data. Finally, unless you have a computer or slave labor (e.g., graduate students), putting a large set of numbers into an array is not easy. Hence, for a large set of numbers the median is not easy to calculate.

4.2(c)   The Mode

The *modal value* of a distribution of numbers is simply the value which occurs most frequently. For grouped data, the *modal class* is the class interval with the highest frequency, and the modal value is often approximated as the midpoint of the modal class.

For example, the modal value in Table 4.2 is 4. The modal class in Table 4.8 is $8,000 to $9,999, and $9,000 could be used as the modal value.

In economics and business, the value of the mode, or the modal value, is decreasing in importance. Of more importance is the *number of modes* for any particular distribution.

In Figure 4.6 the unimodal distribution X may be the income distribution in the United States; a bimodal distribution of grades, Y, is common on the first statistics examination; and a trimodal distribution Z could occur on the load chart of electric energy used in a city.

We shall use the definition of the mode primarily in terms of how many modes a distribution exhibits. Obviously, the mode is also a positional (rather than arithmetic) measure of location. It is not amenable to further calculation.

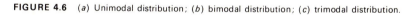

[4] The data in Table 4.5 is hypothetical but based upon experience. Some income distributions are skewed to the left.

FIGURE 4.6   (a) Unimodal distribution; (b) bimodal distribution; (c) trimodal distribution.

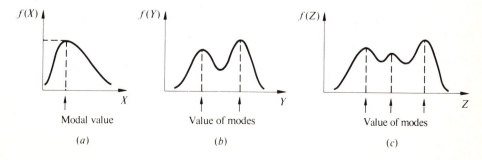

4.2(d) Other Measures of Location

There are at least a couple other measures of location infrequently employed in economics and business. The *harmonic mean*, which is defined as $H = N/\Sigma(1/X_i)$, has only highly specialized uses and will not be discussed further.

The *geometric mean*, which is defined as $G = \sqrt[N]{(X_1)(X_2)\cdots(X_N)}$, or the Nth root of the product of N observations, is more commonly thought of in the context of calculating compound interest. In general, the geometric mean is a more appropriate measure of location for calculating average rates of change and average ratios and for averaging index numbers.

Algebraically, the arithmetic mean is always greater than or equal to the geometric mean. The mathematics of finance and financial management treat these differences, and we shall omit them for the moment.

The special case of unimodal distributions If a distribution is unimodal, then we can say something about the skewness of that distribution by considering the values of the mean, median, and mode. Recall in our introduction to the median that we demonstrated how one extreme value in the data would influence the mean but not the median. Specifically, in our samples of the income of five randomly selected students, we demonstrated: (1) if the distribution is symmetric, then the mean is equal to the median; (2) if the distribution is skewed to the right (e.g., a few very high incomes), then the mean is greater than the median; and (3) if the distribution is skewed to the left (e.g., a few very low incomes), then the median is greater than the mean.

That is, if an extremely high value is not offset by an extremely low value (distribution is not symmetric), then the mean is "pulled toward" the extreme value, or the mean is closer to the "tail" of the distribution. Note that our reasoning followed the order of (1) given a skewed distribution, then (2) the mean will be closer

Figure 4.7 Relation of skewness to mean and median.

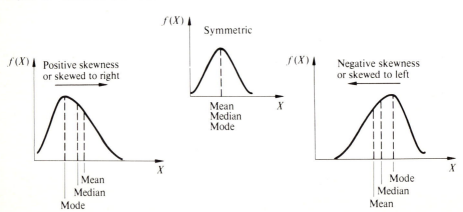

to the tail. Let's reverse our direction on that order. Given that the mean is different from the median, we should be able to tell which way the distribution is skewed.

If, and only if, the distribution is unimodal, then (1) *the median will always lie between the mean and the mode, and* (2) *the mean will point the direction of the distribution's longest tail.* Thus, if the mean is greater than the median, the distribution is skewed to the right, and if the mean is less than the median, the distribution is skewed to the left. Figure 4.7 demonstrates the point.

4.3   MEASURES OF VARIATION

Variation is the name of the game. If there were no variation, scatter, or dispersion in the data of interest, then there would be no need for most of the measures of descriptive statistics. But there is variation in most data, and we need to be able to describe it. Even when we think one particular thing is identical to another, we would probably find, if we measured precisely enough, variation.

Example 4.5   "Getting Down to Nuts and Bolts"

The concept of and the producibility of "standardized parts" have played an important role in the industrial capability of the Western economies. Interchangeable parts lead us to believe that the parts are "identical." In what sense are they identical? In the practical sense that they are interchangeable.

As an experiment, enter a local hardware store. If you can find a bin or box of bolts and corresponding nuts, pick up a nut and a handful of bolts. Check to see if all the bolts will fit that nut. It would be very unusual if they did not. Does that mean that all the bolts have the same (outside) diameter? No! All it means is that whatever variation there is in the diameter of the bolts is small enough to make the bolts interchangeable.

Next, you could take one bolt and a handful of nuts. Do all the nuts, just because they fit, have the same inside diameter? No, not necessarily.

How do we reconcile this variability in diameters, of both the nuts and the bolts, with their demonstrated interchangeability? (*Note:* Engineering students may be shattered by the discussion that follows. The discussion is conceptually correct, but it does not include all the problems in manufacturing nuts and bolts. Nor does this discussion employ the correct engineering terminology.)

With the aid of calipers, let us measure the diameter (Z) of 1,000 bolts and 1,000 nuts to the nearest ten-thousandth of an inch. Our results are graphed in Figure 4.8a where the mean diameter of the bolts is 0.7499 inch and the mean diameter of the nuts is 0.7501 inch. Note that the diameter of the smallest bolt was 0.7498 inch, and the largest was 0.7500 inch. Also, the diameter of the smallest nut was 0.7500 inch, and the largest was 0.7502 inch.

The two frequency distributions in Figure 4.8a appear to be the same except they are *located* at different places on the Z axis; they have the same variation but different means. From the interchangeability viewpoint, it is important to note that these two distributions do not *overlap*. The bolt with the largest diameter would fit (perhaps snugly) into the nut with the smallest diameter.

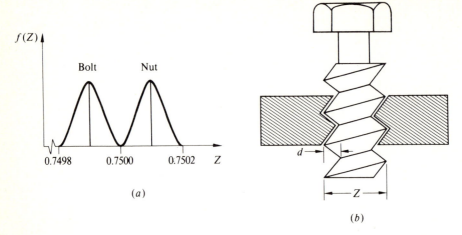

FIGURE 4.8 (*a*) Frequency distribution of 1,000 bolt diameters and 1,000 nut diameters (inches); (*b*) bolt and nut.

To keep the smallest bolt (0.7498 inch) from just falling into the largest nut (0.7502 inch), we would want the width of the "groove," as shown by d in Figure 4.8*b*, to be *greater than* (0.7502 to 0.7498 =) 0.0004 inch.

It is in this context that engineers discuss *tolerances*. If a bolt is manufactured to have a diameter of 0.7499 ± 0.0001 inch (that is, ±0.0001 inch is the tolerance) with a groove of 0.0010 ∓ 0.0005 inch, then all nuts made with a diameter of 0.7501 ± 0.0001 inch will be interchangeable with all the bolts. *Variation exists.* The variations are simply made small enough to "tolerate" one another; the parts are interchangeable.

Granting that variation exists, is it ever important by itself? Subsequent discussion will demonstrate that variation will determine the precision of our statistical estimates and tests. For the moment, however, we can demonstrate that the degree of variation, when we describe data, can influence the types of question we would want to ask.

Consider the mortality data in Figure 4.9. The variable X is the age at death in Bike and Y is the age at death in SMOG. Both countries have the same average age of death: $\mu_X = \mu_Y = 70$ years. Both distributions are symmetric about their mean. Since the data is for 1,000 individuals in each country, the areas under the curves are identical. But the distribution of Y is "flatter" than the distribution of X; the Y values are more spread out than the X values. We say that X exhibits less variation than Y. Even though the average age of death is the same for the two countries, there are more people who die before age 66 in SMOG than in Bike. Also, note that there are more people who live past the age of 74 in SMOG than in Bike. When we look at only the mean age of death, there appears to be no difference; however, when we observe the variation in X and in Y, we might ask: What are the differences between

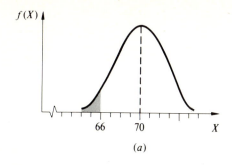
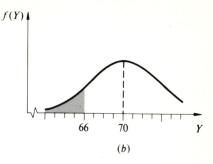

$f(X)$ $f(Y)$

66   70 X 66   70 Y

(a)   (b)

FIGURE 4.9   Hypothetical data on age of death on 1,000 individuals in each of two countries. (a) Republic of Bike; (b) Supreme monachy of Grit (SMOG).

these two countries that would cause a greater dispersion in the age-of-death of SMOG citizens than of Bike citizens?

Since variation frequently occurs and since the extent or degree of variation is important, how do we measure variation in an empirical distribution? We shall consider only three measures of variation: the range, the variance, and the standard deviation.

4.3(a)   The Range

The range is the easiest measure of variation to calculate. For raw or ungrouped data, the *range* is defined as the difference between the highest and lowest values in a set of data. Returning to our number of persons per household in Table 4.1 (or 4.2), we can easily see that the range is (maximum value) − (minimum value) = 7 − 0 = 7 persons per household. From Panel 4.1 we find that the range of income per household is [14,500 − 0 =] $14,500; the range of income per *occupied* house is [14,500 − 3,058 =] $11,442.

With grouped data we do not know the highest and lowest values. If there are no open-ended class intervals, we can approximate the range by using class limits. We approximate the range by taking the *upper class limit of the highest-valued class interval minus the lower class limit of the lowest-valued interval.*

For example, from Table 4.8, the "highest-valued class interval" is $14,000.00 to $15,999.99, and its upper class limit is $15,999.99. In like manner, the "lowest-valued class interval" is $0 to $1,999.99, and its lower class limit is $0. The range is approximately $15,999.99 to $0, which equals $15,999.99, or, to the nearest dollar, $16,000. Note that by using this approximation we are always *equal to or above* the actual value of the range. Further, this approximation is easier to use than to state verbally.

The advantages of the range are: (1) it is easy to calculate, and (2) it is understood by the general populace. An easily identifiable disadvantage is that frequently the range is drastically overstated when approximating from grouped data, and we

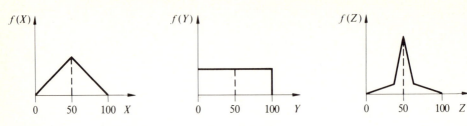

FIGURE 4.10 Three different distributions with the same range.

simply cannot approximate it at all if we have an open-ended class interval (i.e., what is the "upper class limit" of the interval "income of $100,000 or more"?).

A more important disadvantage is that we may wish to know more about the dispersion of the data than can be obtained from the range. Consider Figure 4.10.

All three distributions in Figure 4.10 have a mean of 50 and a range of 100, and all three are symmetric. The dispersions of X, Y, and Z are all quite different, but the range, as a measure of dispersion, does not inform us of those differences. The use of percentiles, deciles, and quartiles can overcome this defect in the range (e.g., the middle 50 percent of the values fall between 35.5 and 64.5 for X, between 25 and 75 for Y, and between 48 and 52 for Z). With regard to statistical inference, which is what interests us, the percentiles, etc., also have disadvantages. Thus, we will proceed to more amenable measures of dispersion.

4.3(b) The Variance

Reconsider Figure 4.10. We observe that the variation or dispersion of Y values is greater than the variation or dispersion of X values. Also, the values of both X and Y are less condensed than the values of Z. It would be convenient, since such variations can be important to us, if our measure of variation reflected those differences. The variance is a measure which provides this convenience.

Since the dispersion of Y is greater than the dispersion of X which is greater than the dispersion of Z, the variance yields the following relationship:

$$\text{Var}(Y) > \text{Var}(X) > \text{Var}(Z)$$

The variance is a measure of dispersion *relative* to some reference point. That reference point is the arithmetic mean of the distribution. More specifically, the variance is a measure of how close, or how far away, the various values are to their own arithmetic mean. In a population (as opposed to a sample), perhaps the easiest measure of *closeness* to the mean is the difference between a particular value, X_i, and the mean, μ_x. Unfortunately, if we find that difference for all N values of X in the population, the sum of the positive differences will *always* just equal the sum of the negative differences. Hence, the net result is always zero![5] That is,

$$\sum_{i=1}^{N} (X_i - \mu_x) \equiv 0$$

If we insist upon using the arithmetic mean as a "reference point"[6] (and we do insist), then this algebraic truth prohibits us from using the deviations about the mean as a *measure* of variation.

We can eliminate this difficulty (of the sum of the positive deviations from the mean being equal to the sum of the negative deviations) in two ways. First, we could just ignore the algebraic sign on each deviation. Mathematicians call this taking the *absolute value* of the difference between any given value of X and the mean of X; symbolically $|X_i - \mu|$ is read "the absolute value of the difference between X_i and μ." If we add all the absolute deviations from the mean

$$\sum_{i=1}^{N} |X_i - \mu_x|$$

and divide that total by the number of items N within the population, then we have "an average of the absolute deviations from the mean." Cleverly enough, we call this the *mean deviation (about the mean)*, or

$$\text{MD} = \text{mean deviation} = \frac{1}{N} \sum_{i=1}^{N} |X_i - \mu_x| \qquad (4.7)$$

This is one solution to our problem of the positive deviations canceling the negative deviations, but, in terms of getting us any closer to statistical inference, it leads us down a blind alley. The mean deviation about the mean is not in common use, and it is not amenable to the further algebraic manipulation currently required for statistical inference.

The second way we can eliminate the algebraic sign on the deviations from the mean is by squaring each deviation. That is, $(X_i - \mu)^2$ is positive if $(X_i - \mu)$ is negative. By adding all the squared deviations from the mean[7] [that is,

---

[5] The proof is simple: $\Sigma(X_i - \mu) = \Sigma X_i - \Sigma \mu = \Sigma X_i - N\mu$; since μ is defined as $(1/N)\Sigma X_i$, then $\Sigma X_i - N(1/N)\Sigma X_i = \Sigma X_i - \Sigma X_i = 0$.

[6] Using the median doesn't help much. If the distribution is symmetric, then the median equals the mean. Referring to Figure 4.10, we can see that the summation of the deviations from the median would be equal to 0 in all three cases. Using the mode as the "reference point" is equally troublesome. What do we do with a multimodal distribution?

[7] Note that there is a big difference between "the square of the sum of the deviations from the mean" and "the sum of the squared deviations from the mean." The *operations* called for are entirely different. That is, $[\Sigma(X - \mu)]^2 \neq \Sigma(X - \mu)^2$, or

$$0 = [(X_1 - \mu) + (X_2 - \mu) + \cdots + (X_N - \mu)]^2 \neq (X_1 - \mu)^2 + (X_2 - \mu)^2 + \cdots + (X_N - \mu)^2$$

$\sum_{i=1}^{N} (X_i - \mu_x)^2]$ and dividing by the total number of items within the population (that is, N), we obtain an average of the squared deviations from the mean. For raw or ungrouped data, we define the *variance of the population of X values* to be

$$\text{Var}(X) = \sigma_x{}^2 = \frac{1}{N} \sum_{i=1}^{N} (X_i - \mu_x)^2 \qquad (4.8)$$

The farther away the X_i's are from the mean, the larger the variance; the closer the X_i's are to the mean, the smaller the variance. Thus, we finally have a measure that would tell us that there is a difference in the variation of Y values and the variations of Z values in Figure 4.10.

Example 4.6   The Variance of Two Simple Distributions

Let the variable X assume the five values 1, 3, 4, 5, and 7 while the variable Y assumes the five values 1, 4, 4, 4, and 7. See Figure 4.11.

There is more variation in the X values than in the Y values. The means are equal to 4 and the ranges are $[7 - 1 =] 6$. What is the variance of X? Of Y? See Table 4.10. Thus, $\sigma_x{}^2 > \sigma_y{}^2$, and there is more variation in the X values than in the Y values. The Y values are more tightly packed about the mean than are the X values.

TABLE 4.10   CALCULATION OF THE MEAN AND VARIANCE

	(a) For X			(b) For Y	
X	$X - \mu$	$(X - \mu)^2$	Y	$Y - \mu$	$(Y - \mu)^2$
1	-3	9	1	-3	9
3	-1	1	4	0	0
4	0	0	4	0	0
5	1	1	4	0	0
7	3	9	7	3	9
20 $= \Sigma X$	0 $= \Sigma(X - \mu)$	20 $= \Sigma(X - \mu)^2$	20 $= \Sigma Y$	0 $= \Sigma(Y - \mu)$	18 $= \Sigma(Y - \mu)^2$

$$\mu_x = \frac{\Sigma X}{N} = \frac{20}{5} = 4 \qquad\qquad \mu_Y = \frac{\Sigma Y}{N} = \frac{20}{5} = 4$$

$$\sigma_x{}^2 = \frac{\Sigma(X - \mu)^2}{N} = \frac{20}{5} = 4 \qquad\qquad \sigma_Y{}^2 = \frac{\Sigma(Y - \mu)^2}{N} = \frac{18}{5} = 3.6$$

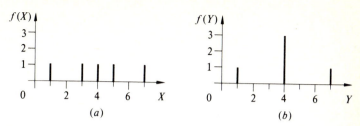

FIGURE 4.11 Distribution of two populations: X and Y.

Example 4.7 **Variance of the Number of Persons per Household on the 2200 Block of Main Street**

Let us return to the data presented in Table 4.2.

X_i	$X_i - \mu$	$(X - \mu)^2$	X_i	$X_i - \mu$	$(X - \mu)^2$
0	−3.5	12.25	4	0.5	0.25
1	−2.5	6.25	4	0.5	0.25
1	−2.5	6.25	4	0.5	0.25
1	−2.5	6.25	4	0.5	0.25
1	−2.5	6.25	4	0.5	0.25
2	−1.5	2.25	4	0.5	0.25
2	−1.5	2.25	4	0.5	0.25
2	−1.5	2.25	5	1.5	2.25
2	−1.5	2.25	5	1.5	2.25
3	−0.5	0.25	5	1.5	2.25
3	−0.5	0.25	5	1.5	2.25
3	−0.5	0.25	6	2.5	6.25
3	−0.5	0.25	6	2.5	6.25
3	−0.5	0.25	7	3.5	12.25
4	0.5	0.25	7	3.5	12.25
			105	0	95.50

$\mu_X = (1/N) \cdot \Sigma X_i = (1/30) \cdot 105 = 3.5$ persons per household.

$\sigma_X^2 = (1/N) \cdot \Sigma(X_i - \mu)^2 = (1/30) \cdot 95.5 = 3.183$ (persons per house)^2

Note the units on the variance.

When there are a small number of items in the data set (that is, $N = 5$ in example 4.6), the definition given by Equation 4.8 is convenient for calculating the variance. When N is large (that is, $N = 30$ in example 4.7), Equation 4.8 becomes more cumbersome. For large data sets it is useful to note the algebraic identity:[8]

---

[8] Proof: $\Sigma(X - \mu)^2 = \Sigma(X^2 - 2\mu X + \mu^2) = \Sigma X^2 - 2\mu\Sigma X + \Sigma\mu^2$

$= \Sigma X^2 - 2\mu N(\Sigma X/N) + N\mu^2 = \Sigma X^2 - 2N\mu^2 + N\mu^2$

$= \Sigma X^2 - N\mu^2 = \Sigma X^2 - (\Sigma X)^2/N = (1/N)[N\Sigma X^2 - (\Sigma X)^2]$

$$\sigma_X{}^2 = \frac{1}{N}\Sigma(X_i - \mu_X)^2 = \frac{1}{N}\left[\Sigma X^2 - \frac{(\Sigma X)^2}{N}\right]$$

$$= \frac{1}{N^2}[N\Sigma X^2 - (\Sigma X)^2] \qquad (4.9)$$

To use Equation 4.9, we do not need to subtract the mean from each observation; instead, we need to calculate only ΣX (which we need to calculate the mean anyway) and ΣX^2.

Example 4.8 Example 4.7 Revisited

X	X²	X	X²	X	X²
0	0	3	9	4	16
1	1	3	9	4	16
1	1	3	9	5	25
1	1	3	9	5	25
1	1	4	16	5	25
2	4	4	16	5	25
2	4	4	16	6	36
2	4	4	16	6	36
2	4	4	16	7	49
3	9	4	16	7	49
				105	463

$$\sigma_X{}^2 = (1/N)[\Sigma X^2 - (\Sigma X)^2/N] = (\tfrac{1}{30})[463 - (105)^2/30]$$
$$= (\tfrac{1}{30})[463 - (11{,}025)/30] = (\tfrac{1}{30})(463 - 367.5)$$
$$= (\tfrac{1}{30})(95.5) = \underline{3.183 \text{ (persons per household)}^2}$$
$$\sigma_X{}^2 = (1/N^2)[N\Sigma X^2 - (\Sigma X)^2] = (\tfrac{1}{900})[(30)(463) - (105)^2]$$
$$= (\tfrac{1}{900})(13{,}890 - 11{,}025) = (\tfrac{1}{900})(2{,}865)$$
$$= \underline{3.183 \text{ (persons per household)}^2}$$

Example 4.7 provided exactly the same result; however, some individuals would find these numbers easier to manipulate.

Example 4.9 The Variance of Income per House on 2200 Main Street

Using Table 4.6, we need to find the square of each income.

Y	Y²	Y	Y²	Y	Y²
$ 0	0	8,045	64,722,025	10,222	104,489,284
3,058	9,351,364	8,500	72,250,000	10,800	116,640,000
4,629	21,427,641	8,726	76,143,076	10,955	120,012,025
5,012	25,120,144	9,000	81,000,000	11,628	135,210,384
6,025	36,300,625	9,200	84,640,000	11,630	135,256,900
6,466	41,809,156	9,500	90,250,000	11,750	138,062,500
7,029	49,406,841	9,500	90,250,000	11,989	143,736,121
7,500	56,250,000	9,750	95,062,500	12,000	144,000,000
7,800	60,840,000	9,826	96,550,276	13,750	189,062,500
8,000	64,000,000	10,000	100,000,000	14,500	210,250,000
				266,790	2,652,093,362

$$\sigma_Y{}^2 = (1/N)[\Sigma Y^2 - (\Sigma Y)^2/N] = (\tfrac{1}{30})[2,652,093,362 - (266,790)^2/30]$$
$$= (\tfrac{1}{30})[2,652,093,362 - (71,176,904,100)/30]$$
$$= (\tfrac{1}{30})(2,652,093,362 - 2,372,563,470)$$
$$= (\tfrac{1}{30})(279,529,892) = 9,317,663.67 \ \$^2$$

You are not expected to crank through all these calculations. The purpose of this example is twofold. First, although these 30 incomes are realistic numbers, their use in calculating the variance welcomes the employment of a computer. Second, we wish to compare this actual variance with a subsequent approximation.

Return to Example 4.7 and note that

$$\sigma_X{}^2 = (1/N) \sum_{i=1}^{N} (X_i - \mu)^2 = (\tfrac{1}{30})[(0 - 3.5)^2$$
$$+ (1 - 3.5)^2 + (1 - 3.5)^2 + \cdots + (7 - 3.5)^2]$$

could be rewritten as

$$\sigma_X{}^2 = (1/N) \sum_{j=1}^{K} [f_j(X_j - \mu)^2]$$
$$= (\tfrac{1}{30}) \cdot [1(0 - 3.5)^2 + 4(1 - 3.5)^2 + 4(2 - 3.5)^2 + 5(3 - 3.5)^2$$
$$+ 8(4 - 3.5)^2 + 4(5 - 3.5)^2 + 2(6 - 3.5)^2 + 2(7 - 3.5)^2]$$

where $X_j = 0, 1, 2, \ldots, 7$ and f_j is how frequently the jth value of X occurred. Again, we will use this concept to develop our approximation of the variance when we have only grouped data. As in the case of approximating the arithmetic mean from grouped data, we will assume that every observation within a class interval can be represented by the midpoint of that class interval. (See footnote 2.) Since we do not know each value of X within a given class interval, we use what we do have: the midpoint of that interval.

Thus, we *approximate the population variance* from

$$\sigma_X{}^2 = \frac{1}{N} \sum_{j=1}^{K} [f_j(M_j - \mu)^2] \qquad (4.10)$$

where K = the number of class intervals

M_j = the midpoint of the jth class interval

f_j = the frequency of the jth class interval

$N = \Sigma f_j$ = the number of items in the population

Since the number of class intervals, K, is usually between 5 and 15, Equation 4.10 is often sufficient for calculational purposes.

Example 4.10 Approximation of the Variance from Grouped Data

Reconsider the income data in Table 4.8 and assume that the cents were omitted in reporting income. Thus, the midpoints of the class intervals are $1,000, $3,000, ..., $15,000.

Midpoints M_j	Frequency f_j	$f_j M_j$	$M_j - \mu$	$(M_j - \mu)^2$	$f_j(M_j - \mu)^2$
$ 1,000	1	1,000	−7,933.33	62,937,724	62,937,724
3,000	1	3,000	−5,933.33	35,204,404	35,204,404
5,000	2	10,000	−3,933.33	15,471,084	30,942,168
7,000	5	35,000	−1,933.33	3,737,765	18,688,825
9,000	10	90,000	66.67	4,449	44,490
11,000	8	88,000	2,066.67	4,271,125	34,169,000
13,000	2	26,000	4,066.67	16,537,804	33,075,608
15,000	1	15,000	6,066.67	36,804,484	36,804,484
		268,000			251,866,703

From Equation 4.4 we can approximate the population mean as

$$\mu = \frac{1}{N} \Sigma M_j f_j = \left(\frac{1}{30}\right)(268,000) = \$8,933.33$$

From the last three columns above, we can approximate the variance by

$$\sigma_Y{}^2 = \frac{1}{N}\Sigma[f_j(M_j - \mu)^2] = \left(\frac{1}{30}\right)(251{,}866{,}703) = 8{,}395{,}557 \text{ (\$)}^2$$

From example 4.9, the actual variance is (\$)² 9,317,663.67, and our estimate reflects a 9.9 percent error relative to the actual variance. What could cause that much error? Equation 4.10 assumes that the midpoint is representative of the observations within that class interval. Since the \$0 to \$1,999.99 interval has only one observation which in fact is 0 but in our approximation is assumed to be \$1,000 (the midpoint), then we have introduced this "error" and other similar errors in approximating the variance by the *choice* of class intervals.

Further, since 1,000 is "closer" to the mean than 0 is, we would expect the influence of the first class interval to make our approximated variance lower than the actual variance. If we had considered the income per household of occupied houses, our error in approximating the variance would have been only about 1 percent even if we did not change the class intervals.

Please note that the careless construction of class intervals when constructing an empirical frequency distribution can lead to error in approximating statistical measures calculated exclusively from that frequency distribution.

The sample variance   The preceding discussion has developed the need for and the ways to calculate the variance of a population. If we wish to take a sample from that population and calculate the variance of the sample, then we employ a slightly different concept of the *average* of the squared deviations from the mean.

The *sample variance* is defined as

$$s_x{}^2 = \frac{1}{n-1} \sum_{i=1}^{n} (X_i - \bar{X})^2 \tag{4.11}$$

where $s_x{}^2$ = sample variance (as opposed to $\sigma_x{}^2$ for a population)

$\bar{X}$ = the sample mean = $(1/n)\Sigma X_i$

n = the sample size or number of observations in the sample

When n is large, there is not much difference in $s_x{}^2$ when n is used instead of $(n-1)$. Thus, the algebraic difference between Equations 4.11 and 4.8 is important only for a small number of observations. Also, note from footnote 8 that

$$s_x{}^2 = \left(\frac{1}{n-1}\right)\left[\Sigma X^2 - \frac{(\Sigma X)^2}{n}\right] = \frac{n\Sigma X^2 - (\Sigma X)^2}{n(n-1)}$$

Later we shall discuss the concepts of degrees of freedom and unbiased estimators, which will clarify the use of $(n - 1)$ instead of n in the denominator of Equation 4.11. For the moment, a certain level of trust is required.

Interpretation of the variance   The purpose of this section was to develop a single number that would tell us how close together or spread apart are the observations in a set of data. Variation must occur relative to something, and we chose the arithmetic mean as a reference point. The easiest measure of distance from the mean [that is, $\Sigma(X_i - \mu)$] was of no assistance because it is always equal to 0. The measure that evolved is an average of squared deviations from the mean, and history has christened that measure the *variance*.

In Chapter 3, we defined the variance of a probability distribution. Are we not using the same word for two different concepts? No! For a probability distribution, we said

$$\sigma_x{}^2 = \sum_{i=1}^{n} \{[X_i - E(X)]^2 P(X_i)\} \qquad \text{(4.12)}$$

First, the expected value of a random variable, $E(X)$, is analogous to the arithmetic mean. Further, if we rewrite Equation 4.10 such that the constant $(1/N)$ is placed within the summation operator,

$$\sigma_x{}^2 = \Sigma \left[\frac{f_j}{N}(M_j - \mu)^2 \right]$$

then all we have to explain is how M_j relates to the value of a random variable and how f_j/N relates to $P(X_i)$. If all the values within a class interval are equal to the value of the midpoint, which is the case with a discrete probability distribution, then M_j is analogous to X_i in Equation 4.12. The term f_j/N is a *relative* frequency, the proportion of the observations that fall within a class interval. Recalling our relative frequency definition of probability from Chapter 2 should complete our analogy. Equation 4.12 for a probability distribution has the same conceptual framework as Equation 4.8 for an empirical distribution.

Finally, reinvestigate the units associated with the variance in examples 4.8 through 4.10. Does (persons per household)^2 or dollars squared mean anything to you? Those units simply are not easy to interpret. This is the major disadvantage of using the variance as a measure of variation. At this point the variance is the most difficult measure to calculate. As a reward for all that calculational effort, surely the result should be easier to interpret! To eliminate the squared units on the variance, we could take the square root.

4.3(c) The Standard Deviation

With all the cleverness and candor of a Perry Mason script, we shall define the *standard deviation* as the positive square root of the variance, or

---

$$\text{Standard deviation} = \sigma_X = \sqrt{\text{Var}(X)}$$

From example 4.6, the standard deviation of X is $\sigma_X = \sqrt{\sigma_X^{\,2}} = \sqrt{4} = 2$, and $\sigma_Y = \sqrt{3.6} = 1.9$. In example 4.7, $\sigma_X = \sqrt{3.183 \text{ (persons per household)}^2} = 1.78$ persons per household; from example 4.9, $\sigma_Y = \sqrt{(\$^2)9{,}317{,}663.67} = \$3{,}052.48$.

The advantages of the variance are not lost by using the standard deviation, and we can understand the units associated with the standard deviation. If

$$\text{Var}(X) > \text{Var}(Y)$$

then

$$\sigma_X > \sigma_Y$$

and there is more variation of the X values about their mean than there is variation of Y values about the mean of Y.

The standard deviation is the principal measure of variation that we will employ in the remaining chapters. Thus, you should become as familiar with interpreting the significance of $\sigma_Y = \$3{,}000$ as you are with interpreting the significance of $\mu_Y = \$9{,}000$. You can relate to a mean income of $\$9{,}000$, but how much income variation is represented by a standard deviation of $\$3{,}000$? Less variation than if σ_Y were $\$4{,}000$ and more variation than if σ_Y were $\$2{,}000$, but so what?

Chebyshev's inequality   In order to obtain a more precise view of how much variation is represented by the value of the standard deviation, we must consider a theorem named after the Russian mathematician Chebyshev (rhymes with "chubby chef"). Chebyshev's inequality is often presented as a probability relationship, but we shall consider it first as a property of numbers.

The proposition that the standard deviation is a measure of how closely the elements of the population are packed around their mean can be viewed in another way. What if we said that there are two numbers $(X_L$ and $X_U)$ such that at least 75 percent of all the numbers within the population will be between those two numbers? Let X_L be the lower limit of X and X_U be the upper limit.

If X_L and X_U are very close to one another and if they include at least 75 percent of all the members of the X population, then the population is not very spread out; the population is rather tightly packed. If X_U and X_L are far apart, then the population is rather spread out.

Now if the population is to be tightly packed around its mean, then μ_X had better be somewhere between X_L and X_U. Let us require X_L and X_U to be limits such that the mean is halfway between them; or, what amounts to the same thing, let $X_L = \mu_X - a$ and $X_U = \mu_X + a$, where a is some constant.

If the standard deviation is a measure of variation, σ should be directly proportional to a. That is, if a is large, then σ should be large; if σ is small, then a should be small. If we want at least 75 percent of the population to lie between X_L and X_U, then

$$a = 2\sigma$$

That is, between $(\mu + 2\sigma)$ and $(\mu - 2\sigma)$ there will be no less than 75 percent of the entire population. Further, a is directly proportional to σ, the population's standard deviation.

That is nice, but where did the $a = 2\sigma$ come from? Chebyshev's inequality can be stated as follows:

Chebyshev's inequality The *proportion* of observations that fall between $\mu - K\sigma$ and $\mu + K\sigma$ (for $K > 1$) is *equal to or greater than* $1 - 1/K^2$.

If $K = 2$, then the proportion of the population that falls between $(\mu - 2\sigma)$ and $(\mu + 2\sigma)$ is at least $1 - 1/2^2 = 1 - \frac{1}{4} = \frac{3}{4}$, or 75 percent. If $K = 3$, then the minimum proportion of observations in the interval $\mu \pm 3\sigma$ is $1 - 1/3^2 = \frac{8}{9}$, or 88.9 percent of the population.

Think about what this implies! By just knowing the mean and standard deviation of the population, we can specify a range of values $(\mu \pm 3\sigma)$ which contains at least 88.9 percent of the population. The entire population may be within three standard deviations of the mean, but there will *never* be less than 88.9 percent of the population within that range. Further, the smaller σ, the smaller the range within which *at least* 88.9 percent of the population resides; the smaller σ, the more closely packed the set of numbers which represent the population.

Finally, note that Chebyshev's theorem does not require that the shape of the population be known, but it does require σ to be known. In later chapters, we shall use Chebyshev's inequality as a last resort.

Proof of Chebyshev's Inequality
Consider a population of X values. Each value of X, say X_i, could occur more than once; hence, let f_i equal the number of times the particular value X_i occurs in the population. Then, $\Sigma f_i = N = $ the size of the population.

Now let us place these N values in an array and subdivide the array into three parts such that

$$-\infty < X_i' < \mu - K\sigma \leq X_i'' \leq \mu + K\sigma < X_i''' < +\infty$$

where $\mu = (1/N)\Sigma(f_i X_i)$, $\sigma = \sqrt{(1/N)\Sigma[(X_i - \mu)^2 f_i]}$, and K is some arbitrarily chosen constant such that $K > 0$.

That is, X'_i represents only values of X which are *less than* $(\mu - K\sigma)$; X''_i represents only those values of X which are *equal to or between* $(\mu - K\sigma)$ and $(\mu + K\sigma)$; and X'''_i represents only those X values which are *greater than* $(\mu + K\sigma)$. There are $\Sigma f'_i$ values of X lower than $(\mu - K\sigma)$, $\Sigma f'''_i$ values of X above $(\mu + K\sigma)$, and $\Sigma f''_i$ values of X between $(\mu - K\sigma)$ and $(\mu + K\sigma)$.

Reconsider the definition of the variance

$$\sigma^2 = \frac{1}{N} \Sigma[(X_i - \mu)^2 f_i]$$

which, by incorporating our three subdivisions, becomes

$$\sigma^2 = \frac{1}{N} \{\Sigma[(X'_i - \mu)^2 f'_i] + \Sigma[(X''_i - \mu)^2 f''_i] + \Sigma[(X'''_i - \mu)^2 f'''_i]\}$$

Consider the right side of this equation. By removing the middle term, what remains must be less than (or equal to) the population's variance. That is,

$$\sigma^2 \geq \frac{1}{N} \{\Sigma[(X'_i - \mu)^2 f'_i] + \Sigma[(X'''_i - \mu)^2 f'''_i]\}$$

By subdividing the population into three classifications, we defined X'''_i such that $X'''_i > \mu + K\sigma$. Thus, $X'''_i - \mu > K\sigma$ and $(X'''_i - \mu)^2 > K^2\sigma^2$. By similar reasoning, $|X'_i - \mu| > K\sigma$ and $(X'_i - \mu)^2 > K^2\sigma^2$. Using these inequalities, we find that

$$\sigma^2 \geq \frac{1}{N} [\Sigma(K^2\sigma^2 f'_i) + \Sigma(K^2\sigma^2 f'''_i)]$$

$$\sigma^2 \geq \frac{1}{N} (K^2\sigma^2)(\Sigma f'_i + \Sigma f'''_i)$$

Dividing both sides by $K^2\sigma^2$ (which requires that both σ and K be greater than 0) yields

$$\frac{1}{K^2} \geq \frac{1}{N} (\Sigma f'_i + \Sigma f'''_i)$$

Recall that $(1/N)(\Sigma f'_i + \Sigma f'''_i)$ represents the proportion of X values that lie *outside* the interval $(\mu \pm K\sigma)$. Therefore,

$$1 - \frac{1}{K^2} \leq \frac{\Sigma f''}{N} = \text{the proportion of } X \text{ values within } \mu \pm K\sigma$$

This result is attributed to Chebyshev (sometimes spelled "Tchebycheff").

SUMMARY

This chapter assumes that data exist. How you get data is a separate question. Existing data are in one of two forms: raw (or original) data or grouped (i.e., a frequency distribution) data. If we have raw data, then we have to consider class limits, class intervals, midpoints, and class frequencies to construct a frequency distribution.

Variation usually exists in a data set. Our basic purpose was to *describe* these data in as concise a manner as possible. A frequency distribution is the most popular method of describing data. The frequency distribution can be presented as a table of numbers or as a diagram (i.e., histogram or polygon).

The measures of location (i.e., mean, median, and mode) and the measures of variation (i.e., range, variance, and standard deviation) allow us to summarize the general nature of a large data set. With just a few numerical measures we could get a reasonably clear picture of a population that could include hundreds of numbers. From the arithmetic mean, the median, and the mode, we have a measure of where the data set is located and whether it is symmetric or skewed. From the standard deviation and the range, we have a measure of the dispersion of the data. Using the mean and the standard deviation, we can say within what range at least, for example, 75 percent of the population resides.

Where we used the expected value for a theoretical probability distribution in Chapter 3, we used the arithmetic mean for an empirical frequency distribution. The relative frequency (f_j/N) of an empirical distribution is similar to the relative-frequency concept of probability. Finally, the variance and standard deviation are measures of the same variation concept for either a probability or an empirical distribution.

When the measures of location and variation are applied to a population, they are called *parameters*. When applied to a portion of (or sample from) the population, they are called sample *statistics*. It is the relationships between these statistics and parameters that consume most of this book.

Postscript on Misrepresentation of Data

Finally, there is the question of numerical and graphical misrepresentation of data. Consider the following statement: We have been selling widgets for 10 years, and over 90 percent of them are still functioning today. This appears to say something about the "useful life" of a widget, but, in that context, it could be a misrepresentation. Consider the total number of widgets sold in the last 10 years. If 1 percent of that total is sold in each of the first 9 years and 91 percent of that total is sold in the tenth year, then the life expectancy of widgets may be only 1 year and the statement is correct. However, any inference concerning the longevity of widgets would require knowledge of the distribution of annual sales.

Frequently, a diagram will be used to dramatize a couple of numbers which someone, or some group, deems important. For example, a school board wishes to

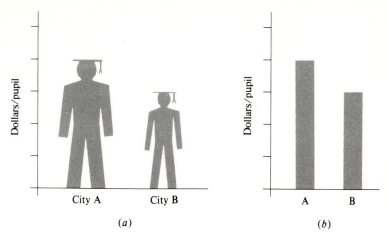

FIGURE 4.12   Expenditure per pupil for city A and city B.

inform the community that their expenditure per pupil is lower than that of a neighboring city. The expenditures per pupil for the two cities could be represented by two numbers, but the school board wants to represent these two numbers in a way that will demand everyone's attention. There is no intention of lying or even of misrepresenting the situation; they just think this comparison is important, and they want everyone to be informed about the difference. As a result, they include Figure 4.12*a* in their annual report, and a local newspaper reprints the diagram.

Let us analyze Figure 4.12*a*. The data used does not say anything about taller and more muscular children in city A; hopefully, most readers would recognize that. The real problem comes from the use of *area* to represent *height*. That is, the ratio of height between person A and person B is 4 to 3; however, the area of A is about twice that of person B. Figure 4.12*b* is a *bar diagram* representing the same data with the same vertical axis; the ratio of areas in bar A and bar B is 4 to 3. In an attempt to emphasize the data, the school board misrepresented the data.

Many more examples are listed in a book, *How to Lie with Statistics,* by Darrell Huff. For over two decades this book has served as an entertaining guide to avoid misrepresenting data; the interested reader will find it delightful.

APPENDIX 4   INDEX NUMBERS

We started the discussion of Chapter 4 by saying that we wanted to take a large set of data and describe it without listing every observation. We wanted to summarize or consolidate information contained in the data. Frequently, data relevant to management and economics refers to many different transactions and/or items. To represent this data, we occasionally employ the concept of an *index number*.

While index numbers are generally regarded as *descriptive statistics*, the main emphasis of this chapter, they are not employed in most of the subsequent chapters. Index numbers are not necessary for understanding statistical inference; however, the basic construction and use of index numbers are important for the student of business and economics. Hence, what follows is a brief introduction to some of the mechanical aspects of index numbers. (The reader may postpone the reading of this appendix without any loss of continuity.)

What are index numbers? How are they constructed? What do they indicate? An *index number* is an aggregate measure of the relative change in a collection of presumably related items. For example, a price index is an attempt to consider a particular market basket of commodities and services, the prices for each item in that market basket at two points in time, the relative importance of each item in the market basket, and to put all that information together in such a way that the general change in price for the total (or aggregate) market basket can be represented by one number. There are several ways to approach this task, and we shall investigate several of the more elementary techniques.

If the price of apricots was 20¢ per pound last week and is 25¢ per pound this week, then you might be willing to say that there has been a 25 percent increase in price. If the price of bread was 30¢ per pound last week and is 33¢ per pound this week, you would be inclined to describe a 10 percent increase in the price of bread. A one-week jump in price from 80¢ to 96¢ per pound for coffee would be described as a 20 percent increase.

The price increase for each individual commodity is unambiguously described as

$$\text{Percent price increase from last week} = \left(\frac{\text{price this week}}{\text{price last week}} - 1\right) \times 100$$

For apricots, $(^{25}\!/_{20} - 1)100 = (1.25 - 1)100 = 25$ percent. These hypothetical prices represent hyperinflation.

The person who eats only bread experiences a 10 percent increase in prices. The apricot connoisseur experiences a 25 percent increase in prices while the coffee clutcher becomes nervous over a 20 percent price increase. But, how many people consume *only* one narrowly defined commodity in the process of living? Most people regularly consume a collection, or market basket, of goods and services during a week. How do we describe the relative increase in "the price of that market basket"?

Aggregate Price Indices

Simple average of price ratios (SAPR)   Assume that we have an individual (presumably in captivity) who consumes only apricots, bread, and coffee.

In Table 4.11 we have calculated the price ratio for each commodity. To describe

TABLE 4.11 WORKSHEET FOR SIMPLE AVERAGE OF PRICE RATIOS

	Price in		Price Ratio
	First Week	**Next Week**	$\dfrac{P_{i,1}}{P_{i,0}}$
Commodity	($t = 0$)	($t = 1$)	
A	20¢ per pound	25¢ per pound	$25/20 = 1.25$
B	30¢ per pound	33¢ per pound	$33/30 = 1.10$
C	80¢ per pound	96¢ per pound	$96/80 = 1.20$
			3.55

the average price increase for this market basket of three commodities, we might calculate a simple average of price ratios (SAPR) from

$$\text{SAPR} = \frac{\text{total of price ratios}}{\text{number of commodities}} = \frac{\sum\limits_{i=1}^{n}(P_{i,1}/P_{i,0})}{n} \times 100 \qquad \textbf{(4.13)}$$

From Table 4.11, SAPR $= (3.55/3)100 = 118.33$. We could interpret this as an 18.33 percent price increase in the market basket over the first week.

The advantage of the SAPR is that the index is not affected by the measurement units. That is, apricots could be measured in cents per pound; bread, in dollars per ounce; coffee, in pesos per gram. The value of the SAPR would be the same.

The disadvantage of the SAPR is that it treats all commodities in the market basket as being equally important. Our single consumer may feel that bread and coffee are relatively more important than apricots, but the SAPR implicitly assigns equal importance to all three commodities.

Of course, a particular individual may assign to each price ratio a weight which reflects her or his preference for that commodity. For example, bread may be 4 times as important to her or him as apricots, and coffee may be 3 times as important as apricots. Hence,

$$\frac{4(1.25) + 3(1.10) + 1(1.20)}{4 + 3 + 1}(100) = \frac{9.50}{8}(100) = 118.75$$

may be a better representation of the price increase for this one individual's market basket.

What if we want a price index for a group of people rather than a single individual? Even for a group of people, one commodity in the market basket may be relatively more important than another commodity. We must find some way to explicitly weigh each price ratio to reflect the varying degree of importance.

The Laspeyres price index (LPI) Consider a group of people who in some *base period* (say, 1970) buy certain quantities of apricots, bread, and coffee at the prevailing prices. For each commodity, the price times the quantity reflects the total expenditure on (or value of) that commodity. We could represent the total value of the market basket as $P_{A, 1970} Q_{A, 1970} + P_{B, 1970} Q_{B, 1970} + P_{C, 1970} Q_{C, 1970}$, or, in general, as

$$\sum_{i=1}^{n} (P_{i, 0} Q_{i, 0})$$

where the subscript 0 represents the base period.

In some other year (for example, 1976), consider only the prices of apricots, bread, and coffee. Now, ask the question: How much would it cost to buy the base-year quantities at the new prices? The answer, of course, is $P_{A, 1976} Q_{A, 1970} + P_{B, 1976} Q_{B, 1970} + P_{C, 1976} Q_{C, 1970}$, or

$$\sum_{i=1}^{n} (P_{i, 6} Q_{i, 0})$$

The ratio of these two values is used to define a *Laspeyres price index (LPI)*:

$$LPI_r = \frac{\Sigma(P_r Q_0)}{\Sigma(P_0 Q_0)} \, 100 \qquad \textbf{(4.14)}$$

where 0 represents the base time period and r represents the time period of interest.

From Table 4.12, assume our population spent 200¢ per week on apricots, 420¢ per week on bread, and 640¢ per week on coffee during 1970. To buy those same quantities at 1976 prices, they would spend 250¢ per week on apricots, 462¢ per week on bread, and 768¢ per week on coffee. The Laspeyres price index,

$$LPI_{76} = \frac{1,480}{1,260} \, (100) = 117.46$$

would show a 17.46 percent increase in the price of a 1970 market basket.

TABLE 4.12 WORKSHEET FOR LASPEYRES PRICE INDEX

Commodity	"Weights" (Pounds per Week) Q_{70}	Prices (Cents per Pound) P_{70}	P_{76}	Calculations (Cents) $P_{70} Q_{70}$	$P_{76} Q_{70}$
Apricots	10	20	25	200	250
Bread	14	30	33	420	462
Coffee	8	80	96	640	768
				1,260	1,480

TABLE 4.13 WORKSHEET FOR PAASCHE PRICE INDEX

Commodity	"Weights" (Pounds per Week) Q_{76}	Prices (Cents per Pound) P_{70}	P_{76}	Calculations (Cents) $P_{70}Q_{76}$	$P_{76}Q_{76}$
Apricots	8	20	25	160	200
Bread	16	30	33	480	528
Coffee	7	80	96	560	672
				1,200	1,400

In the Laspeyres price index the *weights*, which indicate the relative importance of each commodity, are the quantities purchased in the base period.

The Paasche price index (PPI) We could, of course, *weigh* the prices with the *quantities* purchased *in the current time period*. This aggregative price index is called the *Paasche price index (PPI)*:

$$\text{PPI}_r = \frac{P_r Q_r}{P_0 Q_r}(100) \tag{4.15}$$

In Table 4.13 we have used the current-year quantities to weigh the prices for both the current year (1976) and the base year (1970). The resulting Paasche price index is $\text{PPI}_{76} = (1{,}400/1{,}200)(100) = 116.67$, which could be interpreted as a 16.67 percent increase in prices from 1970 to 1976 for a 1976 market basket.

Weighted average of price ratios (WAPR) The SAPR, Equation 4.13, implicitly weighed all commodities equally. We could explicitly weigh each price ratio as follows:

$$\text{WAPR} = \frac{\Sigma[(P_r/P_0)W]}{\Sigma W}(100) \tag{4.16}$$

where W represents the weight for each commodity.

If we let $W = P_0 Q_0$, the value of the commodity purchased in the base period, then WAPR is algebraically identical with the Laspeyres price index. If we let $W = P_0 Q_r$, then WAPR is the Paasche price index. Equation 4.16 may have some calculational advantages for a frequently calculated Laspeyres price index.

Changing the Price Index

Shifting the base of an index number   Consider Table 4.14 where we have a price index with a base year of 1970. Assume that we want the base year to be 1972. That is, we want the index to be equal to 100 for 1972. Table 4.14 demonstrates the rather simple procedure for "shifting" the base from 1970 to 1972.

Splicing of two series of index numbers   Occasionally, we may want to change the items in the market basket, or we may want to alter the weights to better represent current consumption patterns. The result is two slightly different index numbers which individually do not cover the time period of interest. We may "splice" these two indices. See Table 4.15.

The old Laspeyres price index uses 1960 as a base year, and the new index uses 1970 as a base year. First we will investigate how we can extend the old, or 1960 base year, index to 1971 and 1972.

I'_{11}, the new index for 1971, represents a 6.4 percent increase from 1970. A 6.4 percent increase over 127, the value of the old index in 1970, is 8.1; thus, $I_{11} = 127 + 8.1 = 135.1$. That is, 106.4 is to 100 as I_{11} is to 127, or

$$\frac{106.4}{100} = \frac{I_{11}}{127} \qquad \text{or} \qquad I_{11} = 127(1.064) = 135.1$$

For 1972, I_{12} is to 127 as 111.4 is to 100, or

$$\frac{I_{12}}{127} = \frac{111.4}{100} \qquad I_{12} = 127(1.114) = 141.5$$

Of course, we can use the same splicing procedure to extend the new index back to 1969 and 1968. I'_9, the value of the new index for 1969, is to 100 as 121 is to 127, or

$$\frac{I'_9}{100} = \frac{121}{127} \qquad I'_9 = (100)(0.953) = 95.3$$

TABLE 4.14   SHIFTING THE BASE

Year	Price Index $I_{70} = 100$	Calculation	New Index $I_{72} = 100$
1970	100	$(100/120)100 =$	83.33
1972	120	$(120/120)100 =$	100.00
1974	150	$(150/120)100 =$	125.00

TABLE 4.15 TWO OVERLAPPING INDICES

Year	r	Old Index			New Index		
		Value = $\Sigma(P_r Q_0)$	Laspeyres Calculation	Index	Value = $\Sigma(P_r Q_{10})$	Laspeyres Calculation	Index
1960	0	10.00	$\frac{10}{10}(100) = I_0$	100			I'_0
1961	1	10.30	$\frac{10.3}{10}(100) = I_1$	103			I'_1
⋮	⋮	⋮	⋮	⋮			⋮
1968	8	11.80	$\frac{11.8}{10}(100) = I_8$	118			$I'_8 = ?$
1969	9	12.10	$\frac{12.1}{10}(100) = I_9$	121			$I'_9 = ?$
1970	10	12.70	$\frac{12.7}{10}(100) = I_{10}$	127	14.0	$\frac{14.0}{14.0}(100) = I'_{10}$	100
1971	11			$I_{11} = ?$	14.9	$\frac{14.9}{14}(100) = I'_{11}$	106.4
1972	12			$I_{12} = ?$	15.6	$\frac{15.6}{14}(100) = I'_{12}$	111.4

In like manner, I'_8 is to 100 as 118 is to 127, or

$$\frac{I'_8}{100} = \frac{118}{127} \qquad I'_8 = (100)(0.929) = 92.9$$

In general, we can splice an index by solving

$$\frac{\text{Old index for year } r}{\text{Old index for "overlap" year}} = \frac{\text{new index for year } r}{\text{new index for "overlap" year}}$$

Selected Published Indices

All levels of government and several private organizations publish index numbers. The following is a short list of some of the more popular indices.

The consumer price index (CPI)   The CPI is constructed and published by the Bureau of Labor Statistics of the U.S. Department of Labor. The CPI is published monthly in *The Consumer Price Index* (Superintendent of Documents, Government Printing Office).

The CPI is based on the goods and services usually purchased by urban wage earners and clerical workers. Essentially 400 different goods and services are used as weights. Prices are obtained from about 18,000 establishments (i.e., stores, service stations, hospitals, etc.) in over 50 cities in the United States.

The practice of sampling prices changes occasionally, but a typical year might include: (1) determining the prices of selected commodities (e.g., food and fuel) from the more than 50 cities every month; (2) determining the prices of the remaining goods and services monthly from the five largest urban areas and from all sampled cities every 3 months. For 23 of the largest urban areas in the United States, a city CPI is published monthly.

The current base year is 1967, and CPI = 100 for 1967. The base year will soon be changed to 1977.

The CPI is published in both an "unadjusted" and a "seasonally adjusted" framework. For example, the "escalation clause" in union contracts and pension plans is usually tied to the unadjusted CPI.

The construction of the CPI is similar to the Laspeyres price index. Thus, since the CPI for November of 1974 was 154.3, it can be interpreted as a 54.3 percent price increase over 1967 for that market basket usually consumed by urban wage earners and clerical workers.

Wholesale price index (WPI)   In the *Monthly Labor Review* and the *Wholesale Prices and Price Indexes*, the Bureau of Labor Statistics publishes monthly indices. Whereas the CPI attempts to show price changes in the retail market for a particular set of consumers, the WPI attempts to indicate aggregate price movements in all markets except the retail market. The WPI uses over 2,000 items.

Quantity indices   The Federal Reserve Board publishes an index of industrial production (IIP) as an aggregate measure of changes in aggregate manufacturing output. The IIP, a weighted average of quantity (rather than price) relatives, is frequently used as a measure of changes in industrial activity. The IIP is found in the monthly issues of the *Federal Reserve Bulletin* or the *Survey of Current Business* (U.S. Department of Commerce).

Uses of Index Numbers

An index number is an attempt to aggregate a group of variables in one measure. Simply observing how this aggregate changes over time may be very useful for business and economic policy.

Business conditions may be reflected in the IIP, and inflation in the CPI and WPI. As already mentioned, "indexing" a labor contract with the CPI will allow those workers to feel as if they can automatically keep up with inflation.

This suggests one of the more common uses of a price index. A price index can be used with "money income" (or "nominal income") to estimate "real income." Roughly, how has the purchasing power of a person's (or a group's) income changed? In general,

$$\text{Real income} = \frac{(\text{money income})(100)}{\text{relevant price index}}$$

For example, assume that Ms. U. C. Worker earned \$5,000 in 1967 and \$8,000 in 1974. Using the CPI as a *deflator* of her money income, we find that

$$\text{Real income in 1967} = \frac{(5,000)(100)}{(100)} = 5,000 \text{ (in 1967 dollars)}$$

$$\text{Real income in 1974} = \frac{(8,000)(100)}{(154.3)} = 5,184.71 \text{ (in 1967 dollars)}$$

Her money income increased by

$$\frac{8,000 - 5,000}{5,000}(100) = 60 \text{ percent}$$

but because prices increased by 54.3 percent, her real income (or her increase in purchasing power) increased by only

$$\frac{5,184.71 - 5,000}{5,000}(100) = 3.7 \text{ percent}$$

between 1967 and 1974.

SUMMARY

This discussion of the construction and use of index numbers is obviously only an introduction. Especially in the case of price indices, the conceptual framework for the construction (and the resulting interpretation) of the index number requires knowledge of intermediate microeconomic theory.[9] Thus, our discussion is more a description of index numbers than a theoretical basis for their construction and interpretation.

---

[9] For example, see Edwin Mansfield, *Microeconomics: Theory and Applications*, 2d ed., W. W. Norton & Company Inc., New York, 1975, pp. 72–77.

EXERCISES

1. A retail appliance dealer has ten branch outlets in a metropolitan area. In a given week, the owner selected a sample of five outlets and found the number of washing machines sold at each branch was:

Branch:	A	B	C	D	E
Number of washing machines sold:	5	3	6	5	1

a. What are the mean, median, and mode for these five sample values?
b. What are the sample variance and the sample standard deviation?
c. Is this data symmetric about the mean? If not, how is it skewed?
d. What is the range of this data?

2. Consider the data presented in Exercise 1 and assume that the owner also obtained the number of washing machines sold by the remaining five branches:

Branch:	F	G	H	I	J
Number of washing machines sold:	4	7	5	4	0

Consider the ten branches as a population.
a. What are the mean, median, and mode for the ten observations?
b. What are the population variance and standard deviation?
c. What is the range of the ten observations?
d. During the previous week, assume the ten branches had a mean of 3.8 washing machines sold per branch, and assume that there is $100 profit per machine. What is the total profit from selling washing machines that week?

3. The marketing vice president notes that half of the salaried personnel at Baby Food Corporation (BFC) make $18,000 or more per year. The treasurer notes that the 200 salaried personnel have a mean annual salary of $20,000.
a. How could both be correct?
b. If both are correct, what is the total paid for salaries by BFC during a year?
c. If both are correct, could the variance, the standard deviation, or the range of annual salaries be 0?

4. You have five cards; each card has a number on it. Let's put them in an array and call them X_1, X_2, X_3, X_4, X_5.
a. If $\text{Var}(X) = 0$, what can you say about the numbers on those five cards?
b. If the mean value of the five cards is equal to the median value, then is X_3 above, below, or equal to the mean?
c. If the range is 2, if $X_5 > X_1$, and if we knew the value of X_1, what is the value of X_5? What can we say about the values of X_4, X_3, and X_2?
d. If $\sigma_X = \mu_X = 0$, what is the value of X_3?
e. If all five numbers are different, if $X_1 < X_5$, and if $\Sigma X = 5X_4$, then which is larger, the median or the mean?
f. If $\text{Var}(X) \neq 0$, if $X_1 < X_5$, and if $X_3 = X_4 = X_5$, then what can we say about the relative value of μ_X? Which is larger, the mean or the median?
g. In order for the median to equal the mean, *must* $\text{Var}(X)$ be equal to 0? Explain.
h. If $\mu_X = X_2$, ΣX is equal to what?
i. If the median is equal to X_1, what can we say about $\text{Var}(X)$?
j. If $\text{Var}(X) \neq 0$, then what can we say about the relative magnitudes of X_1 and X_5?
k. If $\Sigma(X_i - \mu_X) = \text{Var}(X)$, what can we say about the values of each X_i? If $\Sigma(X_i - \mu_X) \neq \text{Var}(X)$?

5. Consider the following data about the city of Ken.

	Distributions (Annual Income)		
	All Families	**Government Employees**	**Judicial Branch of Government**
Measure	A	B	C
Arithmetic mean	$13,500	$13,000	$15,000
Median	12,000	11,000	16,000
Mode	$9,000 and $14,000	10,000	17,000
Range	$0 to $200,000	$4,500 to $35,500	$10,000 to $35,500
Standard deviation	$3,000	$2,800	$1,000

Sketch all three distributions on the same graph.

6. The totals (dollars per customer) on the cash register at the M & P Grocery are:

3.78	1.02	3.95
5.46	5.09	6.86
1.47	9.89	9.92
7.74	3.74	2.30
7.94	7.16	11.12
2.49	0.39	4.80
10.94	3.79	0.50
3.05	5.62	5.94
1.87	9.18	3.85
5.97	3.83	7.92

Part A   Raw Data (Consider this data as a population)

a. Find the mean and median for these sales.
b. The variance is (^{2}) 9.2337 ($\sqrt{9.2337} = \$3.04$). Find the proportion of observations that are in the interval $\mu \pm (1.5)\sigma$.
c. Does your answer to part *b* obey Chebyshev's inequality for $K = \frac{3}{2}$? What is the range of this data?

Part B   Grouped Data ($2 class intervals)

d. Using class intervals of " $0.00 to under $2.00," " $2.00 to under $4.00," etc., construct a frequency distribution for the sales data.
e. Draw a histogram and a polygon for this data.
f. Approximate the mean and median from your answer to part *d*. How do these answers compare with your answer to part *a*?
g. What is the modal class interval? Is this frequency distribution unimodal? Is this distribution skewed to the left or to the right?
h. Approximate the variance and standard deviation from your answer to part *d*. How does this answer for the variance compare with part *b*?

Part C   Grouped Data ($1 class intervals)

i. Using class intervals of 0.00 to 0.99, 1.00 to 1.99, 2.00 to 2.99, etc., construct a frequency distribution for the sales data.
j. Draw a histogram for this frequency distribution and compare it with part *e*.

k. Approximate the mean and median from your answer to part *i*. How do these answers compare with part *a*?

l. Approximate the variance and standard deviation for the frequency distribution in part *i*. How does this answer compare with the data from part *b*?

Part D   Comparisons

m. If the 30 sales totals were the only thing that interested us, would this data represent a population or a sample? Would the mean, median, variance, and standard deviation be parameters or statistics?

n. If the 30 sales totals were only a portion of the data that interests us, would this data represent a population or a sample? Would the mean, median, variance, and standard deviation be parameters or statistics?

o. If you considered these 30 observations as a sample, how would your answers to parts *b*, *h*, and *l* change?

p. Compare the frequency distributions in parts *d* and *i*. Which do you prefer and why?

7. An instructor gives a quiz to three students, and the resulting scores were: $X_1 = 73$, $X_2 = 75$, $X_3 = 77$.

a. Find the mean, variance, and standard deviation of this population of X values.

b. The room was uncomfortably hot, and there was a bomb scare in the middle of the quiz. The instructor would like to "scale up" the quiz scores to account for these unfortunate environmental conditions. The *first scheme* adds 10 points to each score. Let $Y = X + 10$. Find μ_Y, σ_Y^2, and σ_Y.

c. The *second scheme* increases each score by 10 percent; let $W = (1.1)X$, or $W = \{80.3, 82.5, 84.7\}$. Find μ_W, σ_W^2, and σ_W.

d. The *third scheme* is a combination of the first two. That is, each score is increased by 10 percent and then 10 points are added to it. Let $U = 10 + (1.1)X$, or $U = \{90.3, 92.5, 94.7\}$. Find μ_U, σ_U^2, and σ_U.

e. Check your answers with the following:

(1) $\mu_Y = \mu_X + 10$, $\sigma_Y^2 = \sigma_X^2$, $\sigma_Y = \sigma_X$. Adding 10 points to each score just shifts the distribution over 10 units. It changes the location of the distribution, but it does not change the shape or variation within the distribution.

(2) $\mu_W = (1.1)\mu_X$, $\sigma_W^2 = (1.1)^2\sigma_X^2 = (1.21)\sigma_X^2$, and $\sigma_W = 1.1\sigma_X$. Increasing each score by 10 percent changes the location of the distribution, but it also spreads it out more.

(3) $\mu_U = 10 + 1.1\mu_X$, $\sigma_U^2 = (1.1)^2\sigma_X^2$, $\sigma_U = (1.1)\sigma_X$. Both the 10 percent increase and the additional 10 points affect the location of the distribution; however, only the 10 percent increase affects the shape of the distribution.

Appendix Exercises

8. Consider the Automobile Users' Cost of Operation index from the following hypothetical data:

	Quantities		Prices	
Items	**1971**	**1973**	**1971**	**1973**
Gasoline	800 gallons	760 gallons	30¢ per gallon	50¢ per gallon
Oil	14 quarts	15 quarts	$.90 per quart	$1.40 per quart
Insurance	1	1	$160 per year	$180 per year

a. Find the Laspeyres price index for 1971 and 1973, using 1971 as the base year.

b. Find the Paasche price index for 1971 and 1973, using 1971 as the base year.

c. Find the WAPR index for 1971 and 1973, using total expenditure in 1971 as the weight.

d. Compare and interpret your answers to parts *a*, *b*, and *c*.

9. Consider the student price index for the academic year. This data is bogus, but it represents some of the data collection problems for index numbers.

Items	Quantities 1970–71	Prices ($)		
		1970–71	1971–72	1972–73
Tuition and fees	1.00*	1,000.00	1,000.00	1,100.00
Room or rent	9 months	40.00	43.00	50.00
Meals†	34 weeks	18.20	21.50	23.00
Textbooks	20 books	9.00	10.00	11.50

* Annual data was used because of semester versus quarter problems.
† Prices are a weighted average of on-campus meal ticket prices and off-campus pizza parlor prices.

a. Use 1970–71 as the base (academic) year, and calculate the Laspeyres price index for 1970–71, 1971–72, and 1972–73.
b. Consider the student as consumer. With a relative increase in the price of X, economic theory suggests that the consumer will attempt to consume less of X and more of a substitute. What are the substitutes for (1) tuition and fees, (2) room or rent, (3) meals, (4) books?
c. Assume that the scholarships at Eastwestern University were:

Year:	1970–71	1971–72	1972–73
Scholarship stipend:	$1,300	$1,400	$1,500

Use your answers from part *a* to deflate the scholarship stipend at Eastwestern University. Has the real value of the scholarships changed?

10. You use your car for your work, and your employer paid you $400 car allowance in 1971 and $500 in 1973. Use your answers to Exercise 8, part *a*, to deflate your car allowance. What has happened to the real value of your car allowance?

11. Consider the following values of the CPI and your after-tax income:

Year	Your After-Tax Income	Consumer Price Index* (1967 → 100)
1967	$ 9,000	100.0
1968	9,400	104.2
1969	9,800	109.8
1970	10,200	116.3
1971	10,600	121.3
1972	11,000	125.3
1973	11,400	133.1
1974	11,800	147.7

* Source: *Survey of Current Business.*

Your after-tax income is in "current dollars." Convert it to "constant 1967 dollars" and comment on the result.

12. Consider the following data on gross national product for the United States (in current dollars) and the GNP implicit deflator:

Year	GNP ($ billions)	GNP Implicit Price Deflator (1958 → 100)
1967	793.9	117.59
1968	864.2	122.30
1969	930.3	128.20
1970	977.1	135.24
1971	1,054.9	141.35
1972	1,158.0	146.12
1973	1,294.9	154.31
1974	1,396.7	170.11

Source: *1975 Economic Report of the President, GPO.*

a. Use the GNP deflator as a price index to find real GNP.
b. Why would we not use the CPI to deflate GNP?

CASE QUESTIONS

1. Consider the 30 families in the Oregon sample.
a. Consider V07, V08, and V10.
   (1) By inspection (i.e., you do not have to calculate anything), what are the mean and variance for each variable?
   (2) Use the description of variables section of Appendix B to interpret your results in part 1.
b. Consider V32. (Except for part 7 below, leave all answers in $100 units.)
   (1) Exclude household number 14 (i.e., V01 = 14) and calculate the mean income for the remaining 29 households.
   (2) Why would we exclude household number 14 from part 1?
   (3) Excluding household number 14, calculate the variance and standard deviation.
   (4) Excluding household number 14, place the 29 values for V32 in an array and find the median.
   (5) How would your answer to part 4 change if we included household number 14? Reconsider your answer to part 2. Is there any reason to exclude household number 14 in the calculation of the median?
   (6) Consider your answers to parts 1 and 4. Do the mean and median differ? For that array of 29 numbers, which of these measures, the mean or the median, do you consider to be more representative of "central tendency"?
   (7) Interpret your answers to parts 1 and 4.
   (8) Use the mean, standard deviation, and Chebyshev's inequality to find that income range within which at least 75 percent of the household incomes occur. Now, using the array from part 4, find the proportion of household incomes that actually fall in that range. Was Chebyshev correct?

2. Finding the mean, median, variance, and standard deviation of a set of numbers requires some effort (or cost). The results of these calculations should be of some value (or benefit). Feeding numbers into a formula and letting the formula spit out results does not always provide meaningful informaton.

Part A   Nominal measurement scales

Consider V02, the rural-urban variable.

a. How would you interpret the mean is 2 and the variance is 0? Does the mean and variance provide useful information in this case? Could that same information have been obtained by inspecting the data?

b. Consider the case where 98 households had V02 = 2, one household had V02 = 1, and another had V02 = 3. How would the numerical results of the mean and variance compare with those in part *a*? Given just the mean and variance for this case, how would you interpret them?

c. Consider the case where 50 households have V02 = 0 and an equal number have V02 = 4. How would the numerical results of the mean and variance compare with those same measures in part *b*? How would you interpret the mean and variance in this case?

d. Consider each possible value of V02 as a separate classification (which, in fact, is true). Record (e.g., for the Oregon sample) the number of households that fall into each classification. Does the resulting "frequency table" provide any information? Is it more informative than the raw data?

e. What can you conclude about the methods of "descriptive statistics" presented in this chapter when they are applied to numbers measured on a nominal scale?

Part B   Ordinal measurement scales

Consider V07, plumbing facilities. As reported, V07 assumes values as measured by a nominal scale. By imposing a value judgment on the V07-numbers, we can consider them to represent an ordinal measurement scale. For example,

V07	Code Description	Value Judgment
0	All plumbing facilities	Most desirable situation
1	Lacks hot water only	Next most desirable situation
2	Lacks more than hot water	Least desirable situation
3	NA	Does not apply

By excluding those households where V07 = 3, then the numbers 0, 1, and 2 indicate a rank ordering.

a. For both the Massachusetts and Oregon samples, the mean and variance of V07 are equal to zero. How do we interpret those results?

b. In the Illinois sample, even when we exclude the housing units where V07 = 3, the mean and variance are not exactly equal to zero. Consider (1) the definition of the variance and (2) the difference between two ordinal numbers to discuss why a nonzero variance for V07 has no meaning.

c. Is there a meaningful interpretation to the statement: The median of V07 is zero?

d. What can you conclude about the methods of descriptive statistics when they are applied to numbers measured on an ordinal scale?

Part C   Ratio measurement scale

Consider V03, the number of persons in the household.

a. Construct a frequency distribution using the 30 households in the Oregon sample.

b. Calculate the mean, median, variance, and standard deviation.

c. Interpret the results from part *b*.

d. Building contractors who speculate (i.e., build housing units before they have a buyer) are occasionally criticized for only building units which are amenable to four-person households. If all households with $3 \leq V03 \leq 5$ find the "four-person housing unit" appropriate to their needs, what proportion of the households in the Oregon sample would find such housing units inappropriate for their needs?

3. Consider V10 (value of property) and the Illinois sample. Omit all cases where V10 = 11. Note that V10 = 1, for example, simply tells us the *class interval* (i.e., $5,000–$7,499) in which the housing unit can be classified.

a. Use the code description of V10 to create 11 class intervals. Then, construct the frequency distribution from the Illinois sample.

b. Does household number 36 present any special concerns when we attempt to use this frequency distribution to approximate the mean, the median, and the variance?

c. Assume that the value of the property for household number 36 is $72,000. Calculate the mean, median, variance, and standard deviation from the frequency distribution in part *a*.

d. Again, recall that the values for V10 are "codes" representing class intervals. Excluding those housing units where V10 = 11, consider the mean of these coded V10 values. We could go to the code description of V10 and translate our mean of V10 values into a dollar value. (*Note:* This is not the procedure we followed in parts *a* through *c*.) Why would this technique be inappropriate as an approximation of the mean property value for the Illinois sample?

5

SAMPLING AND SAMPLING DISTRIBUTIONS

INTRODUCTION

Be assured, this chapter concludes our initial "exploration stage" and begins our "discovery" of techniques for statistical inference. In the context of mathematics as a language, Chapters 2 through 4 have provided the vocabulary and grammar for statistical methods. We now have to do something with that language. Idle chatter with this new language may improve our mystique in barroom discussions, but the social significance of this language must provide us with a way to accomplish something in an easier fashion than we are currently capable of doing.

The crucial turning point between the development of the language for statistical methods and the general use of those methods hinges on the difference between *deductive* and *inductive* statistical reasoning. Our reference points are the population and the sample.

With deductive statistical reasoning, we start with a population, about which we already know everything we need or want to know, and we investigate the set of possible samples (and their characteristics) that could be obtained from our known population. That is one basic purpose of this chapter.

Inductive statistical reasoning starts with a known sample and attempts to identify the likelihood that populations with various characteristics generated that sample. More to the point, in many applied economics problems, we already know the population in a general way. What we don't know or what we want to test is a measurement of some specific population characteristic (e.g., the mean). All we have are the characteristics (e.g., mean, standard deviation, etc.) of the sample. We have a sample mean. What further information do we need (and why do we need it?) to infer the value of the population mean? The next several chapters explore these questions.

Why Sample?

Our first question should be, Why sample at all? The descriptive statistics alluded to in Chapter 4 suggest several ways of presenting the results of a census (i.e., a complete enumeration of a population). Except for an independent check on the census of a large population, sampling and statistical inference are not useful in summarizing the results. But, we make this statement under the assumption that the data from a complete enumeration exists.

The importance of sampling and statistical inference is derived from the process of *getting* the data that will provide the desired information about a population. A sample, because only a part of the population is under scrutiny, obviously uses fewer resources than a complete enumeration of the population. Especially for large populations, sampling is simply more economical than a census.

The economic aspects of sampling are most apparent in *destructive sampling*. Consider the producer of incandescent light bulbs who wants to know the mean life (in hours) of the product. This producer could ask the engineering section for the designed mean life. But how long *do* the bulbs last? We could select several bulbs, plug them in, and clock the time it takes for them to burn out. If we took a census, for the random variable " bulb life," we would not have any useful product left. By definition, the population of burned-out bulbs has a mean and standard deviation of 0 hours. Clearly, when the utility of the observed object is destroyed in order to obtain the observation, a census is preposterous.

A population may be so large or its elements so difficult to locate that the cost of a census is prohibitive. The election of the President of the United States is an event ignored by no fewer than 10 percent of the eligible voters. The problems of locating illegal aliens are commonplace in the pursuit of an accurate decennial census of population in the United States. The problem of getting information from such " missing respondents " can be easily misinterpreted, which further increases the cost of complete enumeration.

The ever-popular political polls demonstrate the time utility of sampling versus a census. The extent of changes in opinion should be recognized quickly if policy hopes to accommodate constituents' desires. Dispersed labor union membership, student bodies, urban highway users, etc., are examples of populations where complete enumeration would probably yield useless results because of the time involved to collect the data.

The fewer number of elementary units in a sample can mean that more auxiliary variables can be investigated. In a given locale, for the same amount of money and time, one could obtain either a census of the number of persons per household or a sample which would include persons per household, last year's reported income, type of employment of the head of the household, and years at this address.

Errors

Our second question is, If a sample is taken, then how precise is the resulting information? This question implies a host of other questions.

How was the sample selected? The two basic alternatives are a *judgment sample* and a *random sample*. A judgment sample is one where an "expert" chooses a "representative" sample. Forget about "experts" and consider the impossibility of the "representative" character of any sample. We are taking a sample because we want a more precise estimate of a population characteristic (e.g., the arithmetic mean weight of males enrolled in statistics). In order for the sample to be representative, then either the sample's mean must be close to the population mean or the components or factors which influence the mean in the population must be similar to the composition of those same factors within the sample. In the first case, the "expert" is guessing the value of the mean for both the sample and the population. In the second case, the expert is using a cause-effect model which may or may not be appropriate. Unfortunately, such models are seldom even outlined when they should be explicitly stated. In either case, the expert is biasing the results by biasing the sample selection. We have no way of measuring the extent of that bias.

By the use of several *random* or *probability sampling* methods, we can at least get a measurement of the sampling error involved. A *random sample* is the result of using a procedure where the probability of getting each possible sample is known.

A *simple random sample* is the result of selecting a sample from a population in such a manner that *all* possible samples had the *same probability* of being selected. If there are K possible samples from a particular population, then each and every sample must have a selection probability of $1/K$ in order for the procedure to be considered simple random sampling. Other probability sampling techniques will be discussed later.

In either the judgment or the random sample, the relevant characteristic of the population may be different from the same characteristic in the sample; that error we call *sampling error*. As we shall see later, random sampling permits measurement of the extent of sampling error.

But there is another type of error which has nothing at all to do with sampling. These errors, called *nonsampling*, or *systematic*, *errors*, are the result of incorrect measuring devices, ill-defined questionnaires, etc. Incorrect measurements are not a function of selecting a sample; they could just as easily occur while taking a complete enumeration.

Nonsampling error can be in the form of vague or ambiguous questions when knowledge (e.g., " Your age ?"—At last or closest birthday?) or opinion (e.g., "Is our fiscal policy sound?"—Is the difference between monetary and fiscal policy important?) is requested from persons. Or, when things are measured, systematic (nonsampling) error can occur from incorrectly calibrated measuring devices (e.g., a weighing scale that registers 1 ounce per pound more than what is placed upon it).

For example, a "baker's dozen" is a method of giving a discount; specifically, 13 items are supplied for the price of 12. Suppose that the salesperson at a local bakery tallies the number of "dozens" and "half-dozens" of doughnuts sold to the public before eight o'clock each morning. When a customer orders a dozen doughnuts, a baker's dozen is provided, while a dozen is tallied. When a customer orders a "half-dozen," 6 doughnuts are bagged and a half-dozen is tallied. Assume that on a particular morning there were 10 orders for a dozen and 8 orders for a half-dozen doughnuts. If, each day, the salesperson is asked to report how many dozen doughnuts were ordered, on this particular morning the salesperson would reference the tally sheet and report 14.

If we are interested in the revenue from doughnuts, this " 14 dozen " is accurate and useful data. However, if we are interested in the number of doughnuts supplied, 14 dozen underreports, by 10 doughnuts, the quantity supplied.

Further, assume that we have these morning reports on doughnuts ordered for a period of 6 months. It would not make any difference whether we took a sample of, say, 40 reports or a census of all the reports. From the viewpoint of number of doughnuts supplied, there would be a systematic bias in the measurements.

Thus, each individual measurement has two components of error: sampling error and nonsampling error. Algebraically, this could be stated as:

(Individual measurement) = (true value) + (sampling error) + (nonsampling error)

where all values could be positive or negative.

Accuracy versus Precision

These two components of error yield another semantic distinction found in several statistical studies. Let us consider using a sample mean $\bar{X}$ to provide a point estimate of a population mean μ. Let

$$\bar{X} = \mu + \varepsilon_s + \varepsilon_{ns}$$

where $\bar{X}$ is our estimate, μ the true value, ε_s the error due to sampling, and ε_{ns} the nonsampling error. If we took a census, ε_s would be equal to 0. But, the nonsampling error remains. When the nonsampling error is 0 (or essentially 0), we say the measurement is *accurate*. When we say the measurement is *precise*, then the sampling error is as small as we want it.

The sample itself will provide us with information on the precision aspect only. Statistical inference concentrates on the extent of the sampling error, or precision. Accuracy involves carefully calibrated measuring equipment, concise statements, and logical inference. Henceforth, we shall consider all measurements to be accurate.

5.1 SAMPLING DISTRIBUTION FOR THE MEAN

When we employ a technique that yields a random sample, then there is some known probability associated with each possible sample. From a random sample of n values, $X_1, X_2, \ldots, X_n$, several characteristics could be measured. Using these n values, we could calculate a mean, a median, a range, a variance, a standard deviation, or a total. Any such measure can be called a *statistic*. The sample mean, the sample standard deviation, and the sample proportion (i.e., proportion of n persons who weigh more than 200 pounds, etc.) are the statistics which will receive our primary attention. Remember, the same measured characteristics of the entire population (i.e., the mean, variance, etc.) are called *parameters*.

Consider all possible samples of the same size, n, from a population. Random sampling techniques infer that there is a known probability associated with each sample, and we could, for example, calculate the arithmetic mean of each sample of n observations. Some of these sample means would be the same; other sample means would be different. Grouping all the like sample means and using the probability rule for addition, we could determine the probability of obtaining a sample with that particular sample mean.

Repeating this procedure for each possible sample mean, we construct a probability distribution of the random variable "sample mean." A *sampling distribution* is a *probability distribution* where the random variable is a statistic.

Example 5.1 Ten-Card Deck, V-shaped Population

The Population

Take a standard deck of playing cards and remove three 2s, two 3s, two 5s, and three 6s. Our population consists of these 10 cards, or 2, 2, 2, 3, 3, 5, 5, 6, 6, and 6.

Considered as a sequence of 10 numbers, this population has a mean $\mu_X = (\Sigma X)/N = 40/10 = 4$, and a variance, $\sigma^2 = [\Sigma(X - \mu)^2]/N = [3(4) + 2(1) + 2(1) + 3(4)]/10 = 28/10 = 2.8$. Of course, the standard deviation is $\sqrt{2.8} = 1.673320$.

We could consider the "face value" of a single randomly drawn card to be the random variable X. Then X could assume values of 2, 3, 5, and 6 with probabilities of 0.3,

TABLE 5.1 CALCULATIONS FOR SAMPLE OF SIZE 2

		First Draw			
X		2	3	5	6
	P(X)	(0.3)	(0.2)	(0.2)	(0.3)
2 (0.3)		4 2.0 (0.09)	5 2.5 (0.06)	7 3.5 (0.06)	8 4.0 (0.09)
3 (0.2)		5 2.5 (0.06)	6 3.0 (0.04)	8 4.0 (0.04)	9 4.5 (0.06)
5 (0.2)		7 3.5 (0.06)	8 4.0 (0.04)	10 5.0 (0.04)	11 5.5 (0.06)
6 (0.3)		8 4.0 (0.09)	9 4.5 (0.06)	11 5.5 (0.06)	12 6.0 (0.09)

(Left axis label: Second Draw)

← Total of first and second draws
← Sample mean, $\bar{X}$ = (total)/2
← Probability of this sample mean occurring in this manner

TABLE 5.2 DISTRIBUTION OF SAMPLE MEANS (SAMPLING DISTRIBUTION OF $\bar{X}$)

Total	$\bar{X}$	$P(\bar{X})$
4	2.0	0.09
5	2.5	0.12
6	3.0	0.04
7	3.5	0.12
8	4.0	0.26
9	4.5	0.12
10	5.0	0.04
11	5.5	0.12
12	6.0	0.09
		1.00

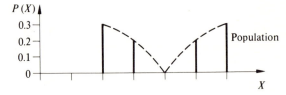

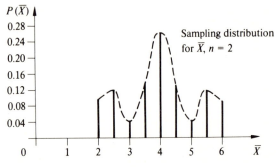

FIGURE 5.1 Sampling distribution for $n = 2$.

0.2, 0.2, and 0.3 respectively. The resulting probability distribution of X has a mean of 4.0 and a standard deviation of 2.8. A graph of X is in Figure 5.1.

The Purpose
We want to find the sampling distribution of the sample means, for three different sample sizes, from this population.

The Experiment
We will take a random sample of two (then three, then four) cards, with replacement, from the deck. (That is, randomly select one card from the well-shuffled deck of 10 cards; observe its face value; return it to the deck; reshuffle the deck; repeat the draw of a single card.)

The Procedure
Enumerate all possible samples resulting from this experiment. For each possible sample, calculate (1) the total of the two (three, or four) cards, (2) the arithmetic mean, and (3) the probability of that sample occurring. Add the probabilities of each of the mutually exclusive samples that have the *same* mean, and construct the sampling distribution.

Table 5.1 enumerates all possible samples and, for each sample, summarizes the total, the mean, and the probability of obtaining that sample. For example, on the first draw a 2 could be selected and a 3 on the second draw. The total is $2 + 3 = 5$, and the mean is $\frac{5}{2} = 2.5$ for that sample. Since we are sampling with replacement, the result of the second draw is independent of what happened on the first draw. Thus, $P(2$ on first draw *and* 3 on second draw$) = P(2)P(3) = (0.3)(0.2) = 0.06$, and this result is shown in parentheses in Table 5.1

The only other way we could get a sample mean $\bar{X}$ of 2.5 is if we got a 3 on the first draw and a 2 on the second. The probability of getting that particular sample is also 0.06. Since the samples $\{2, 3\}$ and $\{3, 2\}$ are mutually exclusive, $P(\bar{X} = 2.5) = 0.06 + 0.06 = 0.12$. In like manner, we can obtain the probability of occurrence for each possible sample mean. Table 5.2 shows the sampling distribution for $\bar{X}$ when samples of size 2 are taken.

$$E(\bar{X}) = \Sigma \bar{X}_i P(\bar{X}_i)$$

$$= 4.0 = \mu_{\bar{X}}$$

$$\sigma_{\bar{X}}^2 = \Sigma[P(\bar{X}_i)(\bar{X}_i - \mu_{\bar{X}})^2]$$

$$= (0.09)(4) + (0.12)(2.25) + (0.04)(1) + (0.12)(0.25)$$

$$+ (0.26)(0) + (0.12)(0.25) + \cdots$$

$$= 1.4$$

or

$$\sigma_{\bar{X}} = \sqrt{1.4} = 1.1832$$

Note that we can consider the sample mean $\bar{X}$ as a new random variable. Our rule for defining this new random variable is $\bar{X} = (\Sigma X)/n$, where ΣX is the total face value of

TABLE 5.3 CALCULATIONS FOR SAMPLE OF SIZE 3

			Total from First and Second Draws								
	X		**4**	**5**	**6**	**7**	**8**	**9**	**10**	**11**	**12**
		$P(X)$	*(0.09)*	*(0.12)*	*(0.04)*	*(0.12)*	*(0.26)*	*(0.12)*	*(0.04)*	*(0.12)*	*(0.09)*
Third Draw	**2**	*(0.3)*	6 2.00 *(0.027)*	7 2.33 *(0.036)*	8 2.67 *(0.012)*	9 3.00 *(0.036)*	10 3.33 *(0.078)*	11 3.67 *(0.036)*	12 4.00 *(0.012)*	13 4.33 *(0.036)*	14 4.67 *(0.027)*
	3	*(0.2)*	7 2.33 *(0.018)*	8 2.67 *(0.024)*	9 3.00 *(0.008)*	10 3.33 *(0.024)*	11 3.67 *(0.052)*	12 4.00 *(0.024)*	13 4.33 *(0.008)*	14 4.67 *(0.024)*	15 5.00 *(0.018)*
	5	*(0.2)*	9 3.00 *(0.018)*	10 3.33 *(0.024)*	11 3.67 *(0.008)*	12 4.00 *(0.024)*	13 4.33 *(0.052)*	14 4.67 *(0.024)*	15 5.00 *(0.008)*	16 5.33 *(0.024)*	17 5.67 *(0.018)*
	6	*(0.3)*	10 3.33 *(0.027)*	11 3.67 *(0.036)*	12 4.00 *(0.012)*	13 4.33 *(0.036)*	14 4.67 *(0.078)*	15 5.00 *(0.036)*	16 5.33 *(0.012)*	17 5.67 *(0.036)*	18 6.00 *(0.027)*

TABLE 5.4 RESULTING DISTRIBUTION OF SAMPLE MEANS

ΣX	$\bar{X}$	$P(\bar{X})$
6	2.00	*0.027*
7	2.33	*0.054*
8	2.67	*0.036*
9	3.00	*0.062*
10	3.33	*0.153*
11	3.67	*0.132*
12	4.00	*0.272*
13	4.33	*0.132*
14	4.67	*0.153*
15	5.00	*0.062*
16	5.33	*0.036*
17	5.67	*0.054*
18	6.00	*0.027*
		1.00

$\mu_{\bar{X}} = 4.0$

$\sigma_{\bar{X}}^2 = 0.9333\ldots$

$\sigma_{\bar{X}} = 0.9661$

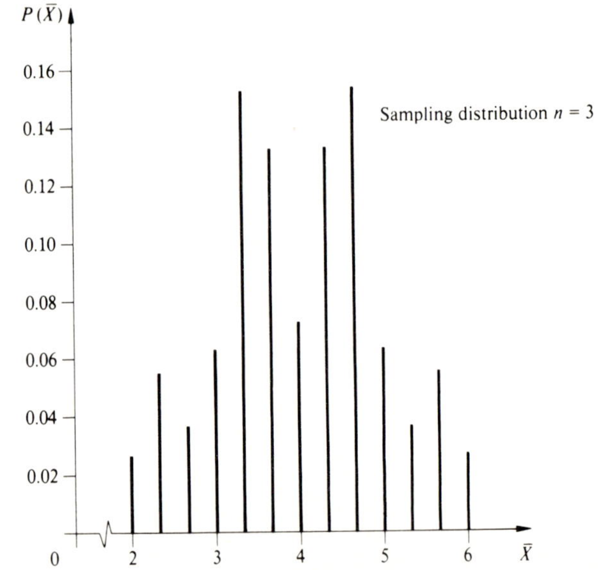

FIGURE 5.2 Sampling distribution for $n = 3$.

TABLE 5.5 CALCULATIONAL TABLE FOR SAMPLE OF SIZE 4

		Total from the First, Second, and Third Draws												
	X	6	7	8	9	10	11	12	13	14	15	16	17	18
	$P(X)$	0.027	0.054	0.036	0.062	0.153	0.132	0.072	0.132	0.153	0.062	0.036	0.054	0.027
Fourth Draw 2	0.3	8 2.00 0.0281	9 2.25 0.0162	10 2.50 0.0108	11 2.75 0.0186	12 3.00 0.0459	13 3.25 0.0396	14 3.50 0.0216	15 3.75 0.0396	16 4.00 0.0459	17 4.25 0.0186	18 4.50 0.0108	19 4.75 0.0162	20 5.00 0.0081
Fourth Draw 3	0.2	9 2.25 0.0054	10 2.50 0.0108	11 2.75 0.0072	12 3.00 2.0124	13 3.25 0.0306	14 3.50 0.0264	15 3.75 0.0144	16 4.00 0.0264	17 4.25 0.0306	18 4.50 0.0124	19 4.75 0.0072	20 5.00 0.0108	21 5.25 0.0054
Fourth Draw 5	0.2	11 2.75 0.0054	12 3.00 0.0108	13 3.25 0.0072	14 3.50 0.0124	15 3.75 0.0306	16 4.00 0.0264	17 4.25 0.0144	18 4.50 0.0264	19 4.75 0.0306	20 5.00 0.0124	21 5.25 0.0072	22 5.50 0.0108	23 5.75 0.0054
Fourth Draw 6	0.3	12 3.00 0.0081	13 3.25 0.0162	14 3.50 0.0108	15 3.75 0.0186	16 4.00 0.0459	17 4.25 0.0396	18 4.50 0.0216	19 4.75 0.0396	20 5.00 0.0459	21 5.25 0.0186	22 5.50 0.0108	23 5.75 0.0162	24 6.00 0.0081

TABLE 5.6 RESULTING DISTRIBUTION OF SAMPLE MEANS

ΣX	$\bar{X}$	$P(\bar{X})$
8	2.00	0.0081
9	2.25	0.0216
10	2.50	0.0216
11	2.75	0.0312
12	3.00	0.0772
13	3.25	0.0936
14	3.50	0.0712
15	3.75	0.1032
16	4.00	0.1446
17	4.25	0.1032
18	4.50	0.0712
19	4.75	0.0936
20	5.00	0.0772
21	5.25	0.0312
22	5.50	0.0216
23	5.75	0.0216
24	6.00	0.0081
		1.000

$\mu_{\bar{X}} = 4$

$\sigma_{\bar{X}}^2 = 0.70$

$\sigma_{\bar{X}} = 0.8366$

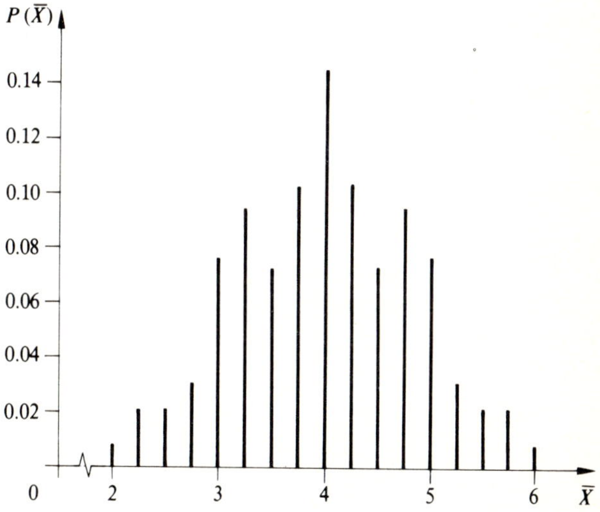

FIGURE 5.3 Sampling distribution for $n = 4$.

TABLE 5.7 SUMMARY OF CHARACTERISTICS OF SAMPLING DISTRIBUTION OF SAMPLE MEANS

Sample Size n	Mean of Sampling Distribution of $\bar{X}$ $E(\bar{X})$	Variance of Sampling Distribution of $\bar{X}$ $\sigma_{\bar{X}}^2$
2	4.0	1.400
3	4.0	0.933
4	4.0	0.700

the n cards in each sample. In Table 5.1, $\bar{X}$ can assume the values of 2.0, 2.5, 3.0, . . . , 5.5, 6.0 (and they are reproduced in Table 5.2). The values of $\bar{X}$ occur by chance because the values of X occur by chance. Note, in Table 5.2, that $P(\bar{X}) \geq 0$ for all values of $\bar{X}$; also, $\Sigma P(\bar{X}) = 1.00$.

Table 5.2 represents a probability distribution for the random variable $\bar{X}$. Since $\bar{X}$ is a *statistic*, the probability distribution of $\bar{X}$ is given a special name: *sampling distribution*. The mean or expected value of the sampling distribution is $E(\bar{X})$; the variance is $\sigma_{\bar{X}}^2$.

From Tables 5.1, 5.3, and 5.5 we can see how a sampling distribution of means can be generated. Once we have the sampling distribution, we calculate its mean and variance. The results of those calculations are given in Table 5.7.

Note that for all three sample sizes, the mean of the sample means is equal to 4.0, which is the mean of the population. This similarity exists even though there is no 4 in the deck of cards.

Since the population variance is 2.8, the variance of the sampling distribution is smaller than 2.8 for all three distributions. Further, in this example at least, it appears that the variance of the sampling distribution of means gets smaller as the sample size (n) increases.

Finally, although the distribution of the population could be described as "V-shaped," the sampling distributions could almost be called "inverted-V-shaped."

The Central Limit Theorem

The results of example 5.1 can be generalized. In formal mathematics, these results are derived from the central limit theorem (henceforth called CLT). The CLT is one of the most powerful (thus, most important) concepts in statistics. This theorem not only justifies but also magnifies the importance of the normal curve. This is a remarkable, even amazing, theorem.

Central limit theorem Let X be a random variable with finite mean μ and a finite variance σ^2. Let $\bar{X}$ be the sample mean of a randomly selected sample of n observations. Then the sampling distribution of sample means will have the following properties:

Property 1. $\mu_{\bar{X}}$ or $E(\bar{X})$ equals μ; that is, the mean of the sampling distribution will always be equal to the population mean.

Property 2. $\sigma_{\bar{X}}^2$ or $\text{Var}(\bar{X})$ equals σ^2/n where sampling with replacement from a finite population (or when sampling from an infinite population, with or without replacement) and

$$\text{Var}(\bar{X}) = (\sigma^2/n)[(N - n)/(N - 1)]$$

when sampling without replacement from a finite population; that is, the variance of the sampling distribution (1) will become smaller as the size of the sample, n, is increased and (2) will always be less than the population variance.

Property 3. Regardless of the shape of the population's frequency distribution, the sampling distribution of the sample means will approach the shape of the normal curve as the size of the sample is increased (usually a sample of 30 or more observations is sufficiently "large").

A mathematical proof of the CLT is beyond the scope of this book; however, a complete understanding of the implications of the CLT is imperative. First, let us investigate the constraints placed upon the population. The random variable may be either discrete or continuous. There may be either an infinite or a finite number of possible values for the random variable X. The population mean μ may be either positive or negative, but it must be finite.

The population variance σ^2 must be finite (and, by definition, it must be positive). Implicitly, the population variance should be greater than 0. [If $\text{Var}(X) = 0$, then all values of X are the same; the sampling distribution of a constant, or a random variable which exhibits no variability, is generally regarded as "uninteresting."] The general shape of the distribution of X is irrelevant to the general conclusions of the CLT.

Next, it is important to reiterate that the CLT makes statements about the *characteristics* of the sampling distribution of sample means. For a given sample size, the CLT does not say anything about *one particular sample mean*, but refers to the distribution of *all possible sample means*. Table 5.2 (also Tables 5.4 and 5.6) is a probability distribution of sample means. If we now consider the sample mean $\bar{X}_i$ as a new random variable, where each value of $\bar{X}_i$ is associated with a known probability of occurrence, then it is for the probability distribution of $\bar{X}$ that the CLT dictates uniform results.

Specifically, the mean of the probability distribution of $\bar{X}$ is always equal to the mean of the population from which the samples were randomly selected. From the three sampling distributions generated in example 5.1, this attribute of the CLT is well demonstrated. In every case, $\mu_{\bar{X}}$ was equal to 4.0, which was the value of the population mean. What does $\mu_{\bar{X}} = \mu$ imply? If we know the mean of the population, then we do *not* have to generate the entire sampling distribution to obtain the value of the mean for that distribution of $\bar{X}$. Calculations similar to those in Tables 5.1 and 5.2 can be eliminated.

Next, the numerical value of the variance for the probability distribution of $\bar{X}$ is always a function of (1) the value of the population variance σ^2 and (2) the number of observations in each sample, n. Note that n, the number of observations in a sample, is not the same thing as the number of possible samples. For completeness, the size of the population may also influence the value of the sampling distribution's variance.

Specifically, $\sigma_{\bar{X}}^2 = (\sigma^2/n)[(N-n)/(N-1)]$, but the term in brackets is frequently equal to, or essentially equal to, 1. If we rewrite $(N-n)/(N-1)$ in the form $(1 - n/N)(1 - 1/N)$, we can see that as N becomes large relative to n, the ratios involving N (that is, n/N and $1/N$) approach 0; thus, $(N-n)/(N-1)$ approaches 1. The term $(N-n)/(N-1)$ is called the *finite population correction factor*, or FPC. Although semantically precise in a mathematical context, in practice the term "finite population correction factor" often leads to confusion.

As we have just seen, if the population is infinitely large, $N = \infty$, then FPC $= 1$, and it can be ignored in calculation. This does not help much since in business and economics problems we seldom have an infinitely large parent population.

Even when the population is finite, the FPC is not always required. When sampling with replacement, the FPC is equal to 1, and it can be ignored. When sampling without replacement from a finite population, the FPC always affects the value of $\sigma_{\bar{X}}^2$; however, the practical question is, How much is $\sigma_{\bar{X}}^2$ affected? There are several rules of thumb concerning when to use the FPC. We will adopt the following: *When sampling without replacement from a finite population, the term $(N-n)/(N-1)$ will be used only when the size of the sample exceeds 10 percent of the population (that is, $n/N > 0.1$). In all other cases, the FPC will be ignored.*

Example 5.2   Ten-Card Deck, Sampling without Replacement

Using the 10-card deck from example 5.1 as the population, consider taking all possible samples of size 2 but using sampling *without replacement*. The probability of getting a particular value of X on the second draw will now depend upon which card was selected on the first draw. The calculations required for generating the probability distribution of sample means (Table 5.9) are summarized in Table 5.8. The mean and variance of the sampling distribution for $\bar{X}$ are calculated in Table 5.10.

TABLE 5.8 CALCULATION FOR SAMPLES OF SIZE 2

Second Draw		First Draw				
		2	3	5	6	X_1
X_2	$P(X_2 \mid X_1)$	3/10	2/10	2/10	3/10	$P(X_1)$
2	3/9 (2/9)	2.0 (6/90)	2.5 6/90	3.5 6/90	4.0 9/90	$\leftarrow \bar{X}$ $\leftarrow P(\bar{X})$
3	2/9 (1/9)	2.5 6/90	3.0 (2/90)	4.0 4/90	4.5 6/90	
5	2/9 (1/9)	3.5 6/90	4.0 4/90	5.0 (2/90)	5.5 6/90	
6	3/9 (2/9)	4.0 9/90	4.5 6/90	5.5 6/90	6.0 (6/90)	

Note: $P(X = 2$ on second draw given $X \neq 2$ on first) is 3/9 and $P(X = 2$ on second draw given $X = 2$ on the first) is (2/9).

TABLE 5.9 SAMPLING DISTRIBUTION

$\bar{X}$	$P(\bar{X})$
2.0	6/90 = 0.0667
2.5	12/90 = 0.1333
3.0	2/90 = 0.0222
3.5	12/90 = 0.1333
4.0	26/90 = 0.2889
4.5	12/90 = 0.1333
5.0	2/90 = 0.0222
5.5	12/90 = 0.1333
6.0	6/90 = 0.0667
	90/90 0.9999

TABLE 5.10 CALCULATIONS FOR $\mu_{\bar{X}}$ AND $\sigma_{\bar{X}}^2$

$\bar{X} \cdot P(\bar{X})$	$(\bar{X} - \mu_X)^2$	$(\bar{X} - \mu_X)^2 P(\bar{X})$
12/90	4.00	24/90
30/90	2.25	27/90
6/90	1.00	2/90
42/90	0.25	3/90
104/90	0.00	0/90
54/90	0.25	3/90
10/90	1.00	2/90
66/90	2.25	27/90
36/90	4.00	24/90
360/90 = 4		112/90 = 1.2444

Thus, $\mu_{\bar{X}} = 4$ and $\sigma_{\bar{X}}^2 = 1.2444$.

Since $n/N = 2/10 = 0.2 > 0.1$ and since we are sampling without replacement, the FPC should be used. From the central limit theorem, $\mu_{\bar{X}} = \mu = 4$ and

$$\sigma_{\bar{X}}^2 = \frac{\sigma^2}{n}\left(\frac{N-n}{N-1}\right) = \frac{2.8}{2}\left(\frac{10-2}{10-1}\right) = 1.2444$$

159

If all we want is the mean and variance of this sampling distribution, then the central limit theorem allows us to calculate them without Tables 5.8, 5.9, and 5.10. What a bargain!

Now reconsider the equation $\sigma_{\bar{x}}^2 = \sigma^2/n$. All possible samples of size 1 ($n = 1$) are no more than a restatement (or census) of the population. Thus, we are usually concerned with samples of size 2 or larger, and $\sigma_{\bar{x}}^2 \leq \frac{1}{2}\sigma^2$. There is always less variation in the sampling distribution than in the parent population! Table 5.7 demonstrates this result from example 5.1. Further, the CLT tells us that the variance of the probability distribution of sample means will always be smaller for large-size samples than for small-size samples.

So far, we have discovered that if we know the mean and variance of the population, the CLT will allow us to calculate the mean and variance of the sampling distribution of sample means without generating the sampling distribution (i.e., the CLT makes all tables in examples 5.1 and 5.2 unnecessary). But what about the shape of the sampling distribution? The CLT has been "friendly" enough up to this point, but in answering this last question the CLT becomes an amazing assistant.

Regardless of how the population is distributed, the probability distribution of sample means will always be similar to the normal distribution if the sample size is "large enough"! In example 5.1, the population had a V shape (which could hardly be confused with the bell-shaped normal curve); yet in Figure 5.3 (with samples of size 4 only) it is not difficult to see how the sampling distribution is rapidly acquiring the shape of a normal curve.

That's nice, but how does it qualify for the title "amazing assistant"? Recall from section 3.6 that a normal distribution is completely defined by its mean and standard deviation (the positive square root of the variance). If we *know* the population's mean and variance, and if we *choose* a "large enough" sample size, a normal curve with mean μ and standard deviation $\sigma/\sqrt{n}$ will be a very close approximation to the actual sampling distribution of sample means. With the relatively simple calculation of $\sigma_{\bar{x}}$, we obtain a very precise picture of any probability distribution of sample means.

The only flaw in this argument involves just how large a sample size is "large enough"? The answer requires judgment and centers on how much difference between the actual sampling distribution and the normal distribution is acceptable. This difference depends, in part, on the shape of the population. If $n = \infty$, there is no difference, but the sample size can assume much more meager values to be acceptable. Since we are attempting to account for all possible shapes of the population, most statisticians consider samples of 30 or more to be large enough.

Figure 5.4 demonstrates the results of the CLT from three different populations. One population has a uniform shape, the next is skewed to the left, and the third is bimodal. The arithmetic mean is denoted with a triangular fulcrum on the horizontal axis; of course, the mean of any given sampling distribution is equal to the mean of the population from which the samples were selected. Also, the variation in the

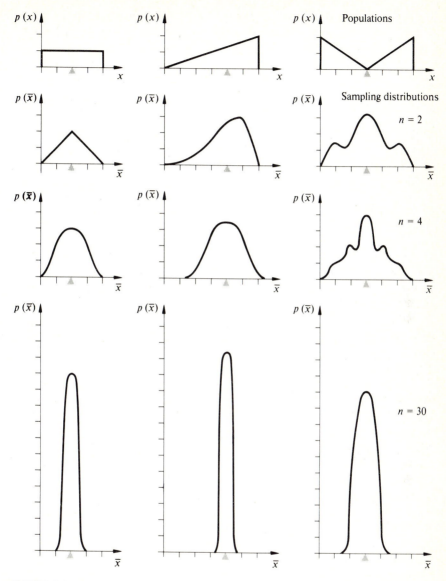

FIGURE 5.4 Selected sampling distributions for three different populations.

sampling distribution becomes less as the sample size increases; that is, $\sigma_{\bar{x}}^2 = \sigma^2/\sqrt{n}$.

The peculiarity of the population's shape affects the shape of the sampling distribution for small samples (i.e., see the cases of $n = 2$ and $n = 4$); however, the influence of the population's shape diminishes rapidly, and for large samples (i.e., see the case of $n = 30$) the sampling distribution appears very much like a normal curve. This amazing similarity allows us to use *only one model*, the normal curve, for the sampling distribution of sample means for large samples.

Use of the Normal-Curve Table

In the last section of Chapter 3 we discovered how to use the z distribution, Appendix Tables IV(A) and IV(B), for a normally distributed continuous random variable X. The central limit theorem gives us reason to use the z table when our random variable is the sample mean $\bar{X}$, which may be either continuous or discrete.

When $\bar{X}$ either is or can be assumed to be a continuous random variable, then the use of the z table is relatively simple. In Chapter 3 we noted that if X was normally distributed with mean equal to μ and standard deviation equal to σ, then $z = (X - \mu)/\sigma$ is normally distributed with mean equal to 0 and standard deviation equal to 1.

Consider a substitution of variables: $\bar{X}$ for X. If $\bar{X}$ is normally distributed (usually requires $n \geq 30$) with mean equal to $\mu_{\bar{X}}$ (which is always equal to μ) and standard deviation $\sigma_{\bar{X}} = \sigma/\sqrt{n}$, then

$$z = \frac{\bar{X} - \mu_{\bar{X}}}{\sigma_{\bar{X}}} = \frac{\bar{X} - \mu}{\sigma/\sqrt{n}} \tag{5.1}$$

is normally distributed with mean equal to 0 and standard deviation equal to 1. Hence,

$$P(\bar{X} \geq k) = P\left(z \geq \frac{k - \mu}{\sigma/\sqrt{n}}\right) \tag{5.2}$$

and

$$P(a \leq \bar{X} \leq b) = P\left(\frac{a - \mu}{\sigma/\sqrt{n}} \leq z \leq \frac{b - \mu}{\sigma/\sqrt{n}}\right) \tag{5.3}$$

Example 5.3 Sampling from the Automatic Cola Dispenser

The automatic dispensing machine in example 3.5 spews forth liquid in amounts which are normally distributed with mean equal to 8.1 ounces and standard deviation of 0.2 ounce. We let X equal the fill, or volume of liquid dispensed, in one try.

Since the *population* is normally distributed, the central limit theorem is valid for any size sample. A sample size of 2 is large enough for the sampling distribution to be normally distributed.

(a) Consider samples of size 4. That is, we get four cups of liquid from this machine and divide the total ounces by 4. What does the sampling distribution look like? From the CLT we know that $\mu_{\bar{X}} = \mu = 8.1$ ounces and $\sigma_{\bar{X}} = \sigma/\sqrt{n} = 0.2/\sqrt{4} = 0.1$. Figure 5.5b shows the probability distribution of all possible sample means, when $n = 4$, from the normally distributed population shown in Figure 5.5a.

FIGURE 5.5

1. What is the probability that a random sample of four filled cups will yield a mean fill of 8.21 ounces or more? The question can be restated as $P(\bar{X} \geq 8.21) = ?$ From Equation 5.2 we find that

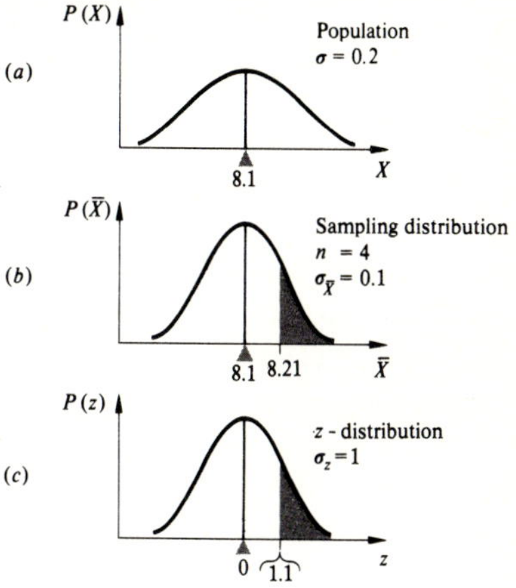

$$P(\bar{X} \geq 8.21) = P\left(z \geq \frac{8.21 - 8.10}{0.2/\sqrt{4}}\right) = P\left(z \geq \frac{0.11}{0.1}\right) = P(z \geq 1.10)$$

Appendix Table IV(B) tells us that

$$P(z \geq 1.10) = 0.1357$$

2. What is $P(8.0 \leq \bar{X} \leq 8.5)$ when a random sample of 4 is selected? From Equation 5.3,

$$P(8.0 \leq \bar{X} \leq 8.5) = P\left(\frac{8.0 - 8.1}{0.1} \leq z \leq \frac{8.5 - 8.1}{0.1}\right)$$

$$= P(-1.00 \leq z \leq +4.00)$$

$$= P(-1 \leq z \leq 0) + P(0 \leq z \leq 4) \qquad \leftarrow \text{Split area at } z = 0$$

$$= P(0 \leq z \leq +1) + P(0 \leq z \leq 4) \qquad \leftarrow \text{Symmetry}$$

$$= 0.3413 + ? \qquad\qquad\qquad \leftarrow \text{Appendix Table IV(A)}$$

The highest value of z provided in Table IV(A) is 3.09. We know that $P(0 \leq z \leq 3.09) = 0.4990$; thus, we have two choices.

Choice 1
First, note that $P(0 \leq z \leq 4) > P(0 \leq z \leq 3.09)$. Given the limitation of Appendix Table IV(A), we could be quite correct to say that

$$P(8.0 \leq \bar{X} \leq 8.5) > 0.3413 + 0.4990 = 0.8403$$

That is, our answer could be that $P(8 \leq \bar{X} \leq 8.5)$ is *greater than* 0.8403.

Choice 2

A more common practice, however, is to round off the probability that z will be between 0 and 4 to three decimal places. That is, $P(0 \le z \le 4) = 0.500$. [If you want a rule of thumb, when $z_1 \ge 3.9$, then $P(0 \le z \le z_1) = 0.5000$; when $3.09 \le z_1 \le 3.9$, then use $P(0 \le z \le z_1) = 0.500$.] Thus

$$P(8.0 \le \bar{X} \le 8.5) = 0.341 + 0.500 = 0.841$$

Compare this answer with part *d* of example 3.5.

(*b*) Assume that we take a random sample of 25 fills and calculate the sample mean. (1) What is the probability that the sample mean will be less than 8 ounces? (2) What is the probability that the sample mean will be between 8.00 and 8.05 ounces?

   First, what does the sampling distribution look like? It is normally distributed with $\mu_{\bar{X}} = \mu = 8.1$ ounces and $\sigma_{\bar{X}} = \sigma/\sqrt{n} = 0.2/\sqrt{25} = 0.04$.

1. $z = (\bar{X}_3 - \mu_{\bar{X}})/\sigma_{\bar{X}}$

$\qquad = (8.0 - 8.1)/0.04$

$\qquad = -2.5$

$\qquad P(\bar{X} \le 8.0) = P(z \le -2.5)$

Again, due to the symmetry of the z distribution,

$\qquad P(z \le -2.5) = P(z \ge +2.5)$

and from Appendix Table IV(B) that probability is 0.0062. Hence, $P(\bar{X} \le 8) = 0.0062$.

2. The question asks for $P(8.00 \le \bar{X} \le 8.05)$. Using Equation 5.3, we get

$$P\left(\frac{8.00 - 8.10}{0.04} \le z \le \frac{8.05 - 8.10}{0.04}\right)$$

$$= P(-2.50 \le z \le -1.25)$$

Carefully observe the area of interest in Figure 5.6*d*.

Method 1 [Appendix Table IV(A)]
Note that $P(-2.50 \le z \le -1.25)$
$= P(-2.50 \le z \le 0) - P(-1.25 \le z \le 0)$
$= 0.4938 - 0.3944 = 0.0994$.

Method 2 [Appendix Table IV(B)]
Also, $P(-2.50 \le z \le -1.25)$ can be written as $P(z \le -1.25) - P(z \le -2.50) = 0.1056 - 0.0062 = 0.0994$.

FIGURE 5.6

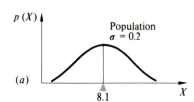

(*a*)

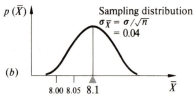

(*b*)

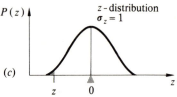

(*c*)

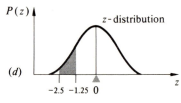

(*d*)

Hence, $P(8.00 \leq \bar{X} \leq 8.05) = 0.0994$, and a diagram similar to Figure 5.6d is always useful.

When the population is normally distributed, we need not be concerned with the size of the sample; the sampling distribution of sample means will be exactly normal, and the CLT allows us to define its mean and standard deviation. However, not all populations of interest in business and economics are normally distributed. When the shape of the population is either not normally distributed or unknown (but $\bar{X}$ is a continuous random variable), then we have to be sure that the sample size is greater than 30 in order for the CLT to be effective. Otherwise, the procedure demonstrated in example 5.3 is the same.

When sampling without replacement from a finite population, the finite population correction factor must be used in calculating the standard deviation of the sampling distribution.

Example 5.4   Distance from "Campus Center" of Off-Campus Students

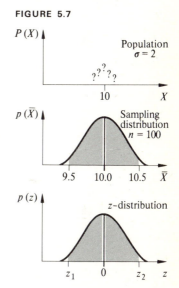

FIGURE 5.7

On a particular campus, assume that there are 500 full-time students who live "off campus." Assume that the variable X is the distance from the center of the campus to place of residence for each of these students. The superaccurate records in the registrar's office and a computer allow us to determine that $\mu = 10$ blocks and $\sigma = 2$ blocks. For some reason (e.g., to investigate the increased commuting costs of a fuel crisis) a random sample of 100 off-campus students is selected without replacement. What is the probability that the mean distance for these 100 students is 10 ± 0.5 blocks? That is, what is the probability that the sample mean $\bar{X}$ will be within one-half block of the population mean?

First, what does the sampling distribution look like for $n = 100$? It will approximate a normal curve with mean equal to $\mu_{\bar{X}} = 10$ blocks. Since $n/N = 100/500 = 0.2$ is larger than 0.1, we must employ the finite population correction factor.

$$\sigma_{\bar{X}} = \frac{\sigma}{\sqrt{n}} \sqrt{\frac{N-n}{N-1}} = \frac{2}{\sqrt{100}} \sqrt{\frac{500-100}{500-1}}$$

$$= 0.2\sqrt{0.8016} = (0.2)(0.8953)$$

$$= 0.1791 \approx 0.18$$

Using Equation 5.3, we find

$$P(9.5 \leq \bar{X} \leq 10.5) = P[(9.5 - 10)/0.18 \leq z \leq (10.5 - 10)/0.18]$$

$$= P(-2.78 \leq z \leq 2.78)$$

$$= P(-2.78 \leq z \leq 0) + P(0 \leq z \leq +2.78)$$

$$= 0.4973 + 0.4973 = 0.9946$$

The only difference between this example and example 5.3 is the inclusion of the FPC to account for smaller variation in sample means when sampling without replacement.

Finally, when the sample mean $\bar{X}$ is a discrete random variable, we may have to consider *continuity correction*. The normal curve is the probability density function for a continuous random variable. As suggested by the central limit theorem, we may want to use the normal curve as a *model* or approximation for an actual sampling distribution of a discrete random variable.

Return to the 10-card deck of cards in example 5.1. The exact sampling distribution, for samples of size 2, is given in Figure 5.1. The normal curve is obviously a poor model to use for that trimodal sampling distribution, but that is not the problem we are trying to investigate. Even if the sampling distribution in Figure 5.1 were bell-shaped, we would have to be careful in using the normal curve as our model. Why?

As shown in Table 5.2 and Figure 5.1, there are only *nine* possible values that $\bar{X}$ can assume (that is, 2.0, 2.5, . . . , 5.5, 6.0). From the actual *discrete* sampling distribution in Table 5.2 we would get the same answer from the two questions $P(\bar{X} \geq 6) = ?$ and $P(\bar{X} \geq 5.75) = ?$ However, from our normal-curve model, a *continuous* probability density function, we would get a different answer to those same two questions. It is this difference which infrequently calls for continuity correction in the use of the normal curve as a model.

Note, for our 10-card deck, it is only if $n \geq 30$ that we would use the normal curve. If we consider the sampling distribution of means for $n = 30$, then there are 121 different possible values of $\bar{X}$ in the range from 2 to 6 (that is, 2.0, 2.033, 2.067, . . . , 5.967, 6.000). Since 5.983 is about halfway between 5.967 and 6.000, we could again ask two questions: $P(\bar{X} \geq 5.983) = ?$ and $P(\bar{X} \geq 6.000) = ?$ Again there would be no difference in the answers if we used the actual sampling distribution, and there would be some difference if we used the normal-curve model. But, the difference in the two answers from the normal-curve model is now slight or insignificant.

We will be concerned only about the continuity correction in our normal-curve model when sampling from a binomial population. (That is, assume a 5-card deck of three 2s and two 6s. On a single trial there are only two possible results: a 2 or a 6.) This case will be discussed in the next section.

5.2   SAMPLING DISTRIBUTIONS FROM THE BINOMIAL

As a discrete probability distribution, the binomial is used to generate two related sampling distributions: the sampling distribution for the *number* of successes in a sample and the sampling distribution for the *proportion* of successes in a sample. Before we consider each of these sampling distributions separately, let us demonstrate their similarities.

We will proceed as we did in the previous section. In an example, we will use only the probability rules (Chapter 2) to generate sampling distributions from a binomial population. We will use the definitions of Chapter 3 to find the mean and variance of these sampling distributions, and we will see if we can generalize the results in order to avoid ever again going through all those calculations.

First, we must refresh our memory of certain definitions. A *binomial population* is any collection of objects where each and every object can be classified as a "success" of a "failure." The size of the population, or total number of objects, is denoted by N, which may or may not be finite. In this population, there are R successes and $(N - R)$ failures (if the population size N is finite). The *population proportion* π is simply defined as the proportion of successes within the population; $\pi = R/N$. (Note that we may know the population proportion even when N is infinite. For example, the proportion of odd-number integers in a population of all positive integers is $\frac{1}{2}$.)

From this population we will randomly select, with replacement, a sample of n objects. The random variable r will refer to the number of successes in this sample; r may assume only integer values. The *sample proportion* $\bar{p}$ (called "p bar") is simply the proportion of successes in the sample; $\bar{p} = r/n$.

If r is the number of successes in a random sample of n objects, then $0 \leq r \leq n$ where r and n are integer values. That is, if we randomly select 10 objects, there cannot be more than 10 successes nor less than 0 successes in that sample. Since a sampling distribution considers *all* possible samples of a given size, in a sampling distribution of the number of successes, the range of values for r must always be $0 \leq r \leq n$.

In like manner, if $\bar{p} = r/n$, then $0/n \leq r/n \leq n/n$, or $\bar{p}$ must always have a value such that $0 \leq \bar{p} \leq 1$. Thus, we know the *range* of the sampling distributions for both the number of successes and the proportion of successes from only the definitions of those random variables. The range of the sampling distribution for the proportion of successes is constant, $0 \leq \bar{p} \leq 1$; however, the range of the sampling distribution for the number of successes is given by the sample size, $0 \leq r \leq n$.

Now, let us resort to a specific example to generate sampling distributions.

Example 5.5   Sampling Distributions of Persons Who Wear Contact Lenses

Suppose that in a community of N persons we know that 30 percent are regular wearers of contact lenses. Let us also define "the wearer of contact lenses" as a success. The population proportion is given as $\pi = 0.30$.

TABLE 5.11 CALCULATIONS OF MEAN AND STANDARD DEVIATION OF A BINOMIAL POPULATION

R	$f(R)$	$R \cdot f(R)$	$R - \mu_R$	$(R - \mu_R)^2$	$f(R) \cdot (R - \mu_R)^2$
0	$(1-\pi)N$	0	$-\pi$	π^2	$\pi^2(1-\pi)N$
1	πN	πN	$1-\pi$	$(1-\pi)^2$	$\pi(1-\pi)^2 N$
	$\overline{N}$	$\overline{\pi N}$			$\overline{\pi(1-\pi)N(\pi+1-\pi)}$
					$= \pi(1-\pi)N$

What does the population look like? It's just a bunch of people; 30 percent wear contact lenses, and 70 percent do not. While this statement is true enough, it will not suffice in a society that denotes everyone with various identification numbers. So, in the spirit of identification numbers but with a desire to maintain some privacy, all wearers of contact lenses will be assigned a 1 and all others will be assigned a 0. Now, our population consists of πN "ones" and $(1 - \pi)N$ "zeros." Let's find the mean and variance of that population. See Table 5.11.

From Chapter 3, $\mu_R = (1/N)\Sigma[R \cdot f(R)] = (1/N)(\pi N) = \pi$, and

$$\sigma_R{}^2 = (1/N)\Sigma[f(R) \cdot (R - \mu_R)^2] = (1/N)[\pi(1-\pi)N] = \pi(1-\pi)$$

Although you can choose some population size, say $N = 100,000$, it should be obvious why it was not necessary to specify N in this example. This population of contact wearers and non-contact-wearers, or 1s and 0s, has a mean of $\pi = 0.30$ and a variance of $\pi(1 - \pi) = (0.3)(1 - 0.3) = 0.21$.

Now, let us consider the sampling distributions from this population. If we select just one person from this community, what is the probability of getting a person who regularly wears contact lenses? $P(\text{success}) = P(\text{a wearer of contact lenses}) = 0.30 = \pi = $ the population proportion.

If $r = $ the number of successes out of a sample of n objects, then when $n = 1$, the variable r can have values of 0 and 1 only. Also, note that P (success) $= P(r = 1) = 0.30$, and $P(\text{failure}) = P(r = 0) = 0.70$. Since all possible samples of size 1 are simply a restatement of the population, our population could be described as a probability distribution as follows:

$$P(r) = \begin{cases} 1 - \pi & r = 0 \\ \pi & r = 1 \end{cases}$$

Using Equations 3.1 and 3.3, you should not be surprised to find that $E(r) = \pi$ and $\text{Var}(r) = \pi(1 - \pi)$. These results are, of course, consistent with our frequency distribution representation of the population.

Now let us consider all possible samples of size 2 (then 3 and then 4), find the probability distribution, and calculate the mean and variance of that sampling distribution.

TABLE 5.12 CALCULATIONAL SUMMARY FOR SAMPLES OF SIZE 2

		First Draw	
r		0	1
	$P(r)$	0.7	0.3

Second Draw				
0	0.7	0 [0.00] 0.49	1 [0.50] 0.21
1	0.3	1 [0.50] 0.21	2 [1.00] 0.09

← Number of persons who wear contact lenses on first draw

← Number of wearers in sample of 2: r
← Sample proportion; $\bar{p} = r/n$
← Probability of this sample occurring

TABLE 5.13 SAMPLING DISTRIBUTIONS FOR r AND $\bar{p}, n = 2$

Random Variable		
Number of Persons Who Wear Contact Lenses (r)	Proportion of Persons Who Wear Contact Lenses $(\bar{p} = r/n)$	Probability $P(r)$ or $P(\bar{p})$
0	0.00	0.49
1	0.50	0.42
2	1.00	0.09
		——
		1.00

TABLE 5.14 CALCULATION OF MEAN AND VARIANCE FOR EACH SAMPLING DISTRIBUTION

Number of Successes			Proportion of Successes		
$r \cdot P(r)$	$(r - \mu_r)^2$	$P(r) \cdot (r - \mu_r)^2$	$\bar{p} \cdot P(\bar{p})$	$(\bar{p} - \mu_{\bar{p}})^2$	$P(\bar{p}) \cdot (\bar{p} - \mu_{\bar{p}})^2$
0.00	0.36	0.1764	0.00	0.09	0.0441
0.42	0.16	0.0672	0.21	0.04	0.0168
0.18	1.96	0.1764	0.09	0.49	0.0441
——		——	——		——
0.60		0.4200	0.30		0.1050
$= E(r)$		$= \text{Var}(r)$	$= E(\bar{p})$		$= \text{Var}(\bar{p})$
$= \mu_r$			$= \mu_{\bar{p}}$		

TABLE 5.15 CALCULATIONAL SUMMARY FOR SAMPLES OF SIZE 3

			First Two Draws		
	r		**0**	**1**	**2**
		$P(r)$	*0.49*	*0.42*	*0.09*
Third Draw **0**	0.7		0 [0.00] *0.343*	1 [0.33] *0.294*	2 [0.67] *0.063*	← Number of successes, r ← Proportion of successes, r/n ← Probability of this sample result
Third Draw **1**	0.3		1 [0.33] *0.147*	2 [0.67] *0.126*	3 [1.00] *0.027*	

TABLE 5.16 SAMPLING DISTRIBUTIONS FOR r AND $\bar{p}$, $n = 3$

Random Variable		
Number of Successes (r)	Proportion of Successes ($\bar{p}$)	Probability $P(r)$ or $P(\bar{p})$
0	0.00	0.343
1	0.33	$0.294 + 0.147 = 0.441$
2	0.67	$0.063 + 0.126 = 0.189$
3	1.00	0.027
		1.000

TABLE 5.17 CALCULATION OF MEAN AND VARIANCE FOR EACH SAMPLING DISTRIBUTION

Number of Successes			Proportion of Successes		
$r \cdot P(r)$	$(r - \mu_r)^2$	$P(r) \cdot (r - \mu_r)^2$	$\bar{p} \cdot P(\bar{p})$	$(\bar{p} - \mu_{\bar{p}})^2$	$P(\bar{p}) \cdot (\bar{p} - \mu_{\bar{p}})^2$
0.000	0.81	0.27783	0.000	0.09000	0.03087
0.441	0.01	0.00441	0.146	0.00111	0.00049
0.378	1.21	0.22869	0.127	0.13444	0.02541
0.081	4.41	0.11907	0.027	0.49000	0.01323
0.900 $= \mu_r$		0.63000 $= \text{Var}(r)$	0.300 $= \mu_{\bar{p}}$		0.07000 $= \text{Var}(\bar{p})$

TABLE 5.18 CALCULATIONAL SUMMARY FOR SAMPLES OF SIZE 4

		First Three Draws				
	r	0	1	2	3	
	$P(r)$	0.343	0.441	0.189	0.027	
Fourth Draw 0	0.7	0 [0.00] 0.2401	1 [0.25] 0.3087	2 [0.50] 0.1323	3 [0.75] 0.0189	←r ←$\bar{p}$ ←Probability
1	0.3	1 [0.25] 0.1029	2 [0.50] 0.1323	3 [0.75] 0.0567	4 [1.00] 0.0081	

TABLE 5.19 SAMPLING DISTRIBUTIONS FOR r AND $\bar{p}$, $n = 4$

Random Variable		Probability
r	$\bar{p}$	$P(r)$ or $P(\bar{p})$
0	0.00	0.2401
1	0.25	$0.3087 + 0.1029 = 0.4116$
2	0.50	$0.1323 + 0.1323 = 0.2646$
3	0.75	$0.0189 + 0.0567 = 0.0756$
4	1.00	0.0081
		1.0000

TABLE 5.20 CALCULATION OF MEAN AND VARIANCE FOR EACH SAMPLING DISTRIBUTION

Number of Successes			Proportion of Successes		
$r \cdot P(r)$	$(r - \mu_r)^2$	$P(r) \cdot (r - \mu_r)^2$	$\bar{p} \cdot P(\bar{p})$	$(\bar{p} - \mu_{\bar{p}})^2$	$P(\bar{p}) \cdot (\bar{p} - \mu_{\bar{p}})^2$
0.0000	1.44	0.345744	0.0000	0.0900	0.021609
0.4116	0.04	0.016464	0.1029	0.0025	0.001029
0.5292	0.64	0.169344	0.1323	0.0400	0.010584
0.2268	3.24	0.244944	0.0567	0.2025	0.015309
0.0324	7.84	0.063504	0.0081	0.4900	0.003969
1.2000 $= \mu_r$		0.840000 $= \text{Var}(r)$	0.3000 $= \mu_{\bar{p}}$		0.052500 $= \text{Var}(\bar{p})$

TABLE 5.21 SUMMARY OF RESULTS

Sample Size (n)	Number of Successes		Proportion of Successes	
	Mean	Variance	Mean	Variance
2	0.60	0.42	0.30	0.1050
3	0.90	0.63	0.30	0.0700
4	1.20	0.84	0.30	0.0525

Figure 5.8 graphs the results of Tables 5.13, 5.16, and 5.19. Recalling that we reasoned that the range of sampling distribution for the number of successes was $0 \leq r \leq n$, we have graphed that sampling distribution, Figure 5.8a, in a rather special way.

In Figure 5.8a, (1) the vertical axes are all connected in the sense that the origin for one distribution is the same point as $P(r) = 1.0$ for the vertical axis for the distribution below it; (2) the horizontal axes are all uniformly scaled; and (3) the succession of sampling distributions is such that the one below the distribution for samples of size n is the distribution for samples of size $(n + 1)$. The "range line" is a straight line connecting the largest possible values of r for each sampling distribution. The "line of means" is a similar line connecting the arithmetic mean of each sampling distribution.

In Figure 5.8b, no special construction is used to graph the sampling distribution of the proportion of success. The range of $\bar{p}$ is always the same (that is, $0 \leq p \leq 1$), and the arithmetic mean of the sampling distribution $\mu_{\bar{p}}$ is always the same (0.3 in this example).

Table 5.21 summarizes the calculations of the mean and variance for each sampling distribution.

First, let us consider the three sampling distributions for the number of successes. Since the upper value of the range is always given by n (and the lower is always 0) and given the particular way Figure 5.8a is constructed, we can easily see how the range line is a *straight line*. But, the line of means is also a straight line! What does that suggest? That the distance along the r axis from 0 to μ_r is always the same proportion of the distance from 0 to n? Yes! That is,

$$\frac{\mu_r}{n} = \text{constant}$$

Returning to Table 5.21, we find that $\mu_r/n = 0.6/2 = 0.9/3 = 1.2/4 = 0.3 = \pi$. The "constant" to which μ_r/n is equal is the population proportion π. Thus, at least for these three sampling distributions of the number of successes, $\mu_r = n\pi$. That is simple enough: the mean of the sampling distribution of the number of successes is the sample size times the population proportion.

We have found a good thing; let us see how far we can stay with it. Could it be that $\text{Var}(r)/n = \text{constant}$? From Table 5.21, $\text{Var}(r)/n = 0.42/2 = 0.63/3 = 0.84/4 = 0.21$! So the ratio of the variance to the sample size is a constant. How would we ever know that that constant was 0.21? Note that $\pi(1 - \pi) = (0.3)(1.0 - 0.3) = (0.3)(0.7) = 0.21$! Since, at least for these three distributions, $\text{Var}(r)/n = \pi(1 - \pi)$, then $\text{Var}(r) = n\pi(1 - \pi)$.

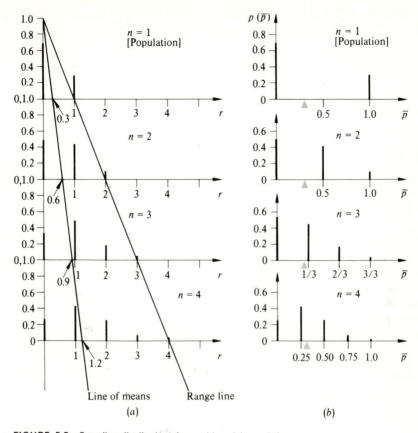

FIGURE 5.8 Sampling distributions from a binomial population.

Now let us turn our attention to the sampling distribution for the proportion of successes. Like the distribution of sample means, Figure 5.8 shows the mean of the sampling distribution of sample proportions to be equal to the mean of the population, which is π; that is, $\mu_{\bar{p}} = \pi = 0.3$.

Let us carry the similarity of the sampling distributions of $\bar{X}$ and $\bar{p}$ one step further. In Table 5.21, $\sigma_{\bar{p}}^2$ decreases as n increases. (This result is visually evident in Figure 5.8b.) From the central limit theorem, we know that $\sigma_{\bar{X}}^2 = \sigma^2/n$. Could it be that $\sigma_{\bar{p}}^2 = \sigma^2/n$? Let us try it. Above we found that the variance of a binomial population is $\pi(1 - \pi)$, which in this case is $\pi(1 - \pi) = (0.3)(0.7) = 0.21$. Thus,

$$\sigma^2/n = 0.21/2 = 0.105$$

$$0.21/3 = 0.07$$

$$0.21/4 = 0.0525$$

These are, of course, the values of the variance we obtained independently from the three sampling distributions.

This example has suggested a string of possible generalizations. How far can we proceed with these generalizations?

5.2(a) The Sampling Distribution for the Number of Successes in n Trials from a Binomial Population

Indeed, the results of example 5.5 can be easily generalized. When sampling from a binomial population with a procedure that satisfies a Bernoulli experiment (see section 3.5), the *exact* sampling distribution of the *number* of successes from a sample of n observations has the following properties:

Property 1. The possible values of r are the integer values between, and including, 0 and n.

Property 2. The probability of each integer value of r occurring is given by

$$p(r = k \mid n \text{ and } \pi) = \frac{n!}{k!(n-k)!}(\pi)^k(1-\pi)^{n-k}$$

For selected values of π and for n up to 16, these binomial probabilities are given in Appendix Table III.

Property 3. The mean is always equal to the sample size times the population proportion

$$E(\text{number of successes}) = \mu_r = n\pi$$

Property 4. The variance is always given by

$$\text{Var(number of successes)} = \sigma_r{}^2 = n\pi(1-\pi)$$

Property 5. The direction and extent of skewness is determined by (1) the size of the sample and (2) how far the population proportion is from $\frac{1}{2}$.

That is, when $\pi = 0.5$, the sampling distribution is symmetric regardless of the size of the sample. (This should not be too surprising; in order for π to equal 0.5, there would have to be an equal number of 0s and 1s within the population. If the population is symmetric and each element in the population has the same probability of being included in a sample, then a symmetric sampling distribution is reasonable.) If $\pi < 0.5$, then the sampling distribution is positively skewed, or *skewed to the right*. Example 5.5 demonstrates the tail of the sampling distribution going to the right when $\pi = 0.3$. If $\pi > 0.5$, then the sampling distribution is negatively skewed, or *skewed to the left*. Further, the greater the absolute difference between π and 0.5 (i.e., the larger $|\pi - 0.5|$), the greater the skewness. On top of all this concern for the difference $(\pi - 0.5)$ is the influence of the sample size n. *Increasing the sample size reduces the skewness of the sampling distribution*. In the hypothetical sampling situation of selecting all possible infinitely large samples, the sampling distribution would be symmetric regardless of the value of π.

Property 6. As the sample size n becomes larger, the sampling distribution approaches the shape of a normal curve.

Ironically, the normal curve reappears. Above, we stated that the sampling distribution becomes symmetric as n increases. We can be even more specific: as n increases, the sampling distribution assumes a bell shape which corresponds to the normal distribution. If π is close to 0.5, then the normal curve may approximate the sampling distribution for small values of n. If π is very, very small (for example, $\pi = 0.05$) or if π is very, very large (for example, $\pi = 0.95$), the sample size may have to be in the hundreds before the normal curve is a good approximation to the exact sampling distribution. Thus, the population proportion π may determine how "large" a "large enough value of n" is, but the normal curve is always the limiting shape of the sampling distribution.

Example 5.6   The Number of Contact Lense Wearers per Sample

Assume we are going to take a random sample, with replacement, of size 4 from the population described in example 5.5. (*a*) Describe the sampling distribution for the number of contact lense wearers per sample. (*b*) What is the exact probability that the one sample we select will have three or more wearers of contact lenses? (*c*) Using the normal curve as an approximation of the exact sampling distribution, what is the probability that we will select a sample with three or more wearers of contact lenses? (*d*) Compare the answers to parts *b* and *c*.

(*a*)   From Appendix Table III, for $n = 4$ and $\pi = 0.3$, we find the following, which are, of course, identical to the results of Table 5.19.

r	$P(r)$
0	0.2401
1	0.4116
2	0.2646
3	0.0756
4	0.0081

The mean is $\mu_r = n\pi = (4)(0.3) = 1.2$ persons; the variance is $\sigma_r^2 = n\pi(1 - \pi) = (4)(0.3) \times$ $(0.7) = 0.84$, and the standard deviation is $\sigma_r = \sqrt{0.84} = 0.91652$. Since $\pi = 0.3 < 0.5$, the sampling distribution is skewed to the right.

(b) $P(r \geq 3) = P[(r = 3) \cup (r = 4)] = P(r = 3) + P(r = 4) = 0.0756 + 0.0081 = 0.0837.$

(c) Now, we want to use the normal-curve approximation to the problem of finding $P(r \geq 3)$. Let us use r to denote the discrete random variable "number of successes" as we have thus far, and let us use x to denote the analogous continuous random variable in our normal-curve approximation.

Figure 5.9 shows the discrete probability distribution for r. We could convert this into a "probability histogram" by noting that class intervals of $r \pm \frac{1}{2}$ yield a class length of 1; thus $P(r) \cdot 1$ equals the area under one bar of the histogram. That is, $P(r - \frac{1}{2} \leq x \leq r + \frac{1}{2}) = P(r)$. Hence, the shaded area in Figure 5.9 is equal to 1. We go one step further and assume that the shaded area can be approximated by a normal curve. Obviously, this is a poor approximation, but it demonstrates both the approach and the problems.

If we want $P(r \geq 3)$, then the appropriate "shaded area" is all that area to the right of $x = 2.5$, or $P(r \geq 3) = P(x \geq 2.5)$. This is the "continuity correction" that was hinted at in the last section.

Now, we know that $z = (x - \mu_x)/\sigma_x$, and we want x to be 2.5. But what are μ_x and σ_x? It is the sampling distribution described in part a which we are assuming is normally distributed. So, $\mu_x = 1.2$ persons and $\sigma_x = 0.91652$ person. Thus, $z = (x - \mu)/\sigma = (2.5 - 1.2)/0.91652 = 1.42$. From Appendix Table IV(B), $P(r \geq 3) \approx P(x \geq 2.5) = P(z \geq 1.42) = 0.0778.$

(d) The actual $P(r \geq 3)$ was given in part b as 0.0837, and our estimate of $P(r \geq 3)$ in part c was 0.0778. That represents a 7 percent error in our estimate. Without the continuity correction, the percent error would have been larger. However, a sample size of 4 could

FIGURE 5.9 Continuous normal-curve approximation to a discrete binomial sampling distribution.

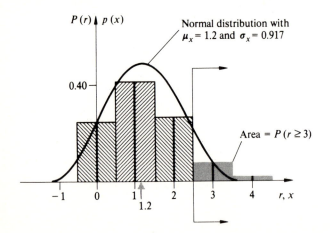

hardly be considered "large." The error in our normal-curve approximation of $P(r \geq 3)$ came from the fact that the actual sampling distribution did not look much like a normal curve. The actual sampling distribution did not look much like a normal curve because $\pi \neq 0.5$ and n was small!

Example 5.6 demonstrates the two-stage decision problem we face when we try to use all the information we have about this sampling distribution of the number of successes. The first decision is, When should we use the exact distribution (e.g., the binomial probabilities in Appendix Table III) and when should we use the normal-curve approximation? The decision rests upon how much error in the approximation is acceptable. We shall use the following rule of thumb: *If $n\pi(1 - \pi) \geq 3$, use the normal-curve approximation; however, if $n\pi(1 - \pi) < 3$, do not use the normal-curve approximation.*[1] Note that, in example 5.6, $n\pi(1 - \pi) = (4)(0.3)(0.7) = 0.84$, which is less than 3; we should not use the normal-curve approximation (a conclusion we have aptly demonstrated).

If we use the binomial probabilities for the exact sampling distribution, there is no reason to be concerned about the continuity correction factor. However, when the normal-curve approximation is called for, how do you use the continuity correction factor? See Table 5.22.

Determining whether the continuity correction factor is a $+\frac{1}{2}$ or a $-\frac{1}{2}$ is a matter of inspecting the particular question asked, but, failing that, Table 5.22 should suffice. The "cookbook" framework of Table 5.22 is indicative only of the very minor importance of the continuity correction.

[1] If π or $(1 - \pi)$ is less than 0.1, there is a good approximation by using the Poisson distribution, which will not be discussed here.

TABLE 5.22 SUGGESTED RULES FOR USING THE CONTINUITY CORRECTION WHEN THE SAMPLE STATISTIC IS THE NUMBER OF SUCCESSES

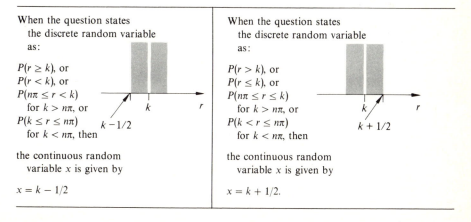

When the question states
  the discrete random variable
  as:

$P(r \geq k)$, or
$P(r < k)$, or
$P(n\pi \leq r < k)$
  for $k > n\pi$, or
$P(k \leq r \leq n\pi)$
  for $k < n\pi$, then

the continuous random
  variable x is given by

$x = k - 1/2$

When the question states
  the discrete random variable
  as:

$P(r > k)$, or
$P(r \leq k)$, or
$P(n\pi \leq r \leq k)$
  for $k > n\pi$, or
$P(k < r \leq n\pi)$
  for $k < n\pi$, then

the continuous random
  variable x is given by

$x = k + 1/2$.

5.2(b) The Sampling Distribution for the Sample Proportion

For the distribution of sample proportions, the results of example 5.5 are easier to generalize. The sample proportion $\bar{p}$ is not just analogous to the sample mean $\bar{X}$; it is a sample mean. The sample proportion is the mean of a sample that contains only 0s and 1s. Assume we have a sample that consists of the following 10 observations: 0, 0, 1, 0, 1, 1, 1, 0, 1, 1. What is the arithmetic mean?

$$(\tfrac{1}{10})(0 + 0 + 1 + 0 + 1 + 1 + 1 + 0 + 1 + 1) = \tfrac{6}{10} = 0.6$$

What is the proportion of 1s (or successes) in this sample? $\tfrac{6}{10} = 0.6$! !

The sampling distribution of sample proportions should exhibit the characteristics of the sampling distribution of sample means. In addition, $\bar{p} = r/n$ is just another random variable to express the events described by the random variable "number of successes," or r. Therefore, the sampling distribution of $\bar{p}$ should have characteristics similar to the sampling distribution of r, "the number of successes."

When sampling from a binomial population with a procedure that satisfies a Bernoulli experiment, the exact sampling distribution of *proportion* of successes within samples of size n has the following properties.

Property 1. The $(n + 1)$ possible values of $\bar{p}$ are always given by

$$0/n, \; 1/n, \; 2/n, \; \ldots, \; (n - 1)/n, \; n/n$$

Property 2. The probability of each value of $\bar{p}$ occurring is given by

$$P(n\bar{p} = k \,|\, n \text{ and } \pi) = \frac{n!}{k!\,(n - k)!}\,\pi^k(1 - \pi)^{n - k}$$

Since $n\bar{p}$ is just the number of successes in a sample of n observations, these are binomial probabilities.

Property 3. The mean is always equal to the population proportion.

$$E(\text{proportion of successes}) = \mu_{\bar{p}} = \pi$$

Property 4. The variance is always given by

$$\text{Var}(\bar{p}) = \sigma_{\bar{p}}{}^2 = \frac{\pi(1 - \pi)}{n}$$

Thus, the standard deviation $\sigma_{\bar{p}}$, must be $\sqrt{[\pi(1 - \pi)]/n}$.

Property 5. The direction and extent of skewness is determined by (1) the size of the sample (n) and (2) how far the population proportion is from $\frac{1}{2}$ (that is, $|\pi - 0.5|$). This is identical to the sampling distribution for the number of successes.

Property 6. As the sample size becomes larger, the sampling distribution approaches the shape of a normal curve.

As in the case of the sampling distribution for the number of successes, for small samples the exact sampling distribution using binomial probabilities is preferred; however, when $n\pi(1 - \pi) \geq 3$ or for large samples, the *normal curve* serves as a very accurate approximation of the exact sampling distribution. Use of the normal-curve approximation also calls for continuity correction. Once again, we shall investigate the normal-curve approximation in the context of an example.

Example 5.7 The Proportion of Contact Lense Wearers in a Sample

Consider selecting, with replacement, a random sample of 16 persons from the population of example 5.5. (*a*) Describe the sampling distribution for the proportion of contact lense wearers per sample. (*b*) What is the probability that the sample we select will have half or more of the persons wearing contact lenses? (*c*) Using the normal-curve approximation, answer part *b*. (*d*) Compare the answers to parts *b* and *c*.

(*a*) From Appendix Table III, the binomial probability distribution for $\pi = 0.30$ and $n = 16$ is given in Table 5.23.

$$\mu_{\bar{p}} = \pi = 0.3 \qquad \sigma_{\bar{p}} = \sqrt{[\pi(1 - \pi)]/n} = \sqrt{(0.3)(0.7)/16}$$

$$= \sqrt{0.21}/4 = 0.4583/4 = 0.1146$$

TABLE 5.23 SAMPLING DISTRIBUTION FOR $\bar{p}$ WHEN $n = 16$, $\pi = 0.30$

$r = n\bar{p}$	$\bar{p}$	$P(\bar{p})$	$r = n\bar{p}$	$\bar{p}$	$P(\bar{p})$
0	0.0000	0.0033	9	0.5625	0.0185
1	0.0625	0.0228	10	0.6250	0.0056
2	0.1250	0.0732	11	0.6875	0.0013
3	0.1875	0.1465	12	0.7500	0.0002
4	0.2500	0.2040	13	0.8125	0.0000
5	0.3125	0.2099	14	0.8750	0.0000
6	0.3750	0.1649	15	0.9375	0.0000
7	0.4375	0.1010	16	1.0000	0.0000
8	0.5000	0.0487			
				Total	0.9999*

* Rounding error; for example, $P(\bar{p} = 0.8125) = 0.00002$.

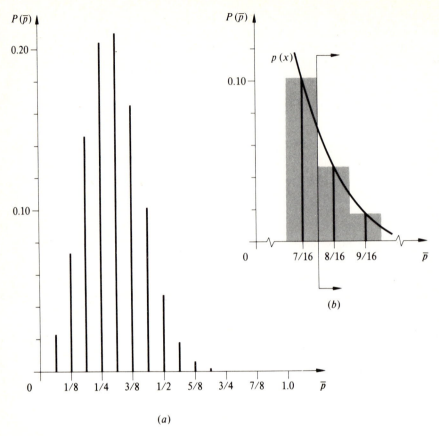

FIGURE 5.10   Sampling distribution of $\bar{p}$ for $n = 16$, $\pi = 0.30$.

Since $\pi = 0.3 < 0.5$, the distribution is skewed to the *right*. Figure 5.10a depicts this sampling distribution of $\bar{p}$.

(b) $P(\bar{p} \geq 0.50) = P(\bar{p} = \frac{8}{16}) + P(\bar{p} = \frac{9}{16}) + P(\bar{p} = \frac{10}{16}) + P(\bar{p} = \frac{11}{16})$

$$+ P(\bar{p} = \frac{12}{16}) + P(\bar{p} = \frac{13}{16}) + P(\bar{p} = \frac{14}{16})$$

$$+ P(\bar{p} = \frac{15}{16}) + P(\bar{p} = \frac{16}{16})$$

$$= (0.0487) + (0.0185) + (0.0056) + (0.0013) + (0.0002) + 4(0.0000)$$

$$= 0.0743$$

(c)   To use the normal-curve approximation, we assume the sampling distribution in Figure 5.10a (which, if we knew it, we would not usually approximate it) is a normal distribution. Since $n\pi(1 - \pi) = (16)(0.3)(0.7) = 3.36 > 3$, the normal distribution should be a good approximation according to our rule of thumb. Without knowing the exact sampling distribution for $\bar{p}$, we do know its mean and standard deviation. From part a, we calculated $\mu_{\bar{p}} = 0.30$ and $\sigma_{\bar{p}} = 0.1146$, and these represent the mean and standard deviation of our normal curve.

Now, consider the continuity correction. For the discrete random variable $\bar{p}$, we are substituting a continuous random variable x. What value of x should we use if we are interested in $P(\bar{p} \geq \frac{1}{2})$? Observe Figure 5.10b where only a small portion of the sampling distribution of $\bar{p}$ is shown. The shaded area of each rectangle represents the probability that $\bar{p}$ will assume the value given by the midpoint of the rectangle's base. Thus, to include *all* the area in the rectangle for $P(\bar{p} = \frac{8}{16} = \frac{1}{2})$, we must go halfway between $\bar{p} = \frac{7}{16}$ and $\bar{p} = \frac{8}{16}$. This continuity correction would mean that $P(x \geq \frac{15}{32})$ is equivalent to $P(\bar{p} \geq \frac{16}{32})$ for the purpose of approximation.

Next, we need to calculate the critical value of the standard normal deviate, z.

$$z = \frac{x - \mu}{\sigma} = \frac{\frac{15}{32} - 0.30}{0.1146} = \frac{0.46875 - 0.3}{0.1146} = 1.47$$

Thus, $P(\bar{p} \geq 0.5) \approx P(x \geq \frac{15}{32}) = P(z \geq +1.47)$ and from **Appendix Table IV(B)** we find that $P(z \geq +1.47) = 0.0708$.

(*d*)   The answer in part *b* was $P(\bar{p} \geq 0.50) = 0.0743$ while the answer from part *c* of 0.0708 represents only a 4.8 percent error. Whether that 4.8 percent error is important depends upon the nature of your problem. In business and economics we would frequently be summarizing the results in something like the following statement: "In only 7 out of 100 cases would we expect to see a random sample of 16 persons with half or more wearing contact lenses." Either part *b* or part *c* would be sufficient for that statement.

TABLE 5.24   SUGGESTED RULES FOR USING
THE CONTINUITY CORRECTION
WHEN THE SAMPLE STATISTIC IS
THE PROPORTION OF SUCCESSES

When the question states the discrete random variable as:	When the question states the discrete random variable as:
$P(\bar{p} \geq k)$, or $P(\bar{p} < k)$, or $P(\pi \leq \bar{p} < k)$ for $k > \pi$, or $P(k \leq \bar{p} \leq \pi)$ for $k < \pi$, then	$P(\bar{p} > k)$, or $P(\bar{p} \leq k)$, or $P(\pi \leq \bar{p} \leq k)$ for $k > \pi$, or $P(k < \bar{p} \leq \pi)$ for $k < \pi$, then
the continuous random variable x is given by	the continuous random variable x is given by
$x = k - 1/(2n)$.	$x = k + 1/(2n)$.

To summarize, if $n\pi(1 - \pi) < 3$, use the exact sampling distribution for $\bar{p}$ (i.e., use binomial probabilities). If $n\pi(1 - \pi) \geq 3$, and the larger $n\pi(1 - \pi)$ the better the approximation, use the normal-curve approximation for the sampling distribution of $\bar{p}$. When the normal-curve approximation is appropriate, continuity correction must be considered. The correct value of x will always be $\bar{p} \pm 1/(2n)$; the only problem is whether to use the plus or the minus sign. Sketching a diagram similar to Figure 5.10b is usually enough guidance; however, Table 5.24 is provided as an added convenience.

Of course, if the sample size is very large (say $n \geq 50$), then the continuity correction factor $1/(2n)$ is insignificant and can be omitted.

5.3 OTHER SAMPLING DISTRIBUTIONS

We have inspected only the normal probability density function and the binomial probability distributions in our quest for sampling distributions. These two probability distributions are the most frequently used (and, presumably, the most useful) distributions in business and economics applications. They serve well as our launching pad into inferential statistics; however, they are not an exhaustive list.

Other useful but, for the purposes of our discussion, avoided *discrete* distributions include the hypergeometric, the Poisson, and the multinomial distributions.

The Hypergeometric Distribution

When a random sample is selected from a small binomial population *without replacement*, the requirements of a Bernoulli experiment are not satisfied and binomial probabilities are not correct. If one success is selected and not replaced, then the probability of a success on the next trial is affected and is not equal to π. The hypergeometric distribution is the correct sampling distribution for random sampling from a binomial population without replacement.

The Poisson Distribution

This is an exponential, but discrete, distribution described in 1837 by Simeon Denis Poisson. Whereas the binomial distribution describes the number of successes in n trials, the Poisson distribution describes the number of successes per unit of time. As such, it has been very useful in operations research. For example, the number of customers arriving at a gasoline pump per hour, the number of telephone calls to the police per minute, and the number of orders for fish sandwiches per hour at a fast-service counter are all problems for which the Poisson distribution is appropriate. Sometimes called the "distribution for rare events," the Poisson is a good approximation for the binomial distribution when π [or $(1 - \pi)$] is less than 0.1.

The Multinomial Distribution

When each element within a population fits into only one of two possible classifications, we called it a binomial population. When the number of classifications is finite but greater than 2, we call it a multinomial population. "First-year student, sophomore, junior, senior" are the four typical classifications of college undergraduates. We could call it a quadnomial population, but we just call it a multinomial population. In sampling from such a population, the multinomial distribution gives us a way of calculating the probability of each different sample. Example 5.1 is an example of sampling from a multinomial population.

In the remaining chapters we will introduce three *continuous* distributions: the *t* distribution, the chi-square distribution, and the *F* distribution. We will develop these distributions as we need them.

Further, the gamma, the beta, the Cauchy, the exponential, the uniform, and the log-normal are all continuous probability distributions that have use in mathematical statistics. Even intermediate statistical methods find it difficult to avoid several of these continuous distributions, but our purpose is more timid.

SUMMARY

We started with general concepts of how a sample might be selected, looked briefly at the population–sampling procedure–sampling distribution process, considered all possible samples of a given size, and viewed in some detail the sampling distributions of the arithmetic mean, the number of successes in a sample, and the sample proportion. In the course of these events, we considered the powerful nature of the central limit theorem and the reoccurring nature of the normal curve. Many of these developments are condensed in Table 5.25.

In general, we can now take a known population and have a reasonably good picture of the probability distribution of all possible samples of size n taken from that population. That is deductive statistical reasoning. We now turn to inductive reasoning.

If we can identify a population from its sampling distribution, and if we can identify a sampling distribution from one sample, then on the basis of one sample (*not* a sample of size 1) we would identify an unknown population. The logic is impeccable Mr. Holmes; however, the premise requires clarification.

What we discovered in this chapter is the probability of the sample statistic (that is, $\bar{X}$, r, or $\bar{p}$) having a particular value when a sample of size n is randomly selected from a known population. Thus, if we randomly select a sample of size n from a population of unknown characteristics (that is, μ or $\pi = ?$), the resulting value of the sample statistic ($\bar{X}$ or $\bar{p}$) would permit only an assessment of the probability that the population parameter (μ or π) has a particular value. Because we know something about sampling distributions, we can, based upon the results of one sample, make a

TABLE 5.25 SUMMARY OF CHARACTERISTICS FOR THE SAMPLING DISTRIBUTIONS OF $\bar{X}$, r, AND $\bar{p}$

Population Parameter	Sample Statistic	Characteristics of Sampling Distribution			
		Mean	Variance	Standard Deviation	Shape
μ *Arithmetic Mean*	$\bar{X}$	$\mu_{\bar{X}} = \mu$	$\sigma_{\bar{X}}^{\,2} =$ $[\sigma^2/n][(N-n)/(N-1)]$	$\sigma_{\bar{X}} =$ $[\sigma/\sqrt{n}]\sqrt{(N-n)/(N-1)}$	If $n < 30$, the shape is determined by the shape of the population. If $n \geq 30$, the shape approaches a symmetric bell-shaped normal curve.
R *The Number of Successes*	r	$\mu_r = n\pi$	$\sigma_r^{\,2} =$ $n\pi(1-\pi)$	$\sigma_r =$ $\sqrt{n\pi(1-\pi)}$	The range is $0 \leq r \leq n$. If $n\pi(1-\pi) \leq 3$, then $\lvert \pi - 0.5 \rvert$ indicates the degree of skewness. If $\pi < 0.5$, the distribution is skewed to the *right*; if $\pi > 0.5$, it is skewed to the *left*. If $n\pi(1-\pi) > 3$, then it approaches a normal curve.
π *The Proportion of Successes*	$\bar{p}$	$\mu_{\bar{p}} = \pi$	$\sigma_{\bar{p}}^{\,2} =$ $\pi(1-\pi)/n$	$\sigma_{\bar{p}} =$ $\sqrt{\pi(1-\pi)/n}$	Same as for r.

"probability statement" about which population(s) it came from. We are not certain from which population the sample came, but we do know how probable each population is. The next several chapters develop techniques for applying this logic.

EXERCISES

1. For most people, the distance between the two joints on their little finger is approximately 1 inch. If you used your little finger as a measuring device to determine the width of a page in this book, are you more likely to introduce "sampling error" or "nonsampling error" when you state your answer in inches?

2. State in your own words the difference between precision and accuracy. Between sampling error and nonsampling error.

3. Suppose your population is the length of 20 pieces of carpet which are all 15 feet wide. Suppose that you carefully measure the length of each piece with a "yardstick" which is, in fact, only 35 inches long. (*Note:* This "yardstick" has 36 intervals of equal length marked on it.)
a. Have you taken a sample or a census?
b. What is the extent of the sampling error?
c. What is the nature of the nonsampling error?
d. Describe how you might sample four pieces of carpet, measure them with a 36-inch "yardstick," and adjust the original 20 measurements (which were to the nearest inch).

4. Consider a population with 10,000 elements in it.
a. If we have a simple random sample of 20 items from that population, what does "simple random sample" indicate?
b. What is the probability that that exact sample of 20 items will be selected?
c. What is the probability that any particular set of 20 items will be selected for a simple random sample? How does this answer differ from that of part b?
d. What does the phrase "all possible samples of size 20" have to do with a simple random sample of size 20?

5. Consider a three-card deck which consists of a 1, a 2, and a 3.

Experiment A (Samples of Size 2—with Replacement) Randomly select two cards, with replacement, from the deck in such a way that trials are independent.
a. List all possible outcomes. Define $\bar{X}$ as the random variable "sample mean" such that $\bar{X} = (\frac{1}{2})$(face value of first draw + face value of second draw).
b. Find the probability of occurrence for each outcome.
c. Construct the probability distribution for $\bar{X}$.
d. From your answer to part c, calculate $E(\bar{X})$ and $\text{Var}(\bar{X})$.
e. Let $X_i = \{1, 2, 3\}$ and find $E(X)$ and $\text{Var}(X)$. Does $\mu_X = \mu_{\bar{X}}$? Does $\sigma_{\bar{X}}^2 = \sigma_X^2/n$?

Experiment B (Samples of Size 2—without Replacement) Randomly select two cards without replacement.
f. List all possible outcomes and define $\bar{X}$ as in part a.
g. Find the probability of occurrence for each outcome, and construct the probability distribution for $\bar{X}$.
h. From the answer to part g, calculate $E(\bar{X})$ and $\text{Var}(\bar{X})$.
i. Does $\mu_{\bar{X}} = \mu_X$? Does $\sigma_{\bar{X}}^2 = \sigma_X^2/n$?

6. Consider a population which consists of 8 five-dollar bills and 2 ten-dollar bills. If we let $X_i = \$5$ bill for $i = 1, \ldots, 8$, and $X_9, X_{10} = \$10$ bill, then show that $E(X) = \$6$ and $Var(X) = \$^2 \ 4$.

Experiment A Place the 10 bills in a hat and randomly select 2 bills, with replacement. Let $\bar{X}$ be the average from each sample.
a. Find the probability distribution for $\bar{X}$.
b. Calculate $E(\bar{X})$ and $Var(\bar{X})$.
c. Is $E(\bar{X}) = \$6$ and $Var(\bar{X}) = \$^2 \ 2$?

Experiment B Repeat experiment A except the two bills are randomly selected without replacement.
d. Find the probability distribution for $\bar{X}$.
e. Calculate $E(\bar{X})$ and $Var(\bar{X})$.
f. How do the answers to part b relate to $E(X)$ and $Var(X)$?

7. A population of 1 million paper clips has a mean of 10 milligrams per paper clip and a standard deviation of 1 milligram per paper clip. The weights are normally distributed.
a. In what weight range about the mean would we expect to find 97.5 percent of the paper clips? That is, for $P(\mu - z\sigma \le \mu \le \mu + z\sigma) = 0.9750$ what is the value of z.
b. Would it be reasonable to say that 5 percent of the paper clips weigh more than 11.645 milligrams? Would it be reasonable to say that 5 percent weigh less than 8.04 milligrams?
c. If a random sample of four paper clips is chosen, then is it reasonable to say that 97.5 percent of the time the mean weight of those four clips will be between 8.88 and 1.12 milligrams?
d. Is it reasonable to assume that only 1.25 percent of the time a sample mean, from a random sample of size 4, would be larger than 11.2 milligrams?
e. If all possible samples of size 100 were selected from this population, then within what range of weights (that is, $\mu_{\bar{X}} \pm z_{\alpha/2}\sigma_{\bar{X}}$) would we find 97.5 percent of the sample means?

8. In Large Apple City (LAC), there are over 1 million employed heads of household. The distribution of their gross annual income is very skewed to the right (or positively skewed). The mean annual income is $16,000, and the standard deviation is $2,000.
a. Would it be reasonable to say that 5 percent of LAC's employed heads-of-household annual incomes are above $19,290?
b. If all possible samples of size 4 were selected, would we expect 97.5 percent of the mean annual incomes to be between $11,520 and $20,480?
c. For the sample means from all possible samples of size 100, would we expect 97.5 percent of the $\bar{X}$ values to lie between $15,552 and $16,448?

9. From past records, a dentist finds that time spent with patients is normally distributed with mean equal to 22 minutes and variance equal to 36 minutes squared. The dentist plans to see 16 patients today. Assume: (1) no patient shows up late; (2) time spent with one patient is independent of time spent with another patient; and (3) these 16 patients represent a random sample from the past experience (the causal system has not changed).
a. What is the probability that the dentist's average time spent will be 25 minutes per patient or more?
b. What is the probability that the first patient will take 20 minutes or less?
c. In order to be on time for a golf date, the dentist must average 20 minutes per patient or less. What is the probability that the dentist will be late to the tee?
d. The dentist starts at 8 A.M. If lunch takes 40 minutes, what time is the golf date?
e. How reasonable are assumptions 1, 2, and 3?

10. A certain grade light bulb has a normally distributed bulb life with mean of 257.1 hours and a standard deviation of 20 hours. A windowless hallway in an apartment house has an unusual light socket which is intended to continuously illuminate the hallway. Four bulbs are plugged into the socket, but only one bulb lights. When it burns out, the next one is automatically switched on. This process continues until all four bulbs are burned out. Every 6 weeks, precisely at noon, the manager comes and replaces all four bulbs. What is the probability that all four bulbs will burn out before the manager returns to replace them?

11. As a "filler" a local newspaper published the fact that of 1,620 persons listed under W in the telephone directory, 567 are musicians. Assume that five of the telephone listings under W are randomly selected.
a. What is the probability that exactly 80 percent will be musicians?
b. What is the probability that exactly 20 percent will be musicians?
c. What is the probability that more than half will be musicians?
d. What is the probability that none will be musicians?

12. A fair coin is flipped n times. What is the probability that the proportion of heads will be exactly $\frac{1}{2}$ for
a. $n = 10$?
b. $n = 100$?
c. $n = 10,000$?
d. Calculate $\mu_{\bar{p}}$ and $\sigma_{\bar{p}}$ for part a.

13. A fair coin is flipped n times. What is the probability that the proportion of heads will be between 0.48 and 0.52 for
a. $n = 10$?
b. $n = 100$?
c. $n = 10,000$?

14. For Exercises 12 and 13:
a. Why is it important to use the continuity correction, $k \pm 1/(2n)$, even for large samples?
b. The classical, or a priori, definition of probability is used to calculate the answers. How do the answers to parts a through c support the relative-frequency concept of probability?

15. A machine packages (i.e., wraps and seals) five sticks of chewing gum. If the label is not on straight, it is considered to be defective; the machine produces 10 percent defective packages. A very large batch of packages have just been produced.

Experiment A   A random sample of five packages is selected from the batch.
a. Find the probability distribution for $\bar{p}$, the proportion of defectives, and plot it.
b. From the probability distribution, find $E(\bar{p})$ and $\text{Var}(\bar{p})$. Do these results correspond to $E(\bar{p}) = \pi$ and $\text{Var}(\bar{p}) = (1/n)\pi(1 - \pi)$?
c. How is this sampling distribution for $\bar{p}$ skewed?
d. Find the probability distribution for r, the number of defectives, and plot it.
e. From part d, find $E(r)$ and $\text{Var}(r)$. Do these results correspond to $E(r) = n\pi$ and $\text{Var}(r) = n\pi(1 - \pi)$?

Experiment B   A random sample of 100 packages is selected from the batch.
f. Find $\mu_{\bar{p}}$ and $\sigma_{\bar{p}}$.
g. Use the normal-curve approximation and omit continuity correction to find:

$P(\bar{p} < 0.050)$ $P(0.100 \le \bar{p} \le 0.125)$

$P(0.050 \le \bar{p} \le 0.075)$ $P(0.125 \le \bar{p} \le 0.150)$

$P(0.075 \le \bar{p} \le 0.100)$ $P(\bar{p} > 0.150)$

h. Plot your results to part *g* on the same scale that was used for part *a* and compare the two. How are they similar? How are they different?

16. An aluminum siding company has two sales representatives, A and B. Although they do not know it, the proportion of houses who will buy aluminum siding is 0.1 in A's territory and 0.2 in B's territory. Each salesperson will call on three customers this morning. What is the probability that A will make more sales than B if both randomly choose houses to call upon and if all trials are independent?

$$P(r_A > r_B) = P(\{r_A = 1, r_B = 0\} \cup \{r_A = 2, r_B = 0\} \cup \{r_A = 2, r_B = 1\} \cup \cdots) = ?$$

17. A city has 10,000 high school pupils. To assist in the city's centennial celebration, the city council will randomly select 100 high school pupils to serve as "official hosts and hostesses."

The height of high school pupils is normally distributed with mean equal to 5 feet 9 inches (69 inches) and standard deviation equal to 3 inches. There are an equal number of male and female pupils.

What is the probability that the city council will select:
a. a sample whose average height is 6 feet or more?
b. pupils whose average height is less than 5 feet, 4 inches?
c. pupils whose average height is between 68 and 70 inches?
d. exactly 50 males and 50 females? (*Hint:* Use normal-curve approximation and continuity correction factor.)
e. 60 or more males?
f. Is this sample selected with or without replacement? Does it make any difference to the answers to parts *a* through *e*? Why or why not?

CASE QUESTIONS

1. In Chapter 3, the first case question asked you to consider the Illinois sample as a population and construct the theoretical probability distribution for all possible samples of size 4. Now we want to take several simple random samples of size 4 from that population, to see if the empirical results duplicate the theoretical model.

Step 1   Selection of simple random samples

There are 100 sequentially numbered households in the Illinois data set. (See V01.) We want each household to have an equal chance of being included in a sample of 4. We could place 100 chips, numbered 1 to 100, in an urn and select 4 chips with replacement. Or, we could use a table of random digits. If we let 00 = 100, we could randomly select *pairs* of digits to represent the household number (V01).

Consider Appendix Table II which includes 5,000 random digits. We should randomly choose a starting point, but, to make the process easier to follow, let us start in the upper left-hand corner of Appendix Table II. The random digits across the first row can be used as follows:

Random digits or household number	1 2 / 1 5 / 9 6 / 6 1 // 4 4 / 0 5 / 0 9 / 1 1 //							
V32 =	500	156	161	109	107	27	171	122
If V32 ≥ 100, enter S; if V32 < 100, enter F.	S	S	S	S	S	F	S	S
Number of successes		4				3		
Sample number		1				2		

Use this strategy to construct 40 samples of size 4. (You will need a string of 320 random digits.)

Step 2   Construct the empirical sampling distribution

If we let X = the number of successes, then the theoretical probability distribution is given by your answer to case question 1a in Chapter 3. From step 1, what percent of your 40 samples contained no successes? How does that compare with $P(X = 0 | \pi = 0.61, n = 4)$ from Chapter 3? Make the same comparison for $X = 1, 2, 3$, and 4.

2.   Consider the Illinois sample as a simple random sample, of size 100, from all households in Illinois.

a.   Assume that half the Illinois households had a family income of $10,000 or more in 1969. What is the probability that a sample of 100 such households would have 61 families with V32 $\geq$ 100? [*Hint*:   Use the normal curve approximation to find $P(60.5 \leq X \leq 61.5 | \pi = 0.5, n = 100)$.]

b.   Repeat part *a* for $\pi = 0.45$. That is, if only 45 percent of all households had income of $10,000 or more.

c.   Repeat part *a* for $\pi = 0.55$.

6

STATISTICAL INFERENCE: ESTIMATION

6.1 THE CONCEPT OF INTERVAL ESTIMATION

The first task, in our examination of estimation techniques, is to use the material from Chapter 5 to develop the concept of "interval estimation." We can then compare and contrast point and interval estimates, design criteria for determining if one estimator is better (and in what sense "better") than another, develop specific estimation techniques, and take another brief look at the data-acquisition process.

Assume that we have a population which is normally distributed with mean equal to μ and standard deviation equal to σ. The sample means, calculated from all possible random samples of size n, will form a probability distribution which is normally distributed with mean equal to μ and standard deviation equal to $\sigma/\sqrt{n}$. As depicted in Figure 6.1b, the $\bar{X}$ axis represents all possible sample means that could come from the population.

Between what two values of $\bar{X}$ do 95 percent of the $\bar{X}$ values *closest* to μ lie? That is, we are primarily interested in values of the sample mean which are close to the population mean. Further, we want some range of values for the sample mean, say $\bar{X}_L$ to $\bar{X}_U$, such that $P(\bar{X}_L \leq \bar{X} \leq \bar{X}_U) = 0.95$.

For reasons which will become obvious later, we want this range of sample means to be as narrow as possible. Since the normal distribution is a symmetric bell-shaped curve, $\bar{X}_L$ and $\bar{X}_U$ must both be the *same distance* from μ if we want their difference to be a minimum (that is, $\bar{X}_U - \bar{X}_L = $ minimum).[1] Hence, $\bar{X}_L = \mu - k$ and $\bar{X}_U = \mu + k$, and the question is to find the appropriate value for k.

Consider the upper limit, $\bar{X}_U$. We know that

$$z_U = \frac{\bar{X}_U - \mu}{\sigma/\sqrt{n}}$$

---

[1] This is a minor point, but it is easily verified. In Figure 6.1b, if we move both the upper limit ($\bar{X}_U$) and the lower limit ($\bar{X}_L$) the same distance (d) to the right, we would give up more area on the left side of the distribution than we gain on the right side. Maintaining the upper limit at $X_U + d$ and preserving 95 percent of the area between the two limits would require that the lower limit be $\bar{X}_L + d'$ where $d' < d$. Doing so, however, would increase the distance between the two limits.

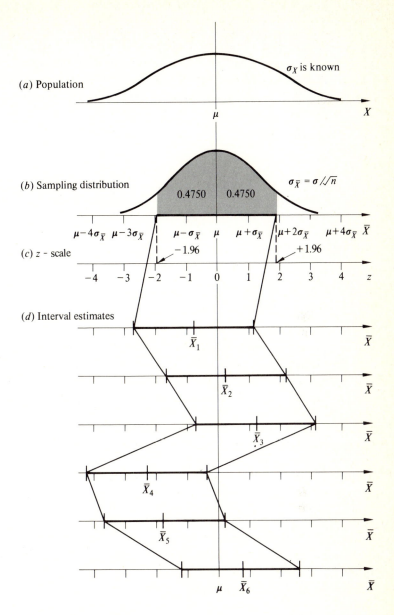

FIGURE 6.1 The concept of "interval estimation" for a population mean.

or $\bar{X}_U = \mu + z_U \sigma/\sqrt{n}$, and we know every value in that expression except z_U. We can determine the value for z_U !

If we want the area under the curve to be 0.95, then the area under each half must be 0.4750, as shown in Figure 6.1b. Since $P(0 \le z \le z_U) = 0.4750$, from Appendix Table IV(B) we find that $P(0 \le z \le +1.96) = 0.4750$. Hence, $z_U = +1.96$ and

$$\bar{X}_U = \mu + \frac{(1.96)\sigma}{\sqrt{n}} \qquad (6.1)$$

The general form of the upper limit was defined as $\bar{X}_U = \mu + k$, and now we have found that $k = (1.96)\sigma/\sqrt{n}$. Since $\bar{X}_L = \mu - k$,

$$\bar{X}_L = \mu - \frac{(1.96)\sigma}{\sqrt{n}} \qquad (6.2)$$

Finally, the probability of a random sample yielding a sample mean in the range $\mu \pm (1.96)\sigma/\sqrt{n}$ is 0.95, or

$$P(\mu - 1.96\sigma/\sqrt{n} \le \bar{X} \le \mu + 1.96\sigma/\sqrt{n}) = 0.95 \qquad (6.3)$$

This range of values for the sample mean is important; let us call it the "interval of means" (IOM). The IOM can be expressed as a straight line between the two endpoints $\mu - 1.96\sigma/\sqrt{n}$ and $\mu + 1.96\sigma/\sqrt{n}$, as shown in Figure 6.1$b$.

We have described both the length and the location of the IOM line by its endpoints. The IOM could be described by specifying (1) its length and (2) the location of its midpoint. The *length* of the IOM line is given by

$$\varepsilon = \bar{X}_U - \bar{X}_L = (\mu + 1.96\sigma/\sqrt{n}) - (\mu - 1.96\sigma/\sqrt{n})$$
$$= (2)(1.96)\sigma/\sqrt{n} \qquad (6.4)$$

The *midpoint* of the IOM is given by μ (along the $\bar{X}$ axis). A line of length ε and midpoint μ is exactly the same line as one with endpoints of $\mu \pm 1.96\sigma/\sqrt{n}$. Carefully note, however, that it is only the *location* or *midpoint*, and *not* the length, of the IOM line that is dependent upon μ.

We will use this information to pursue a sort of logical "gamesmanship" which at the beginning of Chapter 5 we called statistical inference. The length ε of the IOM does not depend upon the population mean; therefore, let us assume that the value of the population mean is *unknown*. Since we now know only the population standard deviation σ and the sample size n, we know the length of the IOM, but we do not know where to place it along the $\bar{X}$ axis; we do not know the midpoint of the IOM.

The interval of means, or IOM, was derived under conditions of certainty; we knew μ, the population mean, and σ. Since we are uncertain about the midpoint or location of the IOM, we will rename this "line" or "range of values" the *confidence interval*. In this case the confidence interval (CI) is a straight line of known length (ε) but uncertain location; the CI is a random variable.

From the population, let us take one random sample of size n and calculate the sample mean $\bar{X}_1$. What would happen if we used $\bar{X}_1$ as the center of our line representing the confidence interval? In Figure 6.1d, $\bar{X}_1$ is assumed to have a value slightly less than μ, and the "line" representing the confidence interval is shifted to the left along the $\bar{X}$ axis. Note that the population mean (μ) falls within this confidence interval; μ is within the range given by $\bar{X}_1 \pm (1.96)\sigma/\sqrt{n}$.

In fact, as we can see from Figure 6.1d the only time that $\bar{X} \pm (1.96)\sigma/\sqrt{n}$ would not include the population mean is when the sample mean is either greater than $(\mu + 1.96\sigma/\sqrt{n})$ or less than $(\mu - 1.96\sigma/\sqrt{n})$. What is the probability that either of those two situations would occur? Well, if $P(\mu - 1.96\sigma/\sqrt{n} \leq \bar{X} \leq \mu + 1.96\sigma/\sqrt{n}) = 0.95$, then $P[(\bar{X} > \mu + 1.96\sigma/\sqrt{n})$ or $(\bar{X} < \mu - 1.96\sigma/\sqrt{n})]$ must be 0.05! That is, if we use $(2)(1.96)\sigma/\sqrt{n}$ to determine the length of our CI and a sample mean to locate the CI along the $\bar{X}$ axis, we will exclude the population mean only 5 percent of the time.

This procedure is the basis for constructing a *confidence-interval estimate of the population mean*. We know that 95 percent of the time a random sample will produce a sample mean whose value lies between $(\mu - 1.96\sigma/\sqrt{n})$ and $(\mu + 1.96\sigma/\sqrt{n})$. If we center the confidence interval on a sample mean, we will include the population mean within the interval $(\bar{X} \pm 1.96\sigma/\sqrt{n})$ 95 percent of the time. That is,

$$P(\bar{X} - 1.96\sigma/\sqrt{n} \leq \mu \leq \bar{X} + 1.96\sigma/\sqrt{n}) = 0.95$$

gives us a probability statement of where the population mean is located. We have estimated the interval within which the value of the population mean lies. It is not a certainty that the population mean lies within the interval (see $\bar{X}_4$ in Figure 6.1d) because we may have randomly selected a rather rare sample mean. However, we can specify the probability (that is, 0.95 in this case) that *this procedure* will establish an interval (that is, $\bar{X} \pm 1.96\sigma/\sqrt{n}$) that includes the value of the population mean.

6.2   ESTIMATES, ESTIMATORS, AND ESTIMATION

Statistical *estimation* is the process by which we attempt to determine the value of a population parameter, without taking a census, from sample information. An *estimate* is the numerical value we think the parameter has, and an *estimator* is the sample statistic used to make an estimate.

For example, the sample mean was used in the previous section as an *estimator* of the population mean. Since specific values for n, σ, and $\bar{X}$ were not used in the previous section, no specific *estimates* of μ were forthcoming.

A *point estimate* is a *single-value* estimate of the parameter of interest. The number of units of product that will be sold during the next year is frequently given as a point estimate. For a variety of planning purposes, a retail paint store may estimate that it will sell 800 gallons of house paint, 1,200 gallons of interior paint, 500 gallons of floor paint, 200 three-inch brushes, and so forth.

An *interval estimate* is the *range of values* within which the value of the parameter is expected to lie. The same retail paint outlet could estimate that it will sell 750 to 850 gallons of house paint, 1,100 to 1,300 gallons of interior paint, and so forth.

Whether a point estimate or an interval estimate is most appropriate depends upon the purpose for which the estimate is to be used. An opinion poll indicating that between 40 and 60 percent of the population will vote for Jones for President is not very useful.

If a purchasing agent is told that the company will need from 6,000 to 7,000 pencils during the next quarter, then (depending upon past data on the supplier's promptness, the inventory of pencils, etc.) she or he can judge how many to order. The purchasing agent may, for example, order 6,300 pencils and discuss with the supplier the possibility of obtaining up to 700 additional pencils on short notice. The interval estimate provides the purchasing agent with an external basis for her or his own contingency planning.

On the other hand, if the purchasing agent is, in fact, a nonjudgmental functionary who just manages the paper work and checks the purchases as they arrive, then a point estimate is more appropriate. Purchase-order forms are seldom amenable to the entry "6,000 to 7,000 pencils." However, if the purchasing agent does exercise some judgment and is provided with a point estimate (say 6,500 pencils), do not be too surprised if she or he superimposes her or his own interval on the estimate. That is, without the external information, the purchasing agent may use a "plus or minus 10 percent" rule of thumb for an interval in the contingency planning.

In investigating criteria for a "good" estimator, we will discuss point estimators. In order to generate point estimators, we would need to understand the "method of maximum likelihood." Unfortunately, this "method" requires the use of at least differential calculus; thus, we will leave that topic for more advanced texts.[2] Most of the remaining topics will employ techniques for interval estimates.

We have briefly considered the sample mean $(\bar{X})$ as a point estimate of the population mean (μ); however, the sample median could also have been used as an *estimator* of the population mean. Which estimator—the sample mean or the sample median—is a better (or best?) estimator of the population mean?

---

[2] The basic concept of maximum likelihood can be found in Ralph E. Beals, *Statistics for Economists: An Introduction*, Rand McNally & Company, Chicago, 1972, pp. 164–169; Thomas H. and Ronald J. Wonnacott, *Introductory Statistics*, 2d ed., John Wiley & Sons, Inc., New York, 1972, chap. 18.

6.3   GUIDELINES FOR CHOOSING AN ESTIMATOR

Consider some population parameter θ, and a set of statistics, $\hat{\theta}_1, \hat{\theta}_2, \ldots$, which can be considered as estimators of θ. (θ is the Greek letter "theta," and $\hat{\theta}$ is read as "theta hat.") We permit θ to be any particular measure of a population because the criteria we discuss should apply to any estimator. For example, θ could be the population mean and $\hat{\theta}_1$ represent the sample mean, $\hat{\theta}_2$ represent the sample median, and $\hat{\theta}_3$ represent the sample mode. Or, we could let θ be the population variance (σ^2), and then consider $\hat{\theta}_1$ to be the sample variance (s^2), $\hat{\theta}_2$ to be the sample range, and so forth.

If we permit $\hat{\theta}_1^*$ to represent one *estimate* from the *estimator* $\hat{\theta}_1$, then the sampling error associated with that estimate is

$$\zeta = \text{error} = \hat{\theta}_1^* - \theta$$

Presumably, any "good" estimator would make that error as small as possible; $\zeta = 0$ is the ideal.

For any given population, the parameter θ is a constant; however, $\hat{\theta}_1^*$ is only one of many possible values that the estimator $\hat{\theta}_1$ could generate. The sampling distribution would tell us what values of $\hat{\theta}_1$ could occur and the probability associated with various ranges of values for $\hat{\theta}_1$. Intuitively, if we want to investigate the distribution of errors (ζ), we should look at the sampling distribution of the estimator.

What properties of the sampling distribution of an estimator $\hat{\theta}$ are desirable for estimating the population parameter θ? For one thing, the estimator's range of possible values should at least include the population parameter. For example, consider a deck of 10 playing cards consisting of five 2s and five 3s from which we will take a random sample of size 2. If we let $\theta = \mu$ (which is 2.5) and $\hat{\theta}_1 = X_1 \cdot X_2$ (the product of the "face value" on each of the two cards), then the smallest possible value for $\hat{\theta}_1$ is $(2)(2) = 4$. As we have defined $\hat{\theta}_1$, it would be a poor estimator for the population mean.

This suggests that a "good" estimator should have a sampling distribution which would have the value of the population parameter somewhat "centrally located" within the sampling distribution. One criterion for a good estimator will account for this property.

Another desirable property of the sampling distribution of the estimator, if we want the error ζ to be as small as possible, is how tightly packed the sampling distribution is in the neighborhood of the parameter θ. The more tightly packed the sampling distribution in the neighborhood of θ, the higher the probability that the estimate would be only a small distance from the parameter and the higher the probability of having a small error. It is in this context that we formally introduce three criteria for a "good" estimator.

Unbiasedness

An estimator $\hat{\theta}$ is said to be *unbiased* if the expected value of the estimator is equal to the population parameter being estimated, or

$$E(\hat{\theta}) = \theta$$

Cleverly, if $E(\hat{\theta}) \neq \theta$, then the estimator $\hat{\theta}$ is said to be *biased*.

Consider all possible samples of size n from a population which has parameter θ. The estimator $\hat{\theta}$ assumes one and only one value for each sample. Take the set of $\hat{\theta}$ values associated with all possible samples of size n and calculate their arithmetic mean $E(\hat{\theta})$. The unbiasedness criterion simply asks that the mean of the sampling distribution, $E(\hat{\theta})$, be equal to that which you are trying to estimate, or θ.

For example, in Chapter 5 we found[3] that $E(\bar{X}) = \mu$; thus, the sample mean is an unbiased estimator of the population mean. If we place the n observations from a randomly selected sample in an array and let m equal the value of the "middle" observation, then m represents the sample median. Although we will not prove it, the expected value of the sample median is the population mean. That is, $E(m) = \mu$; therefore, the sample median is also an unbiased estimator of the population mean.

In Equation 4.11, we defined the sample variance as

$$s_X{}^2 = \frac{\Sigma(X_i - \bar{X})^2}{n-1}$$

and promised an explanation for why we used $(n-1)$ instead of n in the denominator. The time has come! Let us define $*s^2$ to be $(1/n)\Sigma(X_i - \bar{X})^2$. If we wish to use $*s^2$ as an estimator of $\sigma_X{}^2$, the first thing we should note is that it is a biased estimate. That is, we offer without proof that

$$E(*s^2) = \left(1 - \frac{1}{n}\right)\sigma^2$$

Note that as the sample size increases, $*s^2$ approaches the condition of an unbiased estimator of σ^2.

---

[3] The central limit theorem stated that $\mu_{\bar{X}} = E(\bar{X}) = \mu$, which was demonstrated in example 5.1 with the 10-card deck. To prove that $E(\bar{X}) = \mu$, we consider n independent random variables, $x_1, x_2, x_3, \ldots, x_n$, which all have a mean equal to μ. Take a random sample of size 1 from each of the n random variables $X_1, X_2, X_3, \ldots, X_n$. Then

$$E(\bar{X}) = E[(1/n)\Sigma X] = (1/n)[E(X_1 + X_2 + X_3 + \cdots + X_n)]$$
$$= (1/n)[E(X_1) + E(X_2) + E(X_3) + \cdots + E(X_n)]$$

Since $E(X_i) = \mu$ for each i, then $E(\bar{X}) = (1/n)(n\mu) = \mu$.

If we calculate the sample variance according to Equation 4.11, then $E(s^2) = \sigma^2$ and s^2 is an unbiased estimator of σ^2 (regardless of sample size).[4]

One final note on semantics. "Bias" as it is used here means that $E(\hat{\theta}) - \theta \neq 0$, and the "bias" is the amount by which $E(\hat{\theta})$ and θ differ. This "bias" is a mathematical result of statistical theory. The "bias" discussed in Chapter 5 was concerned with the nonsampling (and nonstatistical theory) error in measuring devices (e.g., incorrectly calibrated weighing scales). The same word, unfortunately, is used for two very different concepts.

Unbiasedness is just one criterion for a "good" estimator. Unbiasedness did clarify why we defined the sample variance as we did in Equation 4.11. But, it did not tell us why $\bar{X}$ is a better estimator of μ than m. Both the sample mean and the sample median are unbiased estimators of the population mean.

Efficiency

The second criterion for a "good" estimator deals with how close we can expect $\hat{\theta}$ to be to θ. Remember that $\hat{\theta}$ is an estimator that produces a point estimate of θ. "Efficiency," it is frequently said, "is a relative term." It is no different in this rather technical definition of efficient estimators. In a sense all estimators are efficient; however, some estimators are more efficient than others. Consider two estimators $\hat{\theta}_1$

---

[4] Proof

$$E(s^2) = E\left[\frac{\Sigma(X - \bar{X})^2}{n - 1}\right] = \frac{1}{n - 1} E[\Sigma(X - \bar{X})^2]$$

Note that $X - \bar{X} = X - \bar{X} + (\mu - \mu) = [(X - \mu) - (\bar{X} - \mu)]$ and

$$\Sigma[(X - \mu) - (\bar{X} - \mu)]^2 = \Sigma[(X - \mu)^2 - 2(X - \mu)(\bar{X} - \mu) + (\bar{X} - \mu)^2]$$

where $(\bar{X} - \mu)$ is a constant. Then

$$\begin{aligned}\Sigma(X - \bar{X})^2 &= \Sigma(X - \mu)^2 - 2(\bar{X} - \mu)\Sigma(X - \mu) + n(\bar{X} - \mu)^2 \\ &= \Sigma(X - \mu)^2 - 2(\bar{X} - \mu)(n\bar{X} - n\mu) + n(\bar{X} - \mu)^2 \quad \leftarrow \text{Use } \bar{X} = \Sigma X/n \\ &= \Sigma(X - \mu)^2 - n(\bar{X} - \mu)^2\end{aligned}$$

and

$$\frac{1}{n - 1} E[\Sigma(X - \bar{X})^2] = \frac{1}{n - 1} E[\Sigma(X - \mu)^2 - n(\bar{X} - \mu)^2]$$

$$= \frac{1}{n - 1} \{\Sigma[E(X - \mu)^2] - nE(\bar{X} - \mu)^2\}$$

Since $\text{Var}(X)$, or σ^2, is defined as $E(X - \mu)^2$ and since, from the CLT, $\text{Var}(\bar{X}) = E(\bar{X} - \mu_X)^2 = E(\bar{X} - \mu)^2 = \sigma^2/n$, then

$$E(s^2) = \frac{1}{n - 1}\left[n\sigma^2 - n\left(\frac{\sigma^2}{n}\right)\right] = \frac{1}{n - 1}[\sigma^2(n - 1)] = \sigma^2$$

and $\hat{\theta}_2$ and their respective sampling distributions for the *same* size sample. Further, assume both estimators are unbiased: $E(\hat{\theta}_1) = E(\hat{\theta}_2) = \theta$. If $\text{Var}(\hat{\theta}_2) > \text{Var}(\hat{\theta}_1)$, then the sampling distribution of $\hat{\theta}_2$ is more spread out than the sampling distribution of $\hat{\theta}_1$. Thus, $P(\theta - k \le \hat{\theta}_2 \le \theta + k) < P(\theta - k \le \hat{\theta}_1 \le \theta + k)$—the probability that $\hat{\theta}_1$ will be close to θ is greater than the probability that $\hat{\theta}_2$ will be equally close to θ.

In a more formal manner, one estimator $\hat{\theta}_1$ is said to be *more efficient* than another $\hat{\theta}_2$ if the variance of the first estimator's sampling distribution is less than the variance of the second estimator's sampling distribution. That is, if

$$\text{Var}(\hat{\theta}_1) < \text{Var}(\hat{\theta}_2)$$

for the same size sample in both cases, then $\hat{\theta}_1$ is said to be more efficient than $\hat{\theta}_2$.

The central limit theorem told us that $\sigma_{\bar{x}}^2 = \sigma^2/n$. It can also be shown, when sampling from a symmetric population, that the variance of the distribution of all possible sample medians is $\sigma_m^2 = (1.6)\sigma^2/n$. Thus, $\sigma_{\bar{x}}^2 < \sigma_m^2$ and $\bar{X}$ is said to be a *more efficient* estimator of μ than is the sample median.

Consistency

This closeness of the estimate to the parameter has another dimension. Taking a sample exhausts resources. In general, the larger the sample, the greater the costs. It would be desirable, therefore, for an increase in sample size to cause an increase in the precision of the estimate. Let us define some arbitrary, small number d. We would want $P(|\hat{\theta} - \theta| \le d)$ to approach "certainty" as the sample size increases.

In a formal manner, $\hat{\theta}$ is said to be a *consistent estimator* of θ if $P(|\hat{\theta} - \theta| \le d) \to 1$ as $n \to \infty$. But what does that mean? We measure probabilities as the area under the sampling distribution (or density function) of $\hat{\theta}$. If we want $P(|\hat{\theta} - \theta| \le d)$ to increase as the sample size increases, then we are saying that the variance of the sampling distribution must decrease as the sample size increases. That is $\text{Var}(\hat{\theta}) \to 0$ as $n \to \infty$ (or N for a finite population).[5]

Consider the sample mean as an estimator of the population mean. We know

---

[5] This is always true for *unbiased* estimators. For biased estimators, $\text{Var}(\hat{\theta})$ may approach 0 as the sample size goes to infinity, and the estimator still would not be consistent.

Let $\Psi = \bar{X} + c$, where c is a constant. Then $E(\Psi) = \mu + c$ (Ψ is a biased estimator of μ, and c is the "extent of the bias") and $\text{Var}(\Psi) = \text{Var}(\bar{X}) = \sigma^2/n$. Note that the $\text{Var}(\Psi) \to 0$ as $n \to \infty$, but the estimator Ψ would not necessarily be consistent. Why? In our definition of consistency, we chose "some arbitrary, small number d." If the extent of the bias, c, is greater than d, then $P(|\Psi - \mu| \le d) \to 0$ [*not* 1] as $n \to \infty$. As $n \to \infty$, $\text{Var}(\Psi) \to 0$ and $\Psi \to \mu + c$. If $\Psi - \mu = c$ and if $c > d$, then $P(c \le d) \to 0$ as $n \to \infty$.

If there is very little bias, $c < d$, then Ψ is consistent because $P(|\Psi - \mu| \le d) \to 1$ as $n \to \infty$. If the bias is not a constant (like c), then $\text{Var}(\hat{\theta} - \theta)$ might be equal to $\text{Var}(\text{bias}) + \text{Var}(\hat{\theta})$, and *both* $\text{Var}(\text{bias})$ and $\text{Var}(\hat{\theta})$ must approach 0 as n increases in order for $\hat{\theta}$ to be a consistent estimator.

TABLE 6.1 PROPERTIES OF m (THE MEDIAN) AND $\bar{X}$ (THE MEAN) FOR ESTIMATING μ

Population Parameter (To Be Estimated) θ	Sample Statistic (Estimator) $\hat{\theta}_i$	Unbiasedness	Consistency	Efficiency
μ	$\bar{X}$	Yes $E(\bar{X}) = \mu$	Yes $\text{Var}(\bar{X})\downarrow$ as $n\uparrow$	More so than m $\text{Var}(\bar{X}) < \text{Var}(m)$
	m	Yes $E(m) = \mu$	Yes $\text{Var}(m)\downarrow$ as $n\uparrow$	Less so than $\bar{X}$

that $\mu_{\bar{X}} = \mu$ and $\sigma_{\bar{X}}^2 = \sigma^2/n$. As n increases, the sampling distribution becomes more and more tightly packed about μ. The larger the sample size, the smaller $\sigma_{\bar{X}}^2$ becomes, and $P(|\bar{X} - \mu| \leq d) \to 1$. Thus, $\bar{X}$ is said to be a consistent estimator of μ.

Let us summarize our findings about estimators of the population mean in Table 6.1. In this case, it is easy to choose between the two estimators. The sample mean and the sample median are both unbiased and consistent estimators of the population mean, but the sample mean is a more efficient estimator than the sample median. Although we will not prove it, under certain conditions the sample mean has maximum efficiency in the estimation of the population mean. That is, some other (unspecified) estimator may be *as* efficient as $\bar{X}$, but no estimator will be *more* efficient.

The choice between estimators may not always be so easy. Let us assume that $\hat{\theta}_1$ is a biased, consistent estimator of θ and that $\hat{\theta}_2$ is an unbiased, nonconsistent estimator of θ. Further, let us assume that for $n < 100$, $\text{Var}(\hat{\theta}_2) < \text{Var}(\hat{\theta}_1)$ and for $n \geq 100$, $\text{Var}(\hat{\theta}_2) > \text{Var}(\hat{\theta}_1)$. For $n < 100$, $\hat{\theta}_2$ is more efficient, and for $n \geq 100$, $\hat{\theta}_1$ is more efficient than $\hat{\theta}_2$. Which estimator do we use? If we can estimate the extent of the bias and if we can take samples of $n > 100$, then we might choose $\hat{\theta}_1$; however, those are two important "ifs." There is no easy answer to this "choice-between-alternatives" question.

In introductory economics, there is a principle of "diminishing marginal returns" encountered in the theory of production. This principle has an analogous application in the choice of sample size for a consistent estimator. If $\hat{\theta}$ is a consistent estimator of θ, then the variance of the sampling distribution decreases as the sample size increases. An increase in the size of sample selected increases the "precision" of the estimate, but it also increases the cost of sampling. Is the extra precision worth the extra cost? Consistency is a desirable property for a "good" estimator, but the extent to which we take advantage of a consistent estimator is an economic problem.[6]

Now that we have established that the sample mean is an unbiased, consistent, and efficient estimator of the population mean, let us return to the problem of establishing techniques for an interval estimate of μ.

6.4 CONFIDENCE-INTERVAL ESTIMATES OF THE POPULATION MEAN

Our basic problem is to estimate μ with the information provided by a single sample randomly selected from the population. Section 6.1 introduced the concept of confidence-interval estimates, and the *general form* of an interval estimate of the population mean will always take the form

$$P(\bar{X} - k \leq \mu \leq \bar{X} + k) = 1 - \alpha \qquad \textbf{(6.5)}$$

where $0 < \alpha < 1$. That is, the probability that the population mean will lie in the interval $\bar{X} \pm k$ is $(1 - \alpha)$.

The *interval* given by $\bar{X} \pm k$ is a *random variable*. The probability that μ will *not* be within the interval $\bar{X} \pm k$ is given by α; specifically, if $P(\mu > \bar{X} + k) = P(\mu < \bar{X} - k) = \alpha/2$, then

$$P[(\mu > \bar{X} + k) \text{ or } (\mu < \bar{X} - k)] = \alpha/2 + \alpha/2 = \alpha$$

We could consider α to be the "risk" that the interval $\bar{X} \pm k$ will not include the population mean μ.

In the spirit of positive thinking, the probability that μ *will* lie within the interval $\bar{X} \pm k$ is called the *level of confidence*. Hence, $\bar{X} \pm k$ is called the *confidence-interval* estimate of μ, and $(1 - \alpha)$ specifies the level of confidence.

It would seem reasonable to want a very high level of confidence and a very small interval. Roughly speaking, "we could be very sure, but not certain, that the population mean lies within a very narrow range of values." A "very narrow range of

---

[6] The analogy is formally used in statistical theory in the concept of a *loss function* (L). There is some cost to the investigator if the estimate is in error (i.e., if $\hat{\theta} - \theta \neq 0$); there is some cost associated with selecting the sample and making the estimate (call it C_θ). This could be generalized as $L = l(|\hat{\theta} - \theta|, C_\theta)$. A common specific form for the loss function is given by $L = a(\hat{\theta} - \theta)^2$. This is called a *quadratic loss function*, where a is a constant. As a criterion for deciding which estimator is "best," we might look for the $\hat{\theta}$ which gives the minimum expected loss. If $E(L)$ for $\hat{\theta}_1$ is less than $E(L)$ for $\hat{\theta}_2$, then $\hat{\theta}_1$ is preferred as an estimate of θ. For a more extensive discussion, see Edward J. Kane, *Economic Statistics and Econometrics*, Harper & Row, Publishers, Incorporated, 1968, pp. 182–183.

values" means that k must be small, and we will discuss later what influences the magnitude of k.

Since the level of confidence is a probability, then $0 < (1 - \alpha) < 1$, and to "be very sure" means that $(1 - \alpha)$ should be close to 1. Thus, α (or the "risk" that the interval will not include μ) should be near 0. So how do we determine the value for $(1 - \alpha)$?

In a very real sense, you *choose* what level of confidence $(1 - \alpha)$ that you can accept for any given problem. For practical and pedagogical reasons, the exercises at the end of the chapter and the examples that follow have already specified the level of confidence, but that does not overshadow the fact that the level of confidence is determined by human judgment.

If you are trying to estimate the average amount of toxic material in a prepared medicine for human consumption, you may want the level of confidence to be 0.9990. Economists and business analysts have a fondness for using 0.90, 0.95, or 0.99 for the level of confidence.

In most situations where a numerical value is arbitrarily chosen, certain "standards" or "accepted practices" evolve. Such is the case in applied statistics and economics. As we shall soon observe, the level of confidence $(1 - \alpha)$, the size of the sample (n), and the precision of the estimate (basically, the magnitude of k) are not independent of one another. Thus, the choice of $(1 - \alpha) = 0.999$ may force the selection of a very large sample. The investigator may be willing to give up a little "confidence" in order to select a smaller sample.

Returning to the "precision" of the estimate, we note that the length of the confidence interval is determined by k. In Equation 6.5, we chose the level of confidence $(1 - \alpha)$, $\bar{X}$ is the mean of a randomly selected sample, and finding the value for k is the only quantity that we need in order to calculate the interval estimate of the population mean (μ). In both theory and practice, we always use as much of the information about the *population* as we have available. In general, the more information concerning the population, the smaller the value of k.

Excluding the meaningless case where we know the population mean, we will start with the situation where we have the most population information and progress to the case where we have the least. In all cases, we will attempt to specify the value of k in Equation 6.5.

6.4(a) Population Is Normally Distributed; Population Standard Deviation Is Known

For a given sample size and level of confidence (those are the two variables whose values are chosen by the investigator), we cannot get a smaller value for k than when we know (1) that the population's shape is a normal curve and (2) the population standard deviation (σ). This was the situation presented in section 6.1, and the process of obtaining $k = z_{\alpha/2}\,\sigma/\sqrt{n}$ is summarized in Panel 6.1.

Panel 6.1   Interval Estimate for μ When a Normally Distributed Population Has a Known Standard Deviation

Population
The population is normally distributed. Standard deviation σ is known. Mean μ is unknown.

Sampling Distribution of $\bar{X}$
The shape is that of a normal curve. The standard deviation is $\sigma_{\bar{X}} = \sigma/\sqrt{n}$ and (since we know σ) is known. The value of the mean is unknown, $\mu_{\bar{X}} = \mu = ?$

Probability of Getting Sample Mean
If we knew μ, then $P(\mu - z_{\alpha/2}\sigma/\sqrt{n} \leq \bar{X} \leq \mu + z_{\alpha/2}\sigma/\sqrt{n})$ is equal to $(1 - \alpha)$. The length of the "interval" is known because (1) $\sigma/\sqrt{n}$ is known and (2) $z_{\alpha/2}$ is determined by the chosen value for α. However, the midpoint, or location, of the "interval" is uncertain ($\mu = ?$).

Confidence Interval Estimate of μ
The probability is $(1 - \alpha)$ that we will get a sample mean within the interval $\mu \pm z_{\alpha/2}\sigma/\sqrt{n}$. That interval has a length of $2z_{\alpha/2}\sigma/\sqrt{n}$. If we take that interval and use $\bar{X}$, the sample mean, as its midpoint, then the probability is $(1 - \alpha)$ that μ will be included within the limits $\bar{X} \pm z_{\alpha/2}\sigma/\sqrt{n}$. Thus, for a $(1 - \alpha)$ level of confidence,

$$\bar{X} - z_{\alpha/2}\sigma/\sqrt{n} \leq \mu \leq \bar{X} + z_{\alpha/2}\sigma/\sqrt{n}$$

is the interval estimate of μ.

Interval estimates of μ for five different sample means

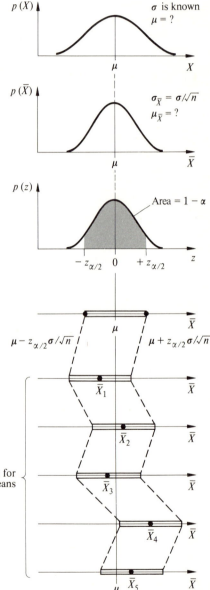

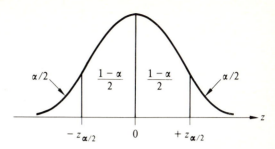

FIGURE 6.2 Determining z value from confidence level.

Perhaps the only remaining mystery is how to determine the value of $z_{\alpha/2}$ from the level of confidence $(1 - \alpha)$. First, recognize that even if we do not know the value of μ, the corresponding value of the standard normal distribution is $z = 0$. Next, once we specify the level of confidence, $(1 - \alpha)$, then we know the appropriate *area* under the z curve (see Figure 6.2), and we must determine the corresponding value of $z_{\alpha/2}$. That is, we can find $z_{\alpha/2}$ from Appendix Table IV(A) by noting that $P(0 \le z \le z_{\alpha/2}) = (1 - \alpha)/2$, or we can find $z_{\alpha/2}$ from Appendix Table IV(B) by noting that $P(z \ge z_{\alpha/2}) = \alpha/2$. You can check your understanding of this procedure by verifying the $z_{\alpha/2}$ values given in Table 6.2.

The larger the area included between $+z_{\alpha/2}$ and $-z_{\alpha/2}$, the larger $z_{\alpha/2}$ must be. Since $z_{\alpha/2}$ increases as $(1 - \alpha)$ increases and since $k = z_{\alpha/2}\, \sigma/\sqrt{n}$, for any given n, an increase in level of confidence will increase the confidence interval (which reduces precision). Also, for any given level of confidence (and thus a specific value of $z_{\alpha/2}$), k decreases as n increases; the interval estimate becomes more precise as the sample size increases.

Finally, note that if we are sampling without replacement from a finite population, the finite population correction factor may be appropriate (see section 3.6). The confidence interval is then given by

$$P\left(\bar{X} - z_{\alpha/2}\frac{\sigma}{\sqrt{n}}\sqrt{\frac{N-n}{N-1}} \le \mu \le \bar{X} + z_{\alpha/2}\frac{\sigma}{\sqrt{n}}\sqrt{\frac{N-n}{N-1}}\right) = 1 - \alpha \qquad \textbf{(6.6)}$$

TABLE 6.2 $z_{\alpha/2}$ FOR SELECTED LEVELS OF CONFIDENCE

Level of Confidence $(1 - \alpha)$	$z_{\alpha/2}$
0.90	1.645
0.95	1.960
0.99	2.575

Example 6.1 Estimating the Mean Fill per Cup from an Automatic Cola Dispenser

Like the frustrated bandit, we return to the automatic dispenser of cola (or coffee, if you prefer). When coin-activated the dispenser releases a volume of liquid which we call a "fill." Due to the particular construction of the dispenser, we know that the fills are normally distributed with a standard deviation of 0.10 ounce regardless of the "fill setting" or the average volume of liquid dispensed per fill.

Now assume that we have such a machine, and we want to know the average volume of liquid per fill, or μ. We take a random sample of 16 cups of liquid from the machine, measure the volume of liquid in each cup, and find the average fill ($\bar{X}$) of the 16 cups to be 8.05 ounces. Construct a "95 percent confidence interval" for the average volume of all fills from this machine.

A 95 percent confidence interval is just a shorter way of saying "an interval estimate with a 0.95 level of confidence." We are sampling without replacement, but $n\,(= 16)$ can be assumed to be less than 10 percent of the total number of times this machine dispenses cola (N); thus, we can assume the finite population correction factor is 1.

From Panel 6.1 or Equation 6.6, the confidence interval is

$$P(\bar{X} - z_{\alpha/2}\sigma/\sqrt{n} \le \mu \le \bar{X} + z_{\alpha/2}\sigma/\sqrt{n}) = 1 - \alpha$$

Since $1 - \alpha = 0.95$, from Table 6.2 we find that $z_{\alpha/2} = 1.96$. The sample information was $n = 16$ and $\bar{X} = 8.05$ ounces. The population standard deviation was given as 0.10 ounce. Thus,

$$8.05 - (1.96)(0.1)/\sqrt{16} \le \mu \le 8.05 + (1.96)(0.1)/\sqrt{16}$$

$$8.05 - (0.196)/4 \le \mu \le 8.05 + (0.196)/4$$

$$8.05 - 0.049 \le \mu \le 8.05 + 0.049$$

$$8.001 \text{ ounces} \le \mu \le 8.099 \text{ ounces}$$

is the 95 percent confidence interval for μ.

Note that the last equation in example 6.1 was not written as a "probability statement." It is incorrect to write $P(8.001 \le \mu \le 8.099) = 0.95$. Why? Because μ is a population parameter which is a constant and not a random variable. Either the true value of μ is in the range of 8.001 to 8.099 or it is not! As soon as we have assigned numbers to the endpoints of the confidence interval, we can no longer write it as a probability statement. In example 6.1, the probability is 0.95 that *this procedure* will generate a specific interval that includes the population mean.

Also, if you found example 6.1 to be easy, do not be too surprised. The actual calculation of a confidence interval is a brief exercise in arithmetic. The only difficulty, as we shall soon discover, is assessing which confidence interval to apply. The last five chapters have been necessary to understand *why* we can construct an interval estimate of the population mean in such an easy fashion.

6.4(b) Population Is Normally Distributed; Population Standard Deviation Is Unknown

In our first case, the population was known to be normally distributed with a known standard deviation. We will now eliminate one of those "known" characteristics of the population. Assume that the population standard deviation, as well as the population mean, is unknown. What damage does this do to our interval estimate?

Recall from our discussion above (and from section 6.1) that the length of the interval was given by $(\mu + z_{\alpha/2}\,\sigma/\sqrt{n}) - (\mu - z_{\alpha/2}\,\sigma/\sqrt{n}) = 2z_{\alpha/2}\,\sigma/\sqrt{n}$. If σ is unknown, then both the "length" and the "midpoint" of the interval are uncertain.

We shall use s, the sample standard deviation (see Equation 4.11) as an estimator of σ. Of course, there may be some sampling error involved in estimating the value of σ with the standard deviation from only one sample.

Let us assume that $\sigma = \delta s$, where δ is a correction factor which would equate σ and s. The length of our interval could then be represented by

$$\frac{2z_{\alpha/2}(\delta s)}{\sqrt{n}}$$

and all we have to do is determine δ. Please note that

$$\frac{2z_{\alpha/2}(\delta s)}{\sqrt{n}} = \frac{2(z_{\alpha/2}\,\delta)s}{\sqrt{n}}$$

If we let $t_{\alpha/2} = (z_{\alpha/2}\,\delta)$, then the length of our interval is

$$\frac{2t_{\alpha/2}\,s}{\sqrt{n}}$$

This provides an intuitive interpretation of the t values; however, it bears absolutely no relationship to the way t is mathematically derived.

The t distribution: the anonymous Mr. Gossett Around the turn of this century, William S. Gossett published the initial development of the t distribution while he was employed by an Irish brewery. Although his results had nothing to do with the secret composition of the brew, there was a company policy against using your real name in published research. Mr. Gossett tried to maintain his anonymity by assuming the pen name "Student." The importance of his research forced revelation of his true identity; however, the t distribution is still often called "Student's distribution."

To investigate Mr. Gossett's t distribution, we have to return to the beginning of section 6.1 where we found that the probability of getting a sample mean was

$$P\left(\mu - \frac{z_{\alpha/2}\sigma}{\sqrt{n}} \le \bar{X} \le \mu + \frac{z_{\alpha/2}\sigma}{\sqrt{n}}\right) = 1 - \alpha$$

from the fact that

$$z = \frac{\bar{X} - \mu}{\sigma/\sqrt{n}}$$

If we let

$$t = \frac{\bar{X} - \mu}{s/\sqrt{n}}$$

note that the only difference between z and t is the substitution of s for σ. We know that if the parent population is normally distributed, then z is normally distributed. But how is t distributed? Remember that σ is a constant and that s is a variable whose value depends upon which particular sample is selected.

If the parent population is normally distributed, then

$$t = \frac{\bar{X} - \mu}{s/\sqrt{n}} \tag{6.7}$$

generates a family of distributions whose characteristics are:

1. The mean of the t distribution is always 0; $E(t) \equiv 0$. [This is similar to the z distribution where $E(z) \equiv 0$.]
2. The variance of the t distribution is a function of "degrees of freedom" which depend upon the sample size; thus, $\text{Var}(t) = f(n)$. [This is *different* from the z distribution where $\text{Var}(z) = 1$, a constant.]
3. The shape of the t distribution is (1) bell-shaped (as is the z curve), (2) symmetric about its mean (as is the z), (3) flatter than the normal curve, and (4) asymptotic to the t axis (as is z).

As shown in Figure 6.3, the only difference between the z curve and the t curve is that the variance (or standard deviation) of the t curve depends upon "degrees of freedom" (or sample size) while the variance of z is always the same.

"Degrees of freedom" are denoted by v (the Greek letter "nu"), but before we find out what "degrees of freedom" are, let us just say that $\text{Var}(t)$ depends upon the

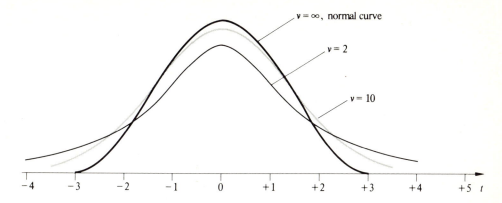

FIGURE 6.3   The t distribution for selected degrees of freedom.

size of sample selected. Why would $\mathrm{Var}(t) = f(n)$? First, it is *not* because of the "square root of n" term in Equation 6.7; note that $\sqrt{n}$ also appears in the definition of z.

Consider taking all possible samples of size 3 $(n = 3)$ from a large but finite population of size N. If we calculate the sample mean $\bar{X}$ and sample standard deviation s from each sample, then s could have many different values. Since s is in the denominator of t (see Equation 6.7), t could assume a wide variety of values (see t for $v = 2$ in Figure 6.3).

Now consider the extreme case for "large samples." If we take the one possible "sample" of size N without replacement (a census), then $s = \sigma$ and the t curve is identical to the z curve. For each sample size n, there is a distribution of possible values for s. When sampling without replacement, it is easy to see that this range of possible values for s *narrows* as we select all possible samples of a larger size. Hence, the t distribution has a larger variance for small samples than it does for large samples. It is the variability in s, not in the $\sqrt{n}$ term, that causes the variance of t to depend upon the size of the sample.

Finally, we should note that *as the sample size increases, the t distribution approaches the shape and characteristics of the standard normal distribution (or z curve)*. Once again, we shall employ that fact and use the z curve as an approximation to the t distribution for "large" samples. How large is "large"? That depends upon how much error we are willing to accept in using the z curve instead of the t curve. For the purposes of estimating the population mean, we shall assume that *for $n > 30$, the z distribution is a close approximation of the t distribution*. (To give you an idea of how "close" an approximation this is, consider the "peak"—or ordinate when $z = t = 0$—of the z and t distributions. For 30 degrees of freedom, the ordinate is 0.396 for the t and 0.399 for the z. At their peak, there is less than 1 percent difference in the height of the two distributions.)

Degrees of freedom   Roughly, the number of "degrees of freedom" refers to the number of values that can be arbitrarily chosen for a particular expression. For example, the "expression"

$$\sum_{i=1}^{3} X_i = X_1 + X_2 + X_3$$

has four terms: the sum on the left side of the expression and the three X values on the right side. You could arbitrarily assign values to any three of those four (for example, $12 = X_1 + 3 + 5$), but the value of the fourth term is given by the expression itself (for example, X_1 must be 4). In this expression there are *three* degrees of freedom.

In Equation 6.7 it is the sample standard deviation s that deserves our attention. From Equation 4.11, we defined $s = \sqrt{\Sigma(X_i - \bar{X})^2/(n - 1)}$. Also, recall that the summation of deviations about their mean is always equal to 0; that is, $\Sigma(X_i - \bar{X}) \equiv 0$. For a sample of size n, only $(n - 1)$ of the X_i values can be arbitrarily specified; the nth value must be such that $\Sigma(X_i - \bar{X}) = 0$. Thus, for the purposes of constructing a confidence-interval estimate of the population mean, *the appropriate t distribution is the one with $(n - 1)$ degrees of freedom*, where n is the size of the sample selected.

Use of the t table   Appendix Table V presents the t values for 34 different degrees of freedom and a few selected probabilities. The left column lists the number of degrees of freedom, the top heading lists six probabilities [or areas in the tail(s) of the t distribution], and the entries in the body of the table are the associated values of t.

For example, for 5 degrees of freedom, what is the value of t^* such that $P(-t^* \le t \le +t^*) = 0.95$? In Table V we move down the left-hand column to degrees of freedom $= 5$ and across to the column for "area in both tails" $= 0.05$ and find that $t^* = 2.571$. Hence, $P(-2.571 \le t \le +2.571) = 0.95$.

For 10 degrees of freedom, what is the value of t^* such that $P(-t^* \le t \le +t^*) = 0.95$? Appendix Table V shows that $t^* = 2.228$ for 10 degrees of freedom. Note that $2.228 < 2.571$ and for larger degrees of freedom the t distribution is more tightly packed about its mean; you do not have to go as far out (in both directions) on the t axis to get 95 percent of the area under the middle of the curve. In the limiting case, for ∞ degrees of freedom, t^* is 1.96; $P(-1.96 \le z \le +1.96) = 0.95$ (see Table 6.2).

6.4(b.1)   Small Sample Interval Estimate of Population Mean; Population Is Normally Distributed

Returning to our original problem, we have a normally distributed population with an unknown mean μ and an unknown standard deviation σ. We want to construct a confidence interval for μ. Following the strategy developed in section 6.1, we need to find the probability that a sample mean will occur between two limits: $P(\bar{X}_L \le \bar{X} \le \bar{X}_U) = 1 - \alpha$.

We now know that $t = (\bar{X} - \mu)/(s/\sqrt{n})$ and solving algebraically for $\bar{X}$ gives

$$\bar{X} = \mu + \frac{ts}{\sqrt{n}}$$

Therefore, $P(\mu - t_{\alpha/2} s/\sqrt{n} \le \bar{X} \le \mu + t_{\alpha/2} s/\sqrt{n}) = 1 - \alpha$, and

$$P\left(\bar{X} - \frac{t_{\alpha/2} s}{\sqrt{n}} \le \mu \le \bar{X} + \frac{t_{\alpha/2} s}{\sqrt{n}}\right) = 1 - \alpha \qquad \textbf{(6.8)}$$

provides the confidence-interval estimate for the $(1 - \alpha)$ level of confidence. This procedure is summarized in Panel 6.2. Since s is also a random variable, both the "location" and the "length" of our interval change from one sample to the next.

Panel 6.2   Interval Estimate of μ When a Small Sample ($n \le 30$) Is Selected from a Normally Distributed Population

Population
It is normally distributed. Mean and standard deviation are unknown; $\mu = ?$, $\sigma = ?$

Sampling Distribution of $\bar{X}$
Shape is that of a normal curve. Standard deviation is unknown; $\sigma_{\bar{X}} = \sigma/\sqrt{n} = ?$ The mean is unknown; $\mu_{\bar{X}} = \mu = ?$

Probability of Getting a Sample Mean
We know that $P(\mu - t_{\alpha/2, v} s/\sqrt{n} \le \bar{X} \le \mu + t_{\alpha/2, v} s/\sqrt{n})$ is equal to $(1 - \alpha)$. The length of this interval is estimated by $2t_{\alpha/2, v} s/\sqrt{n}$. This length can be calculated because (1) s is obtained from the sample and (2) t is determined by $(1 - \alpha)$ and $v = n - 1$ degrees of freedom. Since μ is unknown, the location of the interval is uncertain.

Confidence Interval Estimate of μ
Since the probability is $(1 - \alpha)$ of getting a sample mean within the interval

$$\mu \pm t_{\alpha/2, v} s/\sqrt{n}$$

using the sample mean as the midpoint of the interval indicates that the probability is $(1 - \alpha)$ that $\bar{X} \pm t_{\alpha/2, v} s/\sqrt{n}$ will include μ. Thus, for a $(1 - \alpha)$ level of confidence,

$$\bar{X} - t_{\alpha/2, v} s/\sqrt{n} \le \mu \le \bar{X} + t_{\alpha/2, v} s/\sqrt{n}$$

is the interval estimate of μ.

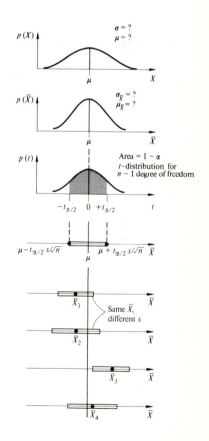

Since the cost of sampling is usually directly proportional to the sample size selected, taking a very small sample has its benefits. Destructive sampling is probably the best example. It is important to remember, however, that in order to use the t distribution you either have to know or be able to assume that the population is a normal distribution.

Example 6.2 The Average Life of Christmas Tree Lights

A producer of Christmas tree lights may know from past experience that the production process always produces lights whose bulb life is normally distributed. The producer wants to estimate the average bulb life (μ) of the latest batch, but also knows from past experience that there is considerable variation in both the mean and standard deviation (σ) from batch to batch. In order to construct a 90 percent confidence interval of the average bulb life of the entire batch, the producer wants to test (burn out) only nine randomly selected lights.

Each of the nine bulbs was plugged in and timed to determine its bulb life. The sample results were $\bar{X} = 8{,}200$ minutes and $s = 420$ minutes.

For $[n - 1 =] 8$ degrees of freedom and $(1 - \alpha) = 0.90$, we find $t_{\alpha/2} = 1.860$ from Appendix Table V. Hence, our interval estimate of μ is

$$\bar{X} - \frac{t_{\alpha/2}\,s}{\sqrt{n}} \le \mu \le \bar{X} + \frac{t_{\alpha/2}\,s}{\sqrt{n}}$$

$$(8{,}200) - (1.860)(420)/\sqrt{9} \le \mu \le (8{,}200) + (1.860)(420)/3$$

$$8{,}200 - (1.860)(140) \le \mu \le 8{,}200 + 260.4$$

$$7{,}939.6 \le \mu \le 8{,}460.4 \text{ minutes}$$

When all we know is that the population is normally distributed and we decide to take a "small" sample, Mr. Gossett's t distribution provides a very convenient way to construct a confidence interval for the population mean. To find the appropriate t value, we must consider both $(1 - \alpha)$ and the number of degrees of freedom $(n - 1)$, and we use the sample standard deviation s to estimate the unknown population standard deviation σ; otherwise the procedure is similar to our first case where σ was known.

6.4(b.2) Large-Sample Interval Estimate of Population Mean; Population Is Normally Distributed

When we decide to take "large" samples ($n > 30$) from a normally distributed population with unknown mean and standard deviation, Equation 6.8 describes the precise form of the confidence interval for the population mean. However, as was previously stated (and can be checked by observing Appendix Table V), the z curve closely approximates the t distribution for large samples.

Thus, we approximate the $t_{\alpha/2}$ value called for in Equation 6.8 with a corresponding value of $z_{\alpha/2}$. When the population assumes the shape of a normal curve (and that is all we know about the population) and the sample size is greater than 30, then our confidence interval for the population mean is given by

$$P\left(\bar{X} - \frac{z_{\alpha/2}\, s}{\sqrt{n}} \leq \mu \leq \bar{X} + \frac{z_{\alpha/2}\, s}{\sqrt{n}}\right) = 1 - \alpha \qquad (6.9)$$

6.4(c) Population Shape Is Unknown; Population Standard Deviation Is Known

We continue our game of checkers concerning what we do or do not know about the population. Consider the case where the shape of the parent population is either unknown or known, but not normally distributed, and the value of the population standard deviation is known. Once again, we have to distinguish between the "large-sample" situation and the "small-sample" situation.

6.4(c.1) Large-Sample Interval Estimate of Population Mean; Population Standard Deviation Known, Population Shape Is Not Known

The central limit theorem again saves those in despair. In section 5.1 the CLT stated that *regardless* of the shape of the parent population, the sampling distribution of sample means approaches a normal curve as the sample size increases. In Figure 5.4 we observed this phenomenon for several populations with very different shapes.

We encounter the reoccurring question of how large must a large sample be? In this situation (that is, σ is known and the population's shape is not), we can *assume the sampling distribution of $\bar{X}$ is normally distributed for samples of size 20 or larger.* When $n \geq 20$, the difference between the actual sampling distribution and the normal curve is slight.

When the population is normally distributed and σ is known (our first case), then

$$P\left(\bar{X} - \frac{z_{\alpha/2}\, \sigma}{\sqrt{n}} \leq \mu \leq \bar{X} + \frac{z_{\alpha/2}\, \sigma}{\sqrt{n}}\right) = 1 - \alpha \qquad (6.10)$$

is the confidence interval *regardless* of sample size. When the population's shape is unknown (or nonnormal) and σ is known, then Equation 6.10 is the confidence interval *only* for samples of size 20 or larger. Panel 6.3 summarizes this logic.

Panel 6.3 **Interval Estimate for μ When a Large $(n \geq 20)$ Sample Is Selected from a Population of Known Standard Deviation but Unknown Distribution**

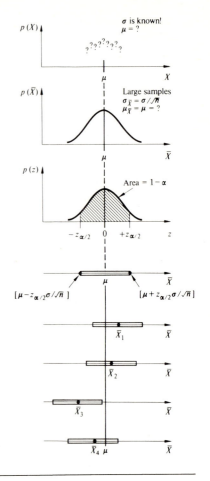

Population
Shape is either unknown or known not to be normally distributed. Standard deviation σ is known; the mean is unknown $(\mu = ?)$.

Sampling Distribution of $\bar{X}$
From the central limit theorem, the shape approaches that of a normal curve as n increases. Samples of size 20 or more are large enough to assume that the sampling distribution is normally distributed.

$$\sigma_{\bar{X}} = \sigma/\sqrt{n} \text{ and is known}$$

$$\mu_{\bar{X}} = \mu = ?$$

Probability of Getting a Sample Mean
As in Panel 6.1, the probability of obtaining a sample mean within the interval $\mu \pm z_{\alpha/2}\sigma/\sqrt{n}$ is $(1 - \alpha)$. The length of the interval is known, $2z_{\alpha/2}\sigma/\sqrt{n}$; the midpoint is uncertain $(\mu = ?)$.

Confidence Interval Estimate of μ
By using the mean of a randomly selected sample as the midpoint of the interval,

$$\bar{X} - \frac{z_{\alpha/2}\sigma}{\sqrt{n}} \leq \mu \leq \bar{X} + \frac{z_{\alpha/2}\sigma}{\sqrt{n}}$$

is the interval estimate of μ for a $(1 - \alpha)$ level of confidence.

6.4(c.2) **Small-Sample Interval Estimate of Population Mean; Population Standard Deviation Known, Population Shape Is Unknown**

The central limit theorem does not assist us for small samples; in Figure 5.4 we can see that the sampling distributions of $\bar{X}$ do not look much like a normal curve for small samples. Furthermore, the t distribution is only valid for small samples from a normally distributed population.

Although we have to slightly alter the character of our interval estimate, we can use Chebyshev's inequality for this situation. In section 4.6, Chebyshev's inequality was stated as a *property of real numbers*. If n_K is the number of observations between

$(\mu - K\sigma)$ and $(\mu + K\sigma)$, where K must be greater than 1, then $n_K/N \geq [1 - (1/K^2)]$, or the proportion of observations between $(\mu - K\sigma)$ and $(\mu + K\sigma)$ is greater than or equal to $1 - (1/K^2)$.

Chebyshev's inequality can also be a *probability statement*. If we randomly select one observation, what is the probability that its value will be between $\mu \pm K\sigma$? If all observations (or values of X) are equally likely, then the a priori definition of probability yields

$$P(\mu - K\sigma \leq X \leq \mu + K\sigma) = \frac{n_K}{N}$$

and Chebyshev's inequality states that $n_K/N \geq 1 - (1/K^2)$. Thus, we do not know the *exact* probability, but we do know the *minimum* probability [that is, $1 - (1/K^2)$].

Instead of the random variable X, with mean μ and standard deviation σ, let us consider the random variable $\bar{X}$, with mean $\mu_{\bar{X}} = \mu$ and standard deviation $\sigma_{\bar{X}} = \sigma/\sqrt{n}$. The probability that a randomly selected sample of size n will have the value of its mean between $\mu \pm K\sigma_{\bar{X}}$ is

$$P\left(\mu - \frac{K\sigma}{\sqrt{n}} \leq \bar{X} \leq \mu + \frac{K\sigma}{\sqrt{n}}\right) \geq 1 - \frac{1}{K^2}$$

If we define $1 - (1/K^2)$ as the "minimum level of confidence," then the length of the "interval," $2K\sigma/\sqrt{n}$, is known because σ is known. If we center the interval on a randomly selected sample mean, then the probability is at least $1 - (1/K^2)$ that the interval will contain the population mean μ. Thus, when (1) the population is either unknown or known but nonnormal, (2) the sample size is less than 20, and (3) the population standard deviation is known, a confidence interval estimate of μ is given by

$$P\left(\bar{X} - \frac{K\sigma}{\sqrt{n}} \leq \mu \leq \bar{X} + \frac{K\sigma}{\sqrt{n}}\right) \geq 1 - \frac{1}{K^2} \tag{6.11}$$

where K must be greater than 1.

First, note that the Chebyshev formulation of a confidence interval does not require the use of auxiliary tables to calculate K. Since you choose the level of confidence $(1 - \alpha)$, we just equate $(1 - \alpha)$ and $[1 - (1/K^2)]$ and solve for K:

$$K = +\sqrt{1/\alpha} \tag{6.12}$$

It is reasonable to ask that α be between 0 and 1; therefore, K will always be greater than 1.

Next, Equation 6.11 could *always* be used when the population standard deviation is known. Why bother with the standard normal distribution and looking up

---

Panel 6.4 Interval Estimate for μ When a Small $(n < 20)$ Sample Is Selected from a Population of Known Standard Deviation but Unknown Distribution

Population
Shape is either unknown or known not to be normally distributed. Standard deviation σ is known; the mean is unknown $(\mu = ?)$.

Sampling Distribution of $\bar{X}$
Shape and mean are unknown; $\mu_{\bar{X}} = \mu = ?$ Standard deviation is known; $\sigma_{\bar{X}} = \sigma/\sqrt{n}$.

Probability of Getting a Sample Mean
From Chebyshev's inequality,

$$P(\mu - K\sigma/\sqrt{n} \le \bar{X} \le \mu + K\sigma/\sqrt{n})$$
$$\ge 1 - (1/K^2)$$

The length of the interval is $2K\sigma/\sqrt{n}$ and is known because K is arbitrary and σ is known; however, the location of the interval is uncertain $(\mu = ?)$.

Confidence Interval Estimate of μ
Since the probability is not less than $1 - (1/K^2)$ that a sample mean will occur within the interval $\mu \pm K\sigma/\sqrt{n}$, using the sample mean as the midpoint of the interval yields a probability of not less than $1 - (1/K^2)$ that the interval $\bar{X} \pm K\sigma/\sqrt{n}$ will include μ. Thus,

$$\bar{X} - \frac{K\sigma}{\sqrt{n}} \le \mu \le \bar{X} + \frac{K\sigma}{\sqrt{n}}$$

or, since $K = \sqrt{1/\alpha}$,

$$\bar{X} - \frac{\sigma}{\sqrt{\alpha n}} \le \mu \le \bar{X} + \frac{\sigma}{\sqrt{\alpha n}}$$

is the interval estimate of μ for a level of confidence of at least $1 - (1/K^2)$.

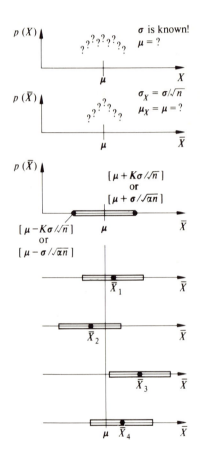

TABLE 6.3 SELECTED COMPARISONS OF z AND K

Level of Confidence $1 - \alpha$	α	$\sqrt{1/\alpha} = K$	$z_{\alpha/2}$
0.90	0.10	$\sqrt{10} = 3.1622$	1.645
0.95	0.05	$\sqrt{20} = 4.4721$	1.960
0.99	0.01	$\sqrt{100} = 10.0000$	2.575

values for $z_{\alpha/2}$ for Equations 6.6 and 6.10? The answer is very practical. For a given sample size and a given level of confidence, the *precision* of our interval estimate is improved when we are able to use z rather than the K in the Chebyshev formulation. The shorter the "length" of the confidence interval, the more "precise" the interval estimate. From Equations 6.6 and 6.10, the length of the interval is $2z_{\alpha/2}\,\sigma/\sqrt{n}$; from Equation 6.11, the length of the interval is $2K\sigma/\sqrt{n}$. The only difference is the value of $z_{\alpha/2}$ versus K. Consider Table 6.3 where three values of $z_{\alpha/2}$ are contrasted with the comparable three values of K. The value of K is always greater than the value of $z_{\alpha/2}$ for the same level of confidence; therefore, when we can use z, the precision of the interval estimate is improved. We introduce the Chebyshev formulation of the interval estimate here because if the sampling distribution is not normally distributed, we cannot use z.

Example 6.3 An Interval Estimate of Electric Fuses

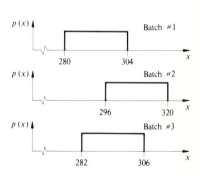

Suppose a producer of specialty electric fuses knows that the fuse-making machine uniformly wears during the production of each batch of (say, 1,000) fuses. The machine can be reset after each batch. The net result is that the producer knows the basic shape of the relevant population, but does not know its location.

That is, if we let $x =$ the disconnect voltage (or the voltage at which the fuse will "blow"), then the three populations above are examples of the producer's model of what happens to each batch. The producer infrequently makes a batch of these fuses, and following the production of each batch he or she wants to estimate the average "disconnect voltage" (μ) of the batch.

Since the sampling procedure destroys the usefulness of the product, the producer wants to take a random sample of only 9 fuses from the batch. For such a small sample from a "uniform" population, the sampling distribution of $\bar{X}$ will not be normally distributed; however, the population variance is known to be 48 volts. Thus, the Chebyshev formulation of the confidence interval is appropriate.

A random sample of 9 fuses exhibits an average disconnect voltage of 290 volts, and the producer wants a 90 percent confidence interval.

Since $(1 - \alpha) = 0.90$, then $\alpha = 0.10$, and $K = +\sqrt{1/\alpha} = \sqrt{10} = 3.1622$.

$$\bar{X} - \frac{K\sigma}{\sqrt{n}} \le \mu \le \bar{X} + \frac{K\sigma}{\sqrt{n}}$$

$$290 - (3.1622)(\sqrt{48})/\sqrt{9} \le \mu \le 290 + (3.1622)(6.928)/3$$

$$290 - (3.1622)(2.309) \le \mu \le 290 + 7.30$$

$$282.7 \text{ volts} \le \mu \le 297.30 \text{ volts}$$

6.4(d) Nothing Is Known about the Population

If we do not know the general shape of the population, the population standard deviation, and the population mean, but we want an estimate of the population mean, then we have to take a sample large enough (1) to have the central limit theorem take effect and (2) to get a precise estimate of σ from the sample standard deviation s. Knowing essentially nothing about the population is a situation that cannot be ignored for real-world applications, but it is also the most difficult situation for which to suggest how "large" the sample size should be.

The problem is complicated by the fact that there is no known general recourse to taking "large" samples. That is, there is no general theory for "small sample interval estimates" when nothing is known about the population.

It may be surprising, but you will seldom encounter a population about which you know absolutely nothing. For example, you will frequently know the lowest possible value of X and the highest possible value of X. This provides the maximum value that the range can assume. We have not proved it, but for any set of real numbers, the standard deviation cannot exceed the range; therefore, we have a maximum value for the standard deviation. Further, you may have a very high subjective probability that a population is symmetric or asymmetric. These simple bits of information can be useful in assessing how large a sample is large enough.

If you have reason to suspect that the population distribution is symmetric, then samples of size 50 or larger may be large enough. If you suspect that the population is very skewed (e.g., income distributions), then sample sizes in excess of 100 are usually necessary.

Exercising some caution with the above statements, we shall adopt the rule that *if the sample size is 50 or larger ($n \ge 50$), then the confidence interval for the population mean is given by*

$$P(\bar{X} - z_{\alpha/2} s/\sqrt{n} \le \mu \le \bar{X} + z_{\alpha/2} s\sqrt{n}) = 1 - \alpha \qquad \textbf{(6.13)}$$

For $n < 50$, there is no general method for establishing an interval estimate of the population mean.

6.4(e) Summary

Building from the basic strategy outlined in section 6.1, we have developed six different interval estimates for the population mean. Each depends upon how much we know about the particular population of interest and, in most cases, the size of the sample taken. Table 6.4 is a "decision tree" which summarizes the various situations and when to use what.

TABLE 6.4 DECISION TREE FOR INTERVAL ESTIMATE OF POPULATION MEAN

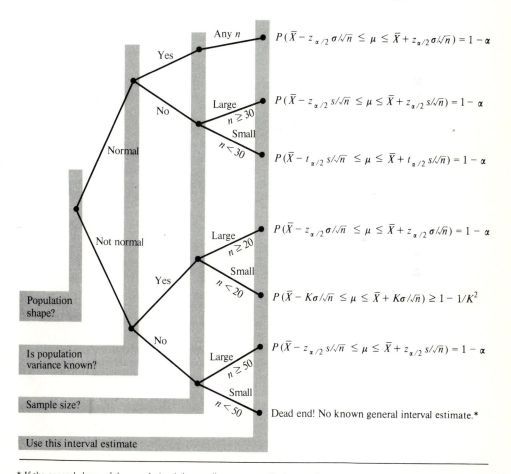

* If the general shape of the population is known (but nonnormal), the sampling distribution is usually derivable.

6.5 CONFIDENCE-INTERVAL ESTIMATE OF THE POPULATION PROPORTION

Section 5.2(b) stated that the sampling distribution for the sample proportion $\bar{p}$ has a mean equal to the population proportion $\mu_{\bar{p}} = \pi$ and a standard deviation $\sigma_{\bar{p}}$ of $\sqrt{(\pi)(1-\pi)/n}$. The sampling distribution of $\bar{p}$ is always a binomial distribution; however, for large samples, the sampling distribution of the sample proportion is approximately a normal distribution.

If we wish to use this information and use the sample proportion $\bar{p}$ to construct a confidence-interval estimate of the population proportion π, then our problem is not very different from using $\bar{X}$ to estimate μ. First, a binomial *population* is never normally distributed (although the sampling distribution, for large samples, from that population does approach a normal curve). Next, since $\sigma = \sqrt{\pi(1-\pi)}$ and π is unknown, the population standard deviation is always unknown.

The analogous situation for establishing a confidence interval of μ (see the decision tree in Table 6.4), when the population shape is not normal and σ is unknown, is that either we take a large sample or we experience a "dead end." A similar situation exists for the confidence interval of the population proportion.

Large-Sample Interval Estimate of π

For large samples, the confidence interval for the population proportion is given by

$$P(\bar{p} - z_{\alpha/2}\,\sigma_{\bar{p}} \le \pi \le \bar{p} + z_{\alpha/2}\,\sigma_{\bar{p}}) = 1 - \alpha \tag{6.14}$$

or, to be more specific,

$$P(\bar{p} - z_{\alpha/2}\sqrt{\bar{p}(1-\bar{p})/n} \le \pi \le \bar{p} + z_{\alpha/2}\sqrt{\bar{p}(1-\bar{p})/n}) = 1 - \alpha \tag{6.15}$$

where $\bar{p}$ is the proportion of "successes" in a randomly selected sample of n observations.

Of course, if we are sampling from a finite population where $n/N > 0.1$, then the "finite population correction factor" should be included:

$$\bar{p} - z_{\alpha/2}\sqrt{\frac{\bar{p}(1-\bar{p})}{n}}\sqrt{\frac{N-n}{N-1}} \le \pi \le \bar{p} + z_{\alpha/2}\sqrt{\frac{\bar{p}(1-\bar{p})}{n}}\sqrt{\frac{N-n}{N-1}} \tag{6.16}$$

Whether designed to evaluate a political candidacy, suggest public policy, or evaluate consumer preference, an opinion poll is one of the common uses of the interval estimate of a population proportion. Estimating the proportion of defects is also of interest to the production manager and the purchasing agent.

Example 6.4   Voter Opinion on Rezoning

Recently, Olympic Refineries proposed an oil refinery in the small (college) town of Durham, New Hampshire. To build the refinery would require a change in the zoning laws, which was to be decided by one of New England's famous town meetings. Clearly a large refinery would induce massive economic change within the town, but with all the claims and counterclaims, it was difficult to tell exactly what the economic and ecological changes would be. Further, the town meeting was to be held during the first united Arab oil boycott of the United States; the affair received the attention of the national news media.

Suppose that prior to the town meeting we wanted to get a reading of the voter's opinion. Let π be "the proportion of voters opposed to the zoning change for an oil refinery." From the town's list of eligible voters we could randomly select (for example) 100 persons. The issue is very emotional within the town, and a very carefully prepared interview procedure yields the following results:

In favor:	35
Opposed:	58
Undecided:	7

What is the 99 percent confidence interval for π? Since there are more than 1,000 eligible voters in the town, we assume the influence of the finite population correction factor is negligible. From Table 6.2, when $1 - \alpha = 0.99$, $z_{\alpha/2} = 2.575$. The sample proportion is $58/100 = 0.58$. Thus,

$$\bar{p} - z_{\alpha/2}\sqrt{\frac{\bar{p}(1-\bar{p})}{n}} \le \pi \le \bar{p} + z_{\alpha/2}\sqrt{\frac{\bar{p}(1-\bar{p})}{n}}$$

$$0.58 - (2.575)\sqrt{\frac{(0.58)(0.42)}{100}} \le \pi \le 0.58 + (2.575)\frac{\sqrt{0.2436}}{10}$$

$$0.58 - (2.575)(0.4936/10) \le \pi \le 0.58 + 0.127$$

$$0.45 \le \pi \le 0.71$$

(For a 95 percent confidence interval, where $z = 1.96$, show that the confidence interval is $0.48 \le \pi \le 0.68$.) Although this interval may have included the proportion of all voters who opposed the zoning change *at the time the poll was conducted*, at the town meeting more than 80 percent of the voters rejected the zoning change and thus the refinery.

Finally, note that the 99 percent confidence interval does not exclude the possibility that less than 50 percent of the entire population would vote against the refinery. A larger sample would be required to increase the precision of the interval estimate of π.

Small-Sample Interval Estimate of π

The previous section, and Table 6.4 in particular, states that when (1) the shape of the population is either unknown or known to be nonnormal and (2) the population standard deviation is unknown, then there is no *general* method for constructing an interval estimate of the population mean. Since we have argued that π is the mean of a "zero-one" binomial population, the matter should be settled; there is no small sample interval estimate for the population proportion. Wrong!

We do know something about the population: it is binomial. With that knowledge and some assumptions (which we will neither develop nor justify), C. J. Clopper and E. S. Pearson developed a graph which can be used to establish a 95 percent confidence interval of π for small samples. Reproduced as Appendix Table VI, this graph is easy to use.

Consider Figure 6.4. For samples of size 20, Figure 6.4 provides a "watermelon-shaped" region of 95 percent confidence intervals for π which depend upon the sample proportion $\bar{p}$. For example, if there are two successes in a sample of 20 observations, then $\bar{p} = r/n = {}^{2}/_{20} = 0.10$, and Figure 6.4 shows the resulting 95 percent confidence interval to be $0.01 < \pi < 0.33$. Also, if there are 16 successes out of 20 randomly selected observations, then $\bar{p} = {}^{16}/_{20} = 0.80$. From Figure 6.4, we find the 95 percent confidence interval to be $0.56 < \pi < 0.94$.

Note that Figure 6.4 is only for $n = 20$ and $(1 - \alpha) = 0.95$. Appendix Table VI includes other sample sizes; however, it is also limited to $(1 - \alpha) = 0.95$.

Finally, we have not discussed how large n must be to invoke the normal-curve attribute of the central limit theorem. Since we do not know π, saying that $n\pi(1 - \pi)$ must be greater than or equal to 3 is not conclusive. But, it is consistent with the rules that follow: *Unless we expect π [or $(1 - \pi)$] to be less than 0.10, samples of size 50 or more are necessary to use the normal-curve approximation of Equation 6.16. If π [or*

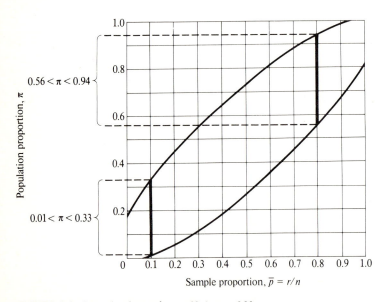

FIGURE 6.4 Interval estimates for $n = 20$, $1 - \alpha = 0.95$.

$(1 - \pi)]$ *is completely unknown or expected to be less than* 0.10, *samples of size 100 or larger are necessary to use Equation 6.16.* Any smaller samples should employ Appendix Table VI to provide a 95 percent confidence interval estimate of π.

6.6 DETERMINATION OF SAMPLE SIZE

What size sample should be selected? Like architecture, *sample survey design* is a curious blend of science and art. In examining the question of sample size, we will touch upon the scientific aspects of sample survey design. Personal experience and the shared experience of others are an important ingredient in the "art" aspects of conducting a sample survey. Although the material presented in this book will serve as a prerequisite, an extensive coverage of sample survey design is beyond the scope of this book.

Sample Size for Estimating μ

First, we must reinvestigate those variables over which the investigator has some control. Let "decision variable" be the title we bestow upon those variables whose value *we* could choose.

Next, recall that in Equation 6.4 we suggested that the "length" of the confidence interval for the population mean could be written as

$$\varepsilon = \frac{2z_{\alpha/2}\sigma}{\sqrt{n}} \tag{6.17}$$

As the length of the interval (ε) increases, the precision of the estimate decreases. The population standard deviation is given by whatever population we are considering; σ is not a decision variable. Recalling the concept of degrees of freedom, we can arbitrarily set the value of any *two* of the following three decision variables:

ε = the "length" of the confidence interval; it should be as small as possible

$1 - \alpha$ = the "level of confidence," which determines z; it should be as large as possible

n = the sample size

Thus, if we know σ, then we could assign values to ε and $(1 - \alpha)$ and solve Equation 6.17 for the sample size that would yield the predetermined precision and level of confidence. That is,

$$n = 4\left(\frac{z_{\alpha/2}\sigma}{\varepsilon}\right)^2 \tag{6.18}$$

Example 6.5   Determining Sample Size for Automatic Cola Dispenser

In example 6.1, the population was known to be normally distributed with $\sigma = 0.10$ ounce. Suppose we wanted the precision of a 95 percent confidence interval to be $\bar{X} \pm 0.01$ ounce. Since we want $(1 - \alpha) = 0.95$, then $z = 1.96$. Given our desired precision, the length of the confidence interval is $\varepsilon = (2)(0.01) = 0.02$.

From, Equation 6.18 the size of the sample should be

$$n = 4\left[\frac{(1.96)(0.10)}{(0.02)}\right]^2$$

$$= 4 \cdot (9.8)^2 = 384.16 \quad \text{or} \quad 385$$

Note that for calculating n from Equation 6.18, we *always round up to the next largest integer*. Also, note that in order to decrease the length of the confidence interval from $[(2)(0.049) \approx] 0.10$ in example 6.1 to the prescribed 0.02 in this example, the sample size has to increase from 16 to 385. The "value" of the additional precision should exceed the "cost" of the additional observations; if it does not, then you have "purchased" (in terms of the number of observations) more precision than you want.

Equation 6.18 is easy to use when the population standard deviation is known. How do we determine the sample size when σ is unknown? Basically, we have to use a nonstatistical "guesstimate" of σ.

Perhaps the easiest guesstimate of σ is when we know (1) that the population is essentially normally distributed, and (2) the range of possible values of X. Since we know that 99.7 percent of the X values lie between $(\mu - 3\sigma)$ and $(\mu + 3\sigma)$ if X is normally distributed, then the range of X values is about equal to 6σ, or $\sigma = $ (the range)/6.

If the shape of the population's distribution is unknown, the range and a process of trial and error are one way to get a "picture" of what size sample to select. For example, guesstimate various values of σ [e.g., (the range)/6, (the range)/4], plug those guesstimates into Equation 6.18, and decide upon some sample size in the neighborhood of those suggested by your guesstimates of σ.

Sample Size for Estimating π

The length of the confidence interval for π is given by

$$\varepsilon = 2z_{\alpha/2}\sqrt{\pi(1 - \pi)/n} \tag{6.19}$$

But π is unknown; π is what we are trying to estimate, and *before* we take the sample, we do not even have an estimate of π. To escape this dilemma, we need to find the maximum value for $\pi(1 - \pi)$.

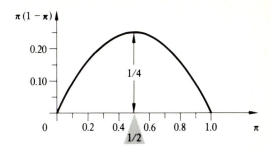

FIGURE 6.5   The relationship between π and $\pi(1 - \pi)$.

We want to choose the value of ε and $(1 - \alpha)$ (the latter determines z). If we find the maximum value for $\pi(1 - \pi)$, then the worst thing that could happen, when we solve Equation 6.19 for n, is that we could take a *larger* sample than was required for our prescribed precision and level of confidence. For given values of ε and z, we would never make the error of taking *too small* a sample.

Since the population proportion must be a fraction between 0 and 1, the maximum value that $\pi(1 - \pi)$ can have is $\frac{1}{4}$, and that occurs when $\pi = \frac{1}{2}$.

Figure 6.5 demonstrates that the product $\pi(1 - \pi)$ decreases as π gets farther away from $\pi = \frac{1}{2}$. You can check this yourself by letting $\pi = \frac{1}{10}, \frac{1}{4}, \frac{1}{2}, \frac{3}{4}, \frac{9}{10}$, etc. If we let $\pi(1 - \pi)$ assume its maximum value, then

$$\varepsilon = 2z\sqrt{\frac{\frac{1}{4}}{n}} = 2z(\tfrac{1}{2})\sqrt{\frac{1}{n}} = z\sqrt{\frac{1}{n}}$$

and solving for n yields

$$n = \left(\frac{z}{\varepsilon}\right)^2 \qquad\qquad \textbf{(6.20)}$$

Although Equation 6.20 is easy to use and completely "safe" in terms of not underestimating the appropriate sample size, Equation 6.20 will drastically overestimate n when π is either close to 1 or close to 0. For example, you may be interested in estimating the proportion of college students whose income exceeds \$20,000 per year. From current information, you are reasonably certain that π is not greater than 0.20. In this situation, there is a practical (rather than a mathematical) maximum for the product $\pi(1 - \pi)$; substitute $\pi = 0.2$ into Equation 6.19 and solve for n.

Example 6.6   Sample Size for Opinion on Oil Refinery

In example 6.4, the interval estimate of π, the proportion of voters who opposed a zoning change for an oil refinery, was inconclusive. What if we wanted our interval estimate to be, at most, ± 5 percent, or $\varepsilon = 0.10$? For a 99 percent confidence interval, $z_{\alpha/2} = 2.575$. Using

Equation 6.20, we find that

$$n = \left(\frac{z}{\varepsilon}\right)^2 = \left(\frac{2.575}{0.10}\right)^2 = (25.75)^2 = 663.0625$$

A sample of the opinion of 664 persons would be required. Recall that example 6.4 was for a small town; let us assume that there are 2,000 eligible voters. One-third of the population is to be sampled.

If we are going to have n/N greater than 10 percent, we should use the finite population correction factor. Equation 6.19 should be amended to include the FPC:

$$\varepsilon = 2z_{\alpha/2}\sqrt{\pi(1-\pi)/n}\ \sqrt{(N-n)/(N-1)}$$

Solving for n, we would obtain

$$n = \frac{4z^2\pi(1-\pi)N}{(N-1)\varepsilon^2 + 4z^2\pi(1-\pi)} \qquad (6.21)$$

Substituting $\frac{1}{4}$ for $\pi(1-\pi)$ yields

$$n = \frac{z^2N}{(N-1)\varepsilon^2 + z^2}$$

Accounting for the fact that we are sampling, without replacement, from a finite population, we find that

$$n = \frac{(2.575)^2(2,000)}{(1,999)(0.1)^2 + (2.575)^2} = \frac{(6.6306)(2,000)}{19.99 + 6.6306} = 498.2$$

499 persons is the appropriate sample size for an interval length of 0.10 and 99 percent level of confidence.

6.7 SURVEYS, SAMPLING, AND SUMMARY

From a single sample, we have inspected the problem of using an estimator $\hat{\theta}$ to estimate the value of the population parameter θ. Unbiasedness, efficiency, and consistency provide a basis for choosing which sample statistic will be a "good" estimator of θ.

A point estimate of θ could be employed, but that does not permit much information about the likelihood of the estimate being "close" to the value of θ. An interval estimate takes the form of $P(\hat{\theta} - K \le \theta \le \hat{\theta} + K) = (1 - \alpha)$ and allows us to say that the probability that θ will be within the limits $\hat{\theta} \pm K$ is $(1 - \alpha)$.

Construct the 88.89 percent confidence interval for the mean "useful life" (in feet) of this production run.

9. In Exercise 8, if 16 randomly selected pens produced the same sample mean:
a. Construct the 88.89 percent confidence interval for μ.
b. How is the interpretation of this confidence interval different from the one in Exercise 8?

10. For storage and ease-of-handling reasons, a moving van company keeps cardboard packing boxes "broken down," or unassembled. The people who do the packing take the unassembled boxes to the site, assemble them, and pack them. The company wants an estimate of the time it takes to assemble a particular type of box. They assume that assembly time is normally distributed, and they randomly select 16 situations in which assembly time is clocked for assembling one such box. The results are $\Sigma X = 336$ seconds and $\Sigma X^2 = 7,116$ (seconds)^2. Find the 99 percent confidence interval for the mean assembly time for putting together this particular type box in all situations.

11. A population is known to be V-shaped with $\sigma = 6$. A random sample of nine observations from this population produced $\Sigma X = 135$ and $\Sigma X^2 = 2,073$. Find the 95 percent confidence interval for μ, the population mean.

12. In Exercise 11, if the population is known to be normally distributed but everything else remains unchanged, then find the 95 percent confidence interval for μ.

13. An oil company allows labor charges (e.g., for putting on snow tires) to be charged on their credit card. The company is interested in the amount of credit extended for local service station labor. The company has over 0.5 million accounts (i.e., credit cards) outstanding. A random sample of 2,025 accounts shows that 1,620 have charged labor services during the past year. The amount of labor charged for each of those 1,620 accounts yielded $\Sigma X = \$81,000$ and $\Sigma X^2 = \$^2 4,051,619$.
a. Construct the 95 percent confidence interval for the proportion of all accounts which charged labor services last year.
b. Construct the 95 percent confidence interval of the mean labor charge for the population of credit card holders who charge labor services.
c. Construct the 95 percent confidence interval for the mean labor charge for all accounts.

14. Comment on the following statement about the interval estimate of a population mean: "For large sample sizes, always use the z distribution (and the population variance, if you know its value). For small samples, you have to know something about the population. If the population is normally distributed, use the t distribution; if the population variance is the only known population characteristic, you can use Chebyshev's inequality. If you are lucky enough to be dealing with a normally distributed population with known variance, then z can be used even for small samples."

15. Because of the rising costs of travel, a citizen suggests that the town's park facilities be expanded. From a list of 2,101 registered voters, the park board randomly selects 400 persons for a survey. All 400 were interviewed, and 240 were in favor of the basic concept of park expansion. Construct a 95 percent confidence interval for the proportion of all voters who would favor park expansion.

16. In a certain hockey league, only a certain amount of curvature is permitted in the blade of the hockey stick. The team has many hockey sticks. A league official takes a random sample of 10 hockey sticks and finds that 2 have too much curvature. What is the 95 percent confidence interval for the proportion of illegal hockey sticks possessed by this team?

17. Reconsider Exercise 13.

a. The oil company wanted a 95 percent confidence interval of the proportion of accounts which included labor charges. Assume they wanted the precision of the estimate to be ± 1 percent. How large must the sample be for a 95 percent confidence interval? (*Hint:* $\bar{p} \pm 0.01$.)

b. Suppose the oil company wants to estimate the average amount of labor charged last year on all its accounts. The company would like a 95 percent confidence interval that would result in an interval something like $\bar{X} \pm \$1$. Further, they guess that the range of labor charges on their credit accounts is between $0 and $150. What size sample should they select?

18. Reconsider Exercise 15. Assume the park board wanted a 95 percent confidence interval estimate somewhat like $\bar{p} \pm 0.01$. Recalling that they may be sampling more than 10 percent of the population, how large a sample should they select to ensure the desired precision?

19. An urban pollution control board wants to estimate the number of times per week that suburban homes use their (inside) incinerators. There are 20,000 homes with incinerators, and a random sample of 400 are interviewed with the following results: $\bar{X} = 5$ times per week; $s = 0.6$ times per week.

a. Construct a 95 percent confidence interval for the average weekly use by all homes. (*Note:* X is a discrete random variable, but it does not represent a binomial population.)

b. If each and every home burned exactly 5 pounds of trash each time they used the incinerator, what is the 95 percent confidence interval for the *total weight* of trash privately burned per week in all suburban home incinerators?

c. What does "95 percent confidence" mean in part *a*? In part *b*?

CASE QUESTIONS

1. Consider the Oregon sample and recall that all households are renters.

a. Construct a 95 percent confidence interval for the mean income of all renters in Oregon. (Be sure to consider case question 1*b* in Chapter 4.)

b. Consider V42. If V42 > 2, the household is above the poverty cutoff. Find the 95 percent confidence interval for the proportion of Oregon renters who are above the poverty cutoff. Use the normal curve approximation.

2. Consider the Illinois sample.

a. Find the 99 percent confidence interval for the proportion of female heads of household in Illinois. Use the normal curve approximation. (*Note:* In Chapter 2, see case question 2A*b*.)

b. Find the 90 percent confidence interval for the proportion of Illinois heads of household who attended college.

7

STATISTICAL INFERENCE: TESTS OF HYPOTHESIS

INTRODUCTION

In Chapter 6, the method of statistical inference was used to propose a value for, or an estimate of, a population parameter. In this chapter, the method of statistical inference will be used to test the validity of a *claim* about the value of a population parameter. The claim, which may be proclaimed by you or someone else, is generally called a "hypothesis"; thus, these methods are denoted "tests of hypothesis."

On a visit to Reno, a squint-eyed, scar-faced gentleman dressed in black with a white silk tie proposes that you indulge in a game of chance with a "fair" coin he discovers behind your right ear. While considering his offer, you take the coin and nervously flip it 10 times. Only twice does a head appear. Is the coin "fair"? Is the probability of getting a head the same as the probability of getting a tail? What can we infer from the sample information (i.e., two heads out of 10 tosses) about his claim that the coin is "fair" (that is, $\pi = 0.5$)?

First, your "degree of suspicion" will influence whether you conduct a statistical test of hypothesis, but it will not affect the statistical results. For example, if the same coin were offered by the Pope, former Attorney General John Mitchell, columnist William Buckley, the Chief Justice of the Supreme Court, or a recent convert to the field of used-car sales persons, your "degree of suspicion" might change, but the coin would not. Either the coin is "fair" or it is not. Your degree of suspicion influences the propensity to conduct a test of hypothesis, and not the results.

Second, it *is* possible that a "fair" coin would turn up heads only twice in 10 tosses. It may not be very probable, but it is possible. With either the conclusion "the coin is fair" or the conclusion "the coin is not fair" there is some uncertainty; there is some risk of being incorrect.

Finally, although a surreptitious character claiming a coin is "fair" may be in the sphere of your experience, those interested in business and economics are more likely to encounter the following claims: "The marginal propensity to consume is 0.85 for the United States," "At least a majority of economists favor floating exchange rates," "Eight out of ten doctors recommend the use of Super Kork for the relief of hemorrhoids," and "This bottled drink is 99.3 percent sugar free."

The Concept of a Statistical Hypothesis

Obviously, a tautological hypothesis requires no statistical testing. "Either the stock market will go up tomorrow or it will not" is neither false nor useful. In general, a *hypothesis* is a proposed explanation (e.g., "George loves me") which may or may not be true.

Our discussion will be limited to statistical hypotheses. A *statistical hypothesis* is a quantitative statement about a population. "Under normal driving conditions, the average life of Brand X tires is 17,000 miles" ($\mu = 17{,}000$ miles) and "No more than 35 percent of our customers prefer returnable glass bottles to all forms of nonreturnable containers" ($\pi \leq 0.35$, where $\pi =$ proportion who prefer returnable containers) are claims where both the population parameter and the quantitative statement concerning the parameter are easily discernible.

Frequently, a claim will have to be carefully dissected in order to isolate the quantitative statement about a population. Consider the statement: "At least a majority of the taxpayers in our fair city pay $800 or more in property tax." Assuming everyone could agree on the particular definitions of "taxpayer" and "property tax" being employed, we still have a problem. The statement does *not* say that the mean property tax payment is $800 or more (that is, $\mu \geq \$800$). The statement could be interpreted as "The median property tax payment is $800 or more," but that is a dubious interpretation because the claim was "At least a majority . . ." and not "Half. . . ."

Consider a binomial population where the two population classifications are "taxpayers who pay less than $800 in property tax" and "taxpayers who pay $800 or more in property tax." If we let π be the proportion of persons who pay $800 or more in property tax, then the "claim" is that $\pi \geq 0.50$. "At least a majority" *infers* a quantitative statement about the population proportion. "$800 or more" is just the means by which the population was separated into two mutually exclusive categories.[1]

The flexibility of the English language permits a variety of phrasings for a claim which contains a statistical hypothesis. Once the statistical hypothesis is detected, it will fall in one of these three classifications:

1. The parameter is *greater* [*less*] *than* some value stated or inferred by the claim (for example, $\mu > 40{,}000$ miles or $\pi > 15$ percent defective).[2]
2. The parameter is *greater* [*less*] *than or equal to* a value stated or inferred by the claim (for examples, $\mu \geq 40{,}000$ miles or $\pi \geq 15$ percent defective).

---

[1] You may wonder why such a claim would be uttered in the first place. Several states either offer or have considered offering a partial refund (usually through a state income tax) on local property tax in excess of some amount (say, $800). It would then be reasonable to ask what proportion of the people pay more than the $800 "cutoff" value.

[2] As we shall soon see, the "claim" in this case will be the "alternative hypothesis"; in the other two cases, the "claim" forms the "null hypothesis."

3. The parameter is *equal to* a value specified or inferred by the claim (for example, $\mu = 40{,}000$ miles or $\pi = 15$ percent).

Although simplistic, these three classifications will assist us in determining how to set up the hypothesis to be tested and indicating whether a one-tail or two-tail test of hypothesis is relevant.

The Null Hypothesis and the Alternative Hypothesis

The *null hypothesis* (which we will abbreviate as H_0 or " H subzero ") and the *alternative hypothesis* (abbreviated H_A or " H sub-A") are simply a mechanism for separating those values of the parameter specified by the claim from those values of the parameter excluded by the claim.[3] Consider example 7.1.

Example 7.1   Sample Claims and Their Null and Alternative Hypotheses

	Claim	*Statistical Hypotheses*
(a)	The average force required to release our ski bindings is 6 pounds.	*H_0: $\mu = 6$ pounds H_A: $\mu \neq 6$ pounds
(b)	Our Super Duper tire averages *at least* 20,000 miles.	*H_0: $\mu \geq 20{,}000$ miles H_A: $\mu < 20{,}000$ miles
(c)	*No more than* 2 percent of the cereal boxes are incorrectly sealed.	*H_0: $\pi \leq 0.02$ H_A: $\pi > 0.02$
(d)	Our "5-pound bags of potatoes" average *more than* 5 pounds per bag.	H_0: $\mu \leq 5$ pounds *H_A: $\mu > 5$ pounds
(e)	*Less than* 90 percent of the full-time students will use the library during the week of final examinations.	H_0: $\pi \geq 0.90$ *H_A: $\pi < 0.90$

In parts *a*, *b*, and *c* of example 7.1, the null hypothesis restates the claim about the population parameter; the alternative hypothesis accounts for all other possible values of the parameter. The feet and ankles must put some stress on the skis to control their direction, but (e.g., in a fall) the skis should not exert an injurious force on the ankles. Thus, the bindings which hold the skis to the boots should release the skis just before the force on the ankles causes injury, but not before that. Parts *b* and *c* are self-explanatory.

---

[3] In later chapters, a parameter may not be involved in the null hypothesis. For example, the null hypothesis may be "events A and B are independent"; the alternative hypothesis may be "A and B are not independent."

Parts *d* and *e* of example 7.1 have the claim in the alternative hypothesis. Why? In order to specify the risk of "judging the H_0 to be false" when in fact the H_0 is true, *the "equals sign" should be associated with the null hypothesis.* For example, if part *d* read: "Our 5-pound bag of potatoes averages 5 pounds *or more*," Then the "claim" would have been in the null hypothesis. (That is, H_0: $\mu \geq 5$ pounds; H_A: $\mu < 5$ pounds.)

Please note that the claims in example 7.1 are provided by someone else. If you originate the claim, you can phrase it anyway you wish. You may phrase the claim in such a way that it is in the null or the alternative hypothesis; however, you should adhere to the equals-sign-in-the-null-hypothesis rule.

Type I and Type II Errors

It is customary to discuss decisions with regard to the null hypothesis. There are three possible decisions:

1. Accept the null hypothesis (or reject the alternative).
2. Reject the null hypothesis (or accept the alternative).
3. Reserve judgment.

In fact, the null hypothesis is either true or false. Only if we accept or reject the null hypothesis can we be correct, and only then can we commit an error. Consider the events described in Table 7.1.

If we accept the null hypothesis when it is true or reject it when it is false, then we have made a correct decision. However, if we reject the null hypothesis when, in fact, it is true, we have made an error; cleverly, we call this event a Type I error. Also, if we

TABLE 7.1 DECISIONS ABOUT THE NULL HYPOTHESIS

We Decide \ Fact	H_0 Is True	H_0 Is False
Accept H_0 H_0 is true	*Bingo!* Correct decision $[1 - \alpha]$	*Zap!* Type II error $[\beta]$
Reject H_0 H_0 is false	*Zap!* Type I error $[\alpha]$	*Bingo!* Correct decision $[1 - \beta]$

accept the null hypothesis when it is false, we have committed a different kind of error; we call it a Type II error. These "errors" are events.

The probability that these errors will occur is given by

$$P(\text{Type I error}) = P(\text{reject } H_0 | H_0 \text{ is true}) = \alpha \qquad \textbf{(7.1)}$$

and

$$P(\text{Type II error}) = P(\text{accept } H_0 | H_0 \text{ is false}) = \beta \qquad \textbf{(7.2)}$$

Of course, we would like α and β to be as small as possible. Recall from Chapter 6 that we called $(1 - \alpha)$ the "level of confidence." In tests of hypothesis, we call α the *significance level.*

We have the option of selecting the significance level, or the value for α, *before* the sample (with which we will test the null hypothesis) is taken and the sample information is known.

General Strategy in the Test of Hypothesis

In the American system of judicial decision making, we have an interesting analogy to statistical hypothesis testing. Compare Table 7.2 with Table 7.1.

Note that the reduction in the probability of convicting an innocent person may increase the probability of acquitting a guilty person. Also, some individuals may judge one "error" to be more important than another. This is one aspect of current legislative discussion concerning Supreme Court decisions which attempt to reduce the probability of convicting the innocent. Further, in the legal oddity of impeachment procedure for public officials, the mechanisms for considering *both* forms of error are not clearly prescribed.

The analogy of judicial decision making and statistical hypothesis testing does not end here. "The accused is presumed innocent until proven guilty" is a common phrase. In statistical hypothesis testing we *assume the null hypothesis is true* until the sample information renders that assumption unsupportable.

TABLE 7.2 DECISIONS ABOUT
THE ACCUSED

Fact / Jury Decides	Accused Is Innocent	Accused Is Guilty
Acquit	Correct	Error
Convict	Error	Correct

"Beyond a reasonable doubt" is another common phrase which infers the existence of "unreasonable doubt." The probability of Type I error (α) determines a "critical value" for a sample statistic which is analogous to the dividing line between reasonable doubt and unreasonable doubt.

With the concepts of (1) ascertaining the null and alternative hypotheses from the claim, (2) specifying the probability of Type I error (α), and (3) assuming the null is true, we are prepared to consider specific techniques for the statistical test of hypothesis.

7.1 TEST OF HYPOTHESIS ABOUT A POPULATION MEAN: POPULATION VARIANCE IS KNOWN

The logic for an interval estimate of a population mean is almost the same as the logic for testing a hypothesis about the population mean. In Chapter 6, the specific formulation of the interval estimate depended upon the answer to the following three questions: (1) Is the population known to be normally distributed? (2) Is the population variance known? (3) What is the size of the sample? This process is summarized in the decision tree provided by Table 6.4.

The same procedure could be employed for developing the specific formulation of tests of hypothesis for the population mean, but such a procedure would consume more of our time than is necessary. Reconsider the decision tree in Table 6.4. When (1) the population variance is known and (2) the sample size is "large," the interval estimate of the population mean is given by $\bar{X} \pm z_{\alpha/2}\, \sigma/\sqrt{n}$ regardless of population shape. In like manner, when (1) σ is unknown and (2) n is large, the confidence interval is $\bar{X} \pm z_{\alpha/2}\, s/\sqrt{n}$ regardless of the population shape. These similarities, which are derived from the central limit theorem, also occur in tests of hypothesis. Thus, in this chapter we will first segregate tests of hypothesis on the basis of whether the population variance is known; then we will consider population shape and sample size.

7.1(a) Population Is Normally Distributed, Any Sample Size *or* a "Large" Sample Is Selected from a Population of Unknown Shape

First, consider a general *claim* that "the population mean is equal to or greater than some specific value μ_0" (for example, "The Super Duper tire averages at least 20,000 miles"). The *null* and *alternative hypotheses* would be $H_0: \mu \geq \mu_0$, and $H_A: \mu < \mu_0$ (for example, $H_0: \mu \geq 20{,}000$ miles, $H_A: \mu < 20{,}000$).

Second, assume that the population is known to have a normal distribution and that the population variance is known (for example, $\sigma^2 = 300$ miles^2). From the central limit theorem, we know that the sample means from all possible samples of size n will exhibit a normal distribution with $\mu_{\bar{X}} = \mu$ and $\sigma_{\bar{X}} = \sigma/\sqrt{n}$ {or $\sigma_{\bar{X}} = (\sigma/\sqrt{n})$ $[\sqrt{(N-n)/(N-1)}]$ if $n/N > 0.10$}. Since σ is known, $\sigma_{\bar{X}}$ can be calculated, but which value of μ do we use?

We assume the null hypothesis is true until we have information to the contrary. The null says $\mu \geq \mu_0$, then $\mu_{\bar{X}} \geq \mu_0$. Consider Figure 7.1a where the shape and standard deviation of the population are known but the location of the population's

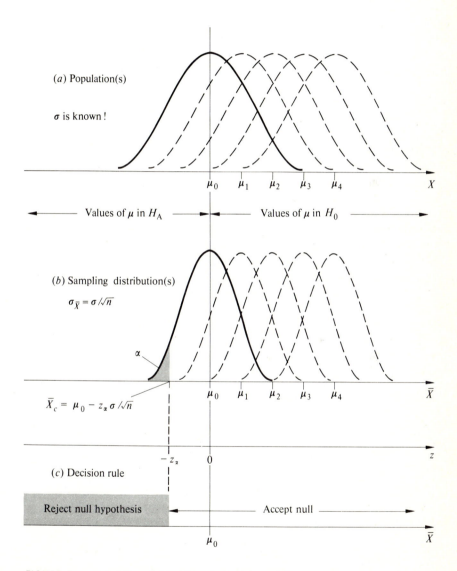

FIGURE 7.1   Formulation of a decision rule for the population mean.

distribution could be anywhere within a *known* range of values (that is, $\mu_0 \leq \mu \leq +\infty$). Figure 7.1*b* demonstrates the resulting situation for the sampling distribution of sample means.

We shall use as a reference point the "finite limit" of the possible values of μ given in the null hypothesis. That is, we shall give special consideration to the population (and the resulting sampling distribution) which has a mean of μ_0.

From a population with mean μ_0 and standard deviation σ ($\sigma > 0$), we know that a random sample of size n could yield a sample mean which is less than the population mean. Specifically, we know that

$$P(\bar{X} < \mu_0 \,|\, \mu_0 \text{ is true mean}) = P(z < 0) = 0.50$$

If the "limiting case" given by the null hypothesis ($\mu = \mu_0$) were true, then we would have a "50 : 50 chance" of selecting a sample mean lower than μ_0.

Remember, we want to maintain a posture of assuming that H_0 is true until evidence dissuades us. So, let us concentrate on the concept of "reasonable doubt." There must be some value of a sample mean that is "small enough" to dissuade us— some $\bar{X}$ is so much less than μ_0 that, while possible, it is not very probable that the sample came from a population whose mean was as large as μ_0.

Thus, we must identify some *critical value of the sample mean*, $\bar{X}_c$, where any observed sample mean equal to or less than $\bar{X}_c$ would cause us to reject the null hypothesis. If the mean calculated from the sample is above $\bar{X}_c$, then we would accept the null hypothesis. For some individuals, the "accept-reject" terminology may be too definite. It is equally acceptable to think of the circumstances where $\bar{X} \leq \bar{X}_c$ as "the sample data does not support the null hypothesis," and $\bar{X} > \bar{X}_c$ as "the sample data does support H_0." Either way, what is the value of $\bar{X}_c$?

Note that rejecting the null hypothesis when $\mu < \mu_0$ is a correct decision, but rejecting H_0 when $\mu \geq \mu_0$ is a mistake. Specifically we called the event {Reject $H_0 \,|\, \mu = \mu_0$} a Type I error, and

$$P(\text{Type I error}) = P(\text{Reject } H_0 \,|\, \mu = \mu_0)$$
$$= P(\bar{X} \leq \bar{X}_c \,|\, \mu = \mu_0) = \alpha \tag{7.3}$$

The probability of committing Type I error (α) is shown as the shaded area in Figure 7.1*b*.

Since the sampling distribution (see heavy-line distribution in Figure 7.1*b*) has a known mean (μ_0 from the H_0) and standard deviation ($\sigma/\sqrt{n}$), then Equation 7.3 would allow us to calculate $\bar{X}_c$ if we specify the value of α, the level of significance. That is,

$$\alpha = P(z \leq -z_\alpha) = P(\bar{X} \leq \mu_0 - z_\alpha \sigma/\sqrt{n}) \tag{7.4}$$

If we *choose* a value for α, then we can find $-z_\alpha$ from Appendix Table IV(B). Since

$$-z_\alpha = \frac{\bar{X}_c - \mu_0}{\sigma/\sqrt{n}}$$

and $\bar{X}_c$ is the only unknown value, then

$$\bar{X}_c = \mu_0 - \frac{z_\alpha \sigma}{\sqrt{n}} \qquad (7.5)$$

(*Note:* The values for $z_{\alpha/2}$ from Chapter 6 cannot be used for z_α.)

Finally, we use this value of $\bar{X}_c$ to establish the *decision rule*. From Equation 7.5, calculate the critical value of the sample mean, $\bar{X}_c$, *then* select a random sample of size n and calculate the sample mean $\bar{X}$.

If $\bar{X} \leq \bar{X}_c$, reject the null hypothesis, H_0; if $\bar{X} > \bar{X}_c$, accept the H_0.

Several aspects of this procedure deserve emphasis. First, when σ is known, the decision rule can be specified *before* the sample is selected. The only sample information required for Equation 7.5, which establishes $\bar{X}_c$, is the size of the sample *to be taken*. This makes this test of hypothesis procedure objective; the rule by which the accused is to be judged is established prior to the collection and knowledge of the evidence.

Second, the reason for the equals-sign-in-the-null-hypothesis rule can be investigated. The "finite limit" from the null hypothesis was used to find the critical value $\bar{X}_c$. Note, when $H_0: \mu \geq \mu_0$, we can use $\mu = \mu_0$ in Equation 7.5. But, if $H_0: \mu > \mu_0$, then what do we use for μ_0 in Equation 7.5? The value of μ_0 would still be used in Equation 7.5, but it would introduce a subtle change in the location of the "equals sign" in our decision rule: If $\bar{X} < \bar{X}_c$, then reject H_0; if $\bar{X} \geq \bar{X}_c$, accept H_0. As we shall discuss in section 7.5, there are situations in which the investigator wants the null hypothesis to read either $\mu > \mu_0$ or $\mu < \mu_0$. The equals-sign-in-the-null-hypothesis rule is simply a temporary pedagogical device to comprehend the procedure for statistical hypothesis testing. Once the procedure is understood, then the subtleties of establishing the null hypothesis can be better appreciated.

Third, the decision rule could have been stated in terms of z, the standard normal deviate:

If $(\bar{X} - \mu_0)/(\sigma/\sqrt{n}) \leq -z_\alpha$, then reject H_0; otherwise, accept H_0.

In fact, only when σ is known can the decision rule be specified in terms of $\bar{X}$ *before* the sample information is known. The next section deals with situations where σ is unknown.

Finally, it is the *choice* of the level of significance (α) that offers the most opportunity for subjective tampering with the results of a statistical test of hypothesis. A larger value for α will reduce the absolute value of z_α which enlarges the rejection region. A particular sample mean may then cause the null to be rejected, where previously that same sample mean suggested the null should be accepted. As a result, it is common practice to specify several values for the level of significance (for example, $\alpha = 0.10, 0.05, 0.01$) and to denote acceptance or rejection of the null at each level. This point will also receive more attention later in this section.

Example 7.2   The Adhesive Strength of Tape

A manufacturer produces a tape that is used to fasten wires to concrete ceilings. The tape is made by applying an adhesive to a superstrong fabric. The adhesive will "break away" from the concrete much quicker than the fabric will break; thus, the "holding power" of the tape is only a function of the "break-away" strength or adhesiveness of the adhesive.

The production process is controlled, but there are slight variations in the amount of adhesive applied (and thus the break-away strength of the tape). The tape is advertised as having an average break-away strength of 14 pounds or more. From past experience, the break-away strength is known to be normally distributed with a standard deviation of 1.2 pounds.

As a means of quality control, for every 10,000 feet of tape produced, 36 randomly selected short strips are tested for adhesive strength. Only 5 times out of 100 does the management want to scrap a production run (10,000 feet of tape) when, in fact, the tape is as strong as advertised.

Construct the decision rule for this producer.

The Claim
The average break-away strength is 14 pounds or more.

Null and Alternative Hypotheses

$$H_0: \mu \geq 14 \text{ pounds}$$

$$H_A: \mu < 14 \text{ pounds}$$

The Population
It is normally distributed with $\sigma = 1.2$ pounds.
See Figure 7.2a.

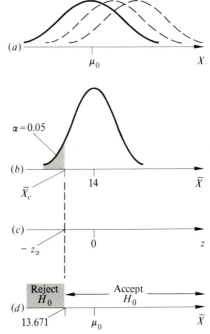

FIGURE 7.2   Decision rule for adhesive tape.

Sampling Distribution
From CLT, $\bar{X}$ is normally distributed with mean $\mu_0 = 14$ and $\sigma_{\bar{X}} = \sigma/\sqrt{n} = 1.2/\sqrt{36} = 0.2$.

Level of Significance
Management wants the risk of Type I error to be 5 out of 100, or $\alpha = 0.05$.

Critical Value of z
From Appendix Table IV(B), $P(z \leq -z_a) = 0.0500$ yields $z_a = 1.645$.

Critical Value of $\bar{X}$
From Equation 7.5,

$$\bar{X}_c = \mu_0 - z_a \sigma/\sqrt{n} = 14 - (1.645)(1.2)/\sqrt{36}$$

$$= 14 - (1.645)(0.2) = 14 - 0.329 = 13.671 \text{ pounds}$$

Decision Rule
If the mean strength of the 36 sample tape strips is less than or equal to 13.671 pounds, then the tape is not as strong as advertised; if $\bar{X} > 13.671$ pounds, then the tape is as strong as advertised.

Thus far we have considered the case for H_0: $\mu \geq \mu_0$ which results in the critical value for $\bar{X}$ being on the left side of the sampling distribution (see Figure 7.1b). For obvious reasons, this is called a "left-tail" test. Panel 7.1 summarizes the logic that provides the decision rule for a "right-tail" test, that is, when the null hypothesis has the general form H_0: $\mu \leq \mu_0$.

Panel 7.1 also illuminates one other point. If the population is normally distributed and σ is known, then the sampling distribution of sample means is normally distributed *regardless* of sample size. If the population shape is unknown (or known but nonnormal), then for large samples the CLT allows the sampling distribution to be approximated by a normal curve. In Chapter 6 we suggested that this approximation is reasonable for $n \geq 20$ when the population standard deviation is known.

Finally, Panel 7.2 summarizes the logic for establishing a decision rule when the null hypothesis assumes the general form H_0: $\mu = \mu_0$ (and H_A: $\mu \neq \mu_0$). In this case, there is only *one* hypothesized value for μ and not a range of values as shown in Figure 7.1a. Thus, a sample mean that is too far from μ_0 in either direction will discredit the null hypothesis. As a result, this is sometimes called a "two-tail" test of hypothesis.

In the "two-tail" test, the risk of committing Type I error is divided equally, and the area in each tail is $\alpha/2$. Thus, the critical values for z, for any given α, are the same as $\pm z_{\alpha/2}$ given in Table 6.2.

Further, the "acceptance region" has finite limits, and the "rejection region" is composed of two parts flanking the acceptance region. This does not present quite the problems encountered in the separation of two Pakistans by 1,000 miles or so.

Panel 7.1 Decision Rule Formulation for Either a Normally Distributed Population with σ Known or a Population of Unknown Shape, σ Known, and Large Samples

Claim
Population mean is less than or equal to some specified value μ_0.

Null and Alternative Hypotheses

$$H_0: \mu \leq \mu_0 \qquad H_A: \mu > \mu_0$$

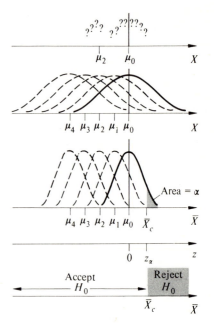

Population
σ is known. Either the population shape is unknown and $n \geq 20$, or the population has a normal distribution and any sample size.

Distribution of Sample Means
Given n, the CLT states that the sampling distribution of $\bar{X}$ is[4] normal with $\mu_{\bar{X}} = \mu$ and $\sigma_{\bar{X}} = \sigma/\sqrt{n}$.
     Since σ is known and, from the H_0, the limiting value of μ is μ_0, the solid-line distribution is the one of interest.

Level of Significance
α is the probability of rejecting H_0 given $\mu = \mu_0$. The H_0 will be rejected if the sample mean $(\bar{X})$ exceeds some critical value $\bar{X}_c$. Thus,

$$P(\text{Reject } H_0 \mid \mu = \mu_0) = P(\bar{X} \geq \bar{X}_c \mid \mu = \mu_0)$$
$$= P(z \geq z_\alpha) = \alpha$$

Given α, z_α is determined by $P(z \geq z_\alpha)$ and Appendix Table IV(B). Hence, the critical value is $\bar{X}_c = \mu_0 + z_\alpha \sigma/\sqrt{n}$.

Decision Rule
At the α significance level,

$$\text{If } \bar{X} \geq \bar{X}_c = \mu_0 + z_\alpha \sigma/\sqrt{n} \qquad \text{reject } H_0$$

$$\text{If } \bar{X} < \bar{X}_c = \mu_0 + z_\alpha \sigma/\sqrt{n} \qquad \text{accept } H_0$$

---

[4] If the population is normally distributed, then the sampling distribution *is* normal; however, if the population shape is unknown, then the sampling distribution *is approximately* normal.

Panel 7.2 Decision Rule Formulation for a Two-Tail Test

Claim
The population mean is equal to some particular value μ_0.

Null and Alternative Hypothesis

$$H_0\colon \mu = \mu_0 \qquad H_A\colon \mu \neq \mu_0$$

Population
σ is known. Either the shape is unknown (and $n \geq 20$), or the population has a normal distribution (and any size n).

Distribution of Sample Means
From the central limit theorem, the sampling distribution of $\bar{X}$ is (or "is approximately") normal with $\mu_{\bar{X}} = \mu$ (and, assuming the H_0 is true, $\mu = \mu_0$) and $\sigma_{\bar{X}} = \sigma/\sqrt{n}$.

Level of Significance
α is the probability of rejecting H_0 given $\mu = \mu_0$. The null will be rejected if the sample mean exceeds some "critical distance" from μ_0 (that is, $|\bar{X} - \mu_0| \geq |\bar{X}_c - \mu_0|$). If $\bar{X}$ is either too much *larger* than μ_0 or too much *smaller* than μ_0, then the null is rejected. Thus, $\alpha/2$ is the area in each tail of the sampling distribution, and

$$P(\text{Reject } H_0|\mu = \mu_0) =$$

$$P(|\bar{X} - \mu_0| \geq |\bar{X}_c - \mu_0|) = P(z \geq z_{\alpha/2}) = \alpha/2$$

Choose a value for α, calculate $\alpha/2$, and find $z_{\alpha/2}$ from Appendix Table IV. Then, $\bar{X}_c = \mu_0 \pm z_{\alpha/2}\,\sigma/\sqrt{n}$.

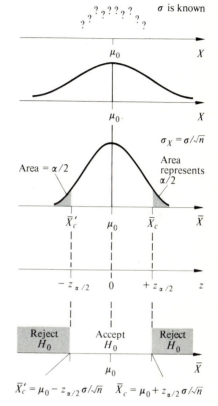

σ is known

$\sigma_{\bar{X}} = \sigma/\sqrt{n}$

Area = $\alpha/2$

Area represents $\alpha/2$

$$\bar{X}_c' = \mu_0 - z_{\alpha/2}\,\sigma/\sqrt{n} \qquad \bar{X}_c = \mu_0 + z_{\alpha/2}\,\sigma/\sqrt{n}$$

Decision Rule
At the α significance level

$$\text{If } \mu_0 - \frac{z_{\alpha/2}\,\sigma}{\sqrt{n}} < \bar{X} < \mu_0 + \frac{z_{\alpha/2}\,\sigma}{\sqrt{n}} \qquad \text{accept } H_0$$

$$\text{If } \bar{X} \geq \mu_0 + \frac{z_{\alpha/2}\,\sigma}{\sqrt{n}}$$

$$\text{or if } \bar{X} \leq \mu_0 - \frac{z_{\alpha/2}\,\sigma}{\sqrt{n}} \qquad \text{reject } H_0$$

Example 7.3   The Release Force on Ski Bindings

Assume the producer of ski bindings claims that the average release force of that equipment is 6 pounds. (Before you panic, these numbers are hypothetical, and most ski bindings have adjustable tension release mechanisms.) The Alpian Juan and Don Ski Shop receives a large shipment the day before their (expected) big weekend rush. They do not have time to test all bindings, so they randomly select 25 bindings and find the average release force to be 6.20 pounds. From past experience, the standard deviation is known to be 0.5 pound. (Can you imagine the image change when a ski bum exclaims "Hey snow buffs, did you know that the standard deviation for tension release in ski bindings is 0.5 pound?" The probability is greater than 0.9 that the reply would be shorter than this example.)

Construct the decision rule for a 1 percent significance level and test the producer's claim.

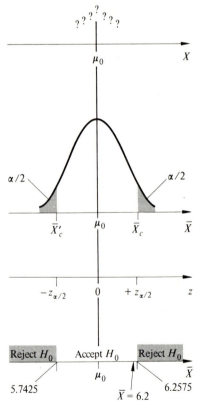

$$H_0: \mu = \mu_0 = 6 \text{ pounds}$$

$$H_A: \mu \neq 6 \text{ pounds}$$

Population shape is unknown, but σ is known and $n > 20$; thus, the sampling distribution is approximately normal. Assuming the H_0 is true, the mean is 6 and $\sigma_{\bar{X}} = \sigma/\sqrt{n} = 0.5/\sqrt{25} = 0.10$. $\alpha = 0.01$ and from Table 6.2, $z_{\alpha/2} = \pm 2.575$.

$$\bar{X}_c = \mu_0 \pm \frac{z_{\alpha/2}\sigma}{\sqrt{n}}$$

$$= 6 \pm (2.575)(0.10)$$

$$= 6 \pm 0.2575, \text{ or } 6.2575 \text{ and } 5.7425$$

The decision rule: If $5.7425 < \bar{X} < 6.2575$, then accept the H_0; if $\bar{X} \le 5.7425$ or if $\bar{X} \ge 6.2575$, then reject H_0.

Since $5.7425 < 6.20 < 6.2575$, we accept the null hypothesis, which means the data supports the claim at the 1 percent level of significance. Would the result be any different at the 5 percent level of significance?

For $\alpha = 0.05$, $z_{\alpha/2} = \pm 1.96$ and

$$\bar{X}_c = 6 \pm (1.96)(0.10) = 6 \pm 0.196, \text{ or } 6.196 \text{ and } 5.804$$

Since $6.20 > 6.196$, we would reject the null hypothesis and assume that the claim is incorrect. (Will the bindings release soon enough?)

7.1(b) Small Sample from Population of Unknown Shape but Known Standard Deviation

For $n < 20$, the central limit theorem is not effective in determining the shape of the sampling distribution of $\bar{X}$ if either the population shape is unknown or the population shape is known but not normally distributed. The shape of the sampling distribution will very much depend upon the shape of the parent population, and in most cases the normal curve would be a very poor approximation. As in Chapter 6, Chebyshev's inequality will provide a decision rule.

Recall the Chebyshev's inequality states that the probability that $\bar{X}$ will fall within the range $\mu_{\bar{X}} \pm K\sigma_{\bar{X}}$ is *not less than* $1 - 1/K^2$, if K is greater than 1. That is, $P(\mu - K\sigma/\sqrt{n} \le \bar{X} \le \mu + K\sigma/\sqrt{n}) \ge 1 - (1/K^2)$. Hence, the probability that a sample mean will fall outside the range is *less than* $1/K^2$. Symbolically,

$$P\left[\left(\bar{X} < \mu - \frac{K\sigma}{\sqrt{n}}\right) \cup \left(\bar{X} > \mu + \frac{K\sigma}{\sqrt{n}}\right)\right] < \frac{1}{K^2} \tag{7.6}$$

Two-tail test If we let the level of significance α be equal to $1/K^2$, then choosing a value for α will allow us to solve for K; if $\alpha = 1/K^2$, then $K = \sqrt{1/\alpha}$. Recall that we are assuming that σ is known. If the claim results in a two-tail test (that is, $H_0: \mu = \mu_0$, $H_A: \mu \neq \mu_0$), then we assume the null is true and let $\mu = \mu_{\bar{X}} = \mu_0$.

The probability of Type I error could be considered as $P(\text{Reject } H_0 | H_0 \text{ is true}) < \alpha$, or

$$P[(\bar{X} < \bar{X}_c | \mu = \mu_0) \cup (\bar{X} > \bar{X}_c | \mu = \mu_0)] < \alpha$$

which, believe it or not, bears a striking resemblance to Equation 7.6. If we let $\bar{X}'_c = \mu_0 - K\sigma/\sqrt{n}$, $\bar{X}_c = \mu_0 + K\sigma/\sqrt{n}$, and $K = \sqrt{1/\alpha}$, then Chebyshev's inequality permits us to state the following decision rule: *For a level of significance less than α,*

If $\bar{X} < \mu_0 - \dfrac{\sigma}{\sqrt{\alpha n}}$ or if $\bar{X} > \mu_0 + \dfrac{\sigma}{\sqrt{\alpha n}}$ then reject H_0

If $\mu_0 - \dfrac{\sigma}{\sqrt{\alpha n}} \le \bar{X} \le \mu_0 + \dfrac{\sigma}{\sqrt{\alpha n}}$ then accept H_0 (7.7)

The critical values of $\bar{X}$ contain only the hypothesized mean (μ_0), the population standard deviation σ, the level of significance α, and the size of sample to be selected (n).

Single-tail test The situation is more complex if the claim results in either a right-tail (that is, $H_0: \mu \le \mu_0$) or a left-tail (that is, $H_0: \mu \ge \mu_0$) test of hypothesis. While Equation 7.6 is a valid statement from Chebyshev's inequality, it is *not* necessarily true that $P(\bar{X} < \mu - K\sigma/\sqrt{n}) < (0.5)(1/K^2)$ or that $P(\bar{X} > \mu + K\sigma/\sqrt{n}) < (0.5)(1/K^2)$. Why? Because the population may not be completely symmetrical. Therefore, the sampling distribution of sample means may not be completely symmetrical.

Consider Figure 7.3. The population is skewed to the right (e.g., an income distribution). Since the sample size is small ($n < 20$), the sampling distribution of sample means is skewed to the right. The mean and standard deviation of the sampling distribution are still $\mu_{\bar{X}} = \mu$ and $\sigma_{\bar{X}} = \sigma/\sqrt{n}$. If we go the same distance above and below the mean ($\mu_0 \pm K\sigma/\sqrt{n}$) to establish the critical values of $\bar{X}$, then there is more area remaining in the right tail than in the left tail. All Chebyshev's inequality tells us is that the total of the areas in both tails is less than $1/K^2$; there is no mention of how much area is in each tail.

Consider the left tail of the sampling distribution in Figure 7.3. If we know the population is skewed to the right *and* if we are conducting a left-tail test (that is, $H_0: \mu \ge \mu_0$), then $P(\bar{X} < \mu_0 - K\sigma/\sqrt{n}) \ll (0.5)(1/K^2)$, where " $\ll$ " is read "much

FIGURE 7.3 Difference of the areas in the tails of a skewed sampling distribution.

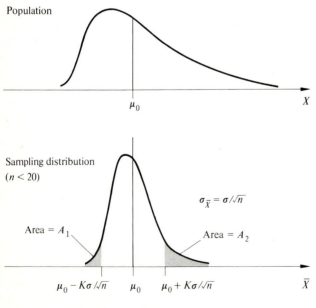

Area A_1 is less than area A_2

TABLE 7.3 VALUE OF K FOR SINGLE-TAIL TEST

Population Symmetry	Left-Tail Test $H_0: \mu \geq \mu_0$ $\bar{X}_c = \mu_0 - K\sigma/\sqrt{n}$	Right-Tail Test $H_0: \mu \leq \mu_0$ $\bar{X}_c = \mu_0 + K\sigma/\sqrt{n}$
Symmetric	$K = 1/\sqrt{2\alpha}$	$K = 1/\sqrt{2\alpha}$
Skewed to right	$K = 1/\sqrt{2\alpha}$	$K = 1/\sqrt{\alpha}$
Skewed to left	$K = 1/\sqrt{\alpha}$	$K = 1/\sqrt{2\alpha}$
Unknown	$K = 1/\sqrt{\alpha}$	$K = 1/\sqrt{\alpha}$

less than." If we let $\alpha = 1/(2K^2)$, then $K = \sqrt{1/(2\alpha)}$, and we could use the following decision rule: *For a level of significance less than α,*

$$\text{If } \bar{X} < \mu_0 - \frac{\sigma}{\sqrt{2\alpha n}} \qquad \text{reject } H_0$$

$$\text{If } \bar{X} \geq \mu_0 - \frac{\sigma}{\sqrt{2\alpha n}} \qquad \text{accept } H_0$$

(7.8)

Consider the right tail in Figure 7.3. If we know that the population is skewed to the right and if we are conducting a right-tail test (that is, $H_0: \mu \leq \mu_0$), then $P(\bar{X} > \mu_0 + K\sigma/\sqrt{n})$ is greater than $(0.5)(1/K^2)$ but less than $1/K^2$. Since it is not very comforting to know that the probability of Type I error is *greater than $1/(2K^2)$*, the only safe way to proceed is to assume that $\alpha = 1/K^2$ (or, $K = \sqrt{1/\alpha}$). Thus, the decision rule would be: *For a level of significance less than α,*

$$\text{If } \bar{X} > \mu_0 + \frac{\sigma}{\sqrt{\alpha n}} \qquad \text{reject } H_0$$

$$\text{If } \bar{X} \leq \mu_0 + \frac{\sigma}{\sqrt{\alpha n}} \qquad \text{accept } H_0$$

(7.9)

The reverse is valid for a population skewed to the left. Thus, the value of K depends upon both the skewness of the population and whether the test is right-tail or left-tail. Table 7.3 summarizes all possible situations.

Example 7.4   The Suspicious Mr. Kool

Suppose that "Conscienceness 30," a semicommune of 200 adults, wishes to acknowledge the relative industry of its members but still maintain the concept of "equal income distribution." Each individual is allowed to make as much money as possible during the year. Everyone submits a statement of income to the Internal Equalization Service. The

IES calculates the mean income and redistributes the income (i.e., taxes the rich, issues subsidies to the poor) until the standard deviation is equal to $500, but maintains the rank-ordering of income earners. (The relative income position is the same before and after taxes.) The records are kept confidential, but the IES publishes the mean income.

Last year, the IES said the mean income was $10,500. Mr. Kool is skeptical of the IES figure, but internal mistrust is rewarded with a form of excommunication. Because of the clandestine nature of his inquiry, Mr. Kool feels that a *random* sample of nine individual incomes is all he dare collect. For a 10 percent significance level, set up a decision rule for Mr. Kool.

The claim is that $\mu = \$10,500$; thus, $H_0: \mu = 10,500$ and $H_A: \mu \neq 10,500$. We do not know the shape of the population, but we do know that $\sigma = 500$. A sample of size 9 is too small for the CLT to be of much assistance; therefore, we are left with the Chebyshev formulation of the decision rule (see Equation 7.7), $\alpha = 0.10 = \frac{1}{10}$.

$$\bar{X}_c = \mu_0 \pm \sigma/\sqrt{\alpha n} = 10,500 \pm (500)/\sqrt{(1/10)(9)}$$

$$= 10,500 \pm (500)\sqrt{1.1111} = 10,500 \pm (500)(1.0541)$$

$$= 10,500 \pm 527.05 \quad (9,972.95 \text{ and } 11,027.05)$$

If $\$9,972.95 \leq \bar{X} \leq \$11,027.05$, accept H_0; if $\bar{X} < \$9,972.95$ or if $\bar{X} > \$11,027.05$, reject H_0.

7.1(c) Type II Error

"'Close' only counts in horseshoes and essay examinations" is a cliché that has enjoyed an enduring popularity with perfectionists. To the regal list of "horseshoes" and "essay examinations," we might add "tests of hypothesis."

Type II error is the acceptance of a null hypothesis which is false. In Equation 7.2, the probability of Type II error was defined as $P(\text{Accept } H_0 | H_0 \text{ is false}) = \beta$. The decision "accept H_0" will depend upon the decision rule; therefore, we always start the process of calculating β with the decision rule. For a left-tail test (that is, $H_0: \mu \geq \mu_0$), β is given by $P(\bar{X} > \bar{X}_c | H_0 \text{ is false})$; for a right-tail test (that is, $H_0: \mu \leq \mu_0$), $\beta = P(\bar{X} < \bar{X}_c | H_0 \text{ is false})$; and for a two-tail test (that is, $H_0: \mu = \mu_0$), $\beta = P(\bar{X}'_c < \bar{X} < \bar{X}_c | H_0 \text{ is false})$.

Consider the left-tail test of hypothesis from a normally distributed population with known standard deviation. In Figure 7.4a, a decision rule, "if $\bar{X} \leq \bar{X}_c$, reject H_0; if $\bar{X} > \bar{X}_c$, accept H_0," has been derived from $\bar{X}_c = \mu_0 - z_\alpha \sigma/\sqrt{n}$. This decision rule *assumes* the null hypothesis ($\mu \geq \mu_0$) is true.

For the sake of argument, assume that the population's mean is μ_1, which is just a little bit less than μ_0; specifically, we want the true mean to lie somewhere between the critical value of the sample mean and the hypothesized mean ($\bar{X}_c < \mu_1 < \mu_0$). Since $\mu_1 < \mu_0$, the null hypothesis is *false*. What is the probability of accepting the null hypothesis when $\mu = \mu_1$?

FIGURE 7.4 Calculation of β, for a left-tail test. (a) Decision rule formulated; (b) Calculation of β, when the true population mean is μ_1.

$$P(\text{Accept } H_0 \mid \mu = \mu_1) = \beta_1$$
$$= P(\bar{X} \geq \bar{X}_c \mid \mu = \mu_1)$$

(c), (d), and (e) Other values of β for various possible *true* values of the population mean. (f) Operating characteristic curve.

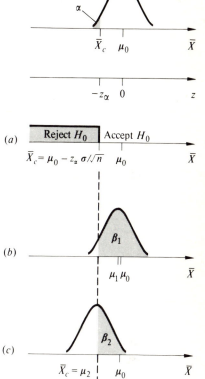

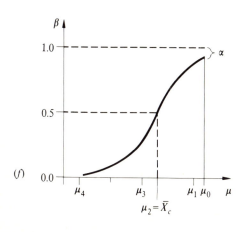

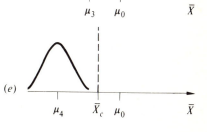

Recall, we know that the population is normal with standard deviation σ; however, we do not know the value of the population's mean. If the population's true mean is μ_1, then the sampling distribution will have a mean of μ_1 and a standard deviation of $\sigma/\sqrt{n}$, as in Figure 7.4b. It looks exactly like the sampling distribution from which the decision rule was generated *except*, since $\mu_1 < \mu_0$, the true sampling distribution is moved slightly to the left.

Given that the true mean is μ_1, we can calculate β. That is, the $P(\text{Accept } H_0 | H_0$ is false) can be written as

$$\begin{aligned}
\beta_1 &= P(\bar{X} > \bar{X}_c | \mu = \mu_1) \\
&= P(\bar{X} > \mu_0 - z_\alpha \sigma/\sqrt{n} | \mu = \mu_1) \\
&= P\left[z > \frac{(\mu_0 - z_\alpha \sigma/\sqrt{n}) - \mu_1}{\sigma/\sqrt{n}} \right] \\
&= P\left(z > \frac{\mu_0 - \mu_1}{\sigma/\sqrt{n}} - z_\alpha \right)
\end{aligned} \tag{7.10}$$

The shaded area in Figure 7.4b is β_1. By inspection we can see that $P[(\bar{X}_c < \mu_1 < \mu_0) | \mu = \mu_1] > 0.5$. That is, for any true population mean value between the hypothesized mean and the critical value of $\bar{X}$, the probability of committing Type II error *exceeds* 0.5.

What happens in the unusual situation where the true value of the population mean μ_2 is equal to the critical value? From Figure 7.4c, we see that one-half the area under the sampling distribution is to the right of $\bar{X}_c$ and one-half is to the left of $\bar{X}_c$. Thus, $\beta = 0.50$. Alternatively,

$$\beta_2 = P(\bar{X} > \bar{X}_c | \mu = \bar{X}_c) = P\left(z > \frac{\bar{X}_c - \bar{X}_c}{\sigma/\sqrt{n}} \right) = P(z > 0) = 0.50$$

By inspection, Figure 7.4d and e shows that β is less than 0.5 if the true value of the mean is less than $\bar{X}_c$. Let μ_T be the "true value of the population mean," which is unknown. By playing the "what if" game (i.e., what if $\mu_T = \mu_1$? what if $\mu_T = \mu_2$? etc.) we can calculate the corresponding value of β.

From Figure 7.4 we can see that when $|\mu_T - \mu_0|$ is small, β is large. When $|\mu_T - \mu_0|$ is large, β is small. When the null hypothesis is "not very wrong" (i.e., when μ_T is very close to μ_0), the probability of Type II error is great. When the null hypothesis is "very wrong" (i.e., when μ_T is far from μ_0), the probability of Type II error is small or close to 0. Thus, the opening remark "'Close' only counts in horseshoes, essay examinations, and hypothesis testing." The relationship between β and the various possible values of μ_T, plotted in Figure 7.4f, is called the *operating characteristic curve*.

Example 7.5   Calculation of β

Consider the producer of Century Light Bulbs, which are advertised to have "an average life of 100 hours or more." "Bulb life" is known to be normally distributed with a standard deviation of 5 hours. A production "run" consists of manufacturing between 7,000 and 8,000 bulbs.

(a)   Construct the decision rule, for a random sample of 25 bulbs, at the 10 percent level of significance. Use $\alpha = 0.10$.

Null and Alternative Hypotheses

$$H_0: \mu \geq 100 \text{ hours} \qquad H_A: \mu < 100 \text{ hours}$$

Population
Bulb life is normally distributed with $\sigma = 5$ hours. Assuming the null is true, the mean is $\mu_0 = 100$ hours.

Sampling Distribution
From CLT, $\bar{X}$ is normally distributed regardless of sample size because the population is normally distributed. The sampling distribution has a mean of 100 hours and a standard deviation of $\sigma/\sqrt{n} = 5/\sqrt{25} = 1$.

Level of Significance
$P(\text{Type I error}) = \alpha = 0.10$.

Critical Value of z
From Appendix Table IV(B), $P(z \leq -z_\alpha) = 0.1000$ yields $-z_\alpha = -1.28$.

Critical Value of $\bar{X}_c$
From Equation 7.5,

$$\bar{X}_c = \mu_0 - z_\alpha \sigma/\sqrt{n} = 100 - [(1.28)(5)]/\sqrt{25}$$
$$= 100 - 1.28 = 98.72 \text{ hours}$$

Decision Rule
If $\bar{X} \leq 98.72$ hours, reject H_0; if $\bar{X} > 98.72$ hours, accept H_0. These results are shown in Figure 7.5a.

(b)   Calculate the probability of accepting the null hypothesis, using the above decision rule, when the "true population mean" is 99 hours.

Assumed "True" Population
Bulb life is normally distributed with $\sigma = 5$ hours, and μ is assumed to be 99 hours.

Assumed "True" Sampling Distribution
It is normally distributed (with mean of 99 and standard deviation of $\sigma/\sqrt{n} = 1$).

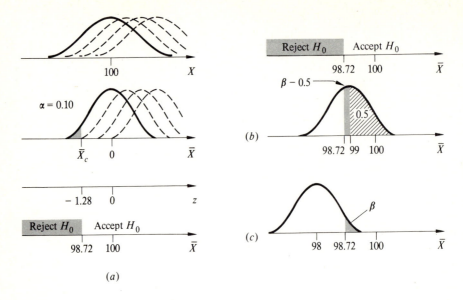

FIGURE 7.5 (*a*) Determination of decision rule; (*b*) and (*c*) Calculation of β.

Probability of Type II Error
Type II error can be committed only if the null hypothesis (that is, $H_0: \mu \geq 100$) is accepted. The "decision rule" in part *a* will suggest the acceptance of the null only if $\bar{X} > 98.72$ hours. Thus, β is the probability of $\bar{X}$ being greater than 98.72 when the sample is taken from a population whose mean is 99. See Figure 7.5*b*.

Method 1 β is the area under the sampling distribution to the right of 98.72. In Figure 7.5*b*, the "cross-hatched" area is the right half of the sampling distribution; the "shaded" area is between 98.72 and 99. Thus,

$$\beta = P(\bar{X} > 98.72 \mid \mu = 99)$$
$$= P(98.72 < \bar{X} \leq 99.00) + P(99.00 < \bar{X} \leq +\infty)$$
$$= P(98.72 < \bar{X} \leq 99.00) + 0.50$$

Since $z = (\bar{X} - \mu)/\sigma_{\bar{X}}$, then

$$P(98.72 < \bar{X} \leq 99.00) = P\left(\frac{98.72 - 99}{1} < z \leq \frac{99 - 99}{1}\right)$$
$$= P(-0.28 < z \leq 0)$$

(because *z* curve is symmetrical, then)

$$= P(0 \leq z \leq +0.28)$$

From Appendix Table IV(A), $P(0 \leq z \leq 0.28) = 0.1103$. Thus, $\beta = 0.1103 + 0.5000 = 0.6103$.

Method 2 The same result is obtained by using Equation 7.10.

$$\beta = P\left(z > \frac{\mu_0 - \mu_1}{\sigma/\sqrt{n}} - z_\alpha\right) = P\left(z > \frac{100 - 99}{1} - 1.28\right)$$

$$= P(z > -0.28) = P(-0.28 < z \leq 0) + P(0 \leq z \leq +\infty)$$

$$= 0.1103 + 0.5000 = 0.6103$$

(c) Using the decision rule from part *a*, calculate the probability of accepting the null hypothesis when the "true" mean is 98 hours. This situation is shown in Figure 7.5c. Using Equation 7.10,

$$\beta = P\left(z > \frac{\mu_0 - \mu_1}{\sigma/\sqrt{n}} - z_\alpha\right) = P\left(z > \frac{100 - 98}{1} - 1.28\right)$$

$$= P(z > +0.72)$$

From Appendix Table IV(B), we find

$$\beta = P(z > +0.72) = 0.2358$$

Note: Parts *b* and *c* demonstrate the results that, for a left-tail test of hypothesis for a population mean, $0.5 \leq \beta < (1 - \alpha)$ when $\bar{X}_c \leq \mu_T < \mu_0$, and $\beta < 0.5$ when $\mu_T < \bar{X}_c$.

In example 7.5, we generalized the use of Equation 7.10. In general, for a *left-tail test of hypothesis* (that is, $H_0: \mu \geq \mu_0$), the value[5] of β is given by

$$\beta = P\left(z > \frac{\mu_0 - \mu_T}{\sigma/\sqrt{n}} - z_\alpha\right) \qquad (7.11)$$

By the same reasoning, it can be shown that, for a *right-tail test of hypothesis* (that is, $H_0: \mu \leq \mu_0$), the value of β is given by

$$\beta = P\left(z < \frac{\mu_0 - \mu_T}{\sigma/\sqrt{n}} + z_\alpha\right) \qquad (7.12)$$

---

[5] The situations where σ is unknown will be considered in the next section. Although β will not be reconsidered in that section, Equations 7.11, 7.12, and 7.13 are valid when s is substituted for σ and when t is substituted for z.

Further, it can be shown that, for a *two-tail test of hypothesis* (that is, $H_0: \mu = \mu_0$; $H_A: \mu \neq \mu_0$), the value of β is given by

$$\beta = P\left(\frac{\mu_0 - \mu_T}{\sigma/\sqrt{n}} - z_{\alpha/2} \leq z \leq \frac{\mu_0 - \mu_T}{\sigma/\sqrt{n}} + z_{\alpha/2}\right) \tag{7.13}$$

and the reader should note the use of $z_{\alpha/2}$ instead of z_α.

Properties of the Operating Characteristic Curve

The procedures described above allow the calculation for β for any particular value of the true mean, μ_T. The *operating characteristic curve* (occasionally abbreviated the "OC" curve) is a plot of the various paired values of β and μ_T. (See Figure 7.4f.)

Statistical studies frequently plot the paired values of $(1 - \beta)$ and μ_T, which is the probability of rejecting a false null hypothesis for different values of μ_T. The relationship between $(1 - \beta)$ and μ_T is called the *power function* or the "power of the test of hypothesis." Since the power curve can be easily derived from the OC curve, only the operating characteristic curve will be discussed.

Instead of plotting a set of paired values of β and μ_T to map out an OC curve, we can obtain a general picture of the OC curve from just three points on it. Table 7.4 presents these three values of β, and Figure 7.6 shows these values on the OC curves.

From Equations 7.11, 7.12, and 7.13, when $\mu_T = \mu_0$, $\beta = (1 - \alpha)$. We have already discussed the fact that when $\mu_T = \bar{X}_c$, $\beta = 0.5$ (i.e., see Figure 7.4c).

In Figure 7.4, we observed that as μ_T moved away from μ_0 and past $\bar{X}_c$, $\beta < 0.5$. How far away from $\bar{X}_c$ do we have to get in order to say that $\beta \approx 0$? From our discussion of the normal curve in Chapter 3, if the true mean of the sampling distribution is three standard deviations away from $\bar{X}_c$, then there is very little area left in the tail beyond $\bar{X}_c$ (see Figure 7.4e). Thus, for a left-tail test, $P(\bar{X} \geq \bar{X}_c | \mu_T = \bar{X}_c - 3\sigma/\sqrt{n}) \approx 0$.

TABLE 7.4 THREE IMPORTANT VALUES OF β

True Value of Mean	Probability of Type II Error
$\|\mu_0 - \mu_T\| \approx 0$	$\beta \approx 1 - \alpha$
$\mu_T = \bar{X}_c$	$\beta = 0.5^*$
$\|\bar{X}_c - \mu_T\| \geq \dfrac{3\sigma}{\sqrt{n}}$	$\beta \approx 0$

* β is exactly equal to 0.5 for a one-tail test. For a two-tail test, $\beta = P(0 < z < 2z_{\alpha/2})$, which is essentially equal to 0.5 for $\alpha \geq 0.12$.

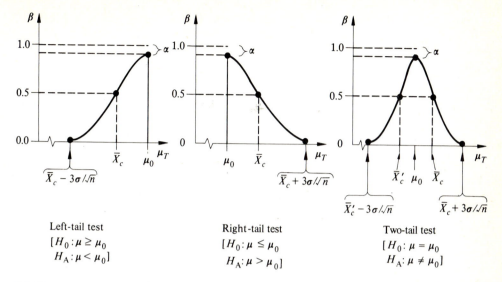

FIGURE 7.6   Properties of the operating characteristic curve.

Decision Maker's Choice of β

There are several ways for a decision maker to approach the problem of what is an "acceptable" OC curve. Only two approaches will be discussed here. The simplest approach is to ignore β, but that is usually unacceptable to everyone involved with the decision.

The next easiest (and probably not much more acceptable) approach requires the decision maker to think in terms of "At what value(s) of μ_T from the alternative hypothesis would I be willing to live with a 50:50 chance of accepting the null hypothesis when it is false?"

Reconsider the producer of Century Light Bulbs in example 7.5. The decision maker was free to choose the values for the level of significance and the sample size. Given the known population information, the critical value of the sample mean was calculated to be 98.72 hours. Thus, $\beta = P(\bar{X} > 98.72 \mid \mu_T = 98.72) = 0.5$. Further, we know that $\beta = P(\bar{X} > 98.72 \mid 98.72 \leq \mu_T \leq 100) \geq 0.5$.

Is the producer willing to accept a 50:50 chance of making a Type II error if the true average bulb life is 98.72 hours? Let us assume the producer decides that she or he wants a decision rule that would yield $\beta = 0.5$ for $\mu_T = 99$ hours. The decision rule calculated in example 7.5 is unacceptable because the risk of Type II error is too large. Since $\beta = 0.5$ for $\mu_T = \bar{X}_c$, then $\bar{X}_c$ must be increased from 98.72 to 99.00.

At this point, the decision maker has three options. First, she or he can increase only the risk of committing Type I error. (That is, increasing α will decrease z_α, which will increase the value of $\bar{X}_c$.) Second, the decision maker could increase only the size of the sample selected (i.e., increasing n will increase $\bar{X}_c$). Finally, she or he could use a combination of the first two.

What is this producer of light bulbs likely to do? First, consider Type I error. Committing Type I error means that the producer scraps (or demotes to a lower price, lower grade) a perfectly good batch of bulbs. We would expect the producer to choose a very small value for α. Further, since the producer immediately absorbs the financial loss, α is sometimes called the *producer's risk*. If the producer commits Type II error, she or he ships bulbs whose average life is not as advertised. (There is no fraud in this situation; the statistical test did not reject the null, and the producer thinks the bulb life is as advertised. Destructive sampling prohibits a census.) Thus, β is sometimes called the *consumer's risk*. The producer could incur costs in committing Type II error either by having the bulbs returned by a retailer (who tested a random sample with contrary results) or by the loss of future sales; however, these "costs" are less certain than those associated with Type I error. Since the sampling procedure destroys the usefulness of the product, there are obvious costs associated with taking a larger sample.

Obviously, anything the producer does to reduce β will increase the direct costs. The equally obvious result is not to advertise the quality of the product. Is there any doubt why (1) consumer advocates want "performance specifications" included with the product and (2) producers are reluctant to comply? Without consumer testing of products, producers are likely to ignore β; with consumer testing, producers are reluctant to advertise performance (because of the increased cost of quality control and internal testing).

Neither the producer nor the consumer are uniquely "the baddie." The existence of "sampling error" requires the possibility of *both* Type I and Type II errors. The producer could commit Type II error, and the consumer could commit Type I error. If the producer and consumer differ about the relative importance of Type I and Type II errors, even a compromise on the proper values of α and β will be difficult.

The interdependency of α, β, and n suggests some form of simultaneous solution for their respective values. "Modern decision theory" is a classification for an area of inquiry which considers these problems. Chapter 14 will provide an introduction to those techniques.

Ignoring β leaves the discussion of tests of hypothesis incomplete. However, primarily for pedagogical reasons, Type II error and β will receive very little additional attention.

7.2 TEST OF HYPOTHESIS ABOUT A POPULATION MEAN: POPULATION VARIANCE IS UNKNOWN

What is the probability that a decision maker would encounter a "real-world" test of hypothesis problem where the population standard deviation is known? Subjective assessments of that probability vary, but they do not include 0 or 1. Further, even many experienced applied statisticians agree that the "σ unknown" situation is more likely.

When the population standard deviation is unknown, the procedures for a test of hypothesis about the population mean conveniently fit into two categories:

1. **Large Samples**
   If the population is normally distributed, then $n \geq 30$ qualifies as "large samples"; if the population has either an unknown shape or a known but nonnormal distribution, then $n \geq 50$ will suffice. (The z distribution and s as an estimate of σ will be used.)
2. **Small Samples**
   Obviously, if $n \geq 30$ is "large," then $n < 30$ is "small" when the sample is selected from a normally distributed population. (The t distribution and s as an estimate of σ will be used.)

   As in Chapter 6, we reach a "dead end" if we attempt to use samples of size smaller than 50 from a population of unknown (or nonnormal) shape.

By using the sample's standard deviation s as an estimate of the unknown population standard deviation σ, it is no longer possible to calculate the critical value of the sample mean $\bar{X}_c$ *before* the sample information is known. The general form of $\bar{X}_c$ in section 7.1 was $\bar{X}_c = \mu_0 \pm z\sigma/\sqrt{n}$; however, if σ is unknown, $\bar{X}_c$ cannot be calculated.

Note that the critical value of z is dependent only upon the value of α. [The critical value of t will depend upon both α and the degrees of freedom $(n - 1)$.] The decision rule can be stated in terms of z (or t) *before* the sample information is known. The "test statistic" is given by

$$z \text{ (or } t) = \frac{\bar{X} - \mu_0}{s/\sqrt{n}} \sqrt{\frac{N - 1}{N - n}} \qquad (7.14)$$

where $\bar{X}$ and s are calculated from the sample of n observations, N is the size of the population, and μ_0 is the hypothesized population mean. Either the value of z (or t), as given by Equation 7.14, falls within the "acceptance region" or it does not.

Stating the decision rule in terms of z (or t) rather than $\bar{X}$ is simply a *change in random variables to describe the same event*. The equation

$$\bar{X}_c = \mu_0 \pm \frac{zs}{\sqrt{n}} \sqrt{(N - n)/(N - 1)}$$

is algebraically identical to Equation 7.14. Thus, if $\bar{X} > \bar{X}_c$, then z will be greater than z_c; and if $\bar{X} < \bar{X}_c$, then $z < z_c$.

Also, there is one practical difference between this section and section 7.1. When

σ is known, the only calculation necessary from the sample data is $\bar{X} = (1/n) \sum X_i$. When σ is unknown, both $\bar{X}$ and s have to be calculated from the sample data (s uses Equation 4.11).

7.2(a) Large Samples

Whether we are considering a normal population with $n \geq 30$ or an unknown population shape with $n \geq 50$, the *raison d'être* may be different, but the results are the same. When the population is normally distributed, the exact sampling distribution

---

Panel 7.3   Decision Rule Formulation When σ Is Unknown: Right-Tail Test for "Large" Samples

Claim
Population mean is less than or equal to some specified value μ_0.

Null and Alternative Hypotheses

$$H_0: \mu \leq \mu_0 \qquad H_A: \mu > \mu_0$$

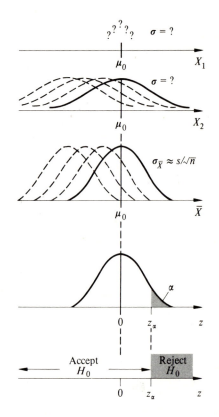

Population
Either X_1 has an unknown distribution (and $n \geq 50$), or X_2 has a normal distribution (and $n \geq 30$).

Distribution of Sample Means
The sampling distribution of $\bar{X}$ is approximately normal with $\mu_{\bar{X}} = \mu$ and $\sigma_{\bar{X}} = \sigma/\sqrt{n}$. The sample standard deviation s is used as an estimate of σ, and the null hypothesis' "limiting value" of μ is μ_0.

Distribution of Test Statistic
If $\bar{X}$ is normally distributed with $\mu_{\bar{X}} = \mu_0$ and $\sigma_{\bar{X}} \approx s/\sqrt{n}$, then $z = (\bar{X} - \mu_0)/(s/\sqrt{n})$ is normally distributed with $\mu_z = 0$ and $\sigma_z = 1$. Given the level of significance α, the critical value of z is z_α. $P(z \geq z_\alpha) = \alpha$.

Decision Rule
For an α significance level,

If $z = \dfrac{\bar{X} - \mu_0}{s/\sqrt{n}} \geq z_\alpha$ reject H_0

If $z = \dfrac{\bar{X} - \mu_0}{s/\sqrt{n}} < z_\alpha$ accept H_0

TABLE 7.5 DECISION RULES FOR LARGE-SAMPLE TEST OF μ WHEN σ IS UNKNOWN

Type of Null Hypothesis	Decision Rule if (1) Population Is Normally Distributed, $n \geq 30$, or (2) Population Is Nonnormally Distributed or Is Unknown $n \geq 50$
$H_0: \mu \geq \mu_0$ $H_A: \mu < \mu_0$ *Left Tail*	If $z = (\bar{X} - \mu_0)/(s/\sqrt{n}) \leq z_\alpha$, reject H_0; otherwise, accept H_0.
$H_0: \mu \leq \mu_0$ $H_A: \mu > \mu_0$ *Right Tail*	If $z = (\bar{X} - \mu_0)/(s/\sqrt{n}) \geq z_\alpha$, reject H_0; otherwise, accept H_0.
$H_0: \mu = \mu_0$ $H_A: \mu \neq \mu_0$ *Two Tail*	If $-z_{\alpha/2} < (\bar{X} - \mu_0)/(s/\sqrt{n}) < +z_{\alpha/2}$, accept H_0; otherwise, reject H_0.

of the sample statistic (Equation 7.14) is the t distribution. But, for $n \geq 30$ the z distribution is a very close approximation of the t (and we do not have to consider further degrees of freedom).

When the population is either of unknown shape or of known but nonnormal shape, then the central limit theorem permits us to assume that $\bar{X}$ is normally distributed for $n \geq 50$. Thus, the "test statistic" (Equation 7.14) has a z distribution.

The development of a right-tail test is outlined in Panel 7.3, which can be compared to Panel 7.1 where σ was known. Table 7.5 presents the decision rules for the large-sample situation when σ is unknown.

Example 7.6 Average Hours Worked per Week

In a community of diverse industrial composition, an economist wants to investigate the proposition that the "average hours worked per week" for wage earners has decreased from the average last year of 43 hours per week. She suspects the population (i.e., hours worked per week for *wage earners*, not the entire community) is skewed, and there is no reason to believe that the standard deviation from the data last year is valid for the situation this year. Let X = hours worked per week, and test the hypothesis that $\mu < 43$ hours per week at the 1 percent level of significance. First,

$$H_0: \mu \geq 43 \text{ hours per week}$$

$$H_A: \mu < 43 \text{ hours per week}$$

Committing Type I error would mean that we would say that the data suggests that the average hours worked per week has decreased when in fact the average has not decreased. For $\alpha = 0.01$, $P(z \leq z_\alpha) = 0.01$ and z_α is determined from Appendix Table IV(B) to be -2.33. Thus, the decision rule is:

$$\text{If } (\bar{X} - 43)/s_{\bar{X}} \le -2.33 \qquad \text{reject } H_0$$

$$\text{If } (\bar{X} - 43)/s_{\bar{X}} > -2.33 \qquad \text{accept } H_0$$

A sample from 64 randomly selected wage earners yields $\bar{X} = 42.6$ hours per week and $s = 0.80$ hour per week. Assume that 64 represents less than 10 percent of the wage earners within the community; the FPC is essentially equal to 1. Hence,

$$s_{\bar{X}} = \frac{s}{\sqrt{n}} = \frac{0.80}{\sqrt{64}} = 0.10$$

and

$$\frac{\bar{X} - 43}{s_{\bar{X}}} = \frac{42.6 - 43}{0.1} = -4$$

Since $-4 < -2.33$, *reject* the null hypothesis. The data supports the hypothesis that the average hours worked per week has decreased from the 43 hours per week recorded last year.

7.2(b)   Small Samples

Either the population of interest is normally distributed, or it is not. If it is, then t can be used as the test statistic, and a test of hypothesis about a population mean is possible.

If the population is not normally distributed, then not knowing σ is a serious problem. For small samples, we simply do not know enough about the sampling distribution to specify a critical value from α, the probability of Type I error. Thus,

TABLE 7.6   DECISION RULES FOR SMALL SAMPLE TEST OF μ WHEN σ IS UNKNOWN **AND** POPULATION IS NORMALLY DISTRIBUTED

Type of Null Hypothesis	Decision Rule
$H_0: \mu \ge \mu_0$ $H_A: \mu < \mu_0$ *Left Tail*	If $t = (\bar{X} - \mu_0)/(s/\sqrt{n}) \le -t_{\alpha,\, n-1}$, reject H_0; otherwise, accept H_0.
$H_0: \mu \le \mu_0$ $H_A: \mu > \mu_0$ *Right Tail*	If $t = (\bar{X} - \mu_0)/(s/\sqrt{n}) \ge t_{\alpha,\, n-1}$, reject H_0; otherwise, accept H_0.
$H_0: \mu = \mu_0$ $H_A: \mu \ne \mu_0$ *Two Tails*	If $-t_{\alpha/2,\, n-1} < (\bar{X} - \mu_0)/(s/\sqrt{n}) < +t_{\alpha/2,\, n-1}$, accept H_0; otherwise, reject H_0.

Panel 7.4 Decision Rule Formulation When σ Is Unknown: Right-Tail Test for Small Samples

Claim
Population mean is less than or equal to some specified value μ_0.

Null and Alternative Hypotheses

$$H_0: \mu \le \mu_0 \qquad H_A: \mu > \mu_0$$

Population
X is normally distributed; σ is unknown.

Distribution of Sample Means
$\bar{X}$ is normally distributed; however, this does not help us very much since $\sigma_{\bar{X}} = \sigma/\sqrt{n}$ is unknown.

Distribution of Test Statistic
If X is normally distributed, then $t = (\bar{X} - \mu_0)/(s/\sqrt{n})$ is t-distributed with $\mu_t = 0$ and $\sigma_t = f(n-1)$. Given the level of significance α and $n-1$ *degrees of freedom*, the critical value of t is $t_{\alpha,\,n-1}$. $P(t \ge t_{\alpha,\,n-1}) = \alpha$.

Decision Rule
For an α significance level,

$$\text{If } t = \frac{\bar{X} - \mu_0}{s/\sqrt{n}} \ge t_{\alpha,\,n-1}, \qquad \text{reject } H_0$$

$$\text{If } t = \frac{\bar{X} - \mu_0}{s/\sqrt{n}} < t_{\alpha,\,n-1}, \qquad \text{accept } H_0$$

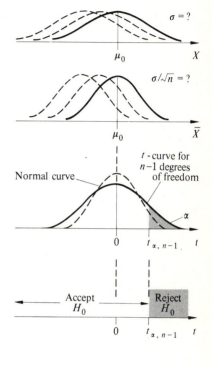

there is no general theory that will allow the construction of a decision rule. In such situations, taking a large sample is a solution.

Panel 7.4 outlines the logic for the decision rule in a right-tail test. Table 7.6 presents the three possible decision rules using the t distribution.

7.3 TESTS OF HYPOTHESIS ABOUT THE POPULATION PROPORTION π

In Chapter 5 we found the sampling distribution of $\bar{p}$ (the proportion of successes in a sample of n tries, r/n) to have a mean $\mu_{\bar{p}} = \pi$ and a standard deviation $\sigma_{\bar{p}} = \sqrt{(1/n)\pi(1-\pi)}$. In Chapter 6, we considered $\bar{p}$ as an estimator of the unknown population proportion π. Since $\sigma_{\bar{p}} = f(\pi)$, then $\sigma_{\bar{p}}$ is always unknown.

In a test of hypothesis about π, the claim leads to the inclusion of some specific population proportion π_0 in the null hypothesis. Further, by assuming the null is true, the decision rule incorporates that value of π_0. Since $\sigma_{\bar{p}} = f(\pi_0)$, a test of hypothesis about the population proportion *always* assumes that the mean and standard deviation of the sampling distribution for $\bar{p}$ are *known*. For confidence-interval estimates of π, $\mu_{\bar{p}}$ and $\sigma_{\bar{p}}$ are always unknown; for tests of hypothesis about π, they are always known.

Let us return to that "squint-eyed, scar-faced gentleman" who was introduced in the introductory section of this chapter. He proposed a game-of-chance with *his* "fair coin." If we define π as the proportion of heads that would appear from the coin, then his claim leads to: $H_0: \pi = 0.5$ and $H_A: \pi \neq 0.5$. For 10 flips ($n = 10$) of a fair coin, we know the sampling distribution for the proportion of successes. Table 7.7 is taken from Appendix Table III for $r = 0, 1, \ldots, 10$; $n = 10$; $\pi = 0.50$.

We still choose a level of significance α; however, *how* we choose α is different. The random variable $\bar{p}$ (or r) is discrete, and the number of possible values for α is also discrete. Note that $P(\bar{p} = 0) = P(\bar{p} = 1) = 0.0010$; it is possible in 10 flips of a fair coin to get all (or no) heads. An "unfair" coin could be one where $\pi < 0.5$ or where $\pi > 0.5$.

Without knowing the details about the game of chance, for the moment, assume that either form of "unfairness" is equally important. When $\pi = 0.5$, the sampling distribution for $\bar{p}$ is symmetric, and $\alpha/2$ represents the probability of Type I error in *each* tail. Table 7.8 presents the five possible values for α; it is from this list of five values for α (thus, five decision rules) that a decision maker must choose.

If we let $\alpha = 0.0020$, then our decision rule would be: "If $0.10 \leq \bar{p} \leq 0.90$, accept

TABLE 7.7 SAMPLING DISTRIBUTION OF $\bar{p}$ FOR $n = 10$, $\pi = 0.50$

Number of Heads (r)	Proportion of Heads ($\bar{p}$)	Probability $P(r)$ or $P(\bar{p})$
0	0.00	0.0010
1	0.10	0.0098
2	0.20	0.0439
3	0.30	0.1172
4	0.40	0.2051
5	0.50	0.2461
6	0.60	0.2051
7	0.70	0.1172
8	0.80	0.0439
9	0.90	0.0098
10	1.00	0.0010
		1.0001

TABLE 7.8 POSSIBLE VALUES FOR α; TWO-TAILED TEST, $n = 10$

Events Causing H_0 to Be Rejected		Probability of Type I Error
r	$\bar{p}$	(α)
0, 10	0.0, 1.0	0.0020
0, 1, 9, 10	0.0, 0.1, 0.9, 1.0	0.0216
0, 1, 2, 8, 9, 10	0, 0.1, 0.2, 0.8, 0.9, 1	0.1094
0, 1, 2, 3, 7, 8, 9, 10	0, 0.1, 0.2, 0.3, 0.7, 0.8, 0.9, 1	0.3438
0, 1, 2, 3, 4, 6, 7, 8, 9, 10	0, 0.1, 0.2, 0.3, 0.4, 0.6, 0.7, 0.8, 0.9, 1	0.7540

H_0; otherwise, reject H_0." There is not much risk of committing Type I error. Except for the cases of a "two-headed" coin (that is, $\pi = 1$) or a "two-tailed" coin (that is, $\pi = 0$), the probability of making a Type II error is quite high.

In Table 7.9, four values of β have been calculated for $\alpha = 0.0020$. Using the above decision rule, consider the probability of accepting the null hypothesis ($\pi = 0.5$) when the coin is weighted such that, in a large number of tosses, a head would appear only 40 percent of the time ($\pi = 0.4$).

$$\beta = P(\text{Accept } H_0 | \pi = 0.4)$$
$$= P(0.1 \leq \bar{p} \leq 0.9 | \pi = 0.4)$$
$$= P(1 \leq r \leq 9 | \pi = 0.4)$$
$$= 1 - P(r = 0 | \pi = 0.4) - P(r = 10 | \pi = 0.4)$$

From Appendix Table III, for $\pi = 0.4$ and $n = 10$, we find that $P(r = 0) = 0.0060$ and $P(r = 10) = 0.0001$; therefore,

$$\beta = 1 - 0.0060 - 0.0001 = 0.9939$$

TABLE 7.9 SELECTED β VALUES FOR EACH OF α; TWO-TAILED TEST, $n = 10$

"True" Values of π	Values of β Given α and the True Value of π				
	α				
	0.0020	0.0216	0.1094	0.3438	0.7540
0.40 (or 0.6)	0.9939	0.9520	0.8205	0.5630	0.2007
0.30 (or 0.7)	0.9717	0.8505	0.6165	0.3398	0.1029
0.20 (or 0.8)	0.8926	0.6242	0.3221	0.1200	0.0264
0.10 (or 0.9)	0.6513	0.2639	0.0702	0.0128	0.0015

which is given in Table 7.9 for $\alpha = 0.0020$ and $\pi = 0.4$. Since the sampling distribution suggested by the null is symmetric (and for small samples the distribution of $\bar{p}$ is only symmetric when $\pi = 0.5$), the β value when $\pi = 0.6$ is the same as when $\pi = 0.4$.

If the decision maker insists on taking a sample of size 10, then Table 7.9 provides the type of information from which a decision rule must be selected. The costs associated with Type I error and Type II error can provide a way of "balancing" α and β, but once α is selected, the decision rule is determined. That is, for each of the five possible values of α, Table 7.8 lists the "rejection region."

For larger sample sizes, there are (1) a larger number of possible values for α, (2) a larger number of possible decision rules, and (3) smaller values of β for a given α. In summary, *for small samples, any hypothesized population proportion (π_0) generates the exact sampling distribution of $\bar{p}$ from the binomial distribution. There are only a finite number of possible values for α; hence there are only a finite number of possible decision rules.* For any given "decision rule," various alternative values of π can be used to generate values of β from binomial distributions. Although this discussion has concentrated on a two-tail test of hypothesis, the same general procedure is followed for a single-tail test.

Large-Sample Tests of π

Most "real-world" situations which are amenable to a test of hypothesis concerning the population proportion can easily be conducted with "large samples." For example, it is usually more expensive to determine (1) the life of a light bulb, (2) the weight of a person, and (3) the gross income of a family than it is to ascertain whether a light bulb is defective (i.e., does it "light"?), whether a person weighs 200 pounds or more, and whether the gross family income exceeds $10,000.

A binomial population consists of two mutually exclusive but exhaustive *classifications.* All observations fit into one classification (a "success") or the other (a "failure"). Identifying the appropriate classification is usually cheaper than taking a measurement (to calculate a mean).

In Chapter 3, we defined a "large" sample as one where $n\pi(1 - \pi) \geq 3$. For $\pi = 0.5$, $n = 12$ is "large enough" with this rule. For, $\pi = 0.1$ (or $\pi = 0.9$) a sample of at least 34 is large enough. Remember that when $\pi = 0.5$, the sampling distribution of $\bar{p}$ is symmetric; therefore, it does not take a very large n in order for the normal curve to be a close approximation of the sampling distribution. The further π is from 0.5, the more skewed the binomial population and the larger the sample required in order to use the normal-curve approximation.

For large samples, the sampling distribution is approximated by a normal distribution with

$$\mu_{\bar{p}} = \pi \quad [= \pi_0 \text{ under the null hypothesis}] \tag{7.15}$$

and
$$\sigma_{\bar{p}} = \sqrt{\frac{\pi_0(1 - \pi_0)}{n}} \tag{7.16}$$

TABLE 7.10 DECISION RULES FOR TEST OF π: LARGE SAMPLES

Type of Null Hypothesis	Critical Value(s) for $\bar{p}$ and Decision Rule
$H_0: \pi \geq \pi_0$ $H_A: \pi < \pi_0$ *Left Tail*	$\bar{p}_c = \pi_0 + (0.5/n) - z_\alpha \sqrt{(1/n)\pi_0(1-\pi_0)}\sqrt{(N-n)/(N-1)}$ If $\bar{p} \leq \bar{p}_c$, reject H_0; otherwise, accept H_0.
$H_0: \pi \leq \pi_0$ $H_A: \pi > \pi_0$ *Right Tail*	$\bar{p}_c = \pi_0 - (0.5/n) + z_\alpha \sqrt{(1/n)\pi_0(1-\pi_0)}\sqrt{(N-n)/(N-1)}$ If $\bar{p} \geq \bar{p}_c$, reject H_0; otherwise, accept H_0.
$H_0: \pi = \pi_0$ $H_A: \pi \neq \pi_0$ *Two Tails*	$\bar{p}_c' = \pi_0 + (0.5/n) - z_{\alpha/2}\sqrt{(1/n)\pi_0(1-\pi_0)}\sqrt{(N-n)/(N-1)}$ $\bar{p}_c = \pi_0 - (0.5/n) + z_{\alpha/2}\sqrt{(1/n)\pi_0(1-\pi_0)}\sqrt{(N-n)/(N-1)}$ If $\bar{p}_c' < \bar{p} < \bar{p}_c$, accept H_0; otherwise, reject H_0.

The (chosen) value of α determines z, and the critical value(s) of the sample proportion $\bar{p}_c$ assumes the familiar general form:

$$\bar{p}_c = \mu_{\bar{p}} \pm z\sigma_{\bar{p}} \qquad (7.17)$$

Panel 7.5 outlines the logic of developing a right-tail test of π, and Table 7.10 summarizes the three possible decision rules. The critical values for the sample proportion in Table 7.10 appear to be considerably more complicated than the "general form" given by Equation 7.17.

First, the values for $\bar{p}_c$ in Table 7.10 contain the finite population correction factor [that is, $\sqrt{(N-n)/(N-1)}$], which is assumed to be equal to 1 if $n/N \leq 0.10$. Second, the continuous random variable z is used as an approximation for the discrete random variable $\bar{p}$. This "continuous-discrete" discrepancy requires the consideration of the continuity correction factor discussed in section 3.5. Thus, the anatomy of the critical value for the sample proportion in a right-tail test could be:

$$\bar{p}_c = \pi_0 - \frac{0.5}{n} + z_\alpha \sqrt{(1/n)(\pi_0)(1-\pi_0)} \, \sqrt{(N-n)/(N-1)}$$

Finite population correction

$\sigma_{\bar{p}}$ when FPC = 1

z-value for α level of significance

Continuity correction factor

Hypothesized value of π

Panel 7.5 **Decision Rule Formulation for a Right-Tail Test of Hypothesis of the Population π: "Large" Samples**

Claim
Population proportion is less than or equal to some specified value π_0.

Null and Alternative Hypotheses

$$H_0: \pi \le \pi_0 \qquad H_A: \pi > \pi_0$$

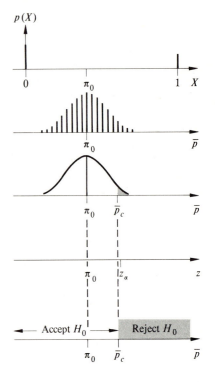

Population
Each success is arbitrarily assigned the value 1, and each failure, the value 0. π_0 is the proportion of successes if the limiting value of the null hypothesis is true.

Exact Sampling Distribution
$\bar{p}$ has a binomial distribution for $\pi = \pi_0$ and n trials.

Approximate Sampling Distribution
If $n\pi_0(1 - \pi_0) \ge 3$, then $\bar{p}$ is closely approximated by a normal curve with $\mu_{\bar{p}} = \pi_0$ and $\sigma_{\bar{p}} = \sqrt{(1/n)\pi_0(1 - \pi_0)}$.

Level of Significance
Since α is chosen, then z_α is determined from $P(z \ge z_\alpha) = \alpha$.

Critical Value of $\bar{p}$
Considering the continuity correction factor $(0.5/n)$, the value of $\bar{p}$ corresponding to z_α is

$$\bar{p}_c = \pi_0 - 0.5/n + z_\alpha\sqrt{(1/n)\pi_0(1 - \pi_0)}$$

Decision Rule
For an α significance level,

$$\text{If } \bar{p} \ge \bar{p}_c \qquad \text{reject } H_0$$

$$\text{If } \bar{p} < \bar{p}_c \qquad \text{accept } H_0$$

7.4 COMPARISON OF TWO POPULATIONS

Occasionally claims and/or decisions are made about the difference in means from two populations. "The mean annual income for female physicians is the same as for male physicians." Or, the proportion of successes from two different populations may be of interest: "The proportion of adult Americans who attend at least one high school athletic contest a year is greater than the proportion who vote in national elections."

Decision problems of this nature require only a little more knowledge of expected values, variances, and sampling distributions. We will start with this "additional knowledge" or "theoretical base."

7.4(a) The Theoretical Base

If X and Y are *independent* random variables, and both are normally distributed, then their difference

$$D = X - Y \tag{7.18}$$

is normally distributed with mean

$$\bar{D} = \mu_X - \mu_Y \tag{7.19}$$

and variance

$$\sigma_D{}^2 = \sigma_X{}^2 + \sigma_Y{}^2 \tag{7.20}$$

Although we will not prove this statement, the validity of Equations 7.19 and 7.20 can be easily demonstrated.

Example 7.7 The Mean and Variance of the Difference of X and Y

Consider $X = \{1, 2, 2, 3\}$ and $Y = \{4, 5, 5, 6\}$.

$$\mu_X = \frac{\Sigma X}{N} = \frac{8}{4} = 2 \quad \text{and} \quad \mu_Y = \frac{\Sigma Y}{N} = \frac{20}{4} = 5$$

$$\sigma_X{}^2 = \frac{\Sigma(X - \mu_X)^2}{N_X} = (1/4)[(1 - 2)^2 + (2 - 2)^2 + (2 - 2)^2 + (3 - 2)^2] = 1/2$$

$$\sigma_Y{}^2 = \frac{\Sigma(Y - \mu_Y)^2}{N_Y} = (1/4)[(4 - 5)^2 + (5 - 5)^2 + (5 - 5)^2 + (6 - 5)^2] = 1/2$$

Now consider $D = X - Y$. We will want to construct the probability distribution for D and then find the mean and variance of that distribution.

X \ Y [P(Y)]	4 [1/4]	5 [2/4]	6 [1/4]	
[P(X)]				
1 [1/4]	-3 [1/16]	-4 [2/16]	-5 [1/16]	$\leftarrow D = X - Y$ $\leftarrow P(D)$
2 [2/4]	-2 [2/16]	-3 [4/16]	-4 [2/16]	
3 [1/4]	-1 [1/16]	-2 [2/16]	-3 [1/16]	

D	$P(D)$	$D \cdot P(D)$	$D - \mu_D$	$(D - \mu_D)^2$	$P(D) \cdot (D - \mu_D)^2$
-1	1/16	- 1/16	+2	4	4/16
-2	4/16	- 8/16	+1	1	4/16
-3	6/16	- 18/16	0	0	0/16
-4	4/16	- 16/16	-1	1	4/16
-5	1/16	- 5/16	-2	4	4/16
	16/16	- 48/16			16/16

$$\bar{D} = \mu_D = E(D) = \Sigma[D \cdot P(D)] = -48/16 = -3$$

$$\sigma_D{}^2 = \Sigma\{P(D) \cdot [D - E(D)]^2\} = 16/16 = 1$$

Note that all these calculations using D were unnecessary. That is,

$$\bar{D} \text{ or } \mu_D = \mu_X - \mu_Y = 2 - 5 = -3$$

and

$$\sigma_D{}^2 = \sigma_X{}^2 + \sigma_Y{}^2 = 1/2 + 1/2 = 1$$

7.4(b) Comparing the Means of Two Populations

One of the more common uses of this test is the use of one population as a "control" group and the other population as an "experimental" group. At this point in the twentieth century, it is common knowledge that the application of fertilizer can increase the yield of a food crop. But how would we test the validity of this "common knowledge"? A control group could be several parcels of land that are planted and allowed to grow without the use of fertilizer. The experimental group could be several different parcels which use fertilizer. We would then want some test of the difference between the average yield (i.e., bushels per acre) of each group.

In the development of pharmaceutical chemicals, a new medicine is administered to the experimental group, and the control group is not so treated. Basically, the "control versus experimental" group technique is useful to determine if some *additive* improves performance.

In contrast, we may also wish to determine if two similar products (or services or "conditions") exhibit the same performance (or results). For example, does the famous "brand X" light bulb (or battery, felt-tip pen, or lawn mower) have the same average life as brand Y?

Whatever the purpose of the test, since you can choose which variable is called X and which variable is called Y, the *claim* can always be rewritten in one of the following two general forms:

1. The mean of X is *greater than or equal to* the mean of Y.
2. The mean of X is *equal to* the mean of Y.

Note that if the claim states "$\mu_X > \mu_Y$," then you can just switch the definitions of X and Y and claim 1 will be valid. Table 7.11 states the null and alternative hypotheses resulting from these two claims.

As suggested by our "theoretical base" stated in section 7.4(a), we want to conduct the test of hypothesis in terms of the single random variable D. If X is normally distributed, then the sampling distribution of $\bar{X}$ is normal with $\mu_{\bar{X}} = \mu_X$ and $\sigma_{\bar{X}} = \sigma_X/\sqrt{n}$. The same is true for Y. Let us define the test statistic d as the difference between the two sample means:

$$d = \bar{X} - \bar{Y} \tag{7.21}$$

From section 7.4(a), d is normally distributed. From Equation 7.19, $\mu_d = \mu_{\bar{X}} - \mu_{\bar{Y}}$ and from the CLT,

$$\mu_d = \mu_X - \mu_Y = \bar{D} \tag{7.22}$$

TABLE 7.11 NULL AND ALTERNATIVE HYPOTHESES

	Null and Alternative Hypotheses for		
Claim	the Random Variables X and Y	the Random Variable $D = X - Y$	
1	$H_0: \mu_X \geq \mu_Y$ $H_A: \mu_X < \mu_Y$	$H_0: \bar{D} \geq 0$ $H_A: \bar{D} < 0$	$(\mu_X - \mu_Y \geq 0)$ $(\mu_X - \mu_Y < 0)$
2	$H_0: \mu_X = \mu_Y$ $H_A: \mu_X \neq \mu_Y$	$H_0: \bar{D} = 0$ $H_A: \bar{D} \neq 0$	$(\mu_X - \mu_Y = 0)$ $(\mu_X - \mu_Y \neq 0)$

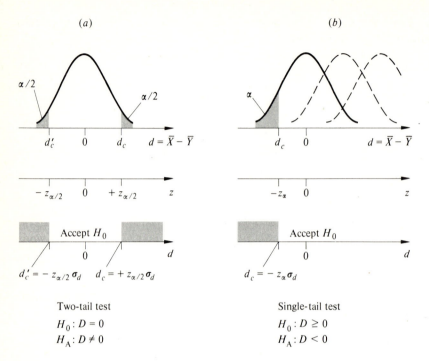

FIGURE 7.7 Decision rule formulation using d.

From Equation 7.20, $\sigma_d^2 = \sigma_X^2 + \sigma_Y^2$ and from the CLT,

$$\sigma_d^2 = \frac{\sigma_X^2}{n_X} + \frac{\sigma_Y^2}{n_Y} \qquad (7.23)$$

where n_X is the number of observations randomly selected from the X population and n_Y is the sample size from Y. (n_X does not have to be equal to n_Y.)

A unique feature of the test of two population means is that the relevant sampling distribution usually has a mean of 0 (that is, $\mu_d = \bar{D} = 0$). The mean of the sampling distribution (μ_d) is equal to the mean of the population ($\bar{D}$) from Equation 7.22. Obviously, for the two-tail test, $\bar{D} = 0$ and the relevant sampling distribution has a mean of 0. From Table 7.11, in the single-tail test the limiting value of $\bar{D}$ within the null hypothesis is $\bar{D} = 0$; therefore, $\mu_d = 0$.

Thus, our problem is to find a critical value of d (or d_c)—a "large enough" difference in the sample means to suggest that the population means are not equal. Consider Figure 7.7 where the sampling distribution is assumed to be normal. The general form of the critical value is

$$d_c = \mu_d \pm z\sigma_d$$

but since $\mu_d = 0$, then $d_c = \pm z \sigma_d$, where σ_d is given by Equation 7.23. Figure 7.7 provides the resulting decision rules.

Both σ_X and σ_Y are known When the standard deviations of both populations are known, then

$$\sigma_d = \sqrt{\frac{\sigma_X{}^2}{n_X} + \frac{\sigma_Y{}^2}{n_Y}} \qquad (7.24)$$

The level of significance α is chosen, and Table 7.12 provides the appropriate decision rule, which also depends upon the population distribution and the sample size.

A random sample of size n_X is selected from X, and the sample mean $\bar{X}$ is calculated. An independent random sample of size n_Y is selected from Y, and $\bar{Y}$ is calculated. The resulting difference in sample means, $d = \bar{X} - \bar{Y}$, is subjected to the appropriate decision rule from Table 7.12.

Both σ_X and σ_Y are unknown: Large samples When the standard deviations of both populations are unknown, then the unbiased estimator $s^2 = [1/(n-1)]\Sigma(X - \bar{X})^2$ is used to provide a point estimate of σ^2. For large samples, the standard deviation of the sampling distribution for the difference between two means is

$$s_d = \sqrt{\frac{s_X{}^2}{n_X} + \frac{s_Y{}^2}{n_Y}} \qquad (7.25)$$

TABLE 7.12 DECISION RULES WHEN σ_X AND σ_Y ARE KNOWN

Is D Normally Distributed*	Sample Size	Decision Rules for	
		Single-Tail Test	Two-Tail Test
Yes	Any n_X Any n_Y	If $\bar{X} - \bar{Y} \le -z_\alpha \sigma_d$, reject H_0; otherwise, accept H_0.	If $-z_{\alpha/2}\sigma_d < \bar{X} - \bar{Y} < +z_{\alpha/2}\sigma_d$, accept H_0; otherwise, reject H_0
No	$n_X \ge 20$ $n_Y \ge 20$	Same as above	Same as above
	$n_X < 20$ $n_Y < 20$	If $\bar{X} - \bar{Y} \le -\sigma_d/\sqrt{\alpha}$, reject H_0; otherwise, accept H_0.	If $-\sigma_d/\sqrt{\alpha} < \bar{X} - \bar{Y} < +\sigma_d/\sqrt{\alpha}$, accept H_0; otherwise, reject H_0.

* If either X or Y is nonnormal, then D is nonnormal; however, if $n \ge 20$ for the nonnormally distributed variable, the sampling distribution of d is approximately normal.

TABLE 7.13 LARGE SAMPLE DECISION RULES WHEN σ_X AND σ_Y ARE UNKNOWN

Is D Normally Distributed?	Sample Size	Decision Rules for	
		Single-Tail Test	Two-Tail Test
Yes	$n_X \geq 30$ $n_Y \geq 30$	If $\bar{X} - \bar{Y} \leq -z_\alpha s_d$, reject H_0; otherwise, accept H_0.	If $-z_{\alpha/2} s_d < \bar{X} - \bar{Y} < +z_{\alpha/2} s_d$, accept H_0; otherwise, reject H_0.
No	$n_X, n_Y \geq 50$		

Of course, if only one population variance is known (i.e., either $\sigma_X{}^2$ or $\sigma_Y{}^2$), that variance is used in Equation 7.25.

The level of significance α is chosen, and Table 7.13 provides the decision rules. Note that s_d, and thus the decision rule, cannot be completely calculated until *after* the sample information is known.

Note that the decision rules could be stated in terms of the standard normal deviate z rather than the random variable d. That is, (1) for a single-tail test, "If $z = (\bar{X} - \bar{Y})/s_d \leq -z_\alpha$, reject H_0; otherwise, accept H_0" and (2) for a two-tail test, "If $-z_{\alpha/2} < (\bar{X} - \bar{Y})/s_d < +z_{\alpha/2}$, accept H_0; otherwise, reject H_0." By using the test statistic z, the decision rule could be stated *before* the sample information is collected.

Example 7.8 Felt-Tip Pens

A large purchaser of felt-tip pens is considering brand A and brand B. Brand A claims that it will write longer, on the average, because it has more ink. Brand B claims that it will write longer because its writing point dispenses ink more evenly and requires less ink. Both pens have the same bid price. The purchaser decides that the risk of Type I error should be 0.05 and wants to test the hypothesis that $\mu_A > \mu_B$ (i.e. the "more ink" argument appeals to the purchaser).

Let X = the length of a continuous line written with a brand B pen and Y = the length of a line with brand A. Then,

$$H_0: \mu_X \geq \mu_Y \to H_0: \bar{D} \geq 0 \qquad \text{where } \bar{D} = \mu_X - \mu_Y$$

$$\mu_B < \mu_A \to H_A: \mu_X < \mu_Y \to H_A: \bar{D} < 0$$

A random sample of pens provided the following:

Brand A (Y):
$\bar{Y} = 1{,}390$ inches
$s_Y{}^2 = 300$ (inches)^2
$n_Y = 60$ pens

Brand B (X):
$\bar{X} = 1{,}380$ inches
$s_X{}^2 = 300$ (inches)^2
$n_X = 75$ pens

From Equation 7.25,

$$s_d = \sqrt{\frac{300}{75} + \frac{300}{60}} = \sqrt{4 + 5} = 3$$

Since $\alpha = 0.05$, then $P(z \leq -z_\alpha) = P(z \geq +z_\alpha) = 0.05$ and Appendix Table IV(A) gives $z_\alpha = 1.645$. Thus, from Table 7.13, the decision rule is. "If $d \leq -z_\alpha s_d = -(1.645)(3) = -4.935$, reject H_0; if $d = \bar{X} - \bar{Y} > -4.935$, accept H_0."

Since $d = \bar{X} - \bar{Y} = 1,380 - 1,390 = -10$, and $-10 < -4.935$, then the H_0 is rejected and the claim by brand A is supported by these test results.

Both σ_X and σ_Y are unknown: Small samples As desirable as small samples may be, the situation where both σ_X and σ_Y are unknown leaves the test of hypothesis of $\bar{D}$ as either very restricted or near impossible. If both X and Y can be assumed to be normally distributed, then the t distribution can be used if we add one further restriction: *both populations have the same standard deviation* $(\sigma_X = \sigma_Y)$. If either X or Y is not normally distributed, then there is no general test procedure available.

Returning to Equation 7.24, if $\sigma_X = \sigma_Y$, then

$$\sigma_d = \sqrt{\frac{\sigma_X{}^2}{n_X} + \frac{\sigma_Y{}^2}{n_Y}} = \sqrt{\sigma^2\left(\frac{1}{n_X} + \frac{1}{n_Y}\right)} = \sigma\sqrt{\frac{1}{n_X} + \frac{1}{n_Y}}$$

It may sound queer, but we assume that $\sigma_X = \sigma_Y$ and that both values are unknown. Thus, we need an estimate for σ. If we pool the information from both samples, then

$$s = \sqrt{\frac{\Sigma(X - \bar{X})^2 + \Sigma(Y - \bar{Y})^2}{(n_X - 1) + (n_Y - 1)}}$$

would serve as an estimate of σ. We can rewrite s as follows:

$$s = \sqrt{\frac{(n_X - 1)s_X{}^2 + (n_Y - 1)s_Y{}^2}{n_X + n_Y - 2}}$$

Hence, the estimate of σ_d becomes

$$s_d = s\sqrt{\frac{1}{n_X} + \frac{1}{n_Y}} = \sqrt{\frac{(n_X - 1)s_X{}^2 + (n_Y - 1)s_Y{}^2}{n_X + n_Y - 2}}\sqrt{\frac{1}{n_X} + \frac{1}{n_Y}} \qquad \textbf{(7.26)}$$

TABLE 7.14 DECISION RULES USING t FOR SMALL SAMPLES

Single-Tail Test	Two-Tail Test
Either	*Either*
If $\bar{X} - \bar{Y} \leq -t_{\alpha, \nu} s_d$, reject H_0;	If $-t_{\alpha/2, \nu} s_d < \bar{X} - \bar{Y} < +t_{\alpha/2, \nu} s_d$
otherwise, accept H_0.	accept H_0; otherwise, reject H_0.
Or	*Or*
If $\dfrac{\bar{X} - \bar{Y}}{s_d} \leq -t_{\alpha, \nu}$, reject H_0;	If $-t_{\alpha/2, \nu} < \dfrac{\bar{X} - \bar{Y}}{s_d} < +t_{\alpha/2, \nu}$,
otherwise, accept H_0	accept H_0; otherwise, reject H_0.

Our next problem is the degrees of freedom for t. In this test, the degrees of freedom are given by

$$\nu = (n_X - 1) + (n_Y - 1) = n_X + n_Y - 2$$

By using Equation 7.26 as an estimate of σ_d and using $t_{\alpha, \nu}$, the decision rules are given in Table 7.14.

Example 7.9 Turning Down the Thermostat at Night

A local oil distributor claims that turning down the thermostat at night makes no difference in how much heating oil is consumed. A local newspaper, skeptical of the distributor's motives, decides to test that claim. A reporter goes to Ticky-Tacky Village, where all the houses have exactly the same floor plan, and obtains the cooperation of 10 homeowners.

Five homes, the control group, will leave their thermostats set in a manner that will keep the house at 72°F all the time. Five other homes, the experimental group, will lower their thermostats to 66°F from 11 P.M. to 7 A.M. every day. The reporter wants a 10 percent level of significance. Oil consumption was metered for the duration of the experiment, and annual rates of oil usage are presented below.

	Control Group (Gallons)				Experimental Group (Gallons)	
X	$X - \bar{X}$	$(X - \bar{X})^2$		Y	$Y - \bar{Y}$	$(Y - \bar{Y})^2$
802	+ 102	10,404		804	+ 124	15,376
749	+ 49	2,401		738	+ 58	3,364
695	− 5	25		677	− 3	9
646	− 54	2,916		611	− 69	4,761
608	− 92	8,464		570	− 110	12,100
3,500		24,210		3,400		35,610

$$\bar{X} = (\Sigma X)/n_X = 3{,}500/5 = 700 \qquad\qquad \bar{Y} = (\Sigma Y)/n_Y = 3{,}400/5 = 680$$

$$s_X{}^2 = \frac{\Sigma(X - \bar{X})^2}{n_X - 1} = \frac{24{,}210}{4} \qquad\qquad s_Y{}^2 = \frac{\Sigma(Y - \bar{Y})^2}{n_Y - 1} = \frac{35{,}610}{4}$$

$$= 6{,}052.5 \qquad\qquad = 8{,}902.5$$

The null and alternative hypotheses are:

$$H_0: \mu_X - \mu_Y = 0$$

$$H_A: \mu_X - \mu_Y \neq 0$$

For $n_X + n_Y - 2 = 5 + 5 - 2 = 8$ degrees of freedom and $\alpha = 0.10$, the t values from Appendix Table V are ± 1.860.

From Table 7.14, the decision rule is

If $-1.86 < (\bar{X} - \bar{Y})/s_d < +1.86$, then accept H_0; otherwise, reject H_0.

From Equation 7.26, s_d is

$$s_d = \sqrt{\frac{24{,}210 + 35{,}610}{5 + 5 - 2}} \sqrt{\frac{1}{5} + \frac{1}{5}} = \sqrt{\frac{59{,}820}{8}} \sqrt{0.4}$$

$$= (86.4725)(0.63246) = 54.69$$

Since $(\bar{X} - \bar{Y})/s_d = (700 - 680)/54.69 = 0.37$, we accept the null hypothesis. Apparently, the oil distributor was correct.

7.4(c)   Matched-Pairs Difference Test

Up to this point, this section has repeatedly required that the two samples be *independently* drawn. One form of *dependence* between X and Y can be considered by using the difference between matched pairs. "Before" and "after" pictures are a common method of advertising the effectiveness of weight-reducing plans and bust-development programs.[6] Let us take a closer look at the implicit claim. If Y_i is the person's weight before starting the weight-reduction plan and X_i is that person's weight after (say) 60 days, then obviously X and Y are not independent. The values X_i and Y_i are a "matched pair." We could then consider the "weight difference"

$$\Delta_i = X_i - Y_i \qquad\qquad\qquad (7.27)$$

---

[6] These examples are subject to cardinal measurement. The "wet look" versus the "dry look" in men's hair would require an ordinal measurement of their "appearance" before and after the alteration in grooming.

and the weight difference for one individual would be independent of the weight difference for another person.

From a randomly selected sample of n matched pairs, the mean weight difference is

$$\bar{\Delta} = \frac{1}{n} \sum_{i=1}^{n} \Delta_i \tag{7.28}$$

and the standard deviation is

$$s_\Delta = \sqrt{\frac{\Sigma(\Delta_i - \bar{\Delta})^2}{n - 1}} \tag{7.29}$$

If the random variable Δ_i is normally distributed, then the sampling distribution of $\bar{\Delta}$ will be normally distributed with mean $\mu_{\bar{\Delta}} = \mu_\Delta$ and $\sigma_{\bar{\Delta}} = \sigma_\Delta/\sqrt{n}$. It is not likely that σ_Δ would be known; therefore, the test statistic would be

$$t = \frac{\bar{\Delta} - \mu_\Delta}{s_\Delta/\sqrt{n}} \tag{7.30}$$

with $(n - 1)$ degrees of freedom.

Now consider the null hypothesis. The "weight before" is greater than the "weight after," or $\mu_Y > \mu_X$ is the implicit claim. To include the equals sign within the null, the null hypothesis would be $\mu_X \geq \mu_Y$, or

$$H_0: \mu_\Delta \geq 0 \qquad (\mu_X - \mu_Y \geq 0, \text{ plan does not work})$$

$$H_A: \mu_\Delta < 0 \qquad (\mu_X - \mu_Y < 0, \text{ average weight decrease})$$

And, for a chosen value of α, the decision rule is given in Table 7.15. Note that $\mu_\Delta = 0$ in Equation 7.30.

TABLE 7.15 DECISION RULES FOR TESTING DIFFERENCE IN POPULATION MEANS FROM MATCHED PAIRS

Null and Alternative Hypotheses	Decision Rule (t Has $n - 1$ Degrees of Freedom)
$H_0: \mu_\Delta \geq 0$ $H_A: \mu_\Delta < 0$ *Single Tail*	If $\dfrac{\bar{\Delta}}{s_\Delta/\sqrt{n}} \leq -t_{\alpha,\,v}$, reject H_0; otherwise, accept H_0.
$H_0: \mu_\Delta = 0$ $H_A: \mu_\Delta \neq 0$ *Two Tails*	If $-t_{\alpha/2,\,v} < \dfrac{\bar{\Delta}}{s_\Delta/\sqrt{n}} < +t_{\alpha/2,\,v}$, accept H_0; otherwise, reject H_0.

First, it is important to note that this test is different. In section 7.4(b) and (c) the statistic d was *the difference between sample means* $(d = \bar{X} - \bar{Y})$; here the statistic $\bar{\Delta}$ is the *sample mean of differences*. The comparison of examples 7.9 and 7.10 will clarify this point. Second, if the number of randomly selected matched pairs is large, then the test statistic is z rather than t.

Finally, the "weight-reducing plan" example is one case in the general category of problems where *the two samples are pairs of observations from the same individuals or objects*. The matched-pair differences method can also be used to account for some other variable that might influence the differences. Example 7.10 demonstrates this.

Example 7.10   The Thermostat Revisited

In example 7.9, we considered the problem of whether the thermostat should be lowered at night. The test procedure used there could not reject the null hypothesis that there was no difference in average oil consumption. By using houses in Ticky-Tacky Village, the same space was being heated in each house, and all houses were subject to the same weather conditions; however, there are other variables which affect heating oil consumption. One of the more obvious is the amount of insulation.

Let us assume that the 10 houses in example 7.9 were matched pairs and that each pair was insulated differently. Now let us test the same hypothesis.

$$H_0: \mu_X - \mu_Y = 0 \qquad H_0: \mu_\Delta = 0$$

$$H_A: \mu_X - \mu_Y \neq 0 \qquad H_A: \mu_\Delta \neq 0$$

For $\alpha = 0.10$ and $(n - 1) = 5 - 1 = 4$ degrees of freedom, the t values from Appendix Table V are ± 2.132. From Table 7.15 the decision rule is

If $-2.132 < \bar{\Delta}/s_{\bar{\Delta}} < +2.132$, accept H_0; otherwise, reject H_0.

Now consider the sample data from example 7.9:

Insulation	X_i	Y_i	Δ_i	$\Delta_i - \bar{\Delta}$	$(\Delta - \bar{\Delta})^2$
None	802	804	-2	-22	484
Light ceiling	749	738	11	-9	81
Heavy ceiling	695	677	18	-2	4
Light ceiling and sidewall	646	611	35	15	225
Heavy ceiling and sidewall	608	570	38	18	324
			100		1,118

$$\bar{\Delta} = (\Sigma \, \Delta_i)/n = 100/5 = 20$$

$$s_{\bar{\Delta}} = \sqrt{\Sigma(\Delta - \bar{\Delta})^2/[n(n-1)]} = \sqrt{1,118/[5(4)]} = \sqrt{55.9} = 7.4766$$

Since $\bar{\Delta}/s_{\bar{\Delta}} = 20/7.4766 = 2.675$, the null hypothesis is rejected; lowering the thermostat at night does make some difference in heating oil consumption when the amount of insulation is considered.

Even this result should be qualified. The sample data suggests that heating oil savings would only occur in well-insulated houses. This data suggests that the heating oil distributor in example 7.9 is only partially incorrect. For poorly insulated houses, there may be savings in heating oil consumption by not lowering the thermostat at night. Thus, the distributor's claim needs to be qualified by the degree of insulation within the house.

Before anyone rushes out to change their thermostat, this data is hypothetical; however, the general conclusions have merit.

7.4(d)   Comparing the Proportions of Two Populations

Consider two populations X and Y and the proportion of successes in each (π_X and π_Y). For example, X may represent females, Y represent males, and "car ownership" is a "success." Again, since X and Y are arbitrarily chosen, the *claim* about two population proportions can always be rewritten in one of the two following general forms:

1. The proportion of successes in X is *greater than or equal to* the proportion of successes in Y.
2. The proportion of successes in X is *equal to* the proportion of successes in Y.

The translation of these claims into a null and an alternative hypothesis is accomplished in Table 7.16.

If we let $d = \bar{p}_X - \bar{p}_Y$, where $\bar{p}$ is the proportion of successes in the sample, then from Equation 7.19

$$\mu_d = \pi_X - \pi_Y = \bar{D} \tag{7.31}$$

TABLE 7.16 NULL AND ALTERNATIVE HYPOTHESES

	Null and Alternative Hypotheses for	
Claim	the Random Variables X and Y	the Random Variable $D = X - Y$
1	$H_0: \pi_X \geq \pi_Y$ $H_A: \pi_X < \pi_Y$	$H_0: \bar{D} \geq 0$ $(\pi_X - \pi_Y \geq 0)$ $H_A: \bar{D} < 0$ $(\pi_X - \pi_Y < 0)$
2	$H_0: \pi_X = \pi_Y$ $H_A: \pi_X \neq \pi_Y$	$H_0: \bar{D} = 0$ $(\pi_X - \pi_Y = 0)$ $H_A: \bar{D} \neq 0$ $(\pi_X - \pi_Y \neq 0)$

With either the single-tail or the two-tail test, $\bar{D} = 0$. Again, the mean of the relevant sampling distribution is always equal to 0 (that is, $\mu_d = 0$).

Further, since $\sigma_{\bar{p}}^2 = (1/n)\pi(1 - \pi)$ and since under the null $\pi_X = \pi_Y$, then $\sigma_{\bar{p}_X}^2 = \sigma_{\bar{p}_Y}^2$. That is, under the null hypothesis, both samples come from identical populations. Hence, if $\pi_X = \pi_Y$, then

$$\sigma_d = \sqrt{\frac{\pi_X(1 - \pi_X)}{n_X} + \frac{\pi_Y(1 - \pi_Y)}{n_Y}} = \sqrt{\pi(1 - \pi)}\sqrt{\frac{1}{n_X} + \frac{1}{n_Y}}$$

If both samples come from the same population, then we can combine the information from both samples to estimate π in the above expression for σ_d. The estimate of π, which we shall call Π, would be the total number of successes $(r_X + r_Y)$ from both samples divided by the total number of observations $(n_X + n_Y)$. Thus, Π is given by

$$\Pi = \frac{r_X + r_Y}{n_X + n_Y} = \frac{n_X\bar{p}_X + n_Y\bar{p}_Y}{n_X + n_Y} \tag{7.32}$$

Finally, the estimate of σ_d is

$$s_d = \sqrt{\Pi(1 - \Pi)} \cdot \sqrt{1/n_X + 1/n_Y} \tag{7.33}$$

After choosing a level of significance α and using Equations 7.32 and 7.33 to obtain s_d, we find the decision rules summarized in Table 7.17.

TABLE 7.17 DECISION RULES FOR TESTING THE DIFFERENCE BETWEEN TWO POPULATION PROPORTIONS

Single-Tail Test $H_0: \pi_X - \pi_Y \geq 0$ $H_A: \pi_X - \pi_Y < 0$	Two-Tail Test $H_0: \pi_X - \pi_Y = 0$ $H_A: \pi_X - \pi_Y \neq 0$
Either If $\bar{p}_X - \bar{p}_Y \leq -z_\alpha s_d$, reject H_0; otherwise, accept H_0. *Or** If $(\bar{p}_X - \bar{p}_Y)/s_d \leq -z_\alpha$, reject H_0; otherwise, accept H_0.	*Either* If $-z_{\alpha/2} s_d < \bar{p}_X - \bar{p}_Y < +z_{\alpha/2} s_d$, accept H_0; otherwise, reject H_0. *Or** If $-z_{\alpha/2} < (\bar{p}_X - \bar{p}_Y)/s_d < +z_{\alpha/2}$, accept H_0; otherwise, reject H_0.

* This statement of the decision rule allows the decision rule to be formulated *before* the sample information is known.

SUMMARY

The techniques discussed in this chapter are generally called *classical tests of hypotheses*. The purpose of these "tests" is to assist someone, or some group, in deciding what is "true" and what is "false." From an almost philosophical viewpoint, it is easier to show that a statistical hypothesis is false. That is, to demonstrate that a hypothesis is true, one has to validate the hypothesis for *all* cases (or situations). But, only *one* case is needed to show that the hypothesis is false.

Thus, the null hypothesis is frequently thought of as the "current opinion of what is true," "the normal circumstance," or "the accepted state of affairs." The alternative hypothesis is "the situation thought to be false," "the abnormal circumstance," or "the unreasonable occurrence." In this context, the rejection of the null hypothesis is of major importance; at least this one case casts doubt upon the "established truth."

Sometimes the original null hypothesis becomes a conditional statement; the hypothesis is refined to include those conditions which would make it true. Other times, a more general theory is developed; the original null hypothesis becomes a special case of a broader explanation of the "truth." In general, there is renewed interest in the *causes* of the unexpected (?) rejection of the null hypothesis; hopefully, there is improved understanding of the factors influencing the hypothesized situation.

As a process of scientific inquiry, these "tests" are abundantly performed today. The concepts of control group and experimental group are derivatives of the same concept of the null and alternative hypotheses. In the so-called "hard" sciences and in most of the academic community, the test of hypothesis is not an *end* in itself, but it is a *means* of investigation.

When we transplant the test-of-hypothesis technique into the disciplines of business and economics, perhaps the closest parallel is quality control on a production line. The null hypothesis reflects some standard of quality. As product comes off the production line, some items are tested. Given the sample information, the decision rule says accept the null (i.e., the quality standards are satisfied) or reject the null (i.e., stop the line). Rejecting the null calls for investigating the causes which are producing an "abnormal" situation.

The hard sciences often have the possibility of conducting controlled experiments to investigate the factors influencing the problems they consider. Most of the problems that concern economists and business managers are not subject to controlled experiments. Even the so-called "controlled economies" are not "controlled" in the context of a laboratory experiment. Experimenting with an economy to determine the factors most influential in economic depression is analogous to using downtown Los Angeles as ground zero for a nuclear explosion in order to determine the radius of fatal shock waves. Extensive study of past "abnormalities" and model building are more common than direct experiments in economics.

In direct experiments and even in model building, the test of hypothesis is just one "tool" the investigators use to get closer and closer to the truth. Rejecting the null hypothesis leads to a new hypothesis or a new experiment or a revised model. There may be considerable time lapse between the first statistical test of hypothesis and some sort of "final judgment."

In many business and some economics problems, the decision maker has, say, 48 hours following the results of the statistical test to reach some sort of final judgment. The decision maker is as close to "the truth" as he or she will get. Thus, the business decision maker is likely to be much more critical of the "test procedure" itself. In the laboratory experiment, committing Type I error (rejecting the null when it is true) in one statistical test will usually be revealed by further experiments and tests. The business decision maker may not have enough time to conduct these "further tests."

If you are going to conduct only one statistical test of hypothesis prior to taking some action, then you are likely to be very concerned about the test procedure. Specifically, the risk of making a mistake must be minimized. In the context of this chapter, given a specific situation, what is the appropriate decision rule?

Picking the correct equation for the critical value is not the answer; it is only a prerequisite for the answer. The question is: What are the appropriate values of α, β, and n? Recall that when either z or t is used as the test statistic, α, β, and n are all dependent on one another. For example, we could choose values of α and β, and then a minimum sample size can be calculated.

If accepting H_0 leads to a particular action (e.g., buy land for industrial development) and rejecting H_0 leads to another action (e.g., do not buy), then there are profits/costs associated with each correct decision and with each mistake. These costs could be measured in either dollars or utilities, but the net result is a measurement of the *relative seriousness* of Type I and Type II errors. As we shall see in Chapter 14, we could use various values of α and β to calculate the expected value of each action. Basically, this procedure could be described as choosing α and β in *dollar* units rather than in *probability* units. For business managers this is a very useful extension of the classical approach of decision rule formulation.

The choice of α and β may lead to such a large value of n that the cost of sampling would exceed the expected profit even when a correct action is taken. Is the value of the sample information greater than its cost? is another important question for the business decision maker.[7]

Finally, this chapter has been loaded with such phrases as "the results *suggest* that the null is true" and "the data fails to *support* the null." This designed vagueness is used to emphasize two points. First, Type I and Type II errors exist with determinable probabilities. Second, as presented here, the tests-of-hypothesis techniques are intended to be *aides* in the decision-making process and not the *decision makers*. Like humans, it is the understanding of both the strengths and the weaknesses of hypothesis testing that makes life interesting.

---

[7] Why should these concerns be limited to "business decision makers"? This question is the subject of considerable debate.

EXERCISES

1. As manager of packaging, you have just received a bin full of transistors for special shipment. The usual production process yields 30 percent of the transistors defective; however, (presumably) the transistors in this bin were produced under stringent quality-control conditions, and only 10 percent are defective.

   The transporter (the person who wheels the bin from the production department to you) is intoxicated, or close to it. Today is the transporter's birthday and his coworkers treated him to lunch. Other than the new dents in the bin you just received, you cannot tell one bin from another. You wonder if you received the specially produced batch or a regularly produced batch of transistors. You establish the following:

$$H_0: \text{The bin has 10 percent defectives } (\pi = 0.10)$$

$$H_A: \text{The bin has 30 percent defectives } (\pi = 0.30)$$

Experiment You take a random sample of 10 transistors from the bin and send them to the laboratory for testing. Meanwhile, you have to decide what you are going to do with the results. Since there are a large number of transistors in the bin, you assume that the sampling procedure is a Bernouli process.

   a. Assume the bin has 10 percent defectives, and find the probability of each possible laboratory result.
   b. Assume the bin has 30 percent defectives, and list the probability that each possible sample result will occur.
   c. Consider the following decision rule: If 2 or more defectives are found in the sample of 10, reject the null hypothesis; if fewer than 2 defectives are found, accept the null hypothesis. Find the probability of Type I and Type II errors.
   d. Find α and β for the decision rule: If one or more are defective, reject H_0; if none are defective, accept H_0.
   e. Find α and β for the decision rule: If three or more are defective, reject H_0; otherwise, accept H_0.
   f. Holding the sample size constant at 10, what happens to β as we decrease α?

2. You are still the manager of the packaging department. In the production department, there is a "scrap bin" for those transistors that do not pass a simple test. The test is not perfect, and about 20 percent of the transistors in the scrap bin are not defective. Later in the afternoon, the celebration-weary transporter delivers a bin of transistors and mumbles something about "I sure hope this isn't the scrap bin." This latest bin is supposed to be a regularly produced batch (that is, 30 percent defective). You establish:

$$H_0: \text{The bin has 30 percent defective } (\pi = 0.30)$$

$$H_A: \text{The bin has 80 percent defective } (\pi = 0.80)$$

The same experiment (as in Exercise 1) is conducted.

   a. Assuming the null is true, find the sampling distribution for r, the number of defectives from a sample of 10.
   b. Assuming the alternative is true, find the sampling distribution for r.
   c. Find α and β for the following decision rules:
   (1) If $r \geq 3$, reject H_0; otherwise, accept H_0.
   (2) If $r \geq 4$, reject H_0; otherwise, accept H_0.
   (3) If $r \geq 5$, reject H_0; otherwise, accept H_0.
   (4) If $r \geq 6$, reject H_0; otherwise, accept H_0.

3. Using the "equals-sign-in-the-null-hypothesis" rule, set up both the null and the alternative hypotheses for the following claims:

a. The Laggard Motor Car averages 32 miles per gallon in highway travel.
b. Except for the Christmas holidays, the average weight of airline baggage per passenger is less than 20 pounds.
c. More than half the adult customers who enter the Scrooge Gift Shop buy something.
d. The average human body temperature is 98.6°F.
e. At least 4 out of 5 homes in this neighborhood have more than two operating television sets.
f. The average annual earnings of "moonlighting" (i.e., employment "off duty") for police officers in Thug Town is no more than $2,000.

4. Using the equals-sign-in-the-null-hypothesis rule, which of the following null and alternative hypotheses are correct interpretations of the claim?

a. *Claim:* At 5 P.M. in downtown Down Town, the average number of particles per cubic meter of air is never more than 2,000.

$$H_0: \mu \geq 2,000 \qquad H_A: \mu < 2,000$$

b. *Claim:* At least 90 percent of our labor force worked 40 or more hours last week.

$$H_0: \mu \geq 40 \text{ hours} \qquad H_A: \mu < 40 \text{ hours}$$

c. *Claim:* Less than half the class earned 80 points or more on the last examination.

$$H_0: \pi \geq 0.5 \qquad H_A: \pi < 0.5$$

d. *Claim:* Electric toasters do not last as long as they used to.

$$H_0: \mu_{1970} - \mu_{1950} \geq 0 \qquad H_A: \mu_{1970} - \mu_{1950} < 0$$

where μ = average number of hours of use before repair.

e. *Claim:* The proportion of people who drive to work is greater than the proportion who do not.

$$H_0: \pi_{\text{drive}} - \pi_{\text{not drive}} > 0 \qquad H_A: \pi_D - \pi_{ND} \leq 0$$

f. *Claim:* As used-car dealers, Ept has a lower percentage of return customers per year than does Louis.

$$H_0: \pi_E - \pi_L \geq 0 \qquad H_A: \pi_E - \pi_L < 0$$

where π = proportion of return customers per year.

5. For each part of Exercise 3, describe in your own words:
a. Type I error
b. Type II error

6. In your own words, describe:
a. acceptance region
b. two-tail test
c. test statistic
d. α
e. one-tail test

f. level of significance
g. $(1 - \beta)$
h. rejection region

7. Establish the decision rule and test the following hypotheses:
a. $H_0: \mu = 13$, $H_A: \mu \neq 13$, $\sigma = 2$, $n = 100$, $\bar{X} = 12.5$, $s = 1.99$, $\alpha = 0.05$, $N = 10$ million
b. $H_0: \mu \leq 200$, $H_A: \mu > 200$, $\alpha = 0.05$, $\sigma = 12.5$, $n = 100$, $\bar{X} = 201.5$, $N = 276$
c. $H_0: \mu \geq 10.12$, $H_A: \mu < 10.12$, $\alpha = 0.10$, $n = 81$, $\Sigma X = 810$, $\Sigma X^2 = 8,180$, $N = $ large
d. $H_0: \mu = 100$, $H_A: \mu \neq 100$, $\alpha = 0.05$, $\sigma^2 = 49$, $\Sigma X = 425$, $\Sigma X^2 = 36,325$, $n = 5$
e. $H_0: \pi \leq 0.8$, $H_A: \pi > 0.8$, $\alpha = 0.05$, $n = 400$, $\bar{p} = 0.84$, $N = 100,000$
f. $H_0: \pi = 0.10$, $H_A: \pi \neq 0.10$, $\alpha = 0.01$, $n = 2,500$, $r = 275$, $N = 4,901$
g. $H_0: \mu_X - \mu_Y \geq 0$, $H_A: \mu_X - \mu_Y < 0$, $\alpha = 0.05$, $n_X = 100$, $\bar{X} = 3,270$, $n_Y = 400$, $\bar{Y} = 3,280$, $\sigma_X = 80$, $\sigma_Y = 120$
h. $H_0: \mu_X - \mu_Y = 0$; $H_A: \mu_X - \mu_Y \neq 0$; $\alpha = 0.05$; $n_X = 400$; $\Sigma X = 1,210,000$; $\Sigma X^2 = 49,370,500$; $n_Y = 900$; $\Sigma Y = 2,700,000$; $\Sigma Y^2 = 132,782,400$.
i. $H_0: \pi_A - \pi_B = 0$, $H_A: \pi_A - \pi_B \neq 0$, $\alpha = 0.10$, $n_A = 100$, $r_A = 19$, $n_B = 80$, $r_B = 17$
j. $H_0: \pi_A - \pi_B \leq 0$; $H_A: \pi_A - \pi_B > 0$; $\alpha = 0.01$; $n_A = 15,000$; $\bar{p}_A = 0.505$; $n_B = 3,000$; $\bar{p}_B = 0.475$

8. Establish the decision rule and test the following hypotheses where all populations are *normally* distributed:
a. $H_0: \mu = 50$, $H_A: \mu \neq 50$, $\alpha = 0.05$, $n = 16$, $\bar{X} = 52$, $s = 4$
b. $H_0: \mu \geq 91.5$, $H_A: \mu < 91.5$, $\alpha = 0.05$, $n = 25$, $\Sigma X = 2,250$, $\Sigma X^2 = 204,900$, $N = 33$
c. $H_0: \mu \leq 35$, $H_A: \mu > 35$, $\alpha = 0.10$, $n = 4$, $\Sigma X = 146$, $\Sigma X^2 = 5,340$, $\sigma = 2$
d. $H_0: \mu_A - \mu_B \geq 0$, $H_A: \mu_A - \mu_B < 0$, $\alpha = 0.05$, $n_A = 20$, $\Sigma X_A = 200$, $\Sigma(X - \bar{X}_A)^2 = 2,299$, $n_B = 5$, $\Sigma X_B = 95$, $\Sigma(X_B - \bar{X}_B)^2 = 1$
e. $H_0: \mu_B - \mu_A \leq 0$, $H_A: \mu_B - \mu_A > 0$, $\alpha = 0.01$, $n_A = 100$, $\bar{X}_A = 200$, $n_B = 100$, $\bar{X}_B = 203$, $\sigma_A = 12$, $\sigma_B = 9$

9. Given that all populations are skewed to the right, establish the decision rule and test the hypotheses for:
a. $H_0: \mu = 300$, $H_A: \mu \neq 300$, $\alpha \leq 0.10$, $n = 10$, $\bar{X} = 288$, $\sigma^2 = 121$
b. $H_0: \mu \geq 500$, $H_A: \mu < 500$, $\alpha \leq 0.05$, $n = 10$, $\Sigma X = 4,940$, $\sigma = 9$
c. $H_0: \mu \leq 500$, $H_A: \mu > 500$, $\alpha \leq 0.10$, $n = 10$, $\Sigma X = 5,100$, $\sigma = 9$
d. $H_0: \mu = 300$, $H_A: \mu \neq 300$, $\alpha = 0.10$, $n = 10$, $\bar{X} = 306$

10. List conditions for the use of:
a. z as a test statistic
b. t as a test statistic
c. Chebyshev inequality as a test statistic

11. Consider the uniform distribution at the right. Assume that this rectangle represents the probability density function for the sample mean, given small samples from a particular population.
a. Let $\alpha = 0.10$ and construct the decision rule, using only Figure A, for $H_0: \mu = 3$ and $H_A: \mu \neq 3$.
b. If the true mean is $\mu_T = 3.5$ (as shown in Figure B), what is the probability of Type II error using the decision rule from part a?
c. If the true mean is $\mu_T = 3.45$ (not shown), what is β?
d. If the true mean is $\mu_T = 4.00$ (not shown), what is β?

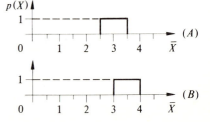

12. For Exercise 7, part *a*, calculate β if the true mean is:

a. 12.5

b. 12.608

c. 12.8

d. 13.392

e. 13.5

13. A producer of quality sneakers produces pink round shoelaces, because no one else will. Today 3,600 inches of pink round string were produced which was presumably cut in lengths of 18 inches per shoe string. A random sample of 25 shoestrings was selected, and the length of each shoelace was measured. The results were $\bar{X} = 18.02$ inches, $s = 0.001$ inch. The producer knows that 200 shoelaces were produced and wants to test the hypothesis that $\mu = 18$ inches. The shoelaces are advertised as being 18 inches long. How would you advise the producer?

14. A producer of floor lamps used in photography studios is concerned about the safety of the product. If someone trips over the light cord, the producer wants the plug at the base of the floor lamp to disconnect before the lamp is pulled down. (The wall-plug end of the cord is not dependable; some wall sockets hold the plug very firmly.) One solution is to make the base of the floor lamp very heavy, but this diminishes the utility of the product for photographers who like to move the floor lamps frequently. The safety engineers decide that the optimal design is for the plug at the base of the floor lamp to disconnect when 20 pounds of pull is placed upon it. If the plug does not disconnect at a 20-pound pull, the light will fall down. If the plug disconnects at less than a 20-pound pull, the plug will fall out in the normal operation of moving the floor lamp around the studio. (This will either result in a loss of sales or inspire the user to tape the plug to the base.)

From a production run of 176 such cord-lamp units, a random sample of 64 units is selected and tested. The results are $\bar{X} = 19.90$ pounds and $s = 1.00$ pound.

a. Test the hypothesis that the mean disconnect pull for the entire production run is 20 pounds per unit at the 5 percent significance level.

b. Given the decision rule used in part *a*, find β for $\mu_T = 21$ pounds per unit.

c. Find β for $\mu_T = 19$ pounds per unit.

15. An official for a local housing authority wants to test the hypothesis, proposed by a local newspaper, that the average annual family income in a public housing project is *greater than* the average family income of local house-trailer occupants. Although they had just published the average and the sample sizes, the newspaper provided the following information from their survey:

Family Annual Income Data for	
Trailer Parks	Public Housing Project
$\bar{X} = \$7,260$	$\bar{Y} = \$7,410$
$s_X = \$800$	$s_Y = \$1,200$
$n_X = 100$	$n_Y = 400$

Assume the samples represent less than 10 percent of their respective populations, and test the hypothesis at the 5 percent significance level.

16. In Exercise 15, if the sample sizes were less than 50 for each group, how would you proceed?

17. A food cooperative is concerned about the quantity of peas in a can. The net weight is consistently 12 ounces, but there seems to be a lot of packing fluid or water in the can. The producer claims that the quantity of peas per can is normally distributed with a mean of 11 ounces. The Coop randomly selects 16 cans, drains the fluid, and weighs the peas. The results are (in ounces):

11.9	11.7	10.7	10.6
10.8	11.2	11.6	10.3
11.4	10.1	11.0	10.4
10.1	10.9	10.3	11.4

a. Calculate $\bar{X}$ and s.
b. For $\alpha = 0.5$, test $H_0: \mu = 11$ ounces, $H_A: \mu \neq 11$.
c. For $\alpha = 0.5$, test $H_0: \mu \geq 11$ ounces, $H_A: \mu < 11$ ounces.
d. For $\alpha = 0.5$, test $H_0: \mu \leq 11$ ounces, $H_A: \mu > 11$ ounces.
e. Which of the hypotheses in parts b through d is most appropriate for this problem?

18. A magazine asserts that if its postage rates go up 20 percent, the magazine will have to increase its price by $.10 per copy to its mail subscribers. The magazine claims that at least one-third of its mail subscriptions will not be renewed if the price is increased by $.10 per copy. A random sample of 2,222 subscribers (out of several million) yields that 711 will cancel or not renew their subscriptions if the proposed price hike is instituted. Is the magazine's claim about the loss of subscribers justified at the 1 percent significance level?

19. During a period of increasing gasoline prices, taxi drivers instituted a 1-week TV advertising campaign extolling their services. The results of a sample of 10 drivers are as follows:

	Gross Income	
Driver	Before	After
1	$132	$142
2	117	119
3	126	121
4	89	109
5	141	144
6	134	123
7	134	149
8	116	97
9	119	129
10	137	143

Test the hypothesis that there was a difference for $\alpha = 0.05$. Assume that each driver worked the same number of hours in the week after as in the week before. Also, assume that gross income per week is normally distributed.

20. A private coeducational college is considering a reduction in the "football budget." The athletic director considers this to be poor business practice and argues that a higher proportion of alumni who hold season tickets to the football games donate $100 or more each year to the college than do the rest of the alumni.

Let population A be living alumni who have season tickets, and let B be the living alumni who do not. Let π be the proportion who gave $100 or more last year. The athletic director's claim is $\pi_A > \pi_B$.

a. What are the null and alternative hypotheses?

b. In the context of this particular problem, what would Type I and Type II errors be?

c. In selecting the samples, what problems would you expect? (*Hint:* What about alumni married to season-ticket holders who did not attend the college?)

d. Would this hypothesis, regardless of the test results, say anything about the relationship of donations to the college if the football budget were reduced?

CASE QUESTIONS

1. At the 5 percent significance level, test the hypothesis that the mean income of Massachusetts renters is equal to the mean income of Oregon renters.

2. At the 5% significance level, test the hypothesis that the proportion of renters above the poverty cutoff in Massachusetts is equal to the proportion of renters above the poverty cutoff in Oregon.

8

TESTS OF ASSUMPTIONS
AND ANALYSIS OF VARIANCE

INTRODUCTION

In the process of establishing a confidence interval and testing a hypothesis, we occasionally made some assumptions concerning our knowledge of the parent population. What assumptions? Recall that Figure 6.4 represented a decision tree for constructing interval estimates of a population mean. The first question suggested by Figure 6.4 concerns our knowledge about the shape of the parent population. Is the parent population normally distributed?

If we have reason to believe that the population is normally distributed, then we are making an assumption. Is there any way to test statistically the assumption that the population is normally distributed? Yes, we can use the information from a randomly selected sample to test the null hypothesis that "the population is normal" *as well as* to establish a confidence interval or test a hypothesis about the population mean. This test is appropriately called the *goodness-of-fit* test, and it uses a new theoretical probability distribution: the chi-square distribution.

Next, observe that the second question suggested by Figure 6.4 concerns our knowledge of the population variance. If we think we know the value of the population variance (see example 6.1), then we are making an assumption about the value of σ^2. With the results of a random sample, we may be able to test the null hypothesis that $\sigma^2 = \sigma_0^2$, where σ_0^2 is the assumed value of the population variance. This test also uses the chi-square distribution.

Frequently we want to know whether two events are independent of (or disassociated from) each other. In economics and business problems, if two events are independent, we can ignore one of them while trying to influence or predict the other. However, if two events are dependent (or associated in some way), we cannot ignore the first event in an attempt to investigate the second.

In Chapter 2 we discussed the concept of independent events [e.g., if $P(A) = P(A|B)$, then A and B are "independent"]. If we had the results of an accurate census, then we could easily determine whether two events are independent. If all we have is the information from one randomly drawn sample, then we need a statistical test of hypothesis about the independence of two events. Once again, the chi-square distribution permits such a test.

Finally, in Chapter 7 we investigated the test of hypothesis about the difference between two population means. What if we had three or more populations and we wanted to test the hypothesis that they all have the same mean? That is, H_0: $\mu_1 = \mu_2 = \mu_3$, etc. The procedure for conducting this test is called "analysis of variance" and uses another theoretical distribution: the F distribution.

The purpose of this chapter is to provide tests of the assumptions used in Chapters 6 and 7 and to extend those concepts of statistical inference. Whereas the normal, t, and binomial probability distributions were used as the theoretical models for the techniques of Chapters 6 and 7, the chi-square and F distributions are used for the theoretical basis of this chapter.

8.1   THE CHI-SQUARE DISTRIBUTION

The mathematical representation of the chi-square probability density function is

$$f(\chi^2) = (k)(\chi^2)^{(v/2)-1}(e^{-\chi^2/2}) \qquad (8.1)$$

where the value of k depends only upon v, v is the "degrees of freedom," e is the base of the Napierian logarithms, a constant, and χ^2 is "chi" squared.

As with the normal and t distributions, we will not be concerned with developing Equation 8.1. Instead, we will attempt only to understand the properties of the distributions that Equation 8.1 represents. Furthermore, we will look only at those properties which will allow us selectively to use the chi-square distribution. Once these properties are understood, applying the chi-square distribution to our "tests of assumptions" is surprisingly easy.

Property 1. χ^2 is a random variable that *cannot* assume negative values.[1] We could let $u = \chi^2$ and call it the u distribution; however, *squaring* "chi" reminds us that χ^2 can only be 0 or greater.

Property 2.   The distribution of χ^2 depends *only* on the *degrees of freedom* v. For a given value of χ^2, everything but v is given on the right-hand side of Equation 8.1 (recall that k is determined only by v).

Property 3.   The chi-square distribution is always *skewed to the right*. The larger v, the *less* the skewness. See Figure 8.1.

Property 4.   The mean of the chi-square distribution is given by the degrees of freedom; $E(\chi^2) = v$.

Property 5.   The variance is twice the degrees of freedom; $\text{Var}(\chi^2) = 2v$.

Property 6.   Equation 8.1 represents a *family* of distributions, a different distribution for each degree of freedom.

---

[1] If Z is normally distributed with $E(Z) = 0$ and $\text{Var}(Z) = 1$, then χ^2 is defined as the sum of n sample values of Z^2. That is, $\chi^2 = Z_1{}^2 + Z_2{}^2 + \cdots + Z_n{}^2$.

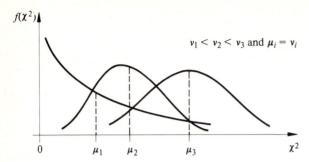

FIGURE 8.1   Chi-square distribution for 3 different degrees of freedom.

Recall that the t distribution is always symmetric with mean equal to 0 $[E(t) = 0]$; however, the variance depends upon the degrees of freedom. Note that the location, dispersion, and degree of skewness of the chi-square distribution are all exclusively given by the degrees of freedom. Thus, if we know the degrees of freedom for any particular problem, we can infer many characteristics of the appropriate chi-square distribution. Check Figure 8.1 for conformity with each of the six properties given above.

8.1(a)   Degrees of Freedom for χ^2

Obviously, determining the degrees of freedom is an all-important step in using the chi-square distribution. The degrees of freedom for the t distribution are determined directly from the sample size n. A similar situation exists for the test of hypothesis about the population variance when the chi-square distribution is employed.

In the "goodness-of-fit test" and the "test of independence," the sample data is recorded in cells or frequency classifications. The *number* of frequency classifications determines the degrees of freedom in these two "tests."

Since *how* we determine degrees of freedom depends upon which test we are going to perform, we will discuss the precise calculation of degrees of freedom with each application of the chi-square distribution.

8.1(b)   Use of the Chi-Square Tables

Since the shape and location of the chi-square distribution is different for each value of v, tables of areas under each curve would be quite lengthy. Thus, Appendix Table VII is similar to Appendix Table V. That is, the left-hand column lists various degrees of freedom, the heading lists several areas under the curve (or probabilities), and the body of the table denotes the upper limit of χ^2.

If we always phrase the question as

$$P(0 \leq \chi^2 \leq k) = 1 - \alpha$$

then for a given number of degrees of freedom v, $(1 - \alpha)$ is the area under the curve between $\chi^2 = 0$ and $\chi^2 = k$, and k is given by the number in the body of Appendix Table VII. For example, for 10 degrees of freedom and $(1 - \alpha) = 0.90$, Appendix Table VII shows that k is 15.99, or

$$P(0 \leq \chi^2 \leq 15.99) = 0.90$$

The probability is 0.90 that a random sample with 10 degrees of freedom will produce a χ^2 value between 0 and 15.99.

8.2   APPLICATION 1: THE GOODNESS-OF-FIT TEST

A classic problem is, How well does the model depict reality? Students bored with theoretical discussions of income distributions or the optimal number of check-out counters are among the first to ask this question. Obviously "models" omit some variables, interrelationships, and considerations. The "ideal" model is perhaps one in which these omissions do not significantly affect the *purpose* of the model. But the answer to the "How well?" question is complex and not necessarily entirely quantitative.

A probability distribution is, in a sense, a model. It is valid to question how good the "fit" of a model is to a particular problem or situation. As a model, a particular probability distribution (e.g., the binomial or the normal) would yield *theoretical frequencies* of occurrence for specific events. The chi-square "goodness-of-fit" test is one technique for comparing these theoretical or expected frequencies with the *actual*, or *observed*, *frequencies*.

The *observed* frequencies are usually not a census of the population; they are only a sample from the possible frequencies. Thus, even if the model (or theoretical distribution) precisely described the actual situation, we would expect the sampling procedure to produce some difference between the observed and theoretical frequencies. On the other hand, if the differences between the observed and theoretical frequencies are very large, it is unlikely that the actual population is distributed according to the theoretical probability distribution that we are using as a model.

For example, consider the several instructors of a multisection economics course who wish to give a common midterm examination. Some sections meet on a Monday-Wednesday-Friday (MWF) schedule; the others meet on a Tuesday-Thursday (TTh) schedule. The total enrollment is large, and the number of students in the MWF sections is approximately equal to the number of students in the TTh sections. Most of the instructors feel that the students, as a group, are equally divided as to which day is chosen for the examination.

Because one instructor seriously doubts the hypothesis, the instructors decide to select at random 60 (of the more than 600 enrolled) students, record their preferences, and test the hypothesis that the five weekdays are equally preferred. That is, H_0: student preferences are *uniformly* distributed, or

$$\pi(M) = \pi(T) = \pi(W) = \pi(Th) = \pi(F) = 0.20$$

where π = the proportion of students preferring the day in parentheses.

If we assume that the null hypothesis is true, then the theoretical or expected frequencies resulting from a sample of 60 are given by πn, or

Day Preferred	Theoretical Frequency (f_t)
Monday	(0.2)(60) = 12
Tuesday	12
Wednesday	12
Thursday	12
Friday	12
	—
	60

As a result of the survey, the observed frequencies are:

Day Preferred	Observed Frequency (f_o)
Monday	6
Tuesday	12
Wednesday	15
Thursday	18
Friday	9
	—
	60

Now, how do we compare these theoretical and observed frequencies?

Under certain conditions,

$$\chi^2 = \sum \left[\frac{(f_0 - f_t)^2}{f_t} \right] \tag{8.2}$$

exhibits a chi-square distribution. We will delay consideration of these conditions until we see how Equation 8.2 is a measure of "closeness."

First, note that we constructed the theoretical frequencies (that is, $f_t = \pi n$) in such a way that $\Sigma f_t = \Sigma f_0 = n$. In this example, $\Sigma f_t = \Sigma f_0 = 60$. This will be important in determining the degrees of freedom.

Second, let us rewrite Equation 8.2 in terms of this example:

$$\chi^2 = (f_{0,M} - f_{t,M})^2/f_{t,M} + (f_{0,T} - f_{t,T})^2/f_{t,T}$$
$$+ (f_{0,W} - f_{t,W})^2/f_{t,W} + (f_{0,Th} - f_{t,Th})^2/f_{t,Th} \tag{8.3}$$
$$+ (f_{0,F} - f_{t,F})^2/f_{t,F}$$

If each observed frequency is equal to its respective theoretical frequency, then the numerator of each ratio on the right side of Equation 8.3 is 0; therefore, $\chi^2 = 0$. That is, if the theoretical model and the sample information are in complete agreement, then χ^2 will equal 0.

Since the difference between each theoretical and observed frequency is squared, each term on the right side of Equation 8.3 must be greater than or equal to 0. Hence, the minimum value of χ^2 is 0.

If the theoretical and observed frequencies are not equal, then $(f_0 - f_t)^2$ is greater than 0 and χ^2 is greater than 0. The greater the difference between the theoretical and observed frequencies, the larger χ^2 becomes.

These algebraic considerations indicate how Equation 8.2 is a measure of "goodness of fit" between a theoretical model and the randomly selected sample observations. *The closer χ^2 is to 0, the better the "fit."* The larger the value of χ^2, the worse (or less "good") the fit.

In reacting to the value of χ^2 as calculated from Equation 8.2, there are two other considerations: one algebraic and one statistical. First, consider the 60 students surveyed. Using the theoretical frequency, we expected 12 to prefer Monday for the examination, but only 6 did. Since $\Sigma f_0 = \Sigma f_t = n = 60$, those 6 persons who were expected to prefer Monday but did not must show up somewhere else (i.e., under Tuesday, Wednesday, etc.). Thus, if $f_0 - f_t \neq 0$ for one day, then there must be at least one other day for which $f_0 - f_t \neq 0$. Equation 8.2 tends to amplify any difference between observed and theoretical frequencies.

In addition, recall that the observed frequencies represent a sample and not a census. Even if the theoretical distribution is a perfect description of the actual population, it is likely that the sampling process will yield some observed frequencies which are different from their respective theoretical frequencies. It is at this point that the chi-square distribution enters.

If our null hypothesis states that the theoretical distribution describes the actual population, then how large must χ^2 be before we reject that null? In previous chapters we have seen, for example, that it is *possible* to obtain a sample mean which is more than three standard deviations from the population mean, but, it is *not very probable*. A similar situation exists in the goodness-of-fit test. When the null hypothesis is true, it is possible to get a very high value of χ^2 from a sample, but it is not very probable. Thus, once again, we *choose* a value for α (i.e., the probability of Type I error) and use α in determining the critical value of χ^2.

Degrees of Freedom v

In our example, a sample of $n = 60$ observations (i.e., student preferences) was tallied into five "frequency classifications" (i.e., Monday, Tuesday, etc.). Also, the theoretical frequencies were derived in a manner that allows $\Sigma f_0 = \Sigma f_t = n$. We can "freely" assign (theoretical) frequencies to any four days, but the frequency for the fifth day is assigned by the relationship $\Sigma f_0 = \Sigma f_t = 60$. Therefore, we have only 4 degrees of freedom for the frequencies which are used in Equation 8.3.

Note that the sample size n only indirectly influences the number of degrees of freedom (that is, $\Sigma f_t = n$). It is the number of "frequency classifications," "classes," or "cells" which directly influences the number of degrees of freedom. Furthermore, it should be obvious that in order for Equation 8.2 to be a meaningful comparison, the frequency classifications for the theoretical frequencies should be the same as those for the observed frequencies. Thus, the *number of degrees of freedom* for the chi-square test of goodness of fit, where the theoretical and observed frequencies share k *frequency classifications*, is given by

$$v = k - 1 \tag{8.4}$$

Although it does not appear in our current example, there may be further restrictions on the calculation of the theoretical frequencies. For example, if the sample mean $\overline{X}$ is used to estimate the population mean in order to calculate theoretical frequencies, then this "restriction" reduces the number of degrees of freedom by 1.

In general, if there are m sample estimates used for m unknown parameters in the calculation of the theoretical frequencies, then the number of degrees of freedom is further reduced by the number m, or

$$v = k - 1 - m \tag{8.5}$$

This calculation of v will be demonstrated in example 8.1.

The Decision Rule

The null hypothesis stated that the proportion of students preferring one weekday is equal to the proportion of students preferring any other weekday. The theoretical frequencies were calculated by assuming that the null is true. Each of the five weekdays represents one frequency classification. From above, the number of degrees of freedom is $v = k - 1 = 5 - 1 = 4$.

In order to use Appendix Table VII to obtain a critical value of χ^2, we must also choose a value for α. Let us choose three values for α: $\alpha = 0.10, 0.05$, and 0.01. Figure 8.2 shows the resulting critical values of χ^2 (from Appendix Table VII).

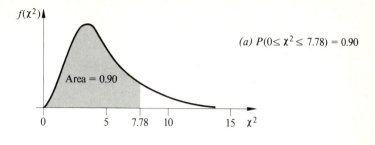

$(a)\ P(0 \le \chi^2 \le 7.78) = 0.90$

Area = 0.90

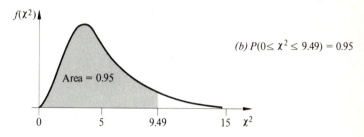

$(b)\ P(0 \le \chi^2 \le 9.49) = 0.95$

Area = 0.95

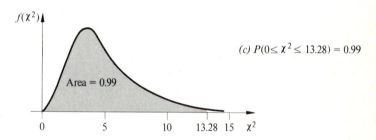

$(c)\ P(0 \le \chi^2 \le 13.28) = 0.99$

Area = 0.99

FIGURE 8.2   Chi-square distribution for 4 degrees of freedom $(v = 4)$.

Using $\alpha = 0.05$ or $(1 - \alpha) = 0.95$, the *decision rule* is:

$$\text{If } \chi^2 > 9.49 \qquad \text{reject } H_0$$
$$\text{If } \chi^2 \le 9.49 \qquad \text{accept } H_0$$

In like manner, for $\alpha = 0.10$ and 0.01, the "rejection region" would be to the right of the critical value given in Figure 8.2*a* and *c*.

Now, let us use Equation 8.2 (or 8.3) to calculate χ^2. Table 8.1 summarizes those calculations. Since $\chi^2 = 7.50 < 9.49$, we accept the null hypothesis. Note that since $7.50 < 7.78$ and $7.50 < 13.28$, the null would be accepted for $\alpha = 0.10$ and $\alpha = 0.01$. Thus, we cannot reject the *uniform distribution* as a model for student preferences.

TABLE 8.1 CALCULATION OF χ^2

Frequency Classification (Weekday Preferred)	Observed Frequency f_0	Theoretical Frequency f_t	$f_0 - f_t$	$(f_0 - f_t)^2$	$\dfrac{(f_0 - f_t)^2}{f_t}$
Monday	6	12	−6	36	3.00
Tuesday	12	12	0	0	0.00
Wednesday	15	12	3	9	0.75
Thursday	18	12	6	36	3.00
Friday	9	12	−3	9	0.75
Total	60	60	0		7.50

Example 8.1 The Automatic Cola Machine Revisited

In example 6.1 the 95 percent confidence interval for the mean fill per cup was calculated for a particular automatic cola dispenser. We assumed (1) that the population is normally distributed and (2) that the population standard deviation is 0.10 ounce.

In example 6.1, a random sample of 16 fills was selected, and a sample mean of 8.05 ounces was calculated. For this example let us assume that a random sample of 49 fills produced the following data (in ounces):

7.85	7.98	8.06	8.11
7.86	7.99	8.06	8.11
7.87	7.99	8.07	8.12
7.87	8.01	8.07	8.16
7.88	8.03	8.07	8.16
7.89	8.03	8.08	8.17
7.92	8.04	8.09	8.19
7.94	8.05	8.09	8.21
7.95	8.05	8.09	8.21
7.96	8.05	8.10	8.22
7.97	8.05	8.10	8.24
7.97	8.05	8.10	8.26
	8.06 ← Median		

From this data, the following can be calculated:

$$\bar{X} = \frac{\Sigma X}{n} = \frac{394.45}{49} = 8.05 \text{ ounces}$$

$$\Sigma(X - \bar{X})^2 = 0.5152 \quad \text{and} \quad s = \sqrt{\frac{\Sigma(X - \bar{X})^2}{n - 1}} = 0.1036 \text{ ounce}$$

Let us use this data to test the assumption that the population is normally distributed. (Later in this chapter we will test the assumption that $\sigma = 0.10$ ounce.) There are two routes that we could follow. One is to assume that $\sigma = 0.10$ ounce; the other is to use the sample standard deviation, $s = 0.1036$, as a point estimate of σ.

(*a*) Assume $\sigma = 0.10$ ounce.

    1. State the null and alternative hypotheses:

H_0: The population is normally distributed

H_A: The normal distribution does not describe the population

    2. Size of the random sample: $n = 49$.
    3. Specify the number of frequency classifications.

The "raw data" given above has already been placed in an array. Using the techniques of Chapter 4, we might use the following class intervals to describe the raw data:

Frequency Classifications	Tally	Observed Frequencies
7.80 to less than 7.90 ounces	𝍅 /	6
7.90 to less than 8.00 ounces	𝍅 ////	9
8.00 to less than 8.10 ounces	𝍅 𝍅 𝍅 ////	19
8.10 to less than 8.20 ounces	𝍅 𝍅	10
8.20 to less than 8.30 ounces	𝍅	5
		——
		49

The observed frequencies are then found from the raw data.

    4. Calculate the theoretical frequencies.

We are assuming that $\sigma = 0.10$, so we do not use a point estimate of the population standard deviation. We do not know the population mean; hence, we will use $\overline{X} = 8.05$ ounces as a point estimate of μ. Thus, assume that we have a normal distribution with $\mu = 8.05$ and $\sigma = 0.10$.

Our first step is to find the probability that a value of X would lie within the class intervals used in step 3 above. Since the normal curve represents the distribution of a continuous random variable which can assume values from $-\infty$ to $+\infty$, we know that $P(X < 7.80)$ and $P(X > 8.30)$ are not 0. Since we want $\Sigma f_t = \Sigma f_0 = n$, we cannot ignore the "tails" of the normal curve. We solve this problem by using the following frequency classifications:

less than 7.90 ounces
7.90 to less than 8.00 ounces
8.00 to less than 8.10 ounces
8.10 to less than 8.20 ounces
8.20 ounces or more

Note that by making the higher and lower class intervals "open-ended," we do not alter the observed frequencies presented in step 3.

Now, we can convert all "class limits" to values of z and use the normal-curve tables in the Appendix to derive the theoretical frequencies.

Lower Class Limit (x_L)	Upper Class Limit (x_U)	Standard Normal Deviate		Area of Class $P(z_L < z < z_U)$	Theoretical Frequency f_t $49 \cdot P(z_L < z < z_U)$
		$z_L = \dfrac{x_L - 8.05}{0.10}$	$z_U = \dfrac{x_U - 8.05}{0.10}$		
$-\infty$	7.90	$-\infty$	-1.50	0.0668	3.2732
7.90	8.00	-1.50	-0.50	0.2417	11.8433
8.00	8.10	-0.50	$+0.50$	0.3830	18.7670
8.10	8.20	$+0.50$	$+1.50$	0.2417	11.8433
8.20	$+\infty$	$+1.50$	$+\infty$	0.0668	3.2732
				1.0000	49.0000

5. Calculate the degrees of freedom.

The degrees of freedom are given by $v = k - 1 - m$, where k is the number of class intervals and m is the number of point estimates of unknown population parameters used in calculating the theoretical frequencies. The number of frequency classifications is 5, and we used $\bar{X} = 8.05$ ounces as a point estimate of the population mean. Thus, there are $v = k - 1 - m = 5 - 1 - 1 = 3$ degrees of freedom.

6. Choose a value for α, the probability of Type I error: Let $\alpha = 0.05$; thus, $(1 - \alpha) = 0.95$.

7. State the decision rule.

For $v = 3$ degrees of freedom and $(1 - \alpha) = 0.95$, we use Appendix Table VII to find the critical value of χ^2 to be 7.81. Thus,

$$\text{If } \chi^2 > 7.81 \quad \text{reject } H_0$$

$$\text{If } \chi^2 \leq 7.81 \quad \text{accept } H_0$$

8. Calculate χ^2.

The following table uses Equation 8.2 to calculate χ^2.

Observed Frequencies (f_0)	Theoretical Frequencies (f_t)	$f_0 - f_t$	$(f_0 - f_t)^2$	$\dfrac{(f_1 - f_t)^2}{f_t}$
6	3.2732	2.7268	7.4354	2.27
9	11.8433	−2.8433	8.0844	0.68
19	18.7670	0.2330	0.0543	0.00
10	11.8433	−1.8433	3.3978	0.29
5	3.2732	1.7268	2.9818	0.91
49	49.0000	0.0000		4.15

Since $\chi^2 = 4.15 < 7.81$, we do not reject the null hypothesis. The sample data supports the assumption that the population is normally distributed.

(b) Use $s = 0.1036$ ounce as an estimate of σ.

In step 4 above, we used $\sigma = 0.10$ to calculate the theoretical frequencies. What if our assumption that $\sigma = 0.10$ is incorrect? Would that not affect our goodness-of-fit test? Later we will test the null that $\sigma = 0.10$, but for now we can use the sample standard deviation s as a point estimate of the population standard deviation. Using s as an estimate of σ will affect both the theoretical frequencies and the degrees of freedom. The former affects our value of χ^2, as calculated from Equation 8.2. The latter affects the critical value of χ^2.

Steps 1, 2, and 3 are identical with part a.

4. Calculate the theoretical frequencies.

We now assume our "model" is a normal distribution with $\mu = 8.05$ and $\sigma = 0.1036$.

Lower Class Limit (χ_L)	Upper Class Limit (χ_U)	Standard Normal Deviate		Area of Class $P(z_L < z < z_U)$	Theoretical Frequency $49 \cdot P(z_L < z < z_U)$
		$z_L = \dfrac{X_L - 8.05}{0.1036}$	$z_U = \dfrac{X_U - 8.05}{0.1036}$		
$-\infty$	7.90	$-\infty$	−1.45	0.0735	3.6015
7.90	8.00	−1.45	−0.48	0.2421	11.8629
8.00	8.10	−0.48	+0.48	0.3688	18.0712
8.10	8.20	+0.48	+1.45	0.2421	11.8629
8.20	$+\infty$	+1.45	$+\infty$	0.0735	3.6015
				1.0000	49.0000

5. Calculate the degrees of freedom.

There are still five frequency classifications; however, s is now being used as an estimate of σ as well as μ being estimated by $\overline{X}$. Thus, $v = k - 1 - m = 5 - 1 - 2 = 2$. There are only 2 degrees of freedom.

6. $\alpha = 0.05$, as in part *a*.

7. State the decision rule.

For $v = 2$ degrees of freedom and $(1 - \alpha) = 0.95$, Appendix Table VII yields 5.99 as the critical value for χ^2.

$$\text{If } \chi^2 > 5.99 \qquad \text{reject } H_0$$

$$\text{If } \chi^2 \leq 5.99 \qquad \text{accept } H_0$$

8. Calculate χ^2:

f_o	f_t	$f_o - f_t$	$(f_o - f_t)^2$	$(f_o - f_t)^2/f_t$
6	3.6015	2.3985	5.7528	1.60
9	11.8629	-2.8629	8.1962	0.69
19	18.0712	0.9288	0.8627	0.05
10	11.8629	-1.8629	3.4704	0.29
5	3.6015	1.3985	1.9558	0.54
49	49.0000	0.0000		3.17

Since $\chi^2 = 3.17 < 5.99$, we accept the H_0 as we did in part *a*.

Words of Warning

The chi-square goodness-of-fit test is not limited to testing the assumptions used in the estimation techniques of Chapter 6. Any time you hypothesize that a population has a particular shape, the goodness-of-fit test is a potential technique for testing that hypothesis.

In the examples, the size of the sample selected, n, does not appear to make much difference. This "appearance" is false. The size of the sample is important; however, this importance is subtle, and the influence of n is indirect.

The chi-square probability distribution (from which Appendix Table VII is calculated) is only an *approximation* to the sampling distribution of χ^2 (as given by Equation 8.2). In general, the larger the sample size, the better the approximation. This is a situation not too different from using the normal curve as an approximation to the binomial (or t) distribution for large samples. Again, the question of how large must n be defies a precise answer. We resort once more to a rule of thumb.

Recall that the sample size n is used to calculate the theoretical frequency in each frequency classification. In general, (theoretical frequency) $= (n)$(probability of observation being in the frequency classification). Our rule of thumb is that *each theoretical frequency should be at least* 5. That is, $f_t \geq 5$ for each theoretical frequency.

Note in example 8.1 that $n = 49$ and in example 6.1 that $n = 16$. In example 8.1, even a sample of 49 observations is not sufficient to have the $f_t \geq 5$ for the open-ended class intervals (that is, $f_t = 3.2732$ for those class intervals in part a). According to our rule, the chi-square distribution is not a good approximation of the sampling distribution of χ^2. What should we do?

The best solution is to take a larger sample. For example, in example 8.1 we might "guess" that the mean fill is 8 ounces. A normally distributed population with $\mu = 8$ and $\sigma = 0.1$ would yield $P(\overline{X} < 7.9) = P[z < (7.9 - 8.0)/0.1] = P(z < 1) = 0.1587$. In order for f_t to be 5 or greater, for that class interval, we would have to take a sample of $(5/0.1587 = 31.5)$ 32 or larger. In example 8.1, we did take a sample larger than 32 (that is, $n = 49$), but our guesstimate of the mean was so bad ($\overline{X} = 8.05$ instead of $\overline{X} = 8.00$) that the open-ended class intervals had theoretical frequencies less than 5.

After the sample is selected, one solution is to group contiguous intervals until the theoretical frequency is at least 5 for each class interval. In example 8.1 we could use the following:

Frequency Classifications	f_0	f_t
Under 8.00 ounces	15	15.1165
8.00 to less than 8.10 ounces	19	18.7670
8.10 ounces or more	15	15.1165
	49	49.0000

By inspection there is a good fit; however, there are now very few degrees of freedom (following part a, there is only 1 degree of freedom).

In example 8.1, we have the luxury of another solution. The fill from an automatic liquid dispenser is a continuous random variable; thus, we could make narrower class intervals and still have $f_t \geq 5$. For example 8.1, consider the following class limits and the associated theoretical and observed frequencies:

Frequency Classifications	Theoretical Frequency	Observed Frequency
Under 7.925 ounces	5.1744	7
7.925 to under 7.975 ounces	5.9290	5
7.975 to under 8.025 ounces	8.5603	4
8.025 to under 8.075 ounces	9.6726	14
8.075 to under 8.125 ounces	8.5603	10
8.125 to under 8.175 ounces	5.9290	3
8.175 ounces or more	5.1744	6
	49.0000	49

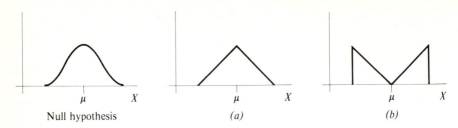

FIGURE 8.3   Three potential populations.

(Following part *a* of example 8.1, there are $v = k - 1 - m = 7 - 1 - 1 = 5$ degrees of freedom. With $\alpha = 0.05$, the critical value of χ^2 is 11.07. It is left to the reader to show that χ^2 is 6.98, which is less than 11.07. We still cannot reject the null hypothesis that the population is normally distributed.) Note that $f_t \geq 5$ for all values of f_t.

The result is that the procedure followed in example 8.1 is not a neat "one-step-after-another" process. Instead, each "step" in example 8.1 is a component in a feedback loop. To ensure that $f_t \geq 5$, after the sample information is available (in example 8.1, after $\bar{X}$ is calculated), you may have to go back to step 3 and either regroup or redefine the class intervals. This feedback process is not intended to alter the results deceptively; it is intended to make the chi-square distribution a better approximation of χ^2 as calculated from Equation 8.2.

Finally, consider Type II error: accepting the null hypothesis when it is false. In the goodness-of-fit test, the null hypothesis refers to a particular theoretical distribution (e.g., the normal). Thus, in order to calculate β, the probability of committing Type II error, you would have to specify some other theoretical distribution. Roughly speaking, if the "other distribution" is very close to the distribution stated in the null, then β is likely to be quite high (e.g., Figure 8.3a). If the other distribution is quite different from the null, then β is likely to be very low (e.g., Figure 8.3b).

8.3   APPLICATION 2: THE TEST OF INDEPENDENCE

"Association" and "independence" are two important concepts in business and economics. If two events are independent, then the occurrence of one event does not affect the probability of the other event occurring. In the chapters on regression we will discuss association in considerably more detail.

Example 8.2   Independence of Purchase Behavior and Consumer Information

Suppose a group of environmentalists challenges an electric utility company about the informational content of the advertisement sent with the monthly electric bill. For example, the environmentalists question how well the utility has informed its public about the fact that "frost-free" refrigerators use considerably more electric energy than the manually defrosted type of unit. They are concerned that, if better informed, the public would purchase fewer of the frost-free units, and the demand for electric energy would be reduced.

From a superbly accurate census, the utility company obtains a complete list of all persons in its service area who purchased electric refrigerators during the past 5 years. With impeccable interviews, the utility company determined whether each purchaser knew, before purchase, that a frost-free refrigerator would use more electric energy. The results are summarized here:

Type of Unit Purchased	Did You Know, before Purchase, that Frost-Free Units Use More Electric Energy?		Total
	Yes (B_1)	No (B_2)	
Frost-free (A_1)	324	108	432
Not frost-free (A_2)	216	72	288
Total	540	180	720

Note that $(324/540 = 0.60)$ 60 percent of those who knew about the difference in relative energy use purchased frost-free units. Also, $(108/180 = 0.60)$ 60 percent of the uninformed purchased frost-free refrigerators; the same percentage $(432/720 = 0.60)$ holds for the total. Since $P(A_1) = P(A_1 | B_1) = P(A_1 | B_2)$, then A and B are independent events. Or, the decision to purchase a frost-free refrigerator is independent of knowledge about the relative energy consumption.

Since this data represents the census of a population, the utility company felt quite comfortable using only this data to support its advertisements. While it may not be the responsibility of the utility company to educate the populace, this data does not answer the question posed by the environmentalists. Consider only those persons who did not know that frost-free refrigerators use more electric energy (i.e., the operating costs are greater). How many of these uninformed would have purchased a different type of refrigerator? This answer is much more difficult to acquire, but at least it is an answer to the question asked.

Assume that a statistics professor argues that afternoon classes do not perform as well as morning classes while the department chairperson argues that between 8 A.M. and 5 P.M. it does not make any difference when the class meets. The instructor locates records for the past semester. The instructor claims that the same lecture was delivered to each of the classes, the same homework and examinations were given, and all students had met the course prerequisites. The grades are presented in Table 8.2.

Tables similar to Table 8.2 are sometimes called *contingency tables*. One form of classification is represented by the columns (the five letter grades), and another classification is denoted by the rows (the two "times of day"). If we use r to denote

TABLE 8.2 LETTER GRADES BY CLASS

	Letter Grade					
Time the Class Met	A	B	C	D	F	Total
Morning	12	18	13	11	6	60
Afternoon	3	7	17	4	9	40
Total	15	25	30	15	15	100

the number of rows and k for the number of columns, then there are $r \times k$ cells in the body of the $r \times k$ *contingency table*. In Table 8.2, there are $r = 2$ rows and $k = 5$ columns; there are $2 \times 5 = 10$ cells.

Each cell represents a cross classification of class time and letter grade. Table 8.2 shows the observed frequency in each cell. There are 12 students from the morning class who earned an A in the course. If we had a theoretical frequency for each cell, then we could use a formula similar to Equation 8.2 to measure how close the observed sample frequencies are to the theoretical frequencies.

Calculation of the Theoretical Frequencies

In the goodness-of-fit test, we calculated the theoretical frequencies from the sample size n and the theoretical probability distribution specified in the null hypothesis. What about our null hypothesis in the "test of independence"? In general, the null and alternative hypotheses are:

$$H_0: \text{Events } A \text{ and } B \text{ are independent}$$

$$H_A: \text{Events } A \text{ and } B \text{ are not independent}$$

Specifically, in our instructor versus chairperson problem,

$$H_0: \text{"Time class meets" and "letter grade" are independent}$$

$$H_A: \text{"Time class meets" and "letter grade" are dependent}$$

Does H_0 suggest a theoretical probability distribution? No, but it does suggest an algebraic relationship within the $r \times k$ contingency table. Refer to example 8.2. Events A and B are independent. The row total for A_1 is 432, the column total for B_1 is 540, and the grand total is 720 persons. If A_1 and B_1 are independent, then

$$\frac{(\text{Total for row } i)(\text{total for column } j)}{(\text{Grand total})} = (\text{frequency for cell } ij) \qquad \textbf{(8.6)}$$

TABLE 8.3 CALCULATION OF THEORETICAL FREQUENCIES

Time	A	B	C	D	F	Total
			Letter Grade			
A.M.	$\dfrac{60 \cdot 15}{100} = 9$	$\dfrac{60 \cdot 25}{100} = 15$	$\dfrac{60 \cdot 30}{100} = 18$	$\dfrac{60 \cdot 15}{100} = 9$	$\dfrac{60 \cdot 15}{100} = 9$	60
P.M.	$\dfrac{40 \cdot 15}{100} = 6$	$\dfrac{40 \cdot 25}{100} = 10$	$\dfrac{40 \cdot 30}{100} = 12$	$\dfrac{40 \cdot 15}{100} = 6$	$\dfrac{40 \cdot 15}{100} = 6$	40
Total	15	25	30	15	15	100

That is, $(432)(540)/720 = 324$, which is the number of persons in example 8.2 that are in both A_1 and B_1 (or the frequency for the event $A_1 \cap B_1$). In like manner, the other cells in example 8.2 can be determined by $(432)(180)/720 = 108$, $(288)(540)/720 = 216$, and $(288)(180)/720 = 72$.

The algebraic relationship suggested by the null hypothesis is Equation 8.6.[2] If two events are independent, then each cell frequency can be calculated from (1) its row total, (2) its column total, and (3) its grand total. For a random sample (e.g., Table 8.2), we use the sample size n as the grand total, and the row and column totals as point estimates for the corresponding population totals. Thus, the calculation of

TABLE 8.4 CALCULATION OF χ^2

Time	Letter Grade	f_o	f_t	$f_o - f_t$	$(f_o - f_t)^2$	$(f_o - f_t)^2/f_t$
	Cell Identification					
A.M.	A	12	9	3	9	1.00
A.M.	B	18	15	3	9	0.60
A.M.	C	13	18	−5	25	1.39
A.M.	D	11	9	2	4	0.44
A.M.	F	6	9	−3	9	1.00
P.M.	A	3	6	−3	9	1.50
P.M.	B	7	10	−3	9	0.90
P.M.	C	17	12	5	25	2.08
P.M.	D	4	6	−2	4	0.67
P.M.	F	9	6	3	9	1.50
Total		100	100	0		11.08

[2] Equation 8.6 is identical to Equation 2.10. If we let $P(A) = a/N$ and $P(B) = b/N$, then $P(A \cap B) = P(A)P(B) = (a/N)(b/N) = ab/N^2$, and $N \cdot P(A \cap B) = ab/N$. Let a = row total, b = column total, N = grand total, and $N \cdot P(A \cap B)$ = cell frequency.

the theoretical frequencies, assuming the null hypothesis is true, is given in Table 8.3. Note that $\Sigma f_t = \Sigma f_0 = n$ for the test of independence as well as the goodness-of-fit test.

Calculation of the Test Statistic χ^2

In the "cells" of Table 8.2 we have the 10 observed frequencies; in the cells of Table 8.3 we have the corresponding 10 theoretical frequencies for independent events. For each of the 10 cells we could use $(f_0 - f_t)^2/f_t$, as in Equation 8.2, as a measure of how close the observed frequencies are to the theoretical frequencies for independent events. Table 8.4 summarizes these calculations.

From Equation 8.2, our value of χ^2 is 11.08. In general, a different notation is used for calculating χ^2 in a test of independence. The resulting equation precisely describes the procedure used in Table 8.4; however, it may appear to be more difficult.

Let i be the subscript for the r rows in a contingency table ($i = 1, 2, ..., r$) and let j be the subscript denoting the k columns ($j = 1, 2, ..., k$). Then

$$\chi^2 = \sum_{i=1}^{r} \sum_{j=1}^{k} \frac{(f_{0,\,ij} - f_{t,\,ij})^2}{f_{t,\,ij}} \tag{8.7}$$

All this says is (1) calculate $(f_0 - f_t)^2/f_t$ for *each cell* and (2) add the results.

Degrees of Freedom

In Table 8.3, the row totals and column totals were used to calculate the theoretical frequencies of independent events. Consider only the morning class. If we calculate the theoretical frequency for the number of A's, B's, C's, and D's, then the number of F's has to be a number such that the row total is 60. Thus, we have only 4 degrees of freedom. Now consider the afternoon class. The theoretical frequency for each letter grade can be determined by subtracting the A.M. theoretical frequency from the column total. That is, the blanks in Table 8.5 are determined by the column and row totals *with no consideration of independent events.*

Therefore, within the 10 cells of Table 8.3 (or 8.5), there are only 4 degrees of freedom. To repeat, calculate any four theoretical frequencies by the technique

TABLE 8.5 DEMONSTRATION OF DEGREES OF FREEDOM

Cell Identification	A	B	C	D	F	Total
A.M.	9	15	18	9	$(60 - 51)$	60
P.M.	$(15 - 9)$	$(25 - 15)$	$(30 - 18)$	$(15 - 9)$	$(15 - 9)$	40
Total	15	25	30	15	15	100

demonstrated in Table 8.3 and the numbers assigned to the other six cells are dictated by the row and column totals. In general, if there are r rows and k columns in a contingency table, then there are

$$(r - 1)(k - 1) = v \text{ degrees of freedom} \tag{8.8}$$

for the test of independence. From Table 8.5, there are $(r - 1)(k - 1) = (2 - 1) \times (5 - 1) = (1)(4) = 4$ degrees of freedom.

Decision Rule for Test of Independence

If each $f_t \geq 5$, then the sampling distribution of χ^2 (i.e., Equation 8.7) is approximated by the chi-square distribution. For α level of significance and $(r - 1)(k - 1)$ degrees of freedom, the critical value of χ^2 is obtained from Appendix Table VII.

In the instructor versus chairperson problem, let $\alpha = 0.05$. With 4 degrees of freedom, the critical value of χ^2 is given by Table VII as 9.49. Hence, the decision rule is:

If $\chi^2 > 9.49$ reject the H_0 ("class time" and "grade" are dependent—the instructor's hypothesis)

If $\chi^2 \leq 9.49$ accept the H_0 ("class time" and "grade" are independent—the chairperson's hypothesis)

From Table 8.4, $\chi^2 = 11.08 > 9.49$; therefore, the null hypothesis is rejected.

Example 8.3   The Shoe Store

For a given price range, a retail shoe store uses four suppliers of men's shoes. Each pair of shoes is inspected by the store before it is placed on the shelves. There are three different defects that would cause the store to return the shoes to the suppliers. The suppliers claim that the defects are endemic to the industry. During the past 3 months the following records were maintained.

		Defects		
Supplier	B_1	B_2	B_3	Total
A_1	17	10	13	40
A_2	10	10	10	30
A_3	18	15	17	50
A_4	15	5	10	30
Total	60	40	50	150

Test the supplier's claim that the type of defect is independent of who supplies the product, using $\alpha = 0.01$.

There are $r = 4$ rows and $k = 3$ columns; there are $(r - 1)(k - 1) = (3)(2) = 6$ degrees of freedom. From Appendix Table VII, the critical value of χ^2, for $(1 - \alpha) = 0.99$ and $v = 6$, is 16.81. The resulting decision rule is:

If $\chi^2 > 16.81$ reject the H_0 (which is that supplier and type of defect are independent)

If $\chi^2 \leq 16.81$ accept the H_0

Using Equation 8.6, calculate the theoretical frequencies for independent events.

		Defects		
Supplier	B_1	B_2	B_3	Total
A_1	$\dfrac{60 \cdot 40}{150} = 16$	$\dfrac{40 \cdot 40}{150} = 10.67$	$\dfrac{50 \cdot 40}{150} = 13.33$	40
A_2	$\dfrac{60 \cdot 30}{150} = 12$	$\dfrac{40 \cdot 30}{150} = 8$	$\dfrac{50 \cdot 30}{150} = 10$	30
A_3	$\dfrac{60 \cdot 50}{150} = 20$	$\dfrac{40 \cdot 50}{150} = 13.33$	$\dfrac{50 \cdot 50}{150} = 16.67$	50
A_4	$\dfrac{60 \cdot 30}{150} = 12$	$\dfrac{40 \cdot 30}{150} = 8$	$\dfrac{50 \cdot 30}{150} = 10$	30
Total	60	40	50	150

Note that each theoretical frequency is greater than 5. Using Equation 8.7 for χ^2, we calculate:

Supplier	Defect	f_0	f_t	$f_0 - f_t$	$(f_0 - f_t)^2$	$(f_0 - f_t)^2/f_t$
A_1	B_1	17	16.00	1.00	1.0000	0.06
A_1	B_2	10	10.67	−0.67	0.4489	0.04
A_1	B_3	13	13.33	−0.33	0.1089	0.01
A_2	B_1	10	12.00	−2.00	4.0000	0.33
A_2	B_2	10	8.00	2.00	4.0000	0.50
A_2	B_3	10	10.00	0.00	0.0000	0.00
A_3	B_1	18	20.00	−2.00	4.0000	0.20
A_3	B_2	15	13.33	1.67	2.7889	0.21
A_3	B_3	17	16.67	0.33	0.1089	0.01
A_4	B_1	15	12.00	3.00	9.0000	0.75
A_4	B_2	5	8.00	−3.00	9.0000	1.12
A_4	B_3	10	10.00	0.00	0.0000	0.00
		150	150.00	0.00		3.23

Since $\chi^2 = 3.23 < 16.81$, we accept the null hypothesis. "Type of defect" and "supplier" are independent. (This is not to say that each supplier ships the same proportion of defects. What it does say is that among the defects, the proportion of B_1-type defects is the same regardless of supplier.)

8.4 APPLICATION 3: SAMPLING DISTRIBUTION OF THE SAMPLE VARIANCE WHEN SAMPLE IS FROM A NORMALLY DISTRIBUTED POPULATION

In the previous two sections, we have used

$$\sum \left[\frac{(f_0 - f_t)^2}{f_t} \right]$$

as a measure of how close the theoretical frequencies are to the observed frequencies. If we consider f_t to be a sort of average of possible f_0 values, then the expression

$$\sum \left[\frac{(X - \mu)^2}{\sigma^2} \right]$$

is analogous to the expression which we used to compare frequencies. Also recall that we stated that Equation 8.2 is *approximately* a chi-square distribution with $(k - 1)$ degrees of freedom (if $f_t \geq 5$).

We shall now state, without proof,[3] that if and only if X is normally distributed with mean μ and a finite variance σ^2, then

$$\chi^2 = \sum \left[\frac{(X - \bar{X})^2}{\sigma^2} \right] \tag{8.9}$$

has a chi-square distribution with $v = n - 1$ degrees of freedom. So what?

First note that σ^2 is a constant; hence

$$\chi^2 = \frac{1}{\sigma^2} \Sigma(X - \bar{X})^2$$

Next, recall that the sample variance was defined as $s^2 = [1/(n - 1)]\Sigma(X - \bar{X})^2$, or $(n - 1)s^2 = \Sigma(X - \bar{X})^2$. Substituting this definition into Equation 8.9, we obtain

$$\chi^2 = \frac{(n - 1)s^2}{\sigma^2} \tag{8.10}$$

---

[3] For a proof, see Lincoln L. Chao, *Statistics: Methods and Analyses*, McGraw-Hill Book Company, New York, 1969, section 12.2.1.

Now that looks more like something we can use. It does? Sure. Equation 8.10 gives us the sampling distribution of all possible values of s^2 when we are taking samples of size n from a normal population with mean μ and variance σ^2.

8.4(a) Test of Hypothesis about a Single Variance

Assume the population is normally distributed and that we want to test the hypothesis that

$$H_0: \sigma^2 = \sigma_0^2 \qquad H_A: \sigma^2 \neq \sigma_0^2$$

where σ_0^2 is some specified value of the population variance. All possible samples of size n would yield a variety of values for s^2, the sample variance. How are these values of s^2 distributed?

Solving Equation 8.10 for s^2 yields

$$s^2 = \left(\frac{\sigma^2}{n-1}\right)\chi^2 \tag{8.11}$$

where n and σ^2 are constants. The sampling distribution for s^2 is an augmented chi-square distribution. See Figure 8.4. The mean of this distribution μ_{s^2} is the population variance, and assuming the null is true, $\sigma^2 = \sigma_0^2$. After choosing a level of significance α, the decision rule is

$$\text{If } s^2 > \left(\frac{\sigma_0^2}{n-1}\right)\chi_U^2 \text{ or if } s^2 < \left(\frac{\sigma_0^2}{n-1}\right)\chi_L^2 \qquad \text{reject the } H_0$$

$$\text{If } \left(\frac{\sigma_0^2}{n-1}\right)\chi_L^2 \leq s^2 \leq \left(\frac{\sigma_0^2}{n-1}\right)\chi_U^2 \qquad \text{accept the } H_0$$

FIGURE 8.4 Sampling distribution for s^2.

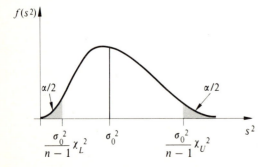

Recall that s^2 can assume only nonnegative numerical values and that χ^2 is not necessarily a symmetric distribution. Thus, our only problem is to determine $\chi_U{}^2$ and $\chi_L{}^2$.

The chi-square distribution is skewed to the right. In Figure 8.4 note that the distance from the mean $\sigma_0{}^2$ to the "lower critical value" (that is, $[\sigma_0{}^2/(n-1)]\chi_L{}^2$) is *less* than the distance from the mean to the "upper critical value." This difference is necessary to ensure that $\alpha/2$ of the area is under *each* tail of the skewed distribution.

When the chi-square distribution is used for the test of hypothesis about a population variance, then the *degrees of freedom* are $n - 1$, and they depend directly upon the size of the sample. In Appendix Table VII, the "area" referred to is the area to the *left* of the tabular value of χ^2. That is, if $\alpha = 0.05$, then $\alpha/2 = 0.025$. For the appropriate degrees of freedom,

$$\chi_L{}^2 = \chi_{0.025}^2 \qquad \text{and} \qquad \chi_U{}^2 = \chi_{0.975}^2$$

Example 8.4 One Last Slice at the Automatic Cola Machine

In example 6.1, 16 randomly selected observations (or cups) gave a mean fill of 8.05 ounces per cup. Also, assume that those 16 measurements gave us the following information:

$$\Sigma(X - \bar{X})^2 = 0.1610 \qquad \text{and} \qquad s = \sqrt{0.1610/15} = 0.1036 \text{ ounce}$$

Example 6.1 assumed that (1) the population of all possible fills is normally distributed and (2) the population standard deviation is 0.10 ounce. Let us assume that the first assumption is true (see example 8.1) and test the second.

1. Null and alternative hypotheses:

$$H_0: \sigma^2 = (0.10)^2 = 0.01$$
$$H_A: \sigma^2 \neq 0.01$$

2. Level of significance: let $\alpha = 0.05$. Then $\alpha/2 = 0.025$ and $(1 - \alpha/2) = 0.975$.
3. Degrees of freedom:

$$v = n - 1 = 16 - 1 = 15$$

4. Critical values of χ^2: For 15 degrees of freedom, Appendix Table VII shows that

$$\chi_L{}^2 = 6.26 \qquad \text{for } \alpha/2 = 0.025$$
$$\chi_U{}^2 = 27.49 \qquad \text{for } 1 - \alpha/2 = 0.975$$

5. Decision rule: Since

$$[\sigma_0^2/(n - 1)]\dot{\chi}_L^2 = (0.01/15)6.26 = 0.004$$

and

$$[\sigma_0^2/(n - 1)]\chi_U^2 = (0.01/15)27.49 = 0.018$$

then

If $s^2 < 0.004$ or if $s^2 > 0.018$ reject H_0

Otherwise   accept H_0

6. Decision: Since $s^2 = (0.1036)^2 = 0.011$, which is between 0.004 and 0.018, we accept (or do not reject) the H_0.

The Large-Sample Case

Appendix Table VII does not list all possible degrees of freedom. For example, in example 8.1 the sample size was $n = 49$. If we wish to test the hypothesis that $\sigma^2 = 0.01$, then the number of degrees of freedom is $n - 1 = 48$. Only $v = 40$ and $v = 60$ appear in Appendix Table VII. How do we determine the critical values of χ^2? There are several solutions.

First, you could locate more detailed tables for the chi-square distribution (e.g., H. Leon Harter, *New Tables of the Incomplete Gamma-Function Ratio and of Percentage Points of the Chi-Square and Beta Distributions*, Aerospace Research Laboratories, Office of Aerospace Research, USAF, 1964, p. 142, which shows $\chi_L^2 = 30.75$ and $\chi_U^2 = 69.02$).

Second, you could "interpolate" from the values given in Appendix Table VII. For the purposes of the problems in this text, linear interpolation is inaccurate but sufficient. Consider the lower critical value of χ^2 for a two-tail test with $\alpha = 0.05$ and 48 degrees of freedom. From Appendix Table VII,

Degrees of Freedom	χ^2 ($\alpha/2 = 0.025$)
40	24.43
60	40.48

From linear interpolation, we estimate the value of χ^2 to be:

$$\chi_{0.025, 48}^2 = 24.43 + \left(\frac{48 - 40}{60 - 40}\right)(40.48 - 24.43) = 24.43 + 6.42 = 30.85$$

Finally, as the degrees of freedom increase, the chi-square distribution approaches the shape of a normal curve with $\mu = v$ and $\sigma = \sqrt{2v}$. Since $z = (X - \mu)/\sigma$, then

$$\chi_L{}^2 = v - z_{\alpha/2}\sqrt{2v} = n - 1 - z_{\alpha/2}\sqrt{2(n-1)}$$

and
$$\chi_U{}^2 = v + z_{\alpha/2}\sqrt{2v}$$

This approximation improves as n increases; however, a common rule of thumb is: *When $v \geq 100$, the chi-square distribution can be approximated with the normal curve.* To demonstrate the accuracy of this approximation, assume $n = 121$ (or $v = 120$). For $\alpha = 0.05$, $z_{\alpha/2} = 1.96$; thus, our approximated values for χ^2 are

$$\chi_L{}^2 = 120 - (1.96)\sqrt{2(120)} = 120 - (1.96)(15.49)$$

$$= 120 - 30.36 = 89.64$$

$$\chi_U{}^2 = 120 + (1.96)\sqrt{240} = 150.36$$

From Appendix Table VII we find the true values:

$$\chi_L{}^2 = 91.57 \quad \text{and} \quad \chi_U{}^2 = 152.2$$

Note that the normal-curve approximation underestimates both values of χ^2—the result you would expect when the chi-square distribution is slightly skewed to the right.

With the exception of the linear-interpolation technique, these techniques regress from the most accurate to the least accurate for degrees of freedom less than 120. The technique employed will depend upon the availability of detailed tables and the type of problem under consideration.

Single-Tail Tests

Occasionally, a single-tail test of a population variance may be appropriate. Consider the introduction of a new harder metal for a die which is used to produce screws. Since the metal should wear at a slower rate than the softer metal dies currently in use, there should be less variation in the diameter of the screws. If we knew the variation in screw diameter was $\sigma_0{}^2$ (inches squared), then we could sample a new die to see if the variation were reduced. The hypotheses would be $H_0: \sigma^2 \geq \sigma_0{}^2$ and $H_A: \sigma^2 < \sigma_0{}^2$. (Using the "equals-sign-in-the-null" rule, the claim is placed in the alternative hypothesis.)

For a given level of significance α, the decision rule would be

$$\text{If } s^2 < [\sigma_0{}^2/(n-1)]\chi^2_{\alpha,\, n-1} \qquad \text{reject } H_0$$

$$\text{If } s^2 \geq [\sigma_0{}^2/(n-1)]\chi^2_{\alpha,\, n-1} \qquad \text{accept } H_0$$

Panel 8.1 Interval Estimate of σ^2

Population
The random variable X *must be normally distributed* with mean μ and variance σ^2.

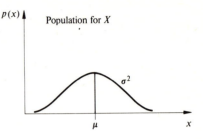

Sampling Distribution
All possible random samples of size n are selected from the population and

$$s^2 = [\Sigma(X - \bar{X})^2]/(n - 1)$$

is calculated for each sample. We know that the ratio $(n - 1)s^2/\sigma^2$ exhibits a chi-square distribution with $(n - 1)$ degrees of freedom.

Level of Confidence
Arbitrarily, $(1 - \alpha)$ is assigned a value. From α and the known degrees of freedom, $\chi_L^2 = \chi^2_{(n-1, \alpha/2)}$ and $\chi_U^2 = \chi^2_{(n-1, 1-\alpha/2)}$ are found in Appendix Table VII.

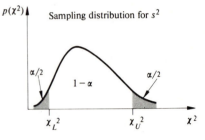

Probability Statement of Ratio
From what we do know,

$$P\left[\chi_L^2 < \frac{(n - 1)s^2}{\sigma^2} < \chi_U^2\right] = 1 - \alpha$$

Algebraic Manipulation
We want something like $P(a < \sigma^2 < b) = 1 - \alpha$.

1. Reciprocal:

$$P\left[\frac{1}{\chi_L^2} > \frac{\sigma^2}{(n - 1)s^2} > \frac{1}{\chi_U^2}\right] = P\left[\frac{1}{\chi_U^2} < \frac{\sigma^2}{(n - 1)s^2} < \frac{1}{\chi_L^2}\right] = 1 - \alpha$$

2. Multiply by $(n - 1)s^2$:

$$P\left[\frac{(n - 1)s^2}{\chi_U^2} < \sigma^2 < \frac{(n - 1)s^2}{\chi_L^2}\right] = 1 - \alpha$$

Confidence-Interval Estimate
For a $(1 - \alpha)$ level of confidence, the interval estimate of the population variance is

$$\frac{(n - 1)s^2}{\chi_U^2} < \sigma^2 < \frac{(n - 1)s^2}{\chi_L^2}$$

Families which are in the process of buying their homes frequently receive an income tax deduction for the interest paid on their mortgages. Let us assume that these tax deductions are normally distributed. A proposed income tax amendment is defended on the basis that the average tax deduction on mortgages will not change. The opposition is more concerned with the increased variation that they suspect the complicated amendment will create. A random sample of n taxpayers (who take this tax deduction) is selected, and their tax deductions are recalculated as if the proposed amendment were in effect.

The opposition proposes that this sample be used to test the following hypothesis:

$H_0: \sigma^2 \leq \sigma_0{}^2$ where $\sigma_0{}^2$ is the variation in the current deduction for interest on mortgage

$H_A: \sigma^2 > \sigma_0{}^2$ (the opposition claim)

For the arbitrarily selected level of significance α, the decision rule is:

$$\text{If } s^2 > [\sigma_0{}^2/(n-1)]\chi^2_{1-\alpha, n-1} \qquad \text{reject } H_0$$
$$\text{If } s^2 \leq [\sigma_0{}^2/(n-1)]\chi^2_{1-\alpha, n-1} \qquad \text{accept } H_0$$

Note that for the left-tail test (e.g., the harder-die example) the decision rule uses $\chi_\alpha{}^2$ for $(n-1)$ degrees of freedom while the right-tail test uses $\chi^2_{1-\alpha}$. This difference occurs because α refers to the area in the appropriate tail of the chi-square distribution and Appendix Table VII is constructed in a manner which considers the area "to the left" of the critical value of χ^2.

8.4(b) Confidence Interval for Population Variance

Recalling Equation 8.10, Panel 8.1 outlines the development of a confidence interval for σ^2.

8.5 TEST OF HYPOTHESIS ABOUT THREE OR MORE MEANS (ANALYSIS OF VARIANCE)

In order to test the hypothesis that three or more means are identical, we must *analyze the variation* both within and between samples. The basic question could be rephrased as, Did all samples come from the same population, or did at least one sample come from a different population?

Example 8.5   Payment Policy for Ambulance Service

A municipally owned hospital is centrally located and purchases its nighttime ambulance services from nearby funeral parlors. The hospital pays for a round-trip emergency ambulance run on the basis of the distance from the funeral home to the patient pick-up to the hospital and back to the funeral home.

There are three "districts" (A, B, and C) within the city. Traffic surveys reveal that nighttime congestion is considerably different in the three districts. The funeral homes want to be paid by the hour rather than according to the current policy of by the mile. They argue that the change would be fairer since it sometimes takes longer to go a mile in district A than in district B.

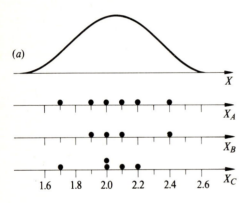

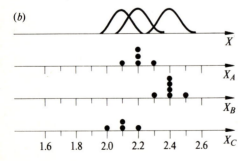

FIGURE 8.5   (*a*) All samples from *one* population? (*b*) Each sample from a *different* population?

TABLE 8.6 WEEKDAY SAMPLE

Observation Number	Minutes per Mile for an Ambulance Run in District		
	A	B	C
1	2.4	2.1	2.2
2	2.0	1.9	2.0
3	2.1	2.0	2.1
4	1.7	2.4	2.0
5	2.2	—	1.7
6	1.9	—	—
Total	12.3	8.4	10.0

Means

$$\bar{X}_A = \frac{12.3}{6} = 2.05 \qquad \bar{X}_B = \frac{8.4}{4} = 2.10 \qquad \bar{X}_C = \frac{10}{5} = 2.00$$

$$\bar{\bar{X}} = \frac{12.3 + 8.4 + 10.0}{6 + 4 + 5} = \frac{30.7}{15} = 2.05 \ (\bar{\bar{X}} = \text{minutes per mile disregarding district})$$

The hospital does not want to change the billing mechanism for several reasons. For example, more traffic congestion means slower-moving traffic, which can mean that traffic more quickly provides a path for emergency vehicles. There are other factors, but the net result is that the hospital argues that the average minutes per mile of an ambulance run is the same for all three districts. That is,

$$H_0: \mu_A = \mu_B = \mu_C$$

The alternative hypothesis is that all three means are not equal (i.e., any two or all three could be different).

If this null hypothesis is true, then it makes no difference whether the hospital continues to pay on a per-mile basis or on a comparable hourly basis. On a randomly selected weekday, the data shown in Table 8.6 was obtained. In addition, Table 8.7 provides a similar sample taken during a weekend (when the congestion is higher in all three districts).

In Figure 8.5, the observations from both tables are plotted. For the weekday sample, the observations from each district appear to have come from a population whose mean is the same as the other two districts. Figure 8.5a appears to indicate a high probability that the null hypothesis is true. However, Figure 8.5b appears to cast some doubt on the null hypothesis.

But how do we statistically test the null?

TABLE 8.7 WEEKEND SAMPLE

Observation Number	Minutes per Mile for an Ambulance Run in District		
	A	B	C
1	2.1	2.4	2.1
2	2.2	2.5	2.0
3	2.2	2.4	2.1
4	2.3	2.4	2.2
5	2.2	2.3	—
6	—	2.4	—
Totals	11.0	14.4	8.4

Means

$$\bar{X}_A = \frac{11}{5} = 2.2 \qquad \bar{X}_B = \frac{14.4}{6} = 2.4 \qquad \bar{X}_C = \frac{8.4}{4} = 2.1$$

$$\bar{\bar{X}} = \frac{11 + 14.4 + 8.4}{5 + 6 + 4} = \frac{33.8}{15} = 2.25$$

Figure 8.5 suggests the procedure for testing the hypothesis concerning the equality of multiple means. Also, it clarifies why the technique is often called "analysis of variance." In Figure 8.5b, it is the amount of variation between districts *relative to* the variation within each district that causes us to suspect that the null hypothesis is not true. Figure 8.6 further generalizes the point.

If the variation *within* each district is small, then it does not take much variation *between* districts to discredit the null hypothesis. However, when the variation *within* each district is large, it takes considerably more variation *between* districts to reject the null hypothesis.

Thus, we are looking for some measure of

$$\frac{\text{Between-district variation}}{\text{Within-district variation}} \qquad (8.12)$$

The larger the value of this ratio, the smaller the probability that the samples came from populations with identical means.

First, let us consider the denominator of this ratio. How are we going to measure within-district variations? For *each* district, we might use the variance, $(1/n) \sum (X_i - \bar{X})^2$, to measure the variation from the mean; however, the "within-district variation" concept reflects a composite (or aggregation) of the within-district

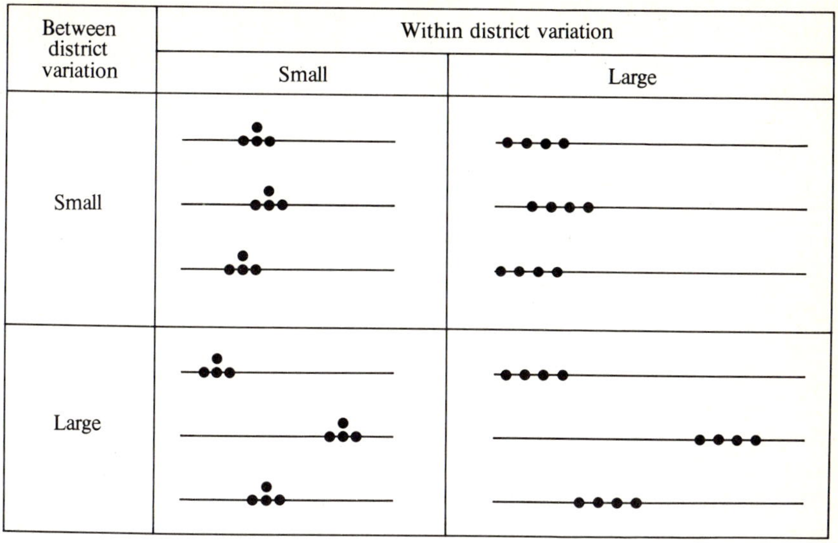

Between district variation	Within district variation	
	Small	Large
Small		
Large		

FIGURE 8.6 Relative magnitude of between- and within-district variation.

variation for *all* districts. For this concept, we shall use the *Sum* of *Squared* deviations from the mean *Within* each district:

$$\text{SSW} = \sum_{i=1}^{n_A} (X_{iA} - \bar{X}_A)^2 + \sum_{i=1}^{n_B} (X_{iB} - \bar{X}_B)^2 + \sum_{i=1}^{n_C} (X_{iC} - \bar{X}_C)^2 \qquad (8.13)$$

In terms of the sample variance,

$$\text{SSW} = (n_A - 1)s_A^2 + (n_B - 1)s_B^2 + (n_C - 1)s_C^2$$

This provides a total of the variation within the various districts. The greater the number of observations, the larger SSW is likely to be; thus an "average" of the variation within the various districts is more suitable for the denominator of the ratio given by Equation 8.12. If we let n represent the total number of observations regardless of district and m the number of districts,[4] then the *Mean* of the *Sum* of *Squared* deviations *Within* the district is defined as

$$\text{MSSW} = \frac{\text{SSW}}{n - m} \qquad (8.14)$$

---

[4] The number of degrees of freedom in calculating SSW is $n - m$. For each district, we lose 1 degree of freedom by using the sample mean to obtain the squared deviations from the mean. Note that $(n_A - 1) + (n_B - 1) + (n_C - 1) = (n_A + n_B + n_C) - 3 = n - m$.

Next, we investigate the "between-district variation" in Equation 8.12. In example 8.5, we calculated the arithmetic mean for each district (that is, $\bar{X}_A, \bar{X}_B, \bar{X}_C$), then we calculated $\bar{\bar{X}}$ (pronounced "X double-bar"). We shall define $\bar{\bar{X}}$ as the *grand mean*.

$$\bar{\bar{X}} = \frac{n_A \bar{X}_A + n_B \bar{X}_B + n_C \bar{X}_C}{n_A + n_B + n_C}$$

$$= \frac{(\Sigma X)_A + (\Sigma X)_B + (\Sigma X)_C}{n_A + n_B + n_C} \tag{8.15}$$

The grand mean is just the sample mean of all the observations, disregarding the "district" classifications. In a sense, it treats the null hypothesis as valid and assumes that all observations came from the same population.

The grand mean $\bar{\bar{X}}$ is used as a reference point, and we look at the variation of the district means relative to the grand mean. Since the deviations about the mean are always equal to 0 [that is, $(\bar{X}_A - \bar{\bar{X}}) + (\bar{X}_B - \bar{\bar{X}}) + (\bar{X}_C - \bar{\bar{X}}) \equiv 0$], we again use the "squared deviations" to measure variation. The *S*um of *S*quared deviations *B*etween districts is defined as

$$\text{SSB} = n_A(\bar{X}_A - \bar{\bar{X}})^2 + n_B(\bar{X}_B - \bar{\bar{X}})^2 + n_C(\bar{X}_C - \bar{\bar{X}})^2 \tag{8.16}$$

Since each district may have a different number of observations, we account for that fact by "weighting" each squared deviation by the sample size.

The SSB reflects the total of the between-district variation. From Equation 8.15 we can see that $\bar{X}_A$, $\bar{X}_B$, and $\bar{X}_C$ are used to calculate $\bar{\bar{X}}$; therefore, there are only 2 degrees of freedom. In general, there are only $(m - 1)$ degrees of freedom in calculating SSB. Hence, the *M*ean of the *S*um of *S*quared deviations *B*etween districts is defined as

$$\text{MSSB} = \frac{\text{SSB}}{m - 1} \tag{8.17}$$

Now, let us define

$$F = \frac{\text{MSSB}}{\text{MSSW}} = \frac{\text{SSB}/(m - 1)}{\text{SSW}/(n - m)} \tag{8.18}$$

as our measure of the concept developed in Equation 8.12.

From example 8.5, let us calculate the F ratio for the weekday sample and the weekend sample. See Tables 8.8 and 8.9.

TABLE 8.8 CALCULATION OF F FOR WEEKDAY SAMPLE

District A		District B		District C	
X_i	$(X - \bar{X}_A)^2$	X_i	$(X - \bar{X}_B)^2$	X_i	$(X - \bar{X}_C)^2$
2.4	0.1225	2.1	0.00	2.2	0.04
2.0	0.0025	1.9	0.04	2.0	0.00
2.1	0.0025	2.0	0.01	2.1	0.01
1.7	0.1225	2.4	0.09	2.0	0.00
2.2	0.0225	—	—	1.7	0.09
1.9	0.0225	8.4	0.14	—	—
—	—			10.0	0.14
12.3	0.2950				

$$n_A = 6, \ \bar{X}_A = 2.05 \qquad n_B = 4, \ \bar{X}_B = 2.1 \qquad n_C = 5, \ \bar{X}_C = 2.0$$

SSW $= 0.295 + 0.140 + 0.140 = 0.575$
MSSW $=$ SSW$/(n - m) = 0.575/(15 - 3) = 0.0479$
$\bar{X} = 2.05$
SSB $= 6(2.05 - 2.05)^2 + 4(2.10 - 2.05)^2 + 5(2.00 - 2.05)^2$
$\quad = 0 + 0.0100 + 0.0125 = 0.0225$
MSSB $=$ SSB$/(m - 1) = 0.0225/(3 - 1) = 0.01125$
$F =$ MSSB/MSSW $= 0.01125/0.0479 = 0.23$

TABLE 8.9 CALCULATION OF F FOR WEEKEND SAMPLE

District A		District B		District C	
X_i	$(X - \bar{X}_A)^2$	X_i	$(X - \bar{X}_B)^2$	X_i	$(X - \bar{X}_C)^2$
2.1	0.01	2.4	0.00	2.1	0.00
2.2	0.00	2.5	0.01	2.0	0.01
2.2	0.00	2.4	0.00	2.1	0.00
2.3	0.01	2.4	0.00	2.2	0.01
2.2	0.00	2.3	0.01	—	—
—	—	2.4	0.00	8.4	0.02
11.0	0.02	—	—		
		14.4	0.02		

$$n_A = 5, \ \bar{X}_A = 2.2 \qquad n_B = 6, \ \bar{X}_B = 2.4 \qquad n_C = 4, \ \bar{X}_C = 2.1$$

SSW $= 0.02 + 0.02 + 0.02 = 0.06$
MSSW $=$ SSW$/(n - m) = 0.06/(15 - 3) = 0.0050$
$\bar{X} = 2.25$
SSB $= 5(2.20 - 2.25)^2 + 6(2.40 - 2.25)^2 + 4(2.10 - 2.25)^2$
$\quad = 0.0125 + 0.1350 + 0.0900 = 0.2375$
MSSB $=$ SSB$/(m - 1) = 0.2375/(3 - 1) = 0.11875$
$F =$ MSSB/MSSW $= 0.11875/0.0050 = 23.75$

If our measure of between-district variations is *less than* our measure of within-district variations, then (1) the F ratio is *less than* 1 and (2) it is quite likely that all districts have the same population mean. In this case the null hypothesis would not be rejected. On the other hand, if the between-district variation is *greater than* the within-district variation, then (1) the F ratio is *greater than* 1 and (2) the equality of district means is dubious.

Note that the weekday sample gave a value of F that was less than 1 (that is, $F = 0.23$), while the weekend sample yields a value of F that is larger than 1 (23.75). Now the question is, How much larger than 1 must F be before we reject the null hypothesis? To answer that question, we must consider another theoretical sampling distribution.

The F Distribution

What is the sampling distribution for the ratio F? The first point to note is that F is the *ratio* of two variables. Next, we will state, without proof, that under specific conditions F is the ratio of two chi-square distributed variables.[5] In Equation 8.9 we said that $(1/\sigma^2)\Sigma(X - \bar{X})^2$ has a chi-square distribution. Recognize that $1/\sigma^2$ can be treated as a constant; then $\Sigma(X - \bar{X})^2$ is the "sum of squared deviations." Both SSW and SSB are analogous to Equation 8.8; however, SSW and SSB may have different degrees of freedom. Thus,

$$F = \frac{\chi_1{}^2/v_1}{\chi_2{}^2/v_2} \tag{8.19}$$

where $\chi_1{}^2$ represents a chi-square distribution with v_1 degrees of freedom and $\chi_2{}^2$ is a chi-square distribution with v_2 degrees of freedom.

Also, recall that the mean, the variance, and the shape of a chi-square distribution depend entirely on the degrees of freedom. Thus, the distribution of this new variable F, which is a ratio of two chi-square distributions, must depend upon the degrees of freedom of *both* the numerator and the denominator. Let v_b (because the between-districts variation is in the numerator) be the degrees of freedom for the numerator and v_w be the degrees of freedom for the denominator of F. Then the sampling distribution of F is completely defined by v_b and v_w.

Since $\chi^2 \geq 0$, then $F \geq 0$. For values of v_b and v_w that are small, the F distribution is skewed to the right (similar to the chi-square). But, as either v_b or v_w increases, the F distribution approaches a symmetric distribution. In addition, we are only

---

[5] The specific conditions are twofold: (1) The population which represents each district is normally distributed. (2) The population variance is the same for all districts. When $\sigma_A{}^2 = \sigma_B{}^2 = \sigma_C{}^2$, the populations are said to be *homoscedastic*. The population mean for each district may be different.

concerned with the extent to which F exceeds 1 (i.e., the between-district variation exceeds the within-district variation); therefore, we will only be concerned with the right tail of the F distribution.[6]

Use of the F Table

If we specify α, the probability of Type I error, and we calculate v_b and v_w, how do we use Appendix Table VIII to find the critical value of F for a decision rule? Appendix Table VIII has been calculated for only two values of α: $\alpha = 0.05$ and $\alpha = 0.01$. For each value of α, there are a variety of values for v_b and v_w.

The heading of Appendix Table VIII provides the number of degrees of freedom for the numerator of F (or, v_b as we have defined it). The number of degrees of freedom in the left-hand column represents v_w, or the degrees of freedom for the denominator. The body of the table presents the (right-tail) critical values of F.

Finally, $v_b = m - 1$ and $v_w = n - m$, as stated in the discussion of Equations 8.14 and 8.17. The number of districts is represented by m, and n ($= n_A + n_B + n_C$) is the total number of observations.

Example 8.6   The Ambulance Service Policy Revisited

Let us use $\alpha = 0.05$. For both the weekday sample and the weekend sample, $n = 15$ and $m = 3$. Thus,

$$v_b = m - 1 = 3 - 1 = 2 \text{ degrees of freedom}$$

and

$$v_w = n - m = 15 - 3 = 12 \text{ degrees of freedom}$$

From Appendix Table VIII, $F(\alpha, v_b, v_w) = F(0.05, 2, 12) = 3.88$. Consequently, our decision rule is:

$$\text{If } F > 3.88 \qquad \text{reject } H_0$$

$$\text{If } F \leq 3.88 \qquad \text{accept } H_0$$

where $H_0: \mu_A = \mu_B = \mu_C$.

---

[6] There are other tests of hypothesis that are concerned with the left tail of the F distribution. Although we will not discuss these tests here, the following relationship allows us to calculate the critical values of F which are less than 1:

$$F(1 - \alpha, v_w, v_b) = \frac{1}{F(\alpha, v_b, v_w)}$$

For the weekday sample, Table 8.7 shows that $F = 0.23$. Since $F = 0.23 < 3.88$, we accept the hospital's hypothesis that there is no difference between districts in the average time it takes per mile.

For the weekend sample, Table 8.8 shows that $F = 23.75$. Since $F = 23.75 > 3.88$, the null hypothesis is rejected and the ambulance owners have a point.

Based upon these results, the hospital may want to continue to pay by the mile. On weekdays there would be a standard mileage rate that does not vary by district. However, a higher rate, which varies by district, would be "fairer" for weekend ambulance runs.

Generalized Notation

The procedure described above is for the ambulance service example. A test of hypothesis for the equality of three or more population means can now be stated in more general terms. Assume that we want to test the equality of m population means. That is,

$$H_0: \mu_1 = \mu_2 = \mu_3 = \cdots = \mu_j = \cdots = \mu_m$$

or

$$H_0: \mu_j = \text{constant} \qquad \text{for } j = 1, 2, \ldots, m$$

From each of the m populations (or districts) we have n_j sample observations. The size of the sample may differ from one population to another; it is not necessary that $n_j = n_{j+1}$.

Table 8.10 presents a general structure for the raw data. Across the top, there is one "classification" for each of the m populations. Each column in the body of the table presents the n_j sample observations for that population.

TABLE 8.10 A GENERAL STRUCTURE FOR THE RAW DATA USED TO TEST THE EQUALITY OF m POPULATION MEANS

Observation Number (i)	Classification (or Population) (j)					
	1	2	$\cdots$	j	$\cdots$	m
1	X_{11}	X_{12}	$\cdots$	X_{1j}	$\cdots$	X_{1m}
2	X_{21}	X_{22}	$\cdots$	X_{2j}	$\cdots$	X_{2m}
$\vdots$	$\vdots$	$\vdots$		$\vdots$		$\vdots$
i	X_{i1}	X_{i2}	$\cdots$	X_{ij}	$\cdots$	X_{im}
$\vdots$	$\vdots$	$\vdots$		$\vdots$		$\vdots$
n	$X_{n_1 1}$	$X_{n_2 2}$	$\cdots$	$X_{n_j j}$	$\cdots$	$X_{n_m m}$
Totals	$\sum\limits_{i=1}^{n_1} X_{i1}$	$\sum\limits_{i=1}^{n_2} X_{i2}$	$\cdots$	$\sum\limits_{i=1}^{n_j} X_{ij}$	$\cdots$	$\sum\limits_{i=1}^{n_m} X_{im}$

Total Number of Observations: $\qquad n = \sum\limits_{j=1}^{m} n_j$

Sample Mean for Each Classification: $\qquad \bar{X}_j = \sum\limits_{i=1}^{n_j} \dfrac{X_{ij}}{n_j}$

Grand Mean: $\qquad \bar{\bar{X}} = \dfrac{\sum\limits_{j=1}^{m} \sum\limits_{i=1}^{n_j} X_{ij}}{n}$

$$SSW = \sum_{j=1}^{m} \left[\sum_{i=1}^{n_j} (X_{ij} - \bar{X}_j)^2 \right] \qquad SSB = \sum_{j=1}^{m} [n_j(\bar{X}_j - \bar{\bar{X}})^2]$$

$$F = \frac{SSB/(m-1)}{SSW/(n-m)}$$

This notation does no more than describe the procedure developed for $m = 3$ in the ambulance service example.

Further Comments

There are several aspects of the analysis-of-variance technique which need to be understood if the diversity and limitations of the technique are to be appreciated.

Segregation of the populations In the ambulance service example, the population was segregated by geographical regions, districts A, B, and C, within the city. Segregating the populations with a *population characteristic* is a common use of analysis of variance. Other examples would include: (1) segregate undergraduates by class standing and test for equality of mean grade-point average; (2) segregate human population by age and test for equality of average wine consumption; and (3) segregate common stocks by industry and test for equality of average rate of return. The population characteristic may be either qualitative or quantitative. If it is quantitative, a regression model for the problem may be more informative; regression techniques are discussed in later chapters.

Another method of segregating the populations falls in the context of experimental design. One (untreated) population may act as a control group, and other populations differ only in the amount (or type) of "treatment" applied. The null hypothesis then implies that the treatment does not make any difference in the average performance of the population. For example, several plots of land may be the various populations. The control plot is unfertilized, and the other plots receive either different amounts or different kinds of fertilizer, and then the equality of average crop yields is tested. Several different approaches to a statistics course could be used, and the average scores on a comprehensive examination could be used to test the differences in approach. The possibilities are extensive. Thus, the type of *treatment* applied serves to segregate the propulations.

More than one classification scheme for segregating populations The techniques outlined in this section are frequently called *one-way analysis of variance*. "One way" refers to the consideration of only one factor (cause or source) that could result in the population means being different. For example, if we had considered "districts" and "time of night" (e.g., before midnight and after midnight), we might have found that "time of night" affects average ambulance time more than either "district" or "day of the week." A two-way or a three-way classification of the data may cause the null hypothesis to be rejected when a one-way classification will not. If the classifications are quantitative, then a procedure for determining their influence will be discussed in the chapter on multiple regression. For qualitative classifications, two-way analysis of variance and its extension are beyond the scope of this text.[7]

Limitations In order to use the theoretical F distribution to test the equality of several population means, the following conditions must be satisfied:

1. Each segregated population must be normally distributed.
2. The variances for all segregated populations must be equal.
3. All sample observations must be independent.

Care in the collection of the sample observations is important to the independence assumption. Testing the equality of the variances is not easy, and we will not discuss it here. If the normality assumption is violated, then a procedure called the Kruskal-Wallis test is available (see Chapter 9).

Further, the probability of Type II error is as difficult to calculate for the F test as it was for the chi-square tests; however, conceptually it is similar to the test of hypothesis about a single mean (discussed in Chapter 7). Using the F distribution as we have described implicitly assumes that the null hypothesis is true (for example, $\mu_A = \mu_B = \mu_C$). Much of the problem of determining β is the multitude of ways in which the null could be false. For example, if the three population means can assume only values between 0 and 10, then all possible combinations of μ_A, μ_B, and μ_C can be represented by a cube with each side 10 units long. The straight line from one corner (0, 0, 0) to the opposite corner (10, 10, 10) represents the possible ways in which the null hypothesis could be true. The rest of the volume of the cube represents alternative hypotheses.

---

[7] For two-way analysis-of-variance technique developed with a business example, see Donald L. Harnett, *Introduction to Statistical Methods*, Addison-Wesley Publishing Company, Inc., Reading, Mass., 1970.

EXERCISES

1. A motel is located in the middle of nowhere. It has 12 units and during the last (100-day) summer season, the occupancy data was represented by:

Number of Rooms Occupied (X)	Number of Days This Number of Rooms Was Occupied (f_o)
0	0
1	3
2	6
3	7
4	10
5	12
6	15
7	19
8	16
9	7
10	5
11	0
12	0
	100

a. Consider the number of rooms occupied as the random variable X and the relative frequency, $f_o/100$, as the probability that each value of X will occur. Find $E(X)$.
b. If X is the number of successes in 12 trials, then $E(X) = n\pi$, and we can estimate the population proportion π from $E(X)/n$. Use your answer from part a to estimate π.
c. Use your estimate of π and Appendix Table III to find the probability that, on *one* given night, X rooms will be occupied.
d. Multiply your answers from part c by 100 days to obtain the theoretical frequencies.
e. For $\alpha = 0.05$, does the above occupancy data provide a good fit for the binomial distribution?

2. Consider the following sales data:

Gross Sales for First Week in September ($)	Number of Retail Stores
0 to under 2,000	10
2,000 to under 4,000	60
4,000 to under 6,000	185
6,000 to under 8,000	165
8,000 to under 10,000	75
10,000 to under 12,000	5
Over 12,000	0
	500

a. Approximate $\bar{X}$ and s.

b. Find the theoretical frequencies for a normal distribution using $\bar{X}$ as an estimate of μ and s as an estimate of σ.

c. Test the goodness of fit of this data to a normal distribution for $\alpha = 0.05$.

3. A large department store has five divisions, one on each floor. A random sample of 1,000 customers provided the following cross classification:

	Division of Purchase				
Type of Payment	B_1 Clothing and Shoes	B_2 Cosmetics and Beauty Care	B_3 Housewares and Furniture	B_4 Jewelry and Gifts	B_5 Toys
A_1: Cash (currency)	20	30	10	10	30
A_2: Cash (check)	100	90	30	20	60
A_3: Store's credit card	40	30	10	30	10
A_4: National credit card	140	150	50	40	100

a. For $\alpha = 0.05$, are A_i and B_j independent?

b. Would any one division be hurt proportionally more than any other division if the store dropped its own credit card (A_3)?

4. A manager of keypunch operators hypothesizes that new employees who know how to type (on a typewriter) are not as satisfactory as those who do not. A random sample of 100 new keypunch operators reveals the following data.

Can They Type?	Job Evaluation	
	Satisfactory	Not Satisfactory
Yes	30	10
No	50	10

For $\alpha = 0.01$, does the data support the manager's hypothesis?

5. In Exercise 3, the data resulted from a random sample of 1,000 customers from the store as a whole unit. If 200 customers were randomly selected from *each* (floor or) division, the results might be as follows.

Type of Payment	Division of Purchase				
	Clothing and Shoes	Cosmetics and Beauty Care	Housewares and Furniture	Jewelry and Gifts	Toys
Currency	15	20	20	20	25
Check	65	40	60	40	95
Store's card	25	20	20	30	5
National card	95	120	100	110	75

a. This data results from a different sampling procedure. Could we use this data for a chi-square test of *independence?*

b. Consider H_0: Each division has the same "type of payment" distribution (i.e., the proportion of customers paying by check is the same for each division) and H_A: The divisions are not *homogeneous* in the form of payment used by customers. Generate a decision rule in the same fashion as you would for a test of independence ($\alpha = 0.05$).

c. Follow exactly the same procedure as if this were a test of independence to: (1) calculate theoretical frequencies, (2) calculate χ^2, and (3) test the null hypothesis.

6. From each of three classifications, which relate to the primary function of the firm, 500 persons in management positions are randomly selected

Sex of Manager	Primary Function of the Firm			
	Manufacturing	Service	Government	Total
Male	300	280	320	900
Female	200	220	180	600
Total	500	500	500	1,500

a. To test the sex composition of each employer classification, would this be a test of independence or a test of homogeneity?

b. For $\alpha = 0.01$, test the hypothesis that the sex composition is the same for all three employer classifications.

7. A manufacturer of heat lamps (for keeping food warm in cafeterias) has been able to produce units with the following characteristics: Bulb life is normally distributed with $\mu = 400$ hours and $\sigma = 40$ hours. While the average life is satisfactory, some customers have complained that replacement scheduling is difficult because of the high variation in bulb life. A supplier of metal alloys suggests that by using a slightly higher-priced alloy as the filament, the standard deviation of bulb life can be reduced. The manufacturer tries the new alloy in the production line, produces 51 bulbs, tests them all, and finds $\bar{X} = 401.2$ hours and $s = 35$ hours. Test the hypothesis that the new alloy produces no less variation than the old alloy at $\alpha = 0.05$.

8. Comment on the following statement: To test the hypothesis that $\sigma = 0$ against the hypothesis that $\sigma \neq 0$, the investigator need not be concerned about the normality of the population or the chi-square distribution; one just samples until $X_i \neq X_j$.

9. From January through March, you conduct extensive research on the television viewing habits of residents of the northern United States. One result is that viewing time per week has $\mu = 20$ hours and $\sigma = 1$ hour. You return to your New York office, and the first question asked is, But, now that it is summer, have viewing habits changed? You rush back to your contacts in the field, take a random sample of 400 viewers, and find $\bar{X} = 19.89$ and $s = 1.2$ hours per week.

a. Assume that the population variance has not changed and, for $\alpha = 0.05$, test the hypothesis that the mean has not changed.

b. Consider the null hypothesis: $H_0: \sigma^2 = 1$ ($H_A: \sigma^2 \neq 1$).
   (1) To test that hypothesis, what assumption is required?
       How would you test that assumption?
   (2) Use a χ^2 test for $\alpha = 0.05$.

10. A random sample of the annual income for 16 families shows $\bar{X} = \$11,500$ and $s = \$800$. The variance in family income is estimated $(\alpha = 0.10)$ as:

$$\frac{(15)(800)^2}{24.996} < \sigma^2 < \frac{(15)(800)^2}{7.261}$$

$$\$384,061 < \sigma^2 < \$1,322,132$$

Comment on the validity of this estimate.

11. How would a chi-square distribution differ from a normal distribution with the same mean if the chi-square distribution has:
a. 1 degree of freedom?
b. 10 degrees of freedom?
c. 600 degrees of freedom?

12. Consider the confidence-interval estimate for σ^2:

$$\frac{(n-1)s^2}{\chi_U^2} < \sigma^2 < \frac{(n-1)s^2}{\chi_L^2}$$

Let $v = n - 1$; recall that $\chi_U^2 = v + z_{\alpha/2}\sqrt{2v}$ and $\chi_L^2 = v - z_{\alpha/2}\sqrt{2v}$. Show that

$$\frac{s^2}{1 + z\sqrt{2}/\sqrt{n-1}} < \sigma^2 < \frac{s^2}{1 - z\sqrt{2}/\sqrt{n-1}}$$

Then observe that the interval gets smaller as n increases.

13. You have recently purchased two plants smaller than your own. You are interested in building a fourth plant, and you are concerned about economies of scale. All three plants produce the same item. You take a random sample of several production runs (or several days, depending upon the nature of the production process) and find the average total cost per unit of output.

Cost per Unit by Size of Plant

Small	Medium	Large
$27	$26	$30
28	24	26
27	25	29
26	24	30
27	25	30
27	26	32
26	25	34
27	25	30
27		29
28		30

a. Test the hypothesis that all three plants have the same mean average total cost for $\alpha = 0.05$.
b. Should you be concerned about scale economies?

14. You do the same type of driving week after week. You keep accurate records of what producer of "regular" gasoline you use and the miles per gallon obtained.

Miles per Gallon by Brand			
Moonco Oil Co.	Sand Oil Co.	Basin Oil Co.	Xnon Oil Co.
18	21	20	17
20	15	19	17
17	17	21	15
21	16	20	16
	15		15
	18		

Do you experience the same mean miles per gallon for all four brands? (Use $\alpha = 0.01$.)

CASE QUESTIONS

1. Reconsider case question 1 in Chapter 5. Use the binomial probabilities to generate "theoretical frequencies" and your 40 empirical values of X, the number of successes, as "observed frequencies." Use a chi-square goodness-of-fit test to determine if the theoretical (binomial) model is a representative model for the empirical sampling distribution.

2. Reconsider case question 1 in Chapter 2. Use only Table 1 of that question (i.e., only gas-heated housing units). Use the chi-square distribution to test the hypothesis that "number of persons" (V03) and "number of bedrooms" (V23) are independent.

3. Consider the Oregon sample. Consider the "highest grade attended" by the head of household (V36). Use V36 to break these 30 households into three groups:

Group	Value of V36	Mean Income
A	V36 < 14	μ_A
B	V36 = 14	μ_B
C	V36 > 14	μ_C

Use V32 to find the incomes for each group. Test $H_0: \mu_A = \mu_B = \mu_C$ at the 5 percent level of significance.

9

NONPARAMETRIC, DISTRIBUTION-FREE, ORDER STATISTICS

INTRODUCTION AND DEFINITIONS

The population mean and the population variance are called *parameters*. Most of what the previous chapters have covered are techniques for making inferences about these parameters. Gauss' (or the normal) distribution is completely defined if we know or can assume the values of μ and σ. *Nonparametric* refers to a technique that makes no assumption about the value of a parameter in a probability density function. For example, in the previous chapter, the chi-square test of independence was not attempting to make any inference about a population parameter; however, it was a statistical inference.

Distribution-free statistical methods are those which do not require any assumptions about the shape or distribution of the parent population. Recall that in order to use the t distribution to provide a confidence-interval estimate of the population mean, the parent population *had* to be normally distributed. Also, remember that the Chebyshev inequality could be used to provide an interval estimate of μ if we know the population variance, but we do not need to know anything about the distribution of the population (e.g., it does not have to be normally distributed). Thus, the Chebyshev inequality was used as a parametric technique, but it was distribution free.

Some data is measured on an ordinal measurement scale (e.g., grades, brand preference) which represent rank-orderings. Finding the mean of the ranks may not mean anything. *Order statistics* are methods for either making inferences from ordinal data or using rank-orderings to investigate data measured on ratio or interval scales. By custom, order statistics and distribution-free statistics are all lumped under the umbrella of "nonparametric methods."

With Chebyshev's inequality and the chi-square distribution we have already treated some nonparametric methods without identifying them as such. Just to be fair, one parametric technique will be slipped into this chapter, but this time it will be so designated.

Finally, you should be aware that "nonparametric methods" include numerous techniques. This chapter serves as a small, nonrandom sample.

Relative Merits of Nonparametric Methods

A major advantage of nonparametric methods is their relative ease of calculation. Frequently, they are called "quick and dirty" methods. If dirty means soiled or obscene, then the point is argumentative; however, if dirty means "contemptibly contrary to honor or rules," a plurality may be possible. At any rate, they are usually much quicker to calculate than their parametric counterpart (if a counterpart exists).

Another major advantage has already been mentioned. Since nonparametric methods are generally distribution free, they are more general in application than the parametric techniques. Also, the ability to deal with ordinal data is especially useful in marketing.

Since this book spends proportionally more coverage on parametric methods and since there are the above-mentioned advantages, there must be some discouraging feature associated with nonparametric methods. Nonparametric techniques have a tendency to be less efficient. For example, for the same sample size and for the same level of confidence, the confidence interval is narrower for the t distribution (knowing the population is normal) than for the Chebyshev approach (knowing the population variance). In contrast, consider the following quote from I. R. Savage:[1]

> Normality is a myth; there never was and never will be, a normal distribution. This is an over-statement from the practical point of view, but it represents a safer initial mental attitude than any in fashion during the past two decades.

Whereas "smoking may be dangerous to your health" is a warning required on cigarette packages, Savage implies that "assuming normality may be dangerous to your statistical inference" should be placed on most of the material in Chapters 6 to 8.

9.1 TESTS OF RANDOMNESS

We have investigated various forms of inferences from a *random* sample. How do we know that a sample is random? Some business and economics data is not randomly selected. For example, data that is collected sequentially through time may have a serial dependency. There is the question of whether the sample results are random. In this section we will look at two techniques for testing:

H_0: the (sample) data is random
H_A: the (sample) data is not random

---

[1] I. R. Savage, "Bibliography of Nonparametric Statistics and Related Topics," *Journal of the American Statistical Association*, vol. 48 (1953), pp. 844–906.

The first, the one sample runs test, is nonparametric and distribution free; the second, the mean-square-of-successive-differences test, is parametric, but it is included here because of its use.

9.1(a) The One Sample Runs Test

To set up a runs test, consider the following steps:

1. The data is ordered chronologically.
2. Two unmistakable, mutually exclusive, and collectively exhaustive categories are established.
3. Each observation, still in chronological order, is replaced by its categorical representation.

For example, consider the first 20 people in a school cafeteria line. The sexes of all individuals, represented by M or F, represent two mutually exclusive and collectively exhaustive categories, which are presumably unmistakable. Then representing all the people, as they appear in that line, by their sex, we could obtain:

Sample 1 / M M M M M M M M M M /   / F F F F F F F F F F /

$$R = 2,\ n_A = 10,\ n_B = 10$$

Sample 2 M F M F M F M F M F M F M F M F M F M F

$$n_A = 10,\ n_B = 10,\ R = 20$$

Sample 3 / F F / / M M M M / / F / / M M / / F / / M / / F F F /

/ M M / / F F F /

$$n_A = 10,\ n_B = 10,\ R = 9$$

Intuitively, you would be inclined to say that samples 1 and 2 are not random. But why? There are an equal number of males and females in all three samples (that is, $n_A = n_B = 10$). Samples 1 and 2 represent a sequence which has a definite pattern; there is a specific *order* to the data in samples 1 and 2. If there is a specific order to the data in sample 3, it is certainly not obvious. The absence of order suggests randomness; the presence of order makes randomness suspicious.

The *number of runs*, designated by R, is the device we shall use, in this test, to investigate randomness. A *run* is a sequence of one or more observations which all fall into the same category. In sample 1, there are ten M's uninterrupted by an F; thus, the underlined set of M's comprises "one run." The ten F's make up another run. Hence, in sample 1, the number of runs is 2.

In sample 2, every run contains just one observation. There are 20 runs. For sample 3, there are nine different groupings, or runs, of M's and F's.

If there are n sample observations, then the maximum number of runs is n (e.g., sample 2). As long as at least one observation falls into both categories, the minimum number of runs is 2 (e.g., sample 1). Therefore, the number of runs will be between 2 and n; $2 \leq R \leq n$. As we have already seen, however, values of R close to 2 and close to n do not encourage the acceptance of the null hypothesis that the sample data is random.

For the null hypothesis, we need some value of R_L and R_U such that

$$R_L + 1 \leq R \leq R_U - 1$$

defines the *acceptance region* for R. Remember that R, the number of runs, must have an integer value. The *rejection region* is $R \leq R_L$ and $R \geq R_U$. If we represent the level of significance by α, then

$$P(R \leq R_L) \leq \alpha/2$$

and
$$P(R \geq R_U) \leq \alpha/2$$

Of course, $P(\{R \leq R_L\} \cup \{R \geq R_U\}) \leq \alpha$. The probability being *less than* or equal to α is necessary because R_L and R_U are integer values. It may be impossible to find integer values for R_L and R_U that make the probability statement "equal to" α.

To test for randomness by using R, the number of runs, we have to choose a value for α and find a way to establish R_L and R_U which will lead to a decision rule (or the specification of the acceptance and rejection regions).

Small samples   For small sample sizes we will use the table published by Eisenhart and Swed, Appendix Table IX, to determine R_L and R_U. As you might suspect, the sampling distribution depends upon the number of sample observations that occur in each of the two categories n_A and n_B. Of course, $n_A + n_B = n$.

In order to use Appendix Table IX, we require only the values of n_A and n_B. Appendix Table IX(L) provides the value of R_L, the lower limit, and Appendix Table IX(U) provides the value for R_U, the upper limit. Please note that these tables are constructed for $\alpha \leq 0.05$ *only!*

Consider sample 3 from the above cafeteria-line example: $n_A = n_B = 10$ (i.e., there are 10 females and 10 males). From Appendix Table IX(L) we find that $R_L = 6$, and from Appendix Table IX(U) we find that $R_U = 16$. Hence, our decision rule is:

If the number of runs R is between $(6 + 1 =)\,7$ and $(16 - 1 =)\,15$, accept the H_0; if R is 6 or less or if R is 16 or more, reject the H_0.

For sample 3, $R = 9$ and we would not reject the null hypothesis that the sample is random at the 5 percent level of significance. For sample 1, where $R = 2$ and for sample 2 where $R = 20$, the null hypothesis would be rejected.

What is a small sample? As suggested by the limits on Appendix Table IX, if *both* n_A and n_B are less than 20, we will consider the situation a small-size sample problem.

Large samples As n_A and n_B increase, the sampling distribution of R approaches a normal distribution[2] with a mean of

$$\mu_R = \frac{2n_A n_B}{n_A + n_B} + 1 \tag{9.1}$$

and a standard deviation of

$$\sigma_R = \sqrt{\frac{2n_A n_B (2n_A n_B - n_A - n_B)}{(n_A + n_B)^2 (n_A + n_B - 1)}} \tag{9.2}$$

For a level of significance α, the decision rule is

If $\mu_R - z_{\alpha/2}\sigma_R < R < \mu_R + z_{\alpha/2}\sigma_R$ accept H_0

If $R \le \mu_R - z_{\alpha/2}\sigma_R$ or if $R \ge \mu_R + z_{\alpha/2}\sigma_R$ reject H_0

Example 9.1   Sales Receipts for Cafeteria

Consider the dollar value of sales for the first 50 customers in the school cafeteria as presented in Table 9.1. Are they random?

If we placed this sales data in an array, we could find the median value by considering the twenty-fifth and twenty-sixth sales slip in that array. For these 50 observations, the median is \$1.03. Define the two mutually exclusive categories as "*Above* \$1.03" and "*Below* \$1.03."

Return to the chronological order of the sales slips, and classify each observation as A or B. By counting the number of A's and B's, we find that $n_A = 25$ and $n_B = 25$. (This is not too surprising since we used the *median* to define the A's and B's.) Finally, the number of runs can be counted: $R = 29$.

Now, we can set up the test of hypothesis.

$$H_0:\text{ The sales are random}$$

$$H_A:\text{ The sales are not random}$$

Let $\alpha = 0.05$ (or any value you choose); then $z_{\alpha/2} = 1.96$. From Equation 9.1,

$$\mu_R = \frac{2n_A n_B}{n_A + n_B} + 1 = \frac{2(25)(25)}{25 + 25} + 1 = 26.00$$

---

[2] Basically, as $n_A, n_B \to \infty$, if n_A/n_B equals a constant, then the sampling distribution of R approaches a normal distribution.

TABLE 9.1 SALES FOR THE FIRST 50 CUSTOMERS

Customer Number	Sales Slip ($)	Above or below Median	Run Number	Customer Number	Sales Slip ($)	Above or below Median	Run Number
1	0.18	B		26	1.21	A	
2	0.98	B	1	27	0.13	B	13
3	0.54	B		28	1.83	A	
4	1.46	A	2	29	1.85	A	14
5	0.50	B	3	30	1.14	A	
6	1.23	A	4	31	0.13	B	15
7	0.63	B	5	32	1.09	A	16
8	0.67	B		33	1.11	A	
9	1.58	A	6	34	0.87	B	17
10	1.00	B	7	35	1.62	A	18
11	1.49	A	8	36	0.94	B	19
12	1.06	A		37	1.40	A	20
13	0.34	B	9	38	0.13	B	21
14	1.19	A	10	39	1.17	A	
15	1.25	A		40	1.25	A	
16	0.15	B		41	1.74	A	22
17	0.94	B		42	1.67	A	
18	0.35	B		43	0.77	B	23
19	0.33	B	11	44	1.65	A	24
20	0.49	B		45	0.33	B	25
21	0.92	B		46	0.51	B	
22	0.89	B		47	1.39	A	26
23	1.34	A		48	0.63	B	27
24	1.30	A	12	49	1.75	A	28
25	1.07	A		50	0.45	B	29

$$\text{Median} = \frac{1.00 + 1.06}{2} = 1.03 \qquad n_A = 25 = n_B \qquad R = 29$$

From Equation 9.2,

$$\sigma_R{}^2 = \frac{2n_A n_B(2n_A n_B - n_A - n_B)}{(n_A + n_B)^2(n_A + n_B - 1)} = \frac{2(25)(25)[2(25)(25) - 25 - 25]}{(25 + 25)^2(25 + 25 - 1)}$$

$$= \frac{1,500,000}{122,500} = 12.2449$$

or $\qquad \sigma_R = \sqrt{12.2449} = 3.50$

Since $\mu_R + z_{\alpha/2}\sigma_R = 26 + (1.96)(3.5) = 26 + 6.86 = 32.86$, and $\mu_R - z_{\alpha/2}\sigma_R = 26 - 6.86 = 19.14$, then the decision rule is:

If $19.14 < R < 32.86$ $\qquad$ accept H_0

If $R \geq 32.86$ or if $R \leq 19.14$ $\qquad$ reject H_0

Since $R = 29$ (and falls in the acceptance region), the hypothesis that this sequence of 50 numbers is random cannot be rejected at the 5 percent level of significance.

Further comments In the previous example, the data is numerical and the two categories were "above" or "below" the median. In the example before that we used the classification "sex" to define the two categories. For other problems we could use defective versus nondefective, owner-occupied home versus not owner-occupied, etc., as classifications. In this context, the one sample runs test of randomness is very flexible.

9.1(b)   The Mean-Square-of-Successive-Differences Test

As previously mentioned, this test is not nonparametric, but it is useful as a test of randomness for data that is numerical (and measured with either a ratio or an interval scale). Also, this test is sometimes referred to as "the von Neumann ratio test for independence."

For example, assume that we have the United States net national annual income for the last 20 years. Net national income in year N may *depend* upon the level of net national income in the previous year. Y_N is not independent of Y_{N-1}; hence, we would expect that the chronological sequence of Y_i values would not be random. They would exhibit some pattern. In a *population*, this dependency is called "autocorrelation"; in a *sample*, "serial correlation." Occasionally, autocorrelation and serial correlation are used interchangeably.

When the observations are chronologically ordered, $(X_i - X_{i-1})$ is defined as the *successive difference*. If X_i is dependent upon X_{i-1}, then the successive differences should be *small*. However, some successive differences may be positive, and some may be negative. To eliminate the sign on each successive difference, square them. The sum of the squared successive differences divided by $(n - 1)$ is the *mean square of successive differences* (δ^2), or

$$\delta^2 = \frac{\sum\limits_{i=2}^{n} (X_i - X_{i-1})^2}{n - 1} \tag{9.3}$$

If the sequence of X_i values is actually random, which is our null hypothesis, then

$$E(\delta^2) = 2\sigma_X{}^2$$

But, we do not usually know σ_X^2, the *population* variance. What we could determine is the *biased* estimate of the *sample* variance,

$$*s^2 = \frac{\Sigma(X_i - \bar{X})^2}{n}$$

Thus, if we define the *von Neumann ratio*[3] (VNR) as

$$\text{VNR} = \frac{\delta^2}{*s^2} \tag{9.4}$$

then the expected value of VNR should be approximately equal to 2.

If X is normally distributed (a restriction we do not have with the runs test) and if $n > 60$, then VNR is approximately a normal distribution with mean

$$\mu_{\text{VNR}} = \frac{2n}{n-1} \tag{9.5}$$

and variance

$$\sigma_{\text{VNR}}^2 = \frac{4n^2(n-2)}{(n+1)(n-1)^3} \tag{9.6}$$

When n is large, $\mu_{\text{VNR}} \simeq 2$ and $\sigma_{\text{VNR}}^2 \simeq 4/n$.

Small samples   For $n \le 60$, the limits for the acceptance region are found in Appendix Table X. The lower tail value is found in Appendix Table X(left tail), and the upper tail value is in Appendix Table X (right tail).

Reconsider example 9.1, where we have the 50 sales in the cafeteria line. The null hypothesis is that the sales are random; the alternative, that they are not random. For $\alpha = 0.02$ (that is, 1 percent of the area in each tail),

$$\text{VNR}(L) = 1.3907 \qquad \text{and} \qquad \text{VNR}(U) = 2.6908$$

---

[3] Von Neumann used the biased estimate of population variance, instead of the unbiased estimate, $s^2 = \Sigma(X - \bar{X})^2/(n-1)$. See John von Neumann, "Distribution of the Ratio of the Mean Square Successive Difference to the Variance," *Annals of Mathematical Statistics*, vol. 12 (1941), pp. 367–395.
    Also, as a hint to the proof that $E(\delta^2) = 2\sigma^2$, note that $(X_i - X_{i-1}) = (X_i - \mu) - (X_{i-1} - \mu)$. If X_i and X_{i-1} are independent, then $E(X_i - X_{i-1})^2 = E(X_i - \mu)^2 + E(X_{i-1} - \mu)^2 = \sigma^2 + \sigma^2$.

TABLE 9.2   CALCULATIONS FOR THE VON NEUMANN RATIO

Sales Slip X_i	First Difference $X_i - X_{i-1}$	Squared Difference $(X_i - X_{i-1})^2$	X_i^2	Sales Slip X_i	First Difference $X_i - X_{i-1}$	Squared Difference $(X_i - X_{i-1})^2$	X_i^2
0.18	0.80	0.6400	0.0324	1.21	-1.08	1.1664	1.4641
0.98	-0.44	0.1936	0.9604	0.13	1.70	2.8900	0.0169
0.54	0.92	0.8464	0.2916	1.83	0.02	0.0004	3.3489
1.46	-0.96	0.9216	2.1316	1.85	-0.71	0.5041	3.4225
0.50	0.73	0.5329	0.2500	1.14	-1.01	1.0201	1.2996
1.23	-0.60	0.3600	1.5129	0.13	0.96	0.9216	0.0169
0.63	0.04	0.0016	0.3969	1.09	0.02	0.0004	1.1881
0.67	0.91	0.8281	0.4489	1.11	-0.24	0.0576	1.2321
1.58	-0.58	0.3364	2.4964	0.87	0.75	0.5625	0.7569
1.00	0.49	0.2401	1.0000	1.62	-0.68	0.4624	2.6244
1.49	-0.43	0.1849	2.2201	0.94	0.46	0.2116	0.8836
1.06	-0.72	0.5184	1.1236	1.40	-1.27	1.6129	1.9600
0.34	0.85	0.7225	0.1156	0.13	1.04	1.0816	0.0169
1.19	0.06	0.0036	1.4161	1.17	0.08	0.0064	1.3689
1.25	-1.10	1.2100	1.5625	1.25	0.49	0.2401	1.5625
0.15	0.79	0.6241	0.0225	1.74	-0.07	0.0049	3.0276
0.94	-0.59	0.3481	0.8836	1.67	-0.90	0.8100	2.7889
0.35	-0.02	0.0004	0.1225	0.77	0.88	0.7744	0.5929
0.33	0.16	0.0256	0.1089	1.65	-1.32	1.7424	2.7225
0.49	0.43	0.1849	0.2401	0.33	0.18	0.0324	0.1089
0.92	-0.03	0.0009	0.8464	0.51	0.88	0.7744	0.2601
0.89	0.45	0.2025	0.7921	1.39	-0.76	0.5776	1.9321
1.34	-0.04	0.0016	1.7956	0.63	1.12	1.2544	0.3969
1.30	-0.23	0.0529	1.6900	1.75	-1.30	1.6900	3.0625
1.07	0.14	0.0196	1.1449	0.45			0.2025
(1.21)							
			Total 48.64			27.3993	59.8628

Thus, the decision rule is:

$$\text{If } 1.3907 < \text{VNR} < 2.6908 \qquad \text{accept } H_0$$

$$\text{Otherwise} \qquad \text{reject } H_0$$

In Table 9.2, the values of X_i have been repeated from Table 9.1, maintaining their chronological order. From Table 9.2,

$$\delta^2 = \frac{\Sigma(X_i - X_{i-1})^2}{n-1} = \frac{27.3993}{49} = 0.5592$$

$$\bar{X} = \frac{\Sigma X}{n} = \frac{48.64}{50} = \$0.9728$$

$${}^*s^2 = \frac{\Sigma X^2 - n\bar{X}^2}{n} = \frac{59.8628 - 50(0.9728)^2}{50} = 0.2509$$

and the von Neumann ratio is

$$\text{VNR} = \frac{\delta^2}{*s^2} = \frac{0.5592}{0.2509} = 2.229$$

Since $1.3907 < 2.229 < 2.6908$, *accept* the null hypothesis that the sales data is random.

Large samples   For $n > 60$, we can use the following normal-curve approximation:

$$\text{VNR}(L) = \mu_{\text{VNR}} - z_{\alpha/2}\sigma_{\text{VNR}} \simeq 2 - z_{\alpha/2}\frac{2}{\sqrt{n}}$$

and $$\text{VNR}(U) = \mu_{\text{VNR}} + z_{\alpha/2}\sigma_{\text{VNR}} \simeq 2 + z_{\alpha/2}\frac{2}{\sqrt{n}}$$

For example, let us use this approximation for the cafeteria-line sales data even though $n = 50 < 60$. For $\alpha = 0.02$, $z_{\alpha/2} = 2.33$, and

$$\text{VNR}(L) = 2 - (2.33)\frac{2}{\sqrt{50}} = 2 - 0.6590 = 1.341$$

$$\text{VNR}(U) = 2 + 0.6590 = 2.6590$$

The resulting decision rule is:

If $1.341 < \text{VNR} < 2.659$ accept H_0

Otherwise   reject H_0

Since $\text{VNR} = 2.229$, we accept the null hypothesis.

As a check on this approximation, we could use Equations 9.5 and 9.6 to obtain

$$\begin{aligned}
\text{VNR}(L) &= \frac{2n}{n-1} - z_{\alpha/2}\sqrt{\frac{4n^2(n-2)}{(n+1)(n-1)^3}} \\
&= \frac{2(50)}{49} - (2.33)\sqrt{\frac{4(50)^2(48)}{(51)(49)^3}} \\
&= 2.041 - (2.33)\sqrt{0.0800} = 2.041 - 0.659 \\
&= 1.382
\end{aligned}$$

(which is closer to the Appendix table value of 1.3907), and $VNR(U) = 2.041 + 0.659 = 2.700$ (which is close to the table value of 2.6908).

Summary

The runs test is easier to apply, and it does not require the assumption that the sample was selected from a normally distributed population. However, when the assumptions for the von Neumann ratio test are met, the mean-square successive-difference test is more efficient.

9.2 INFERENCES ABOUT THE MEDIAN

If a population is symmetric, then the median (ϕ) and the arithmetic mean (μ) have the same value. An interval estimate of the mean, or a test of hypothesis about the mean, would provide similar results for the median. However, if the population is not symmetric, the mean and median have different values, and a separate mechanism for estimating (or testing) the median is necessary.

Interval Estimates for the Population Median

Consider any continuous random variable X, and let ϕ be the median of its probability distribution (or density function). From the definition of the median, the probability that any randomly chosen value of X will be above (below) the median is $\frac{1}{2}$: $P(X_i > \phi) = P(X_i < \phi) = 0.50$. Figure 9.1 demonstrates this.

Define a success as $\{X_i > \phi\}$ and a failure as $\{X_i \leq \phi\}$. Within the population, the probability of a success is always equal to $(\pi =) 0.5$. Let k equal the number of successes.

Select a random sample of five values of X from the population, and place them in an array such that

$$X_1 < X_2 < X_3 < X_4 < X_5$$

FIGURE 9.1 Three distributions and their median.

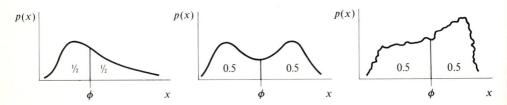

Number of X-values above the median k	$P(k\|n=5,\ \pi=\frac{1}{2})$ $P(k)$	Possible sample results
0	0.0312	
1	0.1562	
2	0.3125	
3	0.3125	
4	0.1562	
5	0.0312	

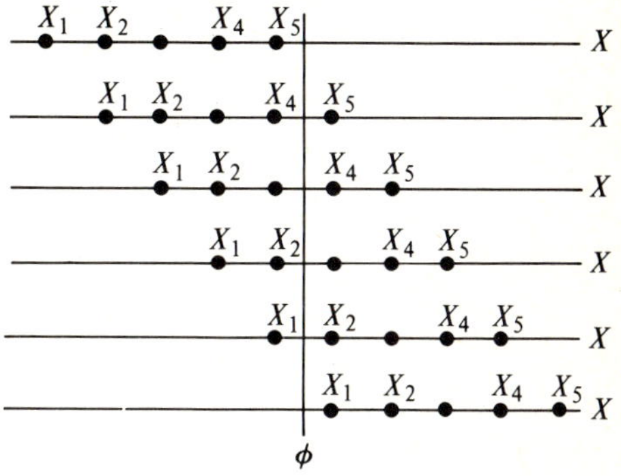

FIGURE 9.2   A random sample of five values of X.

Consider $X_1 \le \phi \le X_5$ as an interval estimate of the median; that is, let X_L equal the lower confidence limit, the smallest value of X, and let X_U equal the upper confidence limit, the highest value of X. What is the probability that the population median will be within the range of the sample?

There are only two ways that the interval from X_1 to X_5 will *not* include the population median: (1) if all five sample values of X are below ϕ, and (2) if all five sample values of X are above ϕ. Consider Figure 9.2.

Let us rephrase the question, using binomial probabilities. Conceptually, randomly selecting five values of X from the population and letting the event $\{X_i > \phi\}$ be a success is no different than flipping a coin 5 times and calling a "head" a success. Thus,

P(all five values of X are *below* the population median)

$$= P(X_5 < \phi) = P(k = 0 \mid n = 5, \pi = 0.5)$$

$$= 0.0312 \qquad \text{(from Appendix Table III)}$$

In like manner,

P(all five values of X are *above* the population median)

$$= P(X_1 > \phi) = P(k = 5 \mid n = 5, \pi = 0.5)$$

$$= 0.0312$$

Thus, the probability that the interval from X_1 to X_5 will *not* include the population median is $0.0312 + 0.0312 = 0.0624$. Consequently, $P(X_1 \leq \phi \leq X_5) = 1 - 0.0624 = 0.9376$. Following the practice of Chapter 6,

$$P(X_L \leq \phi \leq X_U) = 1 - \alpha$$

where $(1 - \alpha)$ is the level of confidence.

To summarize, if we randomly select five values of the continuous random variable X, 93.76 percent of the time the interval from the lowest sample value (X_1) to the highest sample value (X_5) will include the population median.

This 93.76 percent confidence interval might be very large. We might want to increase the precision of the confidence interval; we want to make $X_U - X_L$ smaller without changing the sample size. To do so will reduce the level of confidence.

For example, we could use the second smallest and second largest sample values of X as our interval estimate of the population median. But note from Figure 9.2 that

$$P(X_2 \leq \phi \leq X_4) = P(k = 2) + P(k = 3)$$

$$= 0.3125 + 0.3125 = 0.6250$$

$$= 1 - \alpha$$

Only 62.5 percent of the time would the interval from X_2 to X_4 include the population median when a random sample of 5 is selected.

Small Samples

The size of the sample and the size of the population are relative. The basic concern is: Do we essentially meet all the requirements for a Bernoulli experiment? (See section 3.5.) If the population is infinite, we, of course, have no problem. If the population is finite and we sample with replacement, there is no problem. If the population is finite and we sample without replacement, then the probability of a success changes with each draw. The relevant question is, How much does the probability of a success change? In general, if you reason that

$P($success on nth observation$\mid$all previous $n - 1$ observations were

successes$) \simeq 0.5$ or the $P($success on first sample observation$)$

then the requirements for a Bernoulli experiment are approximately satisfied. In that context, $n < 30$ will generally be considered a small sample.

In general, we want a statement of the form

$$P(X_L \leq \phi \leq X_U) = 1 - \alpha$$

For the left side of this expression:

1. Take a random sample of n values of X from the population.
2. Place these n values of X in an array (or rank-order them), and denote the items in this array as

$$X_1, X_2, X_3, \ldots, X_{n-1}, X_n$$

where X_1 is the smallest sample value of X and X_n is the largest.

3. The *confidence interval* is given by the rth smallest and the rth highest sample value of X:

$$X_r \leq \phi \leq X_{n+1-r}$$

Now consider the right side of the expression, the level of confidence. Recall that for $\pi = 0.5$, the binomial probability distribution is symmetric; therefore,

$$P(k < r) = \alpha/2 = P(k > n + 1 - r)$$

and $\alpha = 2P(k < r)$, where $P(k < r \mid n, \pi = 0.5)$ is obtained from the binomial tables. Finally, the $(1 - \alpha)$ confidence interval for the population median is

$$P(X_r \leq \phi \leq X_{n+1-r}) = 1 - 2P(k < r) \tag{9.7}$$

Example 9.2 An Interval Estimate for the Median Income

We need an estimate of median family income for the residents of some community of several hundred families. Last year's family income is a sensitive topic, so we obtain a small, but random, sample of eight families. Placing these family incomes in an array, we get

X_i	X_1	X_2	X_3	X_4	X_5	X_6	X_7	X_8
Family Income	$9,200	9,750	10,300	11,000	11,400	11,900	12,300	14,700

Using Equation 9.7 for $r = 1$ (where we take the highest and lowest sample values for the confidence limits),

$$P(X_1 \leq \phi \leq X_{8+1-1}) = 1 - 2 \cdot P(k < 1 \mid n = 8, \pi = 0.5)$$
$$P(X_1 \leq \phi \leq X_8) = 1 - 2 \cdot (0.0039)$$
$$P(\$9,200 \leq \phi \leq \$14,700) = 1 - 0.0078 = 0.9922$$

The table (Table 9.3) of all possible confidence intervals for $n = 8$, and their respective levels of confidence, allows you to choose which value of $(1 - \alpha)$ best suits your purpose.

TABLE 9.3 CONFIDENCE INTERVALS FOR MEDIAN INCOME

r	$P(k < r \mid n = 8, \pi = 0.5)$	$2 \cdot P(k < r)$ α	Level of Confidence $1 - \alpha$	Confidence Interval	
				X_r	X_{9-r}
1	0.0039	0.0078	0.9922	\$ 9,200	\$14,700
2	0.0351	0.0702	0.9298	9,750	12,300
3	0.1445	0.2890	0.7110	10,300	11,900
4	0.3633	0.7266	0.2734	11,000	11,400

Large Samples

For large samples (that is, $n > 30$), the normal approximation of the binomial distribution is useful. Figure 9.3 demonstrates the conversion process. The process is as follows:

1. Choose a level of confidence, $1 - \alpha$.
2. From Appendix Table IV find $z_{\alpha/2}$.
3. Find r'. Since $r' = \mu_k - z_{\alpha/2}\,\sigma_k = n\pi - z_{\alpha/2}\sqrt{n\pi(1 - \pi)}$ and $\pi = 0.5$, then

$$r' = \tfrac{1}{2}(n - z_{\alpha/2}\sqrt{n}) \qquad (9.8)$$

4. We must round r' because r must be an integer. If r' is rounded to the lower integer, $(1 - \alpha)$ is increased; if r' is rounded to the higher integer, $(1 - \alpha)$ is decreased. Of course, rounding r' to the nearest integer will not necessarily change the level of confidence by the lesser amount, but this is not much of a problem.
5. Take a random sample of n observations and place them in an array such that X_1 is the smallest and X_n is the largest.

FIGURE 9.3 Normal approximation to the binomial.

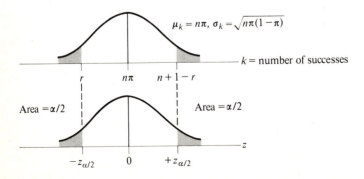

6. Let r be the rounded value of r' from step 4. Count off r observations from each end of the array and use those two values as the confidence interval for the population median:

$$P(X_r \leq \phi \leq X_{n+1-r}) \simeq 1 - \alpha$$

Test of Hypothesis about the Population Median

There are several strategies that may be pursued. First, if the population is known to be symmetric, then test the hypothesis about the population mean, $H_0: \mu = \mu_0 = \phi_0$, where ϕ_0 is the hypothesized value of the population median.

Next, for a two-tail test (that is, $H_0: \phi = \phi_0; H_A: \phi \neq \phi_0$) construct a $(1 - \alpha)$ confidence interval for ϕ as discussed above. If the hypothesized median is within that interval (i.e., if $X_r \leq \phi_0 \leq X_{n+r-1}$), accept H_0; if ϕ_0 is not in that interval, reject H_0.

Finally, for a single-tail test, the techniques discussed in Chapter 7 for a one-tail test of the population proportion are appropriate.

If the null hypothesis is $H_0: \phi \geq \phi_0$ (e.g., "the median income in this city is \$10,000 or more," $\phi_0 = \$10,000$), then we take care to define a success as $\{X_i < \phi_0\}$; that is, a success occurs when a sample value of X is *below* ϕ_0. A large number of successes would discredit the null hypothesis. Let k^* be the number of sample observations (in a sample of size n) which are *below* ϕ_0.

For large samples, the decision rule would be:

$$\text{If } k^* > \tfrac{1}{2}(n + z_\alpha \sqrt{n}) \quad \text{reject } H_0$$

$$\text{Otherwise} \quad \text{accept } H_0$$

For a null hypothesis of the form $H_0: \phi \leq \phi_0$, define a success as $\{X_i > \phi_0\}$. If we define k^* as the number of sample observations that are *above* ϕ_0, then the previous decision rule is also valid for this case.

Matched Pairs

In section 7.4 we discussed the test of hypothesis regarding the difference between two population means. A person's weight before initiating a special diet (X_i) and that same person's weight after ending the diet (Y_i) were an example of a matched pair of observations. The weight change, $D_i = X_i - Y_i$, may not be associated with a symmetric population. Thus, we may want to test the hypothesis that $\phi_D = 0$ against the alternative that $\phi_D \neq 0$.

We could call the event $\{D_i < 0 = \phi_D\}$ a success, which is the same thing as counting the number of *minus signs* associated with D_i. Let k^* equal the number of negative values of D_i. (This is frequently called the *sign test* for the location of two populations.) The values of $k*$ are binomially distributed with $\pi = 0.5$. If $k*$ is very small, or if $k*$ is very large, the sample results would not support the null hypothesis.

For large samples, the decision rule would be:

$$\text{If } \tfrac{1}{2}(n - z_{\alpha/2}\sqrt{n}) \leq k* \leq \tfrac{1}{2}(n + z_{\alpha/2}\sqrt{n}) \quad \text{accept } H_0$$

Otherwise   reject H_0

9.3 ONE-WAY RANK ANALYSIS OF VARIANCE: THE KRUSKAL-WALLIS H TEST

Assume that we have a number of independent samples, say k of them, and we are interested in knowing if they came from populations which all have the same distribution (i.e., the same median, etc.). Consider the following three examples.

Example 9.3   Shelf-Top Ovens

A consumer product-testing agency is interested in the temperature reliability of $(k =)$ 5 brands of portable, shelf-top electric ovens. They obtain $(n_j =)$ 10 ovens of each brand, set all 50 ovens at 400°F, and after 5 minutes measure the temperature inside the oven. Is the distribution of temperatures the same for all five brands?

Example 9.4   Paper Towels

A homemaker is presented with nine unmarked paper towels and asked to perform a standard test with all nine and rank-order preference. Unbeknown to the homemaker, the nine are a random sample of three sheets from $(k = 3)$ three different brands. Does the rank-ordering exhibit any preference for one brand?

Example 9.5   Regional Income Differences

We have $(k =)$ 15 different geographic regions and a random sample of family incomes from each region. Do all 15 regions have the same median family income?

All three examples could use the Kruskal-Wallis H test to test

H_0: All k populations have the same distribution
H_A: Not all k populations have the same distribution

The sample data may be measured with an ordinal, an interval, or a ratio scale. More importantly, there is no assumption about the normality, or even the symmetry, of the parent populations. They can have any shape; the H test is distribution free.

The H test starts by disregarding the k groups or categories and rank-orders *all* the sample observations; then it deals with only those ranks. The ranking process

TABLE 9.4 PLATEAU RANKING EXAMPLE

	Within-Plateau Ranking		
	First	Second	Third
First plateau	1	2	3
Second plateau	4	5	6
Third plateau	7	8	9

tends to diminish the influence of extreme values, which can be a distinct advantage. For example, consider a sample of family income data:

Family Income	$9,600	$10,500	$11,200	$48,300
Rank Order	1	2	3	4

The value of $48,300 will influence the value of the sample mean. Is $48,300 accurate? Is it an exaggeration by the reporting family? Is it a recording mistake by the interviewer? The ranking process tends to diminish its importance regardless of the answer to those questions.

To get an idea of how the H test works, consider the homemaker testing the nine paper towels. In order to rank the nine samples, assume that this person first breaks the samples into three groups (or plateaus). The three towels in the first plateau performed better, in the homemaker's judgment, than the other six. The three towels in the second plateau were better than the remaining three, which end up in the third plateau. Then, the homemaker rank-orders the three paper towels within each plateau. In Table 9.4 the plateau and within-plateau classifications show the final ranking.

Now, the experiment manager decodes the paper towels according to the ranking from 1 to 9.

Case 1

Assume that Table 9.5 shows the results by brand.

The numbers in parentheses are the "ranks within plateaus." Within the first plateau, the brands were ranked A, B, and C respectively. In the second plateau, C, A, and B; and in the third plateau, B, C, and A. Each brand received a "first place" in some plateau; each brand received a "second place" in some plateau; and so forth.

It is more important to note, however, that each brand appeared in each plateau. The rank sum, for each brand, is 15. Next, for each brand, square the rank sum (R^2) and find the mean-square rank sum (R^2/n). Finally, adding the mean-square rank sum across brands gives the sum of R^2/n, which is 225.

Before attempting to interpret that 225, let us consider another possible result of the homemaker's rank-ordering.

TABLE 9.5 CASE 1 RESULTS

	Ranks by Brand				
Brand A		**Brand B**		**Brand C**	
Towel	Rank	Towel	Rank	Towel	Rank
A_1	1 (1)	B_1	2 (2)	C_1	3 (3)
A_2	5 (2)	B_2	6 (3)	C_2	4 (1)
A_3	9 (3)	B_3	7 (1)	C_3	8 (2)
Sum of Ranks (R): 15		15		15	
R^2: 225		225		225	
R^2/n: 225/3 = 75		225/3 = 75		225/3 = 75	
Sum of R^2/n: 3(75) = 225					

Case 2

Table 9.6 demonstrates another possible rank-ordering that the homemaker could have generated.

In case 2, brand A is clearly preferred to brand B, and B is preferred to C. The within-plateau rankings shown in parentheses are no different than they were in case 1; however, only brand A towels show up in the first plateau; only brand B in the second plateau; and only brand C in the third plateau.

Recall the null hypothesis that the k ($= 3$ in this situation) samples come from identical populations. Case 1 supports the null; case 2 does not. We could summarize these two cases as follows:

Case	Apparent Difference between Brands	Sum of R^2/n	H_0 Decision
1	Not much	225	Accept
2	A great deal	279	Reject

TABLE 9.6 CASE 2 RESULTS

	Ranks by Brand				
Brand A		**Brand B**		**Brand C**	
Towel	Rank	Towel	Rank	Towel	Rank
A_1	1 (1)	B_1	4 (1)	C_1	7 (1)
A_2	2 (2)	B_2	5 (2)	C_2	8 (2)
A_3	3 (3)	B_3	6 (3)	C_3	9 (3)
Rank Sum (R): 6		15		24	
R^2: 36		225		576	
R^2/n: 36/3 = 12		225/3 = 75		576/3 = 192	
Sum of R^2/n: 12 + 75 + 192 = 279					

Thus, we will use something like "the sum of R^2/n" as an aggregate measure of whether these samples, as expressed by their rank-ordering, come from identical populations. If the sum of R^2/n is relatively small, the sample results are supportive of the null hypothesis; if the sum of R^2/n is relatively large, the null hypothesis is more dubious.

Specifically, the H *statistic for the Kruskal-Wallis test* is defined as

$$ H = \frac{12}{n(n+1)}\left(\sum_{j=1}^{k} \frac{R_j{}^2}{n_j} \right) - 3(n+1) \tag{9.9} $$

where

k = number of independent samples (or identical populations)

n_j = number of sample observations from the jth population

R_j = sum of ranks in the sample from the jth population (ranks are assigned by grouping all sample observations)

$n = \sum_{j=1}^{k} n_j$ = the total number of observations from the k samples

When each sample has five or more observations (that is, $n_j \geq 5$ for all j), the H approximates a chi-square distribution with $(k-1)$ degrees of freedom. For $k = 3$ and $n_j \leq 5$, special tables must be used.[4]

Note that this is a one-tail test of the null hypothesis; large values of H reject the null. Also note that it is $(k-1)$, *not* $(n-1)$, that determines the degrees of freedom for using the chi-square distribution to obtain a critical value of H for the decision rule.

Example 9.6   Differences in Regional Income

Assume that we have three regions and a random sample of family income from each region (such that $n_j > 5$ for $j = 1, 2$, and 3). Assume that we want to test the hypothesis that the family income distribution is identical in all three regions. Since $k = 3$, we have $k - 1 = 2$ degrees of freedom. Choose a value of α, say $\alpha = 0.05$, and refer to Appendix Table VII to find that $\chi^2 = 5.991$. Hence, the decision rule is:

If $H > 5.991$ reject H_0

If $H \leq 5.991$ accept H_0

---

[4] See W. H. Kruskal and W. A. Wallis, "Use of Ranks in One-Criterion Variance Analysis," *Journal of the American Statistical Association*, vol. 47 (1952), pp. 583–621.

The sample data is given below, and the ranks are assigned by disregarding regions.

Region A		Region B		Region C	
Family Income	Rank	Family Income	Rank	Family Income	Rank
$ 9,100	4	$ 8,700	1	$ 8,900	3
9,300	5	8,800	2	9,900	7
9,400	6	10,400	10	10,000	8
10,500	11	11,100	14	10,100	9
10,800	12	11,600	18	11,200	15
11,000	13	11,800	19	11,300	16
		11,900	20	11,500	17
R_1: 51		R_2: 84		R_3: 75	
R^2: 2,601		7,056		5,625	
R^2/n: 2,601/6 = 433.50		7,056/7 = 1,008		5,625/7 = 803.57	

From Equation 9.9, H is given by

$$H = \frac{12}{(20)(20 + 1)}(433.50 + 1{,}008 + 803.57) - 3(20 + 1)$$

$$= 64.14 - 63 = 1.14$$

Since $H = 1.14 < 5.991$, we accept the null hypothesis.

Ties

As you might suspect, occasionally two or more sample observations may have the same value; there is a tie. If they were all slightly different, a rank could be assigned. To solve this problem, assign the *mean* of the ranks (that they would be assigned if ties were slightly different) to each of the identical values.

Ties can influence the value of H. Call CFT the correction factor for ties, then

$$H' = \frac{H}{\text{CFT}} \tag{9.10}$$

where H is calculated from Equation 9.9 and

$$\text{CFT} = 1 - \frac{\Sigma T}{n^3 - n} \tag{9.11}$$

where $T = t^3 - t$

t = the number of identical observations in a group of tied values

Since CFT is always less than 1, then H' is always greater than H when ties occur. Thus, if H is large enough to reject H_0, there is no need to calculate CFT and H'. If ties do occur and H is not large enough to reject H_0, by all means calculate CFT and H'.

Example 9.7   Regional Income with Ties

Consider the following data:

Region A		Region B		Region C	
Family Income	Rank	Family Income	Rank	Family Income	Rank
$ 9,100	4	$ 8,800	2	$ 8,800	2
9,300	5	8,800	2	9,600	6.5
9,600	6.5	10,400	10	10,000	8
10,500	11	11,100	14	10,100	9
10,800	12	11,600	18	11,200	15
11,000	13	11,800	19	11,300	16
		11,900	20	11,500	17
R:	51.5		85		73.5
R^2:	2,652.25		7,225		5,402.25
R^2/n:	442.04		1,032.14		771.75

$$H = \frac{12}{(20)(21)}(442.04 + 1,032.14 + 771.75) - 3(21)$$

$$= 64.17 - 63 = 1.17$$

For $\alpha = 0.05$, $H = 1.17 < 5.991$. The H_0 is accepted; therefore we should correct H for ties.

Tied Value	Number of Observations with This Value t	t^3	$T = t^3 - t$
$8,800	3	27	24
9,600	2	8	6
			$\Sigma T = 30$

From Equation 9.11,

$$\text{CFT} = 1 - \frac{30}{20^3 - 20} = 1 - \frac{30}{8,000 - 20} = 1 - \frac{30}{7,980} = 0.9962$$

and from Equation 9.10,

$$H' = 1.17/0.9962 = 1.17$$

Hence $H' = 1.17 < 5.991$, and the null hypothesis is accepted. [*Note:* As a calculational check on the rank sums, $\Sigma R_j \equiv (\frac{1}{2})(N)(N + 1)$. For the regional income data, whether or not ties occur, $\Sigma R_j = 210$.]

9.4 SPEARMAN'S RANK CORRELATION COEFFICIENT

In the two chapters that follow, we will investigate the association between two variables when both variables are measured by an interval or a ratio scale. In this section, we will look at a measurement of association for two variables measured on an ordinal scale.

Consider the semester average for five students in a statistics class (X_i) and their mathematics scores on their entrance examination (Y_i):

Student	Course Average Score x_i	Course Average Rank X_i	Entrance Exam Score y_i	Entrance Exam Rank Y_i	Difference in Ranks $d_i = X_i - Y_i$
A	92.7	5	542	5	0
B	89.2	4	507	4	0
C	80.6	3	460	3	0
D	76.3	2	459	2	0
E	70.9	1	456	1	0

Note that the lowest score receives rank 1, etc. Considering the *rank order*, not the magnitude, of the score, there is what is called "a perfect direct rank correlation" between entrance examination results and course average. As the rank of entrance score increases, so does the rank of the course average. Of course, the difference in rank between scores is 0 for each person.

Now consider a group of tennis tournament directors who meet to rank-order (or seed) the six players entered in the tournament according to the player's past performance, etc. Also, assume that they rank-order the six players according to "personality" (perhaps in an attempt to determine who they should have advertising the tournament).

Player	Ability X_i	Personality Rating Y_i	$d_i = X_i - Y_i$	d_i^2
A	6	1	5	25
B	5	2	3	9
C	4	3	1	1
D	3	4	−1	1
E	2	5	−3	9
F	1	6	−5	25
			0	70

The tournament directors rank player A as most likely to win ($X_i = 6$), but they judge that player A has the worst personality ($Y_i = 1$) of the six players. This situation is called a "perfect inverse rank correlation."

So, what are we after? We want a measure of rank correlation, call it r_s, which will distinguish between these two situations. We want to describe "perfect direct rank correlation," the first case, with a measure such that $r_s = +1$. For the "perfect inverse rank correlation" situation, we want that same measure to produce $r_s = -1$. Further, we want $r_s = 0$ when there is "no rank correlation" between the two variables; r_s should be equal to 0 when the ranks X_i and Y_i are mutually independent.

We easily realize that the above two cases are different rank correlation situations, but how do we describe them with a numerical measure? The differences between *ranks* are shown as d_i with both data sets. We encounter an all too familiar problem: $\Sigma d_i = 0$ in both cases. As usual, we solve that problem by squaring each rank difference, d_i^2, and then summing the squared differences, Σd_i^2.

Case	Σd_i^2	Desired Result
1	0	$r_s = +1$
2	70	$r_s = -1$

To obtain the desired result, we define the *Spearman rank correlation coefficient* as

$$r_s = 1 - \frac{6\Sigma d_i^2}{n(n^2 - 1)} \tag{9.12}$$

where n is the number of paired observations.

Note that for the first case

$$r_s = 1 - \frac{6(0)}{5(25 - 1)} = +1$$

and for the second case

$$r_s = 1 - \frac{6(70)}{6(36 - 1)} = 1 - \frac{420}{210} = -1$$

The value of r_s will always be in the range from -1 to $+1$; $-1 \leq r_s \leq +1$.

Testing the hypothesis that r_s is 0 is the same as testing the null hypothesis that X_i and Y_i are mutually independent. When no ties occur, the sampling distribution for r_s is symmetric; however, for small samples the sampling distribution of r_s can assume some rather different (nonsmooth) shapes. For small samples, exact tables should be used.[5]

For large samples, $n > 30$, the sampling distribution of r_s approaches the shape of a normal curve. The two-tail test of hypothesis becomes as follows.

1. H_0: X_i and Y_i (the ranks) are mutually independent.
   H_A: There is some rank correlation.
2. Choose α, the level of significance.
3. Decision rule:

$$\text{If } -z_{\alpha/2} < z < +z_{\alpha/2} \qquad \text{accept } H_0$$

$$\text{If } z \geq +z_{\alpha/2} \text{ or if } z \leq -z_{\alpha/2} \qquad \text{reject } H_0$$

4. Test statistic:

$$z = r_s\sqrt{n-1} \qquad\qquad (9.13)$$

where r_s is given by Equation 9.12 and n is the number of paired observations.

Example 9.8   Aptitude Examination versus Performance

Five years ago a firm built a plant at a new location. At that time 30 new laborers were hired, and they are still with the firm. Management has just finished rating these 30 laborers, and rank-orderings are given in Table 9.7 as X_i. (The laborer with the highest rating is assigned $X = 30$. Also, note that there were some ties.) Before they were hired, each laborer took an aptitude test; the scores have been ranked and listed as Y_i for the appropriate person. (The lowest aptitude score was assigned $Y_i = 1$.)

Management wonders how those old aptitude scores correlate with current performance. See Table 9.7.

The Spearman rank correlation coefficient is

$$r_s = 1 - \frac{6\Sigma d_i{}^2}{n(n^2 - 1)} = 1 - \frac{6(6,961.5)}{30(900 - 1)} = 1 - \frac{41,769}{26,970}$$

$$= 1 - 1.549 = -0.549$$

Let us test the null hypothesis that the ranks are independent with $\alpha = 0.05$. Then, $z_{\alpha/2} = 1.96$, and the decision rule is:

$$\text{If } -1.96 < z < +1.96 \qquad \text{accept } H_0$$

$$\text{If } z \geq +1.96 \text{ or if } z \leq -1.96 \qquad \text{reject } H_0$$

---

[5] See Sidney Siegel, *Nonparametric Statistics for the Behavioral Sciences*, McGraw-Hill Book Company, New York, 1956, pp. 202–213 for discussion of r_s and p. 284 for tables.

TABLE 9.7 APTITUDE AND PERFORMANCE RANKS

X_i	Y_i	d_i	d_i^2	X_i	Y_i	d_i	d_i^2
1	19	-18	324	16	27	-11	121
2	24	-22	484	17	14	3	9
3	15	-12	144	18	22	-4	16
4	30	-26	676	19	4	15	225
5	18	-13	169	20	25	-5	25
7	26	-19	361	21	9	12	144
7	12	-5	25	22	1	21	441
7	17	-10	100	23	23	0	0
9	29	-20	400	24	2	22	484
10	8	2	4	25	10	15	225
11	21	-10	100	26	5	21	441
12.5	16	-3.5	12.25	27	3	24	576
12.5	28	-15.5	240.25	28	13	15	225
14	20	-6	36	29	6	23	529
15	7	8	64	30	11	19	361
							$6{,}961.5 = \Sigma d_i^2$

Since $z = r_s \sqrt{n-1} = (-0.549)\sqrt{29} = (-0.549)(5.385) = -2.96$, we reject the null hypothesis. There does appear to be some inverse rank correlation between those old aptitude scores and current performance.

Do not be too quick with a harsh judgment of the aptitude test. If those hired were only those who "passed" the aptitude test and we do not have any data on how those who "did not pass" might have performed after 5 years, then the aptitude test may have been a very useful hiring device. In addition, in the last 5 years many of those hired with relatively low aptitude scores may have received further training while those hired with relatively high aptitude scores did not.

SUMMARY

These are but a few of the many statistical tests which are either nonparametric or distribution free or which deal with numbers from an ordinal scale. For a more extensive treatment, see Sidney Siegel, *Nonparametric Statistics* (McGraw-Hill Book Company, New York, 1956). To get an idea of the multitude of tests available, see I. R. Savage, *Bibliography of Nonparametric Statistics* (Harvard University Press, Cambridge, Mass., 1962).

Nonparametric methods are not assumption free. Nonparametric methods require a different set of assumptions, but these assumptions seldom require that very much be known about the population. Thus, they are useful for a wide variety of situations.

EXERCISES

1. For the sequence of numbers

$$1, 5, 9, 2, 6, 5, 9, 4, 2, 4, 3, 6, 0, 0, 7, 7, 2, 7, 6, 1$$

test the hypothesis that they are random, using:
a. the runs test ($\alpha = 0.05$)
b. the von Neumann ratio ($\alpha = 0.10$)

2. Consider the production of a soft drink where liquid is dispensed into a bottle, the bottle is capped, dropped into a six-pack, and cased all automatically. As a part of process control, each loaded six-pack slides across a weight-sensitive panel. Assume that under correct production procedures, the weights of six-packs are symmetric with a mean of μ ounces. Let X be the weight of a loaded six-pack, and let the weight-sensitive panel perform as follows:

$$\text{If } X \geq \mu \qquad \text{H is printed on recording paper}$$

$$\text{If } X < \mu \qquad \text{O is printed on recording paper}$$

If there is a random pattern of O's and H's, then the process is correctly self-adjusting. For the following slips of recording paper, would you say the process is in control? (Use $\alpha = 0.05$.)
a. HOHHHOOOHHHHOOOHHOHH $\qquad n = 20$
b. HHHHHOHHHHOHHHHOOHOOHOOHHOHHHHOOHHHOHOOOO-
HOHOHOOOHHOOOHOHHOHHH $\qquad n = 60$
c. Combine parts a and b as one string of 80 observations.

3. If a sequence of data is not random, then there is some pattern within this data. For economic time series, that pattern may be some long-run trend. From the following time-series data, should we look for a trend? (Use $\alpha = 0.02$.)
a. $n = 20$; $\Sigma(X_i - X_{i-1})^2 = 323{,}000$; $\Sigma(X_i - \bar{X})^2 = 85{,}000$
b. $n = 100$; $\Sigma(X_i - X_{i-1})^2 = 1{,}485$; $\Sigma(X_i - \bar{X})^2 = 1{,}000$
c. $n = 2{,}000$; $\Sigma(X_i - X_{i-1})^2 = 1{,}009{,}495$; $\Sigma(X_i - \bar{X})^2 = 500{,}000$

4. Use a runs test to examine randomness for ($\alpha = 0.05$):
a. $n_A = 40$, $n_B = 60$, $R = 36$
b. $n = 200$, $n_A = 50$, $R = 60$

5. Consider Table 9.1 as a population. We know that the median of that population is $\phi = \$1.03$.
a. If we randomly select one sales slip from that population and call the dollar value X_1, what is $P(X_1 > \$1.03)$? What is $P(X_1 < \$1.03)$?
b. Consider randomly selecting two sales slips with replacement. Let $X_1 = $ the lower value and $X_2 = $ the higher value. Find:
    (1) $P(X_1 \cap X_2 > \phi)$
    (2) $P(X_1 \cap X_2 < \phi)$
    (3) α if X_1 and X_2 are used as confidence limits for ϕ
c. Consider randomly selecting three sales slips with replacement. Order them such that $X_1 \leq X_2 \leq X_3$. Find:
    (1) $P(X_1 \cap X_2 \cap X_3 > \phi)$
    (2) $P(X_1 \cap X_2 \cap X_3 < \phi)$
    (3) $1 - \alpha$ if X_1 and X_3 are used as confidence limits for ϕ

6. Consider the following five observations selected at random:

$$\$1,00, \$0.49, \$0.45, \$1.19, \$1.83$$

a. Construct a table, similar to Table 9.3, showing all possible confidence intervals for the population median and their respective levels of confidence.

b. With a table of random numbers, the author picked customers numbered 10, 20, 50, 14, and 28 from Table 9.1 to get this data. If Table 9.1 is considered a population, how many of the intervals from your answer to part *a* actually include the population median?

c. Randomly select five customers from Table 9.1 and order their sales slips as $X_1 \le X_2 \le X_3 \le X_4 \le X_5$.

 (1) Is $X_1 < \$1.03 < X_5$?

 (2) Is $X_2 < \$1.03 < X_4$?

7. Consider the 50 sales slips in Table 9.1 as a random sample of all slips for that day. Find the 95 percent confidence interval for the median sales for that day.

8. A random sample of 625 family incomes has been placed in an array. For a 95 percent confidence interval for the population median income, how many items would be counted off from each end of the array?

9. The mayor said: "In 1970 the median family income was \$9,500 in our city; the median family real income was above that last year." A newspaper reporter takes a random sample of 100 families and finds that 60 of them had incomes above \$9,500 (in 1970 dollars) last year. For $\alpha = 0.01$, does this evidence support the mayor's claim?

10. Last year, the elapsed time of long-distance telephone calls for a national retailer was skewed to the right with a median of 3 minutes 26 seconds. The recession has reduced sales, and the company's treasurer says that the median length of long-distance telephone calls is at least as great this year. A random sample of 5,625 calls was selected, and 2,890 of them were less than 3 minutes 26 seconds. For $\alpha = 0.025$, test the treasurer's claim.

11. Use the data in Exercise 19 of Chapter 7 to test the hypothesis that $\phi_D = \phi_{After} - \phi_{Before} = 0$ against the hypothesis that $\phi_D \ne 0$ for $\alpha = 0.1094$.

12. A typewriter manufacturer has spent a great deal on developing a quieter typewriter. One hundred typists have been given the new machine, and their typing speeds (in words per minute) have been recorded both before and after use of the new typewriter. For $\alpha = 0.10$, set up the decision rule to test $H_0: \phi_D \ge 0$, $H_A: \phi_D < 0$.

13. The same driver's examination is given in five different cities. A random sample of six scores from each city yields the following results:

Driver's Examination Scores by City				
A	B	C	D	E
100	91	92	82	87
98	90	79	78	80
77	100	71	71	75
92	78	91	76	92
99	90	88	71	77
84	83	90	82	98

Use the Kruskal-Wallis H statistic to test the hypothesis that the scores come from identical populations.

14. Reconsider Exercise 13 in Chapter 8. Test the hypothesis that the average costs come from three identical populations, using the H statistic.

15. A group of students take an achievement test, and the results are:

	Scores	
Student	**Mathematics**	**Verbal**
A	79	97
B	93	82
C	89	88
D	80	85
E	89	70
F	71	89

a. Find Spearman's rank correlation coefficient.
b. Is it significantly different from 0 at $\alpha = 0.05$?

16. Six different bond issues are " graded " or ranked by an independent financial institution. The current prices are also available.

Bond	Rank or Grade	Price
1	1 (AAA)	$112
2	2 (AA)	100
3	3 (A)	105
4	4 (BBB)	106
5	5 (BB)	98
6	6 (B)	90

a. Find Spearman's rank correlation coefficient.
b. Is it significantly different from 0 at $\alpha = 0.05$?

CASE QUESTIONS

1. In Chapter 4, case question 1B, you were asked to find the median income for the 30 households in the Oregon sample. Use that median, V32, and the runs test to test the randomness of that sequence of 30 incomes.

2. Construct a 95 percent (or slightly larger) confidence interval for the median income of the renters in Oregon. Compare your result with the interval estimate of the mean income from case question 1a in Chapter 6.

3. Consider the Oregon sample. Construct Spearman's rank correlation coefficient for V11 (gross monthly rent) and V20 (year structure was built). Does the result support or discredit your a priori feelings about the relation between these two variables?

10

SIMPLE LINEAR REGRESSION AND CORRELATION ANALYSIS

10.1 WHY REGRESSION ANALYSIS?

There are two basic reasons why someone would want to consider regression or correlation analysis: (1) to reduce the variation associated with a particular variable, or (2) to know the relationship between two (or more) variables. These two are not exactly mutually exclusive, but they are the essence of the problems most individuals face when they endeavor to construct a regression model.

Consider the manager of a Midwestern restaurant that specializes in fresh Maine lobster meals. An inventory of live lobsters is practically impossible because the requisite salt water is scarce in the Midwest. Thus, at ten o'clock every morning the manager telephones an order to the supplier in Maine, and the lobsters arrive by airplane around four that afternoon. If the manager looks at the number of daily requests for live Maine lobsters from all the records over the past 3 years, there may be $\bar{Y} = 32$ lobsters per day and $s_Y = 20$ lobsters per day. The variation in the number of lobsters per day requested by customers is so large that this data *alone* is not very useful in determining how many lobsters the manager should order today.

You can probably construct a list of things that could cause this variation. Further, you might reason that the act of doing this type of estimate day after day would enable you to acquire a feeling for how many lobsters you should order, a concept which falls under the general heading of "experience." But, can we use this experience as data to construct a model that would reduce the daily variation in the number of lobsters requested? If the *causal factors* that determine this variation are identifiable and consistent, then the answer is: There is hope! For example, if $s_Y = 2$ lobsters per day, then the financial risk in ordering from the Maine supplier is considerably less. (A couple of extra lobsters could make a lot of lobster stew for tomorrow's "luncheon special.")

One way to examine this problem of reducing the variation in the Y values is to look at the "number of reservations before 9:30 A.M." and the "number of lobster meals requested that evening." For example, on only those days when there are 10 reservations before 9:30 A.M., we may find $\bar{Y}_{10} = 11.2$ lobsters requested and $s_Y = 1.3$ lobsters. The number of reservations may be very useful in determining an interval estimate for the number of lobsters requested.

For an example of a case in which you want to know the relationship between two variables, consider the warehouse manager for a company that uses a lot of guy-wire cable. This cable comes on large spools, and you cannot expect the workers to keep accurate records of how much they use; thus, you may have a dozen partially used spools of guy wire and the question is: How much wire is on each spool?

If we knew both how much the spool weighs (call it S) and the number of feet per pound (call it F) of cable, then we could weigh the partially used spool, subtract the spool weight, and calculate the length of cable on the spool. That is, length = $(G - S)F$, where G is the gross weight of spool and cable. Here the relationship between gross weight and cable length is obviously important.

10.2 SOME ELEMENTS OF ANALYTICAL GEOMETRY

A function (or single-valued function) is generally written as $y = f(x)$. This implies that it is possible to map each value of x into y, but the precise form of the mapping is unspecified. For example, some specified *polynomial functions* are:

$$y = \alpha \qquad\qquad\qquad\qquad\qquad constant \text{ function}$$

$$y = \alpha + \beta x \qquad\qquad\qquad\qquad linear \text{ function}$$

$$y = \alpha + \beta_1 x + \beta_2 x^2 \qquad\qquad quadratic \text{ function}$$

$$y = \alpha + \beta_1 x + \beta_2 x^2 + \beta_3 x^3 \qquad cubic \text{ function}$$

The highest exponent, or power, of x is called the *degree* of the polynomial function. Thus, the constant function ($y = \alpha x^0 = \alpha$) is called a zero-degree polynomial function, the linear function ($y = \alpha + \beta x^1$) is called a first-degree polynomial function, etc. The coefficients α, β_1, β_2, and β_3 are constants; they may be negative or positive.

Most of this chapter will be concerned with the linear function; however, the quadratic function and higher-degree polynomial functions are not uncommon in economics and business applications. It is extremely important for your understanding of this chapter that you be able to observe the equation for a straight line and quickly form a mental image of the graphical representation for that equation (or vice versa).

Consider Panel 10.1. For the linear function $y = \alpha + \beta x$, α positions the line and β determines its slope. The quadratic function is one form of representing the traditional U-shaped average-cost curves from the principles of economics. Figure 10.1 summarizes the general shape of this function for various values of the coefficients. Note that if $\beta_2 = 0$, we have a linear function, and if both β_1 and β_2 are 0, we have a constant function.

Higher-degree polynomials are not the only functions that yield "curves" that have application in business and economics. A sample of other functions is shown in Figure 10.2.

Panel 10.1   The Straight Line $y = \alpha + \beta x$

	$\beta > 0$	$\beta = 0$	$\beta < 0$
$\alpha > 0$			
$\alpha = 0$			
$\alpha < 0$			

$\alpha \rightarrow$ When $x = 0$, $y = \alpha$ for finite values of β; thus, α is the y *intercept* or the point where the line crosses the y axis.

$\beta \rightarrow \beta$ is the *slope* of the line, the amount that y will change for a 1-unit change in x. Thus, if $\beta > 0$, an increase in x yields an increase in y; a decrease in x produces a decrease in y. If $\beta < 0$, then x and y move in opposite directions; an increase in x causes a decrease in y. When $\beta = 0$, we have a constant relation; changes in x do not affect y.

Finally, *relations* (or "multivalued functions"), such as $y = \alpha + \beta_1 x + \beta_2 w + \beta_3 z$, will be considered in the next chapter. The four-dimensional linear equation $y = \alpha + \beta_1 x + \beta_2 w + \beta_3 z$ can be "viewed" with the assistance of the *ceteris paribus* assumption so often employed in economics. That is, let us "hold constant" the values of x and w and see what happens to y as z varies. Consider

$$y = (\alpha + \beta_1 x^* + \beta_2 w^*) + \beta_3 z$$

or

$$y = A + \beta_3 z$$

where $A = \alpha + \beta_1 x^* + \beta_2 w^*$. Thus, A is the y intercept and β_3 is the slope of the linear function.

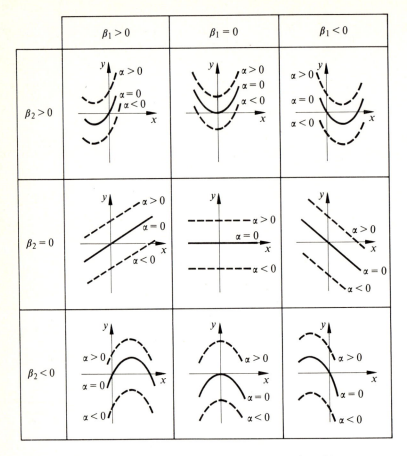

FIGURE 10.1 Graphical representation of the second-degree polynomial.

$$y = a + \beta_1 x + \beta_2 x^2$$

10.3 THE FUNCTIONAL MODEL

In general, the "functional model" takes the form

$$Y = f(X)$$

where Y is called the *dependent* variable and X is called the *independent* variable. If gasoline is \$0.50 per gallon, then

$$Y = (0.50)X$$

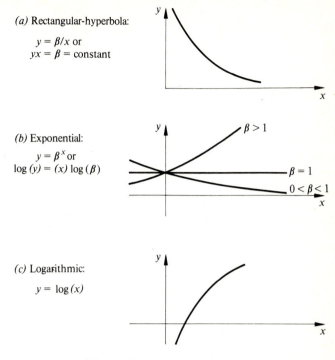

(a) Rectangular-hyperbola:

$$y = \beta/x \text{ or}$$
$$yx = \beta = \text{constant}$$

(b) Exponential:

$$y = \beta^x \text{ or}$$
$$\log (y) = (x) \log (\beta)$$

(c) Logarithmic:

$$y = \log (x)$$

FIGURE 10.2   Selected other curves for $y = f(x)$.

is a functional model for the total cost (Y) and the number of gallons (X) purchased. The value of Y is obviously dependent upon the value of X.

A home service repair person (e.g., TV repair, washing machine repair, etc.) may calculate labor charges (i.e., parts would be extra) as \$10 for the service call and \$5 per hour spent on location. Thus,

$$Y = 10 + 5X$$

is the functional model for the labor charge (Y) and the number of hours (X) used for the repair work. Both of the preceding examples use the linear equation $Y = \alpha + \beta X$ as the functional model.

10.4   THE STATISTICAL MODEL

Unlike the rather precise functional model, the "statistical model" is not a perfect relation between Y and X. For example, consider Table 10.1.

The dozen partially used spools of cable are weighed to get the value of X. Then each spool is unwound, and the length of cable is measured to get the corresponding

TABLE 10.1 CABLE LENGTH AND WEIGHT

Measured Length of Cable (Feet) Y	Gross Weight of Spool and Cable (Pounds) X
70	30
90	40
100	40
120	50
130	50
150	50
160	60
190	70
200	70
200	80
220	80
230	80

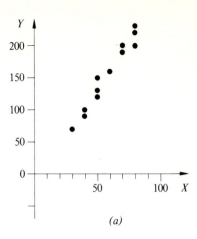

(a)

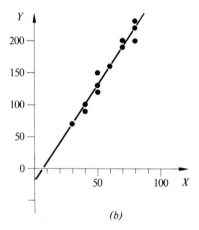

(b)

FIGURE 10.3 (a) Scatter diagram; (b) scatter diagram and linear statistical model.

value of Y. The *paired values* of X and Y from Table 10.1 are plotted in Figure 10.3; the graphical representation of this data is called a *scatter diagram.*

The purpose of a scatter diagram is to suggest the functional form of the statistical model. Note that the scatter diagram in Figure 10.3a suggests a linear relation with the slope β being positive.

Also note that a straight line cannot be placed through all the points in the scatter diagram. Figure 10.3b shows a particular straight-line "statistical model," and most of the points do not actually fall on the line. The relationship between the length of cable and the gross weight of spool and cable is not perfect. Or, not all the variation in Y (cable length) is accounted for by variation in X (gross weight).

In this example there are two possible sources of this variation when we assume that all measurements are correctly performed. First, there may be variation in spool weight. Second, there may be variation in the length of two cables of equal weight (e.g., the diameter of one cable is smaller than the diameter of the other). It is common in economics problems to assume that these variations are random occurrences. Although the statistical relation may not be as exact as the functional model, it is very useful.

In this particular example, if we could estimate the statistical model, or straight line, in Figure 10.3*b* and estimate the amount of variation about that straight line, then we could take another partially used spool of cable, weigh it, and provide an interval estimate of the length of cable remaining on the spool.

Other possible uses of the statistical model are fairly widespread. For example, in macroeconomic theory, we frequently refer to the relation between aggregate consumption and net national product: $Y = \alpha + \beta C$. Estimating α and β are important steps in using that macroeconomic "statistical model."

10.5 THE REGRESSION MODEL

Basic Concepts

The dependent, or "response," variable Y and the independent variable X, in the statistical model, are assumed to have two characteristics:

1. There is some tendency for Y to vary as X varies. It is assumed that this "tendency" is consistent and systematic, but the precise form of this tendency has to be determined. The functional form of the model (that is, $Y = \alpha + \beta X$) must be logically justifiable; however, the parameters (α and β) must be statistically determined.
2. There will be some variation between the observed data and the statistical model; the statistical relation will not be perfect.

From these two characteristics, we construct a regression model with the following assumptions:

1. Our observed "paired values of X and Y" are only samples from a larger population. However, for the moment we are concerned with constructing a model for the population of all possible paired values.
2. For *each* value of X, there are several values of Y that could occur. Further, for *each* value of X there is a probability distribution which represents the relative likelihood of different values (or different *ranges* of values) of Y occurring.
3. The mean of each probability distribution of Y values varies in some systematic fashion with the independent variable X.

The results of these assumptions are represented in Figure 10.4.

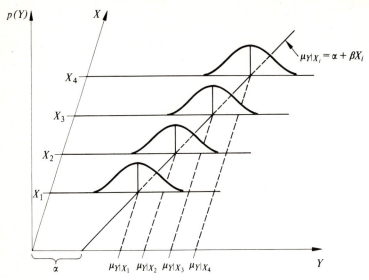

FIGURE 10.4 Linear regression model.

Statement of Model

For the linear regression model, we begin with the statement of the identity

$$Y_i = \alpha + \beta X_i + \xi_i \tag{10.1}$$

where
Y_i = the ith observation of the dependent variable
X_i = the corresponding observation of the independent variable
α and β = *parameters*
ξ_i = the *random error term*

Equation 10.1 is a tautology. The parameters (α and β) are algebraic constants. Although X_i is a particular value of the independent random variable, it can be treated as a known constant. Whatever the observed value Y_i happens to be, ξ_i assumes a value which causes the equality to hold.

Surely, this will not do. The parameters could assume almost any values and Equation 10.1 would be valid. We must impose some further restrictions on the model in order to make it "more meaningful" as an expression of the relation between Y and X.

What criteria would we like to use to make the model "more meaningful"? First, if ξ_i is to represent a "random error" term, then it would be convenient if the average of all errors were 0. That is, $E(\xi) = 0$; over the long haul, the errors cancel each other.

Next, it would be convenient if the variation in the ξ values were as small as possible. Also, if ξ is to be a "random" error term, then the ξ_i occurring in one trial should be independent of the ξ_j occurring in some other trial.

Finally, let us assume that the variation in ξ (and thus the variation in Y) is the *same* regardless of the value of X. This assumption will make it more convenient to measure the variation in ξ.

Restating these criteria, we would like:

1. $E(\xi) = 0$
2. σ_ξ^2 to be as small as possible
3. ξ_i to be independent of ξ_j for all i and j
4. $\sigma_\xi^2 = $ constant for all values of X

To review, we are looking for a model which would describe the linear relation between Y and X. Equation 10.1 perfectly describes any situation, but by itself it does not limit the possible values of the parameters α and β. If the error term ξ_i is relatively small in magnitude and does not exhibit much variation, then the resulting values of α and β closely describe the relation between X and Y.

Implication of Assumptions Concerning the Error Term

First, let us consider the assumption that $E(\xi) = 0$. For a given value of X_i, $\alpha + \beta X_i$ is a constant and

$$
\begin{array}{ccc}
Y_i & = (\alpha + \beta X_i) + & \xi_i \\
\text{random} & \text{constant} & \text{random} \\
\text{variable} & & \text{variable}
\end{array}
$$

The average of the Y values, for a given value of X, is

$$\mu_{Y|X} = E(Y_i) = E[(\alpha + \beta X_i) + \xi_i]$$
$$= E(\alpha + \beta X_i) + E(\xi_i)$$

Since the expected value of a constant is that constant [that is, $E(K) = K$] and since $E(\xi_i) = 0$ by assumption, then

$$\mu_{Y|X} = \alpha + \beta X_i \tag{10.2}$$

Equation 10.2 is our *regression model* of the statistical relation between "the *mean* of the distribution of Y values for a given value of X" and the "X values."

Refer to Figure 10.4. For each value of X there is a distribution of Y values. Figure 10.4 shows these Y values to be normally distributed; the mean of each distribution is given by $\alpha + \beta X_i$. As X changes, so does the mean of the distribution of possible Y values.

Suppose we have four spools of cable which each weigh exactly 40 pounds. Each of the four spools contains a slightly different length of cable. Also, assume that we have four other spools which weigh exactly 50 pounds each, and the lengths of these four cables are slightly different. In the context of this example, Equation 10.2 says simply that the *average length* of cable on the four 50-pound spools is greater than the *average length* of cable on the three 40-pound spools. The gross weight of each spool is represented by X_i, and the mean length of cable is represented by $\mu_{Y|X}$ (i.e., the mean of Y *given* X).

The second implication of our assumptions concerns the variation in the error term. First, we must recall that the variance of a constant is 0 (i.e., if $W_1 = 2$, $W_2 = 2$, and $W_3 = 2$, then $\sigma_W^2 = 0$). Returning to the identity in Equation 10.1, *for a given value* of X,

$$\text{Var}(Y_i) = \text{Var}(\alpha + \beta X_i + \xi_i)$$
$$= \text{Var}(\alpha + \beta X_i) + \text{Var}(\xi_i)$$

Since $\alpha + \beta X_i$ is a constant, then

$$\text{Var}(Y_i) = \text{Var}(\xi_i) \tag{10.3}$$

which we shall call σ^2. Further, we assumed that $\text{Var}(Y_i)$ is equal to σ^2 for *all* values of X. This condition is called *homoscedasticity*. [When $\sigma^2 = f(X)$, this situation is called *heteroscedasticity*.] In Figure 10.4, the mean of the Y values changes as X changes, but $\text{Var}(Y_i)$ is constant and equal to σ^2.

In summary, three of the assumptions we would like to make about the error term ξ_i have led to:

1. The regression model for the population

    $$\mu_{Y|X} = \alpha + \beta X_i$$

    states that the mean of the Y values is given by the value of X and the function $\alpha + \beta X_i$.
2. The variance of the Y values is a constant (σ^2) regardless of the value of X.

Finally, a word of caution about both the homoscedasticity assumption and our desire for $\text{Var}(\xi_i)$ to be as small as possible. Implicit in the simplified notation of Equation 10.3 is the condition that "we know the value of X." The notation may be more cluttered, but Equation 10.3 may be clearer by stating

$$\sigma_{Y|X}^2 = \sigma^2 \tag{10.4}$$

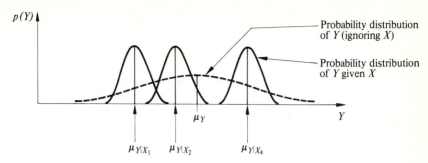

FIGURE 10.5   Comparison of distributions of Y with and without the regression model.

Equations 10.3 and 10.4 are expressions for the variance of a *conditional* probability distribution, the variance of Y given a particular value of X.

Equations 10.2 and 10.3 allude to a regression model for a *population*. Hence, the variance of Y given X is whatever it is; σ^2 is a population parameter, and we cannot arbitrarily change its value. When we said "we would like the variation in ξ_i to be as small as possible," we were stating a desirable characteristic of the actual relation between Y and X. We cannot force σ^2 to be any different than its actual value.

What we *can* do is take care in the selection of our independent variable. In our length-of-cable example, we could measure the spool circumference (i.e., run a tape measure around the partially used spool) and call it X. When X is circumference, σ^2 may be larger than when X is gross weight.

At the beginning of this chapter we stated that one reason for constructing a regression model was to reduce the variation in possible Y values (e.g., How many lobsters do we order today?). Let σ_Y^2 be the variance of the possible Y values for *all* values of X (or, when we ignore the value of X). We would hope that $\sigma_{Y|X}^2 < \sigma_Y^2$; the variance of "$Y$ given a particular value of X" is less than the variance of "Y," as shown in Figure 10.5.

10.6   ESTIMATION OF THE PARAMETERS IN THE REGRESSION FUNCTION

From Equation 10.2 our regression model for the population is

$$\mu_{Y|X} = \alpha + \beta X_i$$

Our next task is to find a technique for estimating the parameters in that model. We estimate the parameters because in essentially all regression model situations we have only a sample of the possible paired values of Y and X. We shall call a the estimate of α and b the estimate of β. Then our *estimate* of the population regression model is

$$\bar{Y}_c = a + bX \tag{10.5}$$

where $\bar{Y}_c$ is the calculated (or estimated) mean of the Y values given the value of X.

How do we estimate a and b? Recall, in Figure 10.3, the scatter diagram represents the observed paired values of Y and X. We want Equation 10.5 to be the straight line that comes "closer" to all those points than any other straight line. That's nice, but how do we define "close"?

In the context of our discussion of the regression model for the population, we continually discussed the distribution of Y *given a particular value of the independent variable X*. Consistent with that discussion, we will define *close* as some measure of the distance between the observed value of Y and the regression-line value $\bar{Y}_c$ for a given value of X. In Figure 10.3, we want some measure of the *vertical* distance between a point and the line.

The Method of Least Squares

Consider one point (Y_i, X_i) on the scatter diagram. For $X = X_i$ the observed value of Y is Y_i. For $X = X_i$ we intend $\bar{Y}_c$ to be the mean of all possible values of Y; therefore,

$$\Sigma(Y_i - \bar{Y}_c) \equiv 0$$

because the sum of the deviations about their mean is always equal to 0. Once again, $(Y_i - \bar{Y}_c)$ will not be a very useful measure of "how close" the Y values are to their mean $\bar{Y}_c$. In previous chapters, we have developed a standard answer to this algebraic nuisance. For a given value of X, we shall use the squared difference between the observed value and the mean, $(Y_i - \bar{Y}_c)^2$, as our measure of "closeness." If we have n paired observations of X and Y, then

$$G = \sum_{i=1}^{n}(Y_i - \bar{Y}_c)^2 = \sum_{i=1}^{n}(Y_i - a - bX_i)^2 \tag{10.6}$$

is our aggregate measure of how close the points are to the regression line. Obviously, we want G to be a minimum.

It can be shown[1] that G will be a minimum when the values of a and b simultaneously satisfy the following equations:

---

[1] To find the values of a and b that minimize G, we take the first partial derivatives of G with respect to a and b, set them equal to 0, and solve for a and b.

$$G = \Sigma(Y^2 - 2aY - 2bXY + 2abX + b^2X^2) + na^2$$

$$\partial G/\partial a = -2\Sigma Y + 2b\Sigma X + 2na = 2(na + b\Sigma X - \Sigma Y) = 0$$

$$\partial G/\partial b = -2\Sigma(XY) + 2a\Sigma X + 2b\Sigma X^2 = 2[a\Sigma X + b\Sigma X^2 - \Sigma(XY)] = 0$$

The terms in brackets must be 0, and they are rewritten above as Equations 10.7a and b.

$$\Sigma Y_i = na + b\Sigma X_i \qquad \text{(10.7a)}$$

$$\Sigma(X_i Y_i) = a\Sigma X_i + b\Sigma X_i{}^2 \qquad \text{(10.7b)}$$

Equations 10.7a and b are called the *normal equations.*
Solving Equations 10.7 for a and b yields

$$b = \frac{n\Sigma(X_i Y_i) - (\Sigma X_i)(\Sigma Y_i)}{n\Sigma(X_i{}^2) - (\Sigma X_i)^2} \qquad \text{(10.8a)}$$

$$a = \frac{1}{n}(\Sigma Y_i - b\Sigma X_i) = \bar{Y} - b\bar{X} \qquad \text{(10.8b)}$$

By using Equation 10.8, a and b are *point estimates* of α and β, respectively, and they
provide a straight line that is "closest" to the observed data.

Example 10.1   The Length of Cable Remaining on a Partially Used Spool

From Table 10.1, we will use the 12 paired values of weight and length to calculate a
and b. (The column of Y^2 values is not needed here, but it will be used later.)

Spool Number i	Length of Cable on Spool i Y_i	Weight of Spool i X_i	XY	X^2	Y^2
1	70	30	2,100	900	4,900
2	90	40	3,600	1,600	8,100
3	100	40	4,000	1,600	10,000
4	120	50	6,000	2,500	14,400
5	130	50	6,500	2,500	16,900
6	150	50	7,500	2,500	22,500
7	160	60	9,600	3,600	25,600
8	190	70	13,300	4,900	36,100
9	200	70	14,000	4,900	40,000
10	200	80	16,000	6,400	40,000
11	220	80	17,600	6,400	48,400
12	230	80	18,400	6,400	52,900
Total	1,860	700	118,600	44,200	319,800

There are $n = 12$ paired observations; there were only 12 partially used spools, but we
took two measurements, Y and X, from each spool. From Equation 10.8a,

$$b = \frac{n\Sigma(X_i Y_i) - (\Sigma X_i)(\Sigma Y_i)}{n\Sigma(X_i{}^2) - (\Sigma X_i)^2} = \frac{(12)(118,600) - (700)(1,860)}{(12)(44,200) - (700)^2}$$

$$= \frac{1,423,200 - 1,302,000}{530,400 - 490,000} = \frac{121,200}{40,400} = 3$$

From Equation 10.8b,

$$a = (1/n)(\Sigma Y - b\Sigma X) = (\tfrac{1}{12})[1,860 - (3)(700)]$$

$$= (-240)/12 = -20$$

Thus, the least-squares regression line is

$$\bar{Y}_c = -20 + 3X_i$$

Standard Error of the Estimate

In the regression model we acknowledged a distribution of Y values for a given value of X. In Equations 10.3 and 10.4 we called the variance of error term (which is equal to the variance of Y for a given X) $\sigma_{Y|X}{}^2$. We will need an estimate of $\sigma_{Y|X}{}^2$ to complete the model.

If we ignore the X values and consider only the Y values, we have previously used $\bar{Y} = (\Sigma Y)/n$ as a point estimate of μ_Y and

$$s_Y{}^2 = \frac{\Sigma(Y_i - \bar{Y})^2}{n - 1}$$

as a point estimate of $\sigma_Y{}^2$. Our consideration of $\mu_{Y|X}$ and $\sigma_{Y|X}{}^2$ is analogous to our treatment of a single variable Y.

In keeping with the above analogy, the numerator of our estimate of $\sigma_{Y|X}{}^2$ would be $\Sigma(Y_i - \bar{Y}_c)^2$. It is important to remember that each value of Y, say Y_j, is associated with a particular value of X, say X_j. In our regression model, Y_j is assumed to come from a distribution whose mean is estimated to be $(\bar{Y}_c)_j = a + bX_j$. It is the squared deviation of Y_j from its own mean, $(\bar{Y}_c)_j$, that is considered in the numerator of the point estimate of $\sigma_{Y|X}{}^2$. Thus,

$$\sum_{i=1}^{n} [Y_i - (\bar{Y}_c)_i]^2 = \sum_{i=1}^{n} (Y_i - a - bX_i)^2$$

is a more complete representation of the numerator.

Finally, note that in order to obtain $\bar{Y}_c$, we had to estimate both α and β (with a and b, respectively). Since there were *two* parameters estimated to obtain the numerator, there are $(n - 2)$ degrees of freedom. And,

$$s_{Y|X}{}^2 = \frac{\Sigma[Y_i - (\bar{Y}_c)_i]^2}{n - 2} \tag{10.9}$$

serves as our point estimate of $\sigma_{Y|X}{}^2$. The square root of Equation 10.9, $s_{Y|X}$, is called the *standard error of the estimate* for our regression model.

For the purpose of calculation, an algebraically identical expression is preferred:

$$s_{Y|X} = \sqrt{\frac{\Sigma Y^2 - a\Sigma Y - b\Sigma X Y}{n - 2}} \qquad \textbf{(10.10)}$$

Example 10.2 Simple Mechanics of Estimating the Regression Model

The purpose of this example is to demonstrate (1) the manipulations required to estimate the regression model and (2) some of the characteristics of that model. Consider the following five paired observations of X and Y:

Y	X	Point
5	4	A
9	2	B
3	4	C
6	3	D
7	2	E

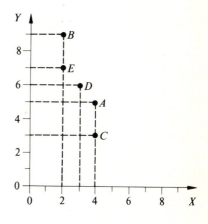

FIGURE 10.6a Scatter diagram.

(a) The Scatter Diagram
Be sure you can plot each point as shown in Figure 10.6a. Observing the scatter diagram, we would expect a to be positive (since the regression line will intersect the Y axis at a positive value of Y) and b to be negative (because the "tendency" is for Y to become smaller as X increases).

(b) Calculation of a and b
In order to determine $\bar{Y}_c = a + bX$, our estimate of the population regression model, we must calculate a and b. The framework of Table 10.2 is all that is ever necessary to provide the data base for calculations used in a linear regression model.

Since there are five "paired observations," $n = 5$. From Equation 10.8a,

$$b = \frac{n\Sigma X Y - (\Sigma X)(\Sigma Y)}{n\Sigma X^2 - (\Sigma X)^2} = \frac{(5)(82) - (15)(30)}{(5)(49) - (15)^2} = \frac{410 - 450}{245 - 225} = \frac{-40}{+20} = -2$$

TABLE 10.2 GENERAL WORKSHEET

Y	X	XY	Y²	X²	
5	4	20	25	16	
9	2	18	81	4	
3	4	12	9	16	
6	3	18	36	9	
7	2	14	49	4	
30	15	82	200	49	←Totals

From Equation 10.8b,

$$a = \frac{\Sigma Y - b\Sigma X}{n} = \frac{(30) - (-2)(15)}{5} = \frac{60}{5} = +12 \cdot$$

The least-square regression line is given by $\bar{Y}_c = a + bX_i = 12 - 2X_i$. As we suspected in part a, $a\ (= +12) > 0$ and $b\ (= -2) < 0$.

(c) Graphing the Regression Line
It takes only two points to plot a straight line, but as a safeguard against making a calculational error it is useful to calculate three points, plot them, and make sure that all three lie on the line. Further, as a check, the point $(\bar{Y}, \bar{X})$ must always lie on the regression line. (From Equation 10.8b, $a = \bar{Y} - b\bar{X}$ or $\bar{Y} = a + b\bar{X}$; therefore, $\bar{Y}_c = a + b\bar{X}$ must equal $\bar{Y}$ when $X = \bar{X}$.) Finally, look at the observed values of X and choose a convenient small value of X and a convenient large value.

For example, the observed X values were 2, 3, and 4. Let us use $X = 0, 2,$ and 4 as X values in the regression line.

X	$12 - 2(X) = \bar{Y}_c$	
0	12 − 2(0)	12
2	12 − 2(2)	8
4	12 − 2(4)	4

These paired values of $\bar{Y}_c$ and X, and the regression line they map, are plotted in Figure 10.6b.

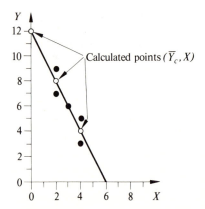

FIGURE 10.6b Regression line.

Note that $\bar{Y} = (\Sigma Y)/n = {}^{30}\!/_5 = 6$, $\bar{X} = (\Sigma X)/n = {}^{15}\!/_5$, and note that ($\dot{Y} = 6$, $X = 3$) is a point on the regression line.

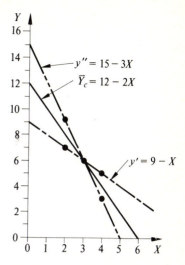

FIGURE 10.6c   The least-squares aspect of the regression line.

(d) The "Least-Squares" Nature of the Regression Line

In Figure 10.6b, the regression line $\bar{Y}_c = 12 - 2X$ goes through only one point out of the five points in the data. In Figure 10.6c we have two other straight lines which go through three of the five data points. Why is our regression line a "better" representation of those five points?

With Equation 10.6, we argued that the line of "best fit" would be the line which minimized the sum of the squared vertical distances between the point and the line. Consider the following calculations:

Y	X	$12 - 2(X) = \bar{Y}_c$	$(Y - \bar{Y}_c)^2$	$9 - X = y'$	$(Y - y')^2$	$15 - 3X = y''$	$(Y - y'')^2$
5	4	$12 - 2(4) = 4$	1	$9 - 4 = 5$	0	$15 - 3(4) = 3$	4
9	2	$12 - 2(2) = 8$	1	$9 - 2 = 7$	4	$15 - 3(2) = 9$	0
3	4		4		5	4	3	0
6	3	$12 - 2(3) = 6$	0	$9 - 3 = 6$	0	$15 - 3(3) = 6$	0
7	2		8	1		7	0	9	4
Total			4		8		8

For our regression line, $\Sigma(Y - \bar{Y}_c)^2 = 4$. However, for the line $y' = 9 - X$ (which goes through three points), $\Sigma(Y - y')^2 = 8$; the same is true for $y'' = 15 - 3X$. For those five points, *no straight line*, say $y^* = A + BX$, will produce a value of $\Sigma(Y - y^*)^2$ which is less than 4; hence, using Equations 10.8a and b to generate the line $\bar{Y}_c = 12 - 2X_i$ is called the "method of least squares." It is our use of Equation 10.6 as a measure of "best fit" that yields our regression line a "better" representation of those five points.

(e) The "Algebraic Irreversibility" of the Regression Line

If we considered the "algebraic" line $y = a + bx$, then we could solve for x in terms of y. Or,

$$x = -(a/b) + (1/b)y = c + dy$$

where $c = -a/b$ and $d = 1/b$. For example, $y = 12 - 2x$ is the same as $x = 6 - \frac{1}{2}y$.

In general, this algebraic manipulation is *not valid* for a least-squares regression line. (One special case where it is valid occurs when all the observed points lie *exactly* on a straight line; there is a perfect relation between Y and X.) In terms of our regression model, we might want to define X as the dependent variable and Y as the independent variable. The least-squares regression line could be expressed as $\bar{X}_c = c + dY_i$. Using the data from part b, we find

$$d = \frac{n\Sigma XY - (\Sigma X)(\Sigma Y)}{n\Sigma Y^2 - (\Sigma Y)^2} = \frac{5(82) - (15)(30)}{5(200) - (30)^2} = \frac{410 - 450}{1,000 - 900} = \frac{-40}{100} = -0.40$$

$$c = \frac{\Sigma X - d\Sigma Y}{n} = \frac{(15) - (-0.4)(30)}{5} = \frac{27}{5} = 5.4$$

Thus, $\bar{X}_c = 5.4 - (0.4)Y$ and $\bar{X}_c$ is *not* equal to $6 - \frac{1}{2}Y$! Why?

Figure 10.6d helps show us why. The least-squares regression line $\bar{Y}_c = a + bX$ minimizes the *vertical* distances between the data points and the line while $\bar{X}_c = c + dY$ minimizes the *horizontal* distances.

The moral of this story is: If you have $\bar{Y}_c = a + bX$ and you want $\bar{X}_c = c + dY$, *calculate* it and do *not* try to solve $\bar{Y}_c = f(X)$ for X!

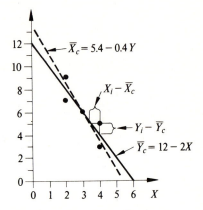

FIGURE 10.6d The algebraic irreversibility of the regression line.

(f) Calculation of the Standard Error of Estimate

We can calculate $s_{Y|x}$ from its definition, from the square root of Equation 10.9, or from Equation 10.10. From part d above, we found that $\Sigma(Y - \bar{Y}_c)^2 = 4$; therefore,

$$s_{Y|x} = \sqrt{\frac{\Sigma(Y - \bar{Y}_c)^2}{n - 2}} = \sqrt{\frac{4}{5 - 2}} = \sqrt{\frac{4}{3}}$$

Using Equation 10.10,

$$s_{Y|x} = \sqrt{\frac{\Sigma Y^2 - a\Sigma Y - b\Sigma XY}{n - 2}} = \sqrt{\frac{(200) - (12)(30) - (-2)(82)}{5 - 2}}$$

$$= \sqrt{\frac{200 - 360 + 164}{3}} = \sqrt{\frac{4}{3}} = 1.15$$

(g) Comparison of s_Y and $s_{Y|x}$

If we ignore the X values, then s_Y can be calculated as follows:

Y	$Y - \bar{Y}$	$(Y - \bar{Y})^2$
5	-1	1
9	3	9
3	-3	9
6	0	0
7	1	1
30	0	20

$$\bar{Y} = \Sigma Y/n = 30/5 = 6$$

$$s_Y = \sqrt{\frac{\Sigma(Y - \bar{Y})^2}{n - 1}}$$

$$= \sqrt{20/4} = \sqrt{5} = 2.236$$

From part f, $s_{Y|X} = 1.15 < s_Y = 2.236$. Knowing the value of X_i and the regression line reduces the variability in Y. We will use this result in later sections.

Example 10.3   A Cost Study of a Hosiery Mill

To provide some perspective on both the vintage and the economic application of the simple linear regression model, consider the published cost study performed by Professor Joel Dean over 30 years ago.[2] He considered a silk-hosiery knitting mill which, at that time, manufactured an intermediate product (dyeing and finishing occurred at other plants) with "highly mechanized equipment and skilled labor."

The dependent variable (Y) was total cost in dollars per month, and the independent variable (X) was output in dozens of pairs of stockings per month. One of the results was:

$$\bar{Y}_c = \$2,935.59 \text{ per month} + (\$1.998 \text{ per dozen pairs})X_i$$

$$s_{Y|X} = \$6,109.83 \text{ per month}$$

$$r = 0.973$$

The significance of r, the correlation coefficient, will be discussed later. The regression line's Y intercept, $\$2,935.59$ per month, is an estimate of the "fixed-costs" component of this production process. The slope, $\$1.998$ per dozen pairs, times output (X_i) is an estimate of the "variable-cost" component.

The standard error of the estimate, $\$6,109.83$ per month, may seem high. The missing ingredient is the range of X; that is, X ranged from 4,000 (dozen pairs) to 44,000. Within

---

[2] Joel Dean, *Statistical Cost Functions of a Hosiery Mill*, University of Chicago Press, Chicago, Ill., 1941. For other examples, see Edwin Mansfield (ed.), *Elementary Statistics for Business and Economics: Selected Readings*, W. W. Norton & Company, Inc., New York, 1970.

this range of output, the regression equation is a very useful model for decision making within this plant.

Interpretation of the Model

With the estimates of the regression line and the standard error of estimate completed, we can now interpret our model of the population. Figure 10.4 depicts the model and we will attempt to view Figure 10.4 from "above," using only two-dimensional diagrams.

For a given value of X, we want to make a statement about the probability of Y being within some interval: $P(Y_L \leq Y \leq Y_U)$. As in previous chapters, we will view Y_L and Y_U as being so many standard deviations (or standard errors) away from the mean. The mean of the Y values depends upon which value of X we are considering. The standard deviation of Y given X depends upon the variation in the error term, which we have assumed is independent of X. In general,

$$Y_L = \mu_{Y|X} - k\sigma_{Y|X} = (\alpha + \beta X) - k\sigma_{Y|X}$$

$$Y_U = \mu_{Y|X} + k\sigma_{Y|X} = (\alpha + \beta X) + k\sigma_{Y|X}$$

The value of k and the $P(Y_L \leq Y \leq Y_U)$, or the proportion of Y values between Y_L and Y_U, will also depend upon the *shape*, or distribution, of the Y values for a given X. But, the distribution of Y values for a given X is the distribution of the error term ξ_i. If we can assume that the error term is the result of minor random influences, it is not difficult to *assume that the error term is normally distributed.*

How important is the assumption that ξ_i, and thus Y, is normally distributed? The answer is noteworthy, but repetitious. For a large sample size, this assumption is usually not critical; for a small sample size, it is essential.

Figure 10.7 denotes the population model for $k = 1$, 2, and 3. If we know the

FIGURE 10.7   Proportion of Y values in particular intervals.

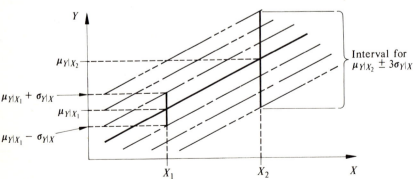

error term is normally distributed, then $\mu_{Y|X} \pm 2\sigma_{Y|X}$ includes 95.5 percent of the Y values. Since $\bar{Y}_c$ and $s_{Y|X}$ are estimates of their respective population parameters, $\mu_{Y|X}$ and $\sigma_{Y|X}$, our estimated regression model gives us some idea of how the population of Y values migrate for particular values of X.

10.7 INFERENCES ABOUT THE LINEAR MODEL

The method of least squares has permitted us to use a as a point estimate of α and b as a point estimate of β. The next question is: How do we construct interval estimates of α and β? Only occasionally will the particular problem under consideration require inferences about α, but we will *always* be concerned about an inference about β. Why?

If $\beta = 0$, then we have a trivial regression line, $\mu_{Y|X} = \alpha$. Knowledge of X does not help us determine Y, so why bother with X? Thus, we are interested in whether β is equal to 0.

Whereas β is the slope of the regression line, α is the parameter that positions the line on the graph. As such, α may not have any special meaning. In the spool-of-cable example, $\bar{Y}_c = -20 + 3X_i$. When $X_i = 0$ (an imaginary spool that weighs nothing?), $\bar{Y}_c = -20$. It is not very meaningful to say that a cable spool of zero weight has -20 feet of cable on it. Of course, the $a/b = {}^{20}\!/_3$ is an estimate of the weight of an empty spool. In the Joel Dean cost-function example, a is an estimate of fixed costs, and it is meaningful to say that you will have to pay $2,936 per month even if you produce nothing.

Interval Estimate of β

Recall from Chapter 6 that if X is normally distributed but σ_X^2 is unknown, then the t distribution is appropriate, where

$$t = \frac{\bar{X} - \mu}{s_X} = \frac{\bar{X} - \mu}{s/\sqrt{n}}$$

In general, we could state

$$t = \frac{(\text{sample statistic}) - (\text{population parameter})}{(\text{standard deviation of sampling distribution})}$$

It is in this context that we approach the sampling distribution of b. We shall state without proof that, if the "Y given X" values are normally distributed, then

$$\mu_b = E(b) = \beta$$

$$\sigma_b = \frac{\sigma_{Y|X}}{\sqrt{\Sigma X^2 - n\bar{X}^2}}$$

Since we do not know $\sigma_{Y|X}$, we use the estimate $s_{Y|X}$ and

$$t = \frac{b - \beta}{s_b} = \frac{b - \beta}{s_{Y|X}/\sqrt{\Sigma X^2 - n\bar{X}^2}}$$

In a fashion analogous to constructing a confidence interval for the population mean in Chapter 7, we have

$$P(b - t_{\alpha/2}s_b \leq \beta \leq b + t_{\alpha/2}s_b) = 1 - \alpha$$

or
$$b \pm t_{\alpha/2}s_{Y|X}/\sqrt{\Sigma X^2 - n\bar{X}^2}$$

provides the $(1 - \alpha)$ confidence interval for β.

If n is large, then the central limit theorem permits us to say that the sampling distribution of b is approximately normal even when the conditional probability distributions for the Y values are not normally distributed. Thus, for large sample sizes, $t_{\alpha/2}$ can be replaced by $z_{\alpha/2}$, and the assumption that Y is normally distributed becomes less critical.

Recall that $s_{Y|X}$ has $(n - 2)$ degrees of freedom (because a and b are estimates from the sample data); therefore, *in the interval estimate of β, $t_{\alpha/2}$ has $(n - 2)$ degrees of freedom.*

From example 10.2, $b = -2$, $s_{Y|X} = \sqrt{4/3} = 1.15$, $\Sigma X^2 = 49$, $n = 5$, and $\bar{X} = (\Sigma X)/n = 3$. For a 95 percent confidence interval, $n - 2 = 3$ degrees of freedom and $t_{\alpha/2} = 3.182$; hence

$$b \pm t_{\alpha/2}s_{Y|X}/\sqrt{\Sigma X^2 - n\bar{X}^2}$$
$$(-2) \pm (3.182)(1.15)/\sqrt{49 - (5)(3)^2}$$
$$(-2) \pm (3.66)/\sqrt{4}$$
$$-2 \pm 1.83 \quad \text{or} \quad -3.83 \leq \beta \leq -0.17$$

Note that the interval does not include $\beta = 0$.

Test of Hypothesis about β

Although you may wish to test the hypothesis that β is equal to a particular nonzero value, the common interest is the null hypothesis that $\beta = 0$. That is,

$$H_0: \beta = 0$$
$$H_A: \beta \neq 0$$

The decision rule is:

$$\text{If } -t_{\alpha/2,\, n-2} \leq t \leq +t_{\alpha/2,\, n-2} \qquad \text{accept } H_0$$
$$\text{Otherwise} \qquad \text{reject } H_0$$

For $\beta = 0$, the test statistic t is

$$t = b/s_b = (b\sqrt{\Sigma X^2 - n\bar{X}^2})/s_{Y|X}$$

Again, from example 10.2,

$$t = (-2)(\sqrt{4})/1.15 = -3.478$$

Since $t_{\alpha/2,\, n-2} = 3.182$, we reject the null hypothesis that $\beta = 0$.

Interval Estimate of α

If the conditional probability distribution of Y is normally distributed, then the sampling distribution for a is normal with

$$\mu_a = E(a) = \alpha$$

$$\sigma_a{}^2 = \sigma_{Y|X}{}^2 \left[\frac{\Sigma X^2}{n\Sigma(X_i - \bar{X})^2} \right]$$

Since $\sigma_{Y|X}{}^2$ is usually unknown, then

$$s_a{}^2 = s_{Y|X}{}^2 \left[\frac{\Sigma X^2}{n\Sigma(X_i - \bar{X})^2} \right]$$

and

$$t = \frac{a - \alpha}{s_a}$$

Hence, the probability statement

$$P(a - t_{\alpha/2} s_a \leq \alpha \leq a + t_{\alpha/2} s_a) = 1 - \alpha$$

leads to

$$a \pm t_{\alpha/2} s_{Y|X} \sqrt{\frac{\Sigma X^2}{n\Sigma(X_i - \bar{X})^2}}$$

or

$$a \pm t_{\alpha/2} s_{Y|X} \sqrt{1/n + \bar{X}^2/[\Sigma(X - \bar{X})^2]}$$

[where $t_{\alpha/2}$ has $(n - 2)$ degrees of freedom] as the $(1 - \alpha)$ confidence interval for α.

In example 10.2, $a = +12$, $s_{Y|X} = 1.15$, $n = 5$, $\bar{X} = 3$, and $\Sigma(X - \bar{X})^2 = \Sigma X^2 - n\bar{X}^2 = 49 - (5)(3)^2 = 4$. For a 95 percent confidence interval, $t_{\alpha/2,\, n-2} = 3.182$, and

$$12 \pm (3.182)(1.15)\sqrt{\tfrac{1}{5} + (3)^2/4}$$

$$12 \pm (3.66)\sqrt{0.20 + 2.25}$$

$$12 \pm (3.66)(1.565) = 12 \pm 5.73$$

or

$$6.27 \leq \alpha \leq 17.73$$

Test of Hypothesis about α

For a test of the hypothesis:

$$H_0: \alpha = \alpha_0$$

$$H_A: \alpha \neq \alpha_0$$

the decision rule is:

$$\text{If } t_{\alpha/2,\, n-2} \leq t \leq t_{\alpha/2,\, n-2} \qquad \text{accept } H_0$$

$$\text{Otherwise} \qquad \text{reject } H_0$$

and the test statistic is

$$t = \frac{a - \alpha_0}{s_{Y|X}\sqrt{1/n + \bar{X}^2/(\Sigma X^2 - n\bar{X}^2)}}$$

Example 10.4 Length-of-Cable Model

We wish to test the hypothesis that $\beta = 0$ and provide an interval estimate for α. From example 10.1, the relevant data is:

$$\bar{Y}_c = -20 + 3X \qquad s_{Y|X} = 10.9545$$

$$\bar{X} = \Sigma X/n = 700/12 = 58.3333 \qquad \Sigma X^2 = 44{,}200$$

(a) $H_0: \beta = 0$; $H_A: \beta \neq 0$. Use a 5 percent level of significance; for $n - 2 = 10$ degrees of freedom, $t = 2.228$. The decision rule is:

$$\text{If } -2.228 \leq t \leq +2.228 \qquad \text{accept } H_0$$

$$\text{Otherwise} \qquad \text{reject } H_0$$

The test statistic is:

$$t = b/s_b = (b\sqrt{\Sigma X^2 - n\bar{X}^2})/s_{Y|X}$$

$$= [(3)\sqrt{44,200 - (12)(58.3333)^2}]/10.9545$$

$$= [(3)\sqrt{3,366.715}]/10.9545$$

$$= [(3)(58.02)]/10.9545 = 15.89$$

Since $15.89 > +2.228$, we *reject* the H_0 that $\beta = 0$.

(b)   The 95 percent confidence interval for α would yield $t = 2.228$ from part *a*.

$$a \pm t_{\alpha/2} s_{Y|X} \sqrt{1/n + \bar{X}^2/(\Sigma X^2 - n\bar{X}^2)}$$

$$-20 \pm (2.228)(10.9545)\sqrt{\frac{1}{12} + (58.3333)^2/[44,200 - (12)(58.3333)^2]}$$

$$-20 \pm (24.4066)\sqrt{0.0833 + 1.0107}$$

$$-20 \pm (24.4066)(1.046) = -20 \pm 25.53$$

or
$$-45.53 \le \alpha \le 5.53$$

This is a very large range for the interval estimate of α. The precision of this estimate could be improved with a larger-sized sample.

10.8   INFERENCES WITH THE REGRESSION MODEL

Finally, we reach the point where we can use the regression model for prediction. Given a particular partially used spool of cable, we weigh it and use the model to determine an interval estimate for the length of cable remaining on the spool. Or, the restaurant manager takes the number of reservations for the evening meal, uses the regression model to construct an interval estimate of the number of lobsters that will be requested, and decides what quantity to order.

For a given value of X, we will consider three types of predictions: (1) the *mean* value of Y, (2) the *next* value of Y, and (3) the *mean* of the *next m* values of Y. These *prediction intervals* will look very much like *confidence intervals*. However, in order for these prediction intervals to be trustworthy, we must make one more assumption.

The regression model, as we have discussed it, is *descriptive* of the relationship between Y and X; it neither proves nor disproves *causation*. Changes in Y and X may be caused (or forced to occur) by one or more other variables. In this context, the method of least squares is stupid! The method of least squares finds the "best" straight line that represents the relationship between Y and X as paired values are "fed" to it. It is the investigator who chooses which variable is called Y and which other variable is called X. The causal chain of events that links changes in Y (or the probability distribution of Y values) to the various values of X is extraneous to the statistical regression model.

Of course, we assume that the causal chain of events did not change during the period of gathering the data from which the regression model was created. Now, if we wish to use the regression model to predict "future" values relating to Y for a given value of X, then *we must assume that the causal chain of events has not changed.* Although the magnitude of X may be different, we must assume that *the parameters that describe the relation between Y and X have not changed.*

Finally, we shall assume that X_j is some "new" value of the independent random variable. It is "new" in the sense that it was not necessarily a value of X that was used to develop the estimated regression line.

Interval Estimation of the Conditional Mean

The regression line $(\bar{Y}_c)_j = a + bX_j$ provides a point estimate of $\mu_{Y|X}$ for the new value of X. The value of $(\bar{Y}_c)_j$ may be in error because of the sampling error involved in estimating α and β. It is our custom to search for a probability statement of the form

$$P[(\bar{Y}_c)_{\text{Lower}} \le \mu_{Y|X_j} \le (\bar{Y}_c)_{\text{Upper}}] = 1 - \alpha$$

in order to provide a confidence interval. Further, the confidence limits have assumed the form

(Point estimate) $\pm (t_{\alpha/2}$ or $z_{\alpha/2})$(standard deviation of sampling distribution for estimator)

For example, $\bar{X} \pm t_{\alpha/2} s_{\bar{X}}$. Our past procedures would lead us to expect

$$(\bar{Y}_c)_j \pm t_{\alpha/2} s_{\bar{Y}_c}$$

to be the limits for a $(1 - \alpha)$ level of confidence.

This expectation will be realized, but it includes an important new twist. That "new twist" is the fact that $s_{\bar{Y}_c}$ is *not* a constant. Specifically, $s_{\bar{Y}_c} = f(X_j - \bar{X})$, where X_j, the "new" value of X, is not necessarily a value of X that was used to compute $\bar{X}$. Before we state exactly how the value of $s_{\bar{Y}_c}$ varies with the difference between X_j and the mean of the X values used to construct the regression line, let us intuitively investigate why that relationship exists.

First, recall that the regression line $\bar{Y}_c = a + bX$ is an estimate of the true regression line $\mu_{Y|X} = \alpha + \beta X$, and the method of least squares was used to provide a and b. Next, assume that the point $(\bar{Y}, \bar{X})$ lies on the true regression line. Since we have treated X values as known constants throughout this discussion, then $\bar{X}$ is just a

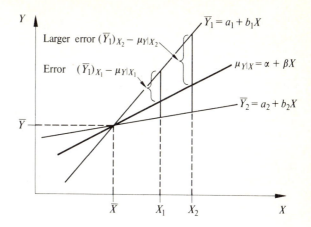

FIGURE 10.8   Influence of sample error in b.

representative value of X. We will soon relax the assumption that, given $X = \bar{X}$, $Y = \mu_{Y|X}$, but the point $(\bar{Y}, \bar{X})$ is a useful reference point.

The third point concerns only the sampling error involved in using b as an estimator for β. In the previous section, a technique for constructing an interval estimate of β was introduced. In Figure 10.8 we have the presumed population regression line, $\mu_{Y|X} = \alpha + \beta X$, and two regression lines estimated from different samples of that population. The estimate b_1 is greater than β while the estimate b_2 is less than β. Note that X_1 is closer to $\bar{X}$ than is X_2. Also note that the difference between $\mu_{Y|X_1}$ and $(\bar{Y}_1)_{X_1}$ [or $\mu_{Y|X_1}$ and $(\bar{Y}_2)_{X_1}$] is *less* than the difference between $\mu_{Y|X_2}$ and $(\bar{Y}_1)_{X_2}$ [or between $\mu_{Y|X_2}$ and $(\bar{Y}_2)_{X_2}$]. *The greater the distance* $|X_j - \bar{X}|$, *the greater the error* $[(\bar{Y}_c)_{X_j} - \mu_{Y|X_j}]$ *when $b \neq \beta$. Therefore, as $|X_j - \bar{X}|$ increases, we would expect $s_{\bar{Y}_c}$ to increase.* This is the point we intended to establish intuitively, but there are some "loose ends."

The first question may be: What about a as an estimate of α? The answer is quite simple. From Equation 10.8b, $a = \bar{Y} - b\bar{X}$, we can see that any error in b will be directly imposed upon the value of a. The estimates of a and b are *not* independent, and, for that reason, we used $(\bar{Y}, \bar{X})$ as a reference point [rather than the Y intercept, $(a, 0)$, or $(\alpha, 0)$].

Finally, when $X = \bar{X}$, the Y values are normally distributed with mean $\mu_{Y|X}$ and standard deviation σ. Sampling from this distribution of Y values, we would expect to obtain a sample mean $(\bar{Y})$ within the interval $\mu_{Y|X} \pm 1.96\sigma/\sqrt{n}$ about 95 percent of the time. Thus, for any given sample, the sample mean $\bar{Y}$ may not be equal to $\mu_{Y|X}$. But, the general relationship between $|X_j - \bar{X}|$ and $s_{Y|X}$ would still be valid when $b \neq \beta$.

From above, we would expect the standard deviation of the sampling distribution of $(\bar{Y}_c)_j$ to be a function of $|X_j - \bar{X}|$, and from the central limit theorem, we would expect it to be a function of $\sigma_{Y|X}/\sqrt{n}$. The estimated standard deviation[3] of $(\bar{Y}_c)_j$ is

$$S_{(\bar{Y}_c)_j} = s_{Y|X}\sqrt{\frac{1}{n} + \frac{(X_j - \bar{X})^2}{\Sigma X_i^2 - n\bar{X}^2}} \qquad (10.11)$$

Hence, the $(1 - \alpha)$ confidence interval for $\mu_{Y|X_j}$ is

$$(\bar{Y}_X)_j \pm t_{\alpha/2,\, n-2}\, s_{Y|X}\sqrt{1/n + (X_j - \bar{X})^2/(\Sigma X^2 - n\bar{X}^2)} \qquad (10.12)$$

Note that the confidence interval is *smaller* (or the estimate is more precise) when n is large and when X_j is close to $\bar{X}$.

Example 10.5   Interval Estimate for the Average Length of Cable

(a)   With reference to example 10.1, assume that we want to know the average length of cable remaining on spools whose gross weight is 60 pounds. Using a 95 percent confidence interval, $\alpha/2 = 0.025$ and there are $n - 2 = 10$ degrees of freedom; $t = 2.228$.
The regression model is $\bar{Y}_c = -20 + 3X$ and $s_{Y|X} = 10.9545$. Other pertinent data from example 10.1 includes $\bar{X} = 58.3333$ pounds, and $\Sigma X_i^2 = 44{,}200$.
Since

$$(\bar{Y}_c)_j = -20 + 3X = -20 + 3(60) = 160$$

and

$$\frac{(X_j - \bar{X})^2}{\Sigma X_i^2 - n\bar{X}^2} = \frac{(60 - 58.3333)^2}{44{,}200 - (12)(58.3333)^2}$$

$$= \frac{2.7778}{3{,}366.715} = 0.000825$$

then the 95 percent confidence limits for $\mu_{Y|60}$ are

$$(\bar{Y}_c)_j \pm t_{\alpha/2}\, s_{Y|X}\sqrt{1/n + (X_j - \bar{X})^2/(\Sigma X_i^2 - n\bar{X}^2)}$$

$$160 \pm (2.228)(10.9545)\sqrt{\tfrac{1}{12} + 0.000825}$$

$$160 \pm (24.4066)\sqrt{0.0842}$$

$$160 \pm 7.08$$

or

$$152.92 \le \mu_{Y|60} \le 167.08 \text{ feet}$$

Note the width of the interval is $2(7.08) = 14.16$ feet.

---

[3] This result is not very difficult to obtain. $\bar{Y}_c = a + bX = (\bar{Y} - b\bar{X}) + bX$, or $\bar{Y}_c = \bar{Y} + (X - \bar{X})b$. Since $\bar{Y}$ and b are independent, then

$$\text{Var}(\bar{Y}_c) = \text{Var}[\bar{Y} + (X - \bar{X})b] = \text{Var}(\bar{Y}) + (X - \bar{X})^2\, \text{Var}(b)$$

$\text{Var}(\bar{Y}) = \sigma_{Y|X}^2/n$ and from the previous section, $\text{Var}(b) = \sigma_{Y|X}^2/(\Sigma X^2 - n\bar{X}^2)$. Thus $\text{Var}(\bar{Y}_c) = \sigma_{Y|X}^2[1/n + (X_j - \bar{X})^2/(\Sigma X_i^2 - n\bar{X}^2)]$.

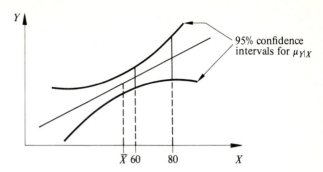

FIGURE 10.9   The 95 percent confidence-interval band.

(*b*)   Now consider the 95 percent confidence interval for the average length of cable on all spools whose gross weight is 80 pounds. The value of t is still 2.228.

$$(\bar{Y}_c)_j = -20 + 3X = -20 + 3(80) = 220 \text{ feet}$$
$$(X_j - \bar{X})^2/(\Sigma X_i{}^2 - n\bar{X}^2) = (80 - 58.3333)^2/(3{,}366.715)$$
$$= 0.1394$$

The 95 percent confidence limits for $\mu_{Y|80}$ are

$$220 \pm (2.228)(10.9545)\sqrt{0.0833 + 0.1394}$$
$$220 \pm 11.52$$

or
$$208.48 \le \mu_{Y|80} \le 231.52 \text{ feet}$$

Note that the width of the interval is $2(11.52) = 23.04$ feet. The width of the interval is greater because 80 pounds is not as close to $\bar{X}$ ($= 58.33$ pounds) as is 60 pounds.

Figure 10.9 shows the 95 percent confidence limits for $\mu_{Y|x}$; these limits diverge from the regression line as $|X_j - \bar{X}|$ increases.

Prediction Interval for the "Next" Value of Y

The warehouse manager who receives a partially used spool of cable is now concerned with how much cable is on that particular spool. If it weighs 60 pounds, the manager is not interested in the *mean* length of cable on *all* spools that weigh 60 pounds; what is needed is an estimate of the cable length on this one particular spool. The 60 pounds is the "new" value of X, X_j, and this one particular cable length is called Y_{next}.

For an intuitive picture of the procedure used to obtain the prediction interval for the "next" value of Y, let us assume that we know the population mean of Y for

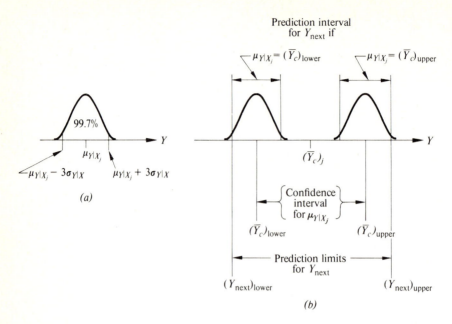

FIGURE 10.10 (a) Prediction interval for Y_{next} when the population parameters are known; (b) prediction interval for Y_{next} when population parameters are unknown.

X_j: $\mu_{Y|X_j}$. Also, assume that we know the population standard deviation $\sigma_{Y|X}$ and that the Y values are normally distributed. For $X = X_j$, we would expect 99.7 percent of the Y values to be in the interval $\mu_{Y|X_j} \pm 3\sigma_{Y|X}$, as shown in Figure 10.10a.

But let us be more realistic about what we know. We do not know $\mu_{Y|X_j}$; instead we have a confidence-interval estimate from Equation 10.12. The confidence limits for $\mu_{Y|X_j}$ are given as $(\bar{Y}_c)_{lower}$ and $(\bar{Y}_c)_{upper}$ in Figure 10.10b.

Assuming we know $\sigma_{Y|X}$, then we could center the distribution in Figure 10.10a on $(\bar{Y}_c)_{lower}$ and $(\bar{Y}_c)_{upper}$ in order to obtain a prediction interval for Y_{next} from both of those limits (as shown at the top of Figure 10.10b). Our prediction interval for Y_{next} extends from $(Y_{next})_{lower}$ to $(Y_{next})_{upper}$. Of course, we do not know $\sigma_{Y|X}$ but will use $s_{Y|X}$ as a point estimate.

In summary, for the same level of confidence $(1 - \alpha)$, *the prediction interval for a particular value of Y, Y_{next}, must be larger than the prediction interval for the mean of Y, $\mu_{Y|X}$.* Thus, we will use

$$s_{Y_{next}} = s_{Y|X} \sqrt{\frac{1}{n} + \frac{(X_j - \bar{X})^2}{\Sigma X_i^2 - n\bar{X}^2} + 1} \qquad \textbf{(10.13)}$$

as the estimate of the standard deviation of the sampling distribution of Y_{next}. Note the similarity with Equation 10.11; Equation 10.13 has an additional $+1$ under the radical. Since the expression under the radical sign is multiplied times the standard error of the estimate, this $+1$ term accounts for the variation in Y around its conditional mean (which is provided by the regression line).

As a result, the $(1 - \alpha)$ prediction interval for Y_{next}, given X_j, is

$$(\bar{Y}_c)_j \pm t_{\alpha/2, n-2} S_{Y|X} \sqrt{\frac{1}{n} + \frac{(X_j - \bar{X})^2}{\Sigma X_i^2 - n\bar{X}^2} + 1} \qquad (10.14)$$

In example 10.5a, we calculated the interval estimate of the mean length of cables from all spools of 60 pounds gross weight. If we have one spool of 60 pounds gross weight, we can provide a prediction interval of its length of remaining cable by using Equation 10.14. Recalling that Equations 10.11 and 10.13 differ only by the $+1$ term, we can use the calculations in example 10.5 as follows:

$$160 \pm (2.228)(10.9545)\sqrt{0.0842 + 1}$$

$$160 \pm (24.4066)\sqrt{1.0842}$$

$$160 \pm (24.4066)(1.0412)$$

$$160 \pm 25.41$$

or
$$134.59 \le Y_{next} \le 185.41 \text{ feet}$$

The length of this confidence interval, 50.82 feet, is considerably larger than the 14.16 feet in example 10.5 for the 95 percent confidence interval of $\mu_{Y|60}$. But that is what we expect when we construct a prediction interval for the cable length on a *particular* spool as compared to the mean length of cable on *all* spools of the same gross weight.

Prediction Interval for Mean of m "Next" Values of Y Given X_j

Finally consider the situation where our cable keeper has three partially used spools, each of which weighs exactly 60 pounds, and would like an estimate of the total length of cable on the three spools. If there were an interval estimate of the mean length of the three spools, then those confidence limits could be multiplied by 3 in order to have an interval estimate of the total length of cable on the three spools.

It is important to recognize that this is *not* the same thing as taking 3 times the confidence interval provided by Equation 10.12. The interval estimate of $\mu_{Y|X_j}$ is based on samples of size n. In like manner, the prediction interval for Y_{next} is analogous to samples of size 1. From the central limit theorem, we know that the standard deviation for the sampling distribution of the sample means becomes *smaller* as the sample size increases. If we assume that $n > m > 1$, then the analogous situation for the regression line would be

$$S_{(\bar{Y}_c)_j} < S_{(\bar{Y}_m)_j} < S_{Y_{next}}$$

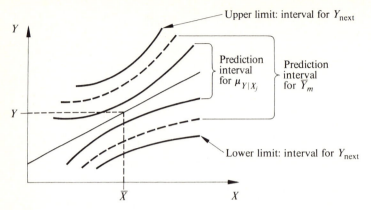

FIGURE 10.11 95 percent confidence bands.

and we would expect the prediction interval for $(\bar{Y}_m)_j$ to be smaller than the prediction interval for Y_{next} but larger than the prediction interval for $\mu_{Y|X_j}$. Figure 10.11 demonstrates the point.

For $X = X_j$, we shall define $\bar{Y}_m$ as the conditional mean of the m "next" values of Y, or $\bar{Y}_m = (\sum_{j=1}^{m} Y_j)/m$. The standard deviation of the sampling distribution of $\bar{Y}_m$ is

$$s_{(\bar{Y}_m)_j} = s_{Y|X} \sqrt{\frac{1}{n} + \frac{(X_j - \bar{X})^2}{\Sigma X_i^2 - n\bar{X}^2} + \frac{1}{m}} \qquad (10.15)$$

In contrast with $s_{Y_{\text{next}}}$, the $+1$ has been replaced with the $1/m$ term.

The resulting $(1 - \alpha)$ prediction interval is

$$(\bar{Y}_c)_j \pm t_{\alpha/2,\, n-2} s_{Y|X} \sqrt{\frac{1}{n} + \frac{(X_j - \bar{X})^2}{\Sigma X_i^2 - n\bar{X}^2} + \frac{1}{m}} \qquad (10.16)$$

where $(\bar{Y}_c)_j = a + bX_j$.

For example, consider the 95 percent confidence interval of the mean length $(\bar{Y}_m)$ of the $(m =)$ three 60-pound spools of cable using the regression model in example 10.1.

$$160 \pm (2.228)(10.9545)\sqrt{0.0842 + \tfrac{1}{3}}$$

$$160 \pm (24.4066)\sqrt{0.4175}$$

$$160 \pm 15.77$$

or

$$144.23 \leq \bar{Y}_m \leq 175.77 \text{ feet}$$

Or, the 95 percent prediction interval for the "total length" is

$$3(144.23) \le m\bar{Y}_m \le 3(175.77)$$

$$432.7 \text{ feet} \le \Sigma Y_j \le 527.3 \text{ feet}$$

Real-world problems may not require the prediction interval for $\bar{Y}_m$ as frequently as the prediction interval for either $\mu_{Y|X_j}$ or Y_{next}.

Summary

We start with the assumption that the basic causal system has not changed. Our description of the relation between Y and X, the regression model, is then useful for prediction purposes. A "new" value of X is observed: call it X_j. Given X_j, a $(1 - \alpha)$ (confidence or) prediction interval can be constructed for: (1) the mean of all possible Y values, $\mu_{Y|X}$, (2) the next single value of Y, Y_{next}, or (3) the mean of the next m values of Y, $\bar{Y}_m$. The main difference between these prediction intervals and the confidence intervals considered in previous chapters is the fact that the length of the prediction intervals also depends upon "how far" the new value of X is from the mean of the X values used in constructing the regression line.

10.9 THE CORRELATION MODEL

Throughout this chapter, we are concerned with the *association* between Y and X. The regression model, as presented in prior sections, first considers the values of X to be a set of known constants and then tries to specify the *form* of the association between Y and X.

Our first task in this section will be to use the correlation model to measure the *degree* of association between Y and X. The standard error of the estimate, $s_{Y|X}$, is a measure of how close the "points" (i.e., the paired observations of Y and X) are to the regression line. Although $s_{Y|X}$ is a measure of the *degree* of association between Y and X, it is an imperfect measure because the magnitude of $s_{Y|X}$ depends upon the units used to measure Y. In the length-of-cable example, if Y were measured in inches rather than feet, then $s_{Y|X}$ would have been much greater. But, the degree of association between Y and X would be no different. Thus, we search for a "unitless" index for the *degree* of association. The correlation model will provide better (though not necessarily "perfect") measures for the degree of association.

Our second task is to consider what happens to (1) our regression model and (2) the results of the correlation techniques when X is a *random variable* rather than a set of known constants. For example, consider the farmer's problem when attempting to find the regression line between $Y =$ bushels of corn per acre and $X =$ inches of rain during the growing season. Even with recent advances in meteorology, classifying the independent variable, inches of rainfall, as a set of known constants is dubious. The classification "random variable" is certainly more appropriate for this independent variable. Under certain conditions, X can be a random variable, and the results of the first eight sections in this chapter are still valid.

10.9(a) Reference Point: The Regression Line and $\overline{Y}$

Our goal is to measure the *degree* of association between Y and X. The regression line measures the *form* of the association, and we are looking for a unitless measure of how close the points are to the regression line. We shall use three extreme cases to demonstrate the nature of the problem.

Case 1:   Zero Correlation
Y is not associated with X; Y is not influenced by anything. There is no variation in Y; the problem of predicting Y is trivial.

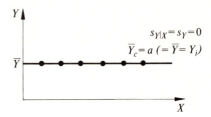

Case 2:   Zero Correlation
Y is not associated with X. There is variation in Y; predicting Y is not trivial, but knowing X does not reduce the variation in Y.

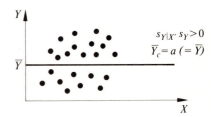

Case 3:   Perfect Correlation
The association between Y and X maps a straight line. All the points lie on a line. There is variation in Y, but knowing X tells us exactly what Y will be.

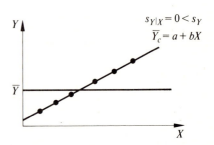

FIGURE 10.12   The extreme cases of correlation.

In case 2 there is "no association," or what we shall call "zero correlation." In case 3, a straight line (with $b \neq 0$) perfectly describes the association of Y and X, or "perfect correlation." Case 1 has the disturbing characteristic that a straight line (with $b = 0$) perfectly describes the points, but there is "no association" because Y is independent of X. Case 1 is "uninteresting" from the viewpoint of this discussion because there is no variation in Y.

For our first measure of degree of association, we would like a measure that assigns the value 0 to case 2 and a value 1 to case 3. On a scale between 0 and 1, the measure would indicate how close the points are to the regression line, and the measure would be independent of the units with which X and Y are measured.

First, let us ignore X and consider the degree of scatter about $\bar{Y}$. The term

$$\Sigma(Y - \bar{Y})^2$$

is a familiar term for measuring the variation of Y about its mean. Now consider the scatter about the regression line. Considering the vertical distance between each point and the regression line yields

$$\Sigma(Y - \bar{Y}_c)^2$$

as a measure of the scatter about the regression line. Both are in the (squared) units of Y, so let us take the ratio of the two:

$$\frac{\Sigma(Y - \bar{Y}_c)^2}{\Sigma(Y - \bar{Y})^2}$$

If the denominator is 0, then we have a situation similar to case 1, which means we would never consider regression or correlation analysis anyway. Only if $\Sigma(Y - \bar{Y})^2$ were greater than 0 would we be interested in a measure of the degree of association.

If $\bar{Y}_c = \bar{Y}$ (i.e., case 2), then the ratio is 1; if the numerator is 0 (i.e., case 3), then the ratio is 0. The ratio gives a measure of degree of association, but the scale from 0 to 1 represents the reverse of what we were seeking. We can rectify that by considering

$$r^2 = 1 - \frac{\Sigma(Y - \bar{Y}_c)^2}{\Sigma(Y - \bar{Y})^2} \tag{10.17}$$

Since we would never be interested in such a measure if $\Sigma(Y - \bar{Y})^2 = 0$, we will assume that $\Sigma(Y - \bar{Y})^2 > 0$. The only way that $\Sigma(Y - \bar{Y}_c)^2$ will equal 0 is if there is "perfect correlation;" hence, $\Sigma(Y - \bar{Y}_c)^2 = 0$ causes r^2 to be equal to 1. Bingo!

If $\bar{Y}_c = \bar{Y}$ (i.e., case 2), then $\Sigma(Y - \bar{Y}_c)^2 = \Sigma(Y - \bar{Y})^2$ and $r^2 = 0$. Bingo again! Further, it can be shown that $\Sigma(Y - \bar{Y}_c)^2 \leq \Sigma(Y - \bar{Y})^2$; therefore, r^2 will never be negative.

The term r^2 is called the *sample coefficient of determination* and has the following properties:

1. $r^2 = 0$ reflects "no correlation"
2. $r^2 = 1$ reflects "perfect correlation"
3. $0 < r^2 < 1$ reflects the "degree of correlation"

Equation 10.17 is a relatively easy way to interpret the meaning of the sample coefficient of determination, but it is not usually the easiest way to calculate r^2. There are several equivalent algebraic expressions which make calculation of r^2 easier.

If the regression line has been calculated, then

$$r^2 = \frac{a\Sigma Y + b\Sigma XY - n\bar{Y}^2}{\Sigma Y^2 - n\bar{Y}^2} \tag{10.18}$$

is often easier. If the regression line has not been calculated, then

$$r^2 = \frac{(\Sigma XY - n \cdot \bar{X} \cdot \bar{Y})^2}{(\Sigma X^2 - n\bar{X}^2)(\Sigma Y^2 - n\bar{Y}^2)} \tag{10.19}$$

and

$$r^2 = \frac{[n\Sigma XY - (\Sigma X)(\Sigma Y)]^2}{[n\Sigma X^2 - (\Sigma X)^2][n\Sigma Y^2 - (\Sigma Y)^2]} \tag{10.20}$$

Example 10.6 Coefficient of Determination for Length of Cable

(a)
$$r^2 = \frac{a\Sigma Y + b\Sigma XY - n\bar{Y}^2}{\Sigma Y^2 - n\bar{Y}^2}$$

$$= \frac{(-20)(1,860) + (3)(118,600) - (12)(155)^2}{319,800 - (12)(155)^2}$$

$$= \frac{-37,200 + 355,800 - 288,300}{31,500}$$

$$= \frac{30,300}{31,500} = 0.9619$$

(b) Using Equation 10.20,

$$r^2 = \frac{[n\Sigma XY - (\Sigma X)(\Sigma Y)]^2}{[n\Sigma X^2 - (\Sigma X)^2][n\Sigma Y^2 - (\Sigma Y)^2]}$$

$$= \frac{\{(12)(118,600) - (700)(1,860)\}^2}{\{(12)(44,200) - (700)^2\}\{(12)(319,800) - (1,860)^2\}}$$

$$= \frac{\{1,423,200 - 1,302,000\}^2}{\{530,400 - 490,000\}\{3,837,600 - 3,459,600\}}$$

$$= \frac{(121,200)^2}{(40,400)(378,000)} = \frac{14,689,440,000}{15,271,200,000}$$

$$= 0.9619$$

An Alternative Interpretation of the Coefficient of Determination

In constructing the definition of r^2 that led to Equation 10.17, we used the extreme cases (cases 2 and 3) in Figure 10.12. In case 2, none of the variation in Y could be described by the regression line, while in case 3 all the variation in Y could be described by the regression line. Apparently, when $0 < r^2 < 1$, *some* of the variation in Y can be described by the regression line, but, how do we interpret "some"?

What we will now show is that r^2 *is the proportion of the total variation in Y which is explained by the regression line*. In the previous example, $r^2 = 0.9619$. We could say that 96.19 percent of the variation in Y is explained by (or "described by" or "accounted for") the regression line.

Recall, in example 10.1, that we unwound the 12 partially used spools of cable and measured the length of each cable. The length ranged from 70 to 230 feet with a mean of $(1,860/12 =)$ 155 feet. The term "total variation" refers to the variation of these 12 lengths from 155 feet, their mean. Then we found the regression line, $\bar{Y}_c = -20 + 3X$, using the gross weight (spool plus cable) and cable length. Since $r^2 = 0.9619$, we can say that 96.19 percent of the total variation in cable lengths can be described by the statistical relation between gross weight and cable length.

On one of those spools, we found $X = 80$ pounds and $Y = 230$ feet of cable. Since the mean length of the 12 cables was 155 feet, then $Y - \bar{Y} = 230 - 155 = 75$ feet is what we will call the *total deviation* of the observed value from the mean. For $X = 80$ pounds, the regression line yields a value of $\bar{Y}_c = -20 + 3(80) = 220$ feet. We shall call $\bar{Y}_c - \bar{Y} = 220 - 155 = 65$ feet the *explained deviation*, or the deviation of $\bar{Y}_c$ from the mean as described by the regression line. The vertical distance between the observed value and the regression line, $Y - \bar{Y}_c = 230 - 220 = 10$ feet, is that component of the total deviation *not explained* by the regression line. Figure 10.13

FIGURE 10.13   Total, explained, and unexplained deviation.

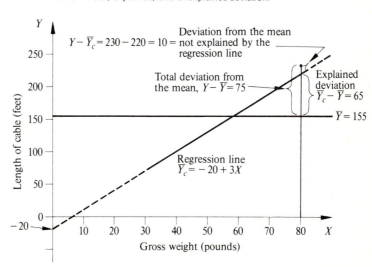

demonstrates the concept that

$$(\text{Total deviation}) = (\text{unexplained deviation}) + (\text{explained deviation})$$

or
$$(Y - \bar{Y}) = (Y - \bar{Y}_c) + (\bar{Y}_c - \bar{Y})$$

In this context, we switch from the *deviation* of one point about the mean to the *variation* of all points about their mean. To be consistent with our past measurements of variation, *total variation* is measured by $\Sigma(Y - \bar{Y})^2$ and *unexplained variation* is measured by $\Sigma(Y - \bar{Y}_c)^2$. The proportion of unexplained variation is

$$\frac{\text{Unexplained variation}}{\text{Total variation}} = \frac{\Sigma(Y - \bar{Y}_c)^2}{\Sigma(Y - \bar{Y})^2}$$

Thus, the proportion of explained variation should be

$$\frac{\text{Explained variation}}{\text{Total variation}} = 1 - \frac{\text{unexplained variation}}{\text{total variation}}$$

$$= 1 - \frac{\Sigma(Y - \bar{Y}_c)^2}{\Sigma(Y - \bar{Y})^2}$$

$$= r^2$$

The proportion of the total variation explained by the regression line is given by the sample coefficient of determination.

Although this is a common interpretation of r^2, the adjective "explained" should be carefully employed. The ability of the regression line to *describe* $r^2(100)$ percent of the total variation does not necessarily imply that $r^2(100)$ percent of the variation in Y is *caused* by the variation in X as expressed in the regression line. The following example demonstrates the point.

Example 10.7 Spurious Correlation

From the records of the Bureau of Motor Vehicles, we select four automobile registration forms and record the curb weight (Y) and the last three digits of the license plate number (X). From the resulting data, we calculate the regression line and the sample coefficient of correlation as follows:

Curb Weight (Pounds) Y	Last Three Digits on License Plate X	XY	Y^2	X^2
4,000	624	2,496,000	16,000,000	389,376
2,750	424	1,166,000	7,562,500	179,776
2,650	408	1,081,200	7,022,500	166,464
2,250	344	774,000	5,062,500	118,336
11,650	1,800	5,517,200	35,647,500	853,952

$$b = \frac{n\Sigma XY - (\Sigma X)(\Sigma Y)}{n\Sigma X^2 - (\Sigma X)^2} = \frac{(4)(5,517,200) - (1,800)(11,650)}{(4)(853,952) - (1,800)^2} = \frac{1,098,800}{175,808}$$

$$= 6.25$$

$$a = \frac{\Sigma Y - b\Sigma X}{n} = \frac{11,650 - (6.25)(1,800)}{4} = \frac{400}{4} = 100$$

$$\bar{Y}_c = 100 + 6.25X$$

Since $\bar{Y} = 11,650/4 = 2,912.5$,

$$r^2 = \frac{a\Sigma Y + b\Sigma XY - n\bar{Y}^2}{\Sigma Y^2 - n\bar{Y}^2} = \frac{(100)(11,650) + (6.25)(5,517,200) - (4)(2,912.5)^2}{(35,647,500) - (4)(2,912.5)^2}$$

$$= \frac{1,716,876}{1,716,876} = 1.00$$

There is perfect correlation between X and Y. Certainly you would not say that variation in the last three digits on the license plate causes variation in curb weight of the automobile. Nor is the reverse true; license plate numbers are not assigned according to curb weight.

The Population Coefficient of Determination

Our regression model of the population is such that, for each value of X, there is a distribution of Y values. The mean of each conditional distribution of Y is given by $\mu_{Y|X} = \alpha + \beta X$ and the variance is $\sigma_{Y|X}^2$. Disregarding X, the variance of the Y values is σ_Y^2, which is greater than 0. (If $\sigma_Y^2 = 0$, we are not interested in regression analysis, as in case 1 in Figure 10.12.) If $\sigma_{Y|X}^2$ is small relative to σ_Y^2, then we have a "high degree" of association between Y and X. If $\sigma_{Y|X}^2$ is about the same as σ_Y^2, then we have a "low degree" of association. Hence, the *coefficient of determination for the population*, denoted ρ^2 (where ρ is the Greek letter rho), is defined as

$$\rho^2 = 1 - \frac{\sigma_{Y|X}^2}{\sigma_Y^2}$$

As defined in Equation 10.17, r^2 is a point estimate of ρ^2.

10.9(b)   Reference Point: $\bar{Y}$ and $\bar{X}$

The regression model when X is a random variable   Now consider the degree of association when both Y and X are considered to be random variables. Our discussion of the regression model has explicitly treated the values of X as a set of known constants. What happens to that model when X is a random variable?

It can be shown that all the preceding discussion of the regression model is valid for a random independent variable when the following two conditions are satisfied:

1. The conditional probability distributions of Y are *normally* distributed with mean $\mu_{Y|X} = \alpha + \beta X$ and a constant variance $\sigma_{Y|X}^2$.
2. The values of X_i are independent of one another, and the probability distribution for X does not include any of the parameters: α, β, or $\sigma_{Y|X}^2$.

In general, if X is a random variable, it poses no serious problem in our regression model.

The sample correlation coefficient   Let us consider the scatter diagrams for two hypothetical examples: (1) the bushels of corn per acre and the inches of rainfall per growing period on a farm, and (2) the person-hours per week spent on municipal tennis courts and inches of rainfall per week.

Consider Figure 10.14. The first step is to shift the origin from $Y = 0$, $X = 0$ to $Y = \bar{Y}$, $X = \bar{X}$.

In Figure 10.14a there is a general tendency for Y to increase as X increases. With $(\bar{Y}, \bar{X})$ as the origin, most of the points fall in the upper right-hand quadrant and lower left-hand quadrant. Consider the algebraic sign on the product $(Y_i - \bar{Y}) \times (X_i - \bar{X})$ and the location of the point (Y_i, X_i); the results are summarized in Table 10.3.

Considering all the points in Figure 10.14a, we would expect $\Sigma[(Y - \bar{Y}) \times (X - \bar{X})]$ to be greater than 0. In like manner, in Figure 10.14b, we would expect $\Sigma[(Y - \bar{Y})(X - \bar{X})]$ to be less than 0 (or negative). Furthermore, if the points were rather evenly distributed in all four quadrants (indicating little or no correlation), the points where the product $(Y - \bar{Y})(X - \bar{X})$ is negative would roughly equal the

FIGURE 10.14   Scatter diagram. (a) The farm; (b) municipal tennis courts.

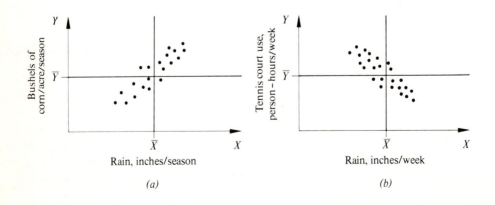

TABLE 10.3 SIGN ON CROSS-PRODUCT

Quadrant for (Y_i, X_i)	Algebraic Sign on		
	$(Y - \bar{Y})$	$(X - \bar{X})$	$(Y - \bar{Y})(X - \bar{X})$
Upper left	+	−	−
Upper right	+	+	+
Lower left	−	−	+
Lower right	−	+	−

points where the product is positive, and we would expect $\Sigma[(Y - \bar{Y})(X - \bar{X})]$ to be close to 0.

Therefore, if $\Sigma[(Y - \bar{Y})(X - \bar{X})]$ is positive, there is a *direct* correlation (i.e., increases in X are associated with increases in Y); if it is negative, there is an *inverse* correlation (i.e., increases in X are associated with decreases in Y); if it is 0, there is no correlation between Y and X. We set out to find a measure of the *degree* of association; the aforementioned describes a measure for the *nature* of the association between two random variables. Could we use the magnitude or absolute value of $\Sigma[(Y - \bar{Y})(X - \bar{X})]$ as a measure for the degree of association? The larger $|\Sigma[(Y - \bar{Y})(X - \bar{X})]|$, the greater the degree of association? With some qualification, yes!

The term $|\Sigma[(Y - \bar{Y})(X - \bar{X})]|$ has two deficiencies: (1) it is measured in "units of Y" times "units of X" (and we want a unitless measure), and (2) the magnitude of this term depends upon the size of the sample. To account for the first deficiency, we shall divide $|\Sigma[(Y - \bar{Y})(X - \bar{X})]|$ by $s_Y \cdot s_X$; hence, the ratio is a unitless measure of relative variation. A remedy for the second deficiency is to use an average concept; divide the ratio by $(n - 1)$.

We will define[4] the *sample correlation coefficient r*, as

$$r = \frac{\Sigma[(Y - \bar{Y})(X - \bar{X})]}{s_Y s_X (n - 1)} \tag{10.21}$$

The correlation coefficient has the following properties:

Property 1. $r = -1$ when there is a perfect inverse association between Y and X (i.e., all the sample points will fall on a straight line which has a *negative* slope: $b < 0$).

---

[4] Please note that

$$\frac{\Sigma\left[\left(\frac{Y - \bar{Y}}{s_Y}\right)\left(\frac{X - \bar{X}}{s_X}\right)\right]}{n - 1}$$

is the same as Equation 10.21.

Property 2. $r = 0$ when there is no correlation between Y and X.

Property 3. $r = +1$ when there is a perfect direct association between Y and X (i.e., all the sample points will fall on a straight line which has a *positive* slope: $b > 0$).

Property 4. $0 < |r| < 1$ represents the degree of association.

Undoubtedly, denoting the sample correlation coefficient with the symbol r and denoting the sample coefficient of determination with the symbol r^2 are no accidents. Although r^2 was developed using X as a set of known constants and $\bar{Y}$ and $\bar{Y}_c$ (from the regression line) as reference points, while r was developed using both X and Y as random variables and $(\bar{Y}, \bar{X})$ as a reference point, they are related by

$$r = \pm \sqrt{r^2} \tag{10.22}$$

The obvious question is: How do we know which root of r^2 to use for r? *The algebraic sign of r is always the same as the algebraic sign of b, the slope of the regression line.*

Equation 10.18, 10.19, or 10.20 can be used to calculate r^2. To get r, take the square root of r^2 and impose the sign of b. If the regression line has not been calculated, then

$$r = \frac{n\Sigma XY - (\Sigma X)(\Sigma Y)}{\sqrt{[n\Sigma X^2 - (\Sigma X)^2][n\Sigma Y^2 - (\Sigma Y)^2]}} \tag{10.23}$$

will usually be easier to calculate than Equation 10.21.

Inferences about the correlation model The most common inference about ρ, the population correlation coefficient is $H_0: \rho = 0$, $H_A: \rho \neq 0$. Fortunately, for a given sample size n and level of significance α, the results of testing $\rho = 0$ will always be the same as testing the null hypothesis that $\beta = 0$. The test of hypothesis discussed in section 10.7 is sufficient for both cases.

In order to test the hypothesis that $\rho = \rho_0$ where ρ_0 is a hypothetical value of importance to a particular problem, we must make further assumptions about the probability distribution of Y and X. An example is given in Exercise 9.

A "YELLOW SIGN" SUMMARY

On the highways of the United States, yellow signs are used to caution the motorist of possible hazards or road conditions. Experienced drivers almost subconsciously react to these signs; the novice may take more time to read and interpret these roadside messages. This section is a list of "caution messages" for those who construct or use regression and correlation methods.

Does the Model Make Sense?

This is not an entirely facetious question. Usually this is a question of cause and effect. If we want to predict Y, then the variable that most affects Y should logically be chosen as X. Cause and effect usually require a specified network of interrelationships that are logically derived. Regression and correlation techniques measure the association between two variables; however, they do not *ensure* the existence of a cause-and-effect relationship.

For example, consider the high school principal who wants to predict the number of incoming first-year students (Y) for next fall. The data on past sizes of the incoming class could be plotted against time; time could be used as the independent variable X. This might plot a diagram that appears to be a straight line with positive slope. In Chapter 12 we will discuss time-series analysis.

Time-series analysis is a very useful method for *describing* what has happened in the "time frame" of the past, but it has one important defect. With a few exceptions, time seldom *causes* anything. The "growing time" for a tomato, the "drying time" for glue or paint, the "travel time" between two geographic locations are all descriptive measures of a cause-effect sequence that involves other variables. Growing time, drying time, and travel time are more likely to be considered the dependent variable. The "passing of time" as measured in calendar years is not likely to be the *cause* of the size of the incoming class.

The high school principal would more logically look at the variable "populations" for X. In many cases, a public high school will have a number of middle or elementary schools which will "feed" students to that institution. The use of X = "number of eighth graders in feeder schools as of some date" would have a more logical influence on Y than would calendar time. Changes of net migration into or out of the area, changing school district boundaries, etc., will affect the relationship, but, at least initially, the model makes sense.

Can we tie Y logically (as well as statistically) to X? is just one important facet of this discussion. The second question asks: Has the cause-effect system changed from the time frame of the data (from which the statistical model is constructed) to the present or future time? Answers to this question employ the concept of subjective probability.

Prediction Constraints on the Regression Model

The data used to construct the regression model will also provide a range of X values (the observed maximum and minimum of X). For "new" values of X within that range, the techniques of section 10.8 can be used for prediction intervals. For "new" values of X outside that range, we can "extrapolate" the line and construct prediction intervals, but the extrapolation process introduces new hazards. Outside the

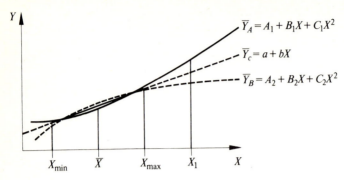

FIGURE 10.15 Danger of extrapolating regression line.

range of X values used to construct the regression line, we have no data on the form or the degree of association between Y and X.

Figure 10.15 demonstrates the point. The regression line $\bar{Y}_c$ may be very representative of the association between X_{min} and X_{max}. For a new value X_1, we could extrapolate $\bar{Y}_c$ (dashed portion of regression line), but such extrapolation is only a guess. Given data for Y and X from X_{min} to X_1, we may find a curvilinear relation is more representative—for example, $\bar{Y}_A$ or $\bar{Y}_B$.

EXERCISES

1. Let Y be the dependent variable and X the independent variable.

Case A		Case B		Case C	
Y	X	Y	X	Y	X
3	4	1	1	3	3
5	7	6	3	1	7
1	3	2	5	4	1
3	6	5	7	2	5

For each case:
a. Plot the scatter diagram. What algebraic sign do you expect for a and b?
b. Estimate the regression line.
c. Calculate the standard error of estimate.
d. Calculate the coefficient of determination and interpret your answer.
e. Test the hypothesis that $\beta = 0$. What does this say about $\rho = 0$? (In the remaining parts of this question, use your regression line regardless of these results.)
f. For $X = 4$, calculate the 95 percent prediction interval for the mean of the Y values.
g. For $X = 4$, calculate the 95 percent prediction interval for the mean of the next four values of Y.
h. For $X = 4$, calculate the 95 percent prediction interval for the next value of Y.

i. Compare the "length" of the 95 percent prediction intervals with their respective:
  (1) scatter diagrams
  (2) standard errors of estimate
  (3) r^2
j. Calculate $\bar{X}_c = c + dY$. Do $c = -a/b$ and $d = 1/b$ as they would if you solved $y = a + bx$ for x?

2. Ten applicants for employment were given an aptitude test and hired, and a year later their job performances were rated.

Job Rating Y_i	Aptitude Test Score X_i
8	7
7	6
6	5
6	4
7	5
10	8
9	7
9	8
10	9
8	6

a. Estimate the regression line $\bar{Y} = a + bX$.
b. How do you interpret a? When $X = 0$, $\bar{Y} = a$; but does that indicate that all applicants who receive a 0 on the aptitude test will earn an average job rating of a?
c. How do you interpret b, the slope coefficient?
d. What is the degree of association between these two sets of scores? (Find r^2.)
e. Test the hypothesis that: $H_0: \beta = 0$, $H_A: \beta \neq 0$ at the 5 percent significance level.
f. What does the result from part e say about the test $H_0: \rho = 0$, $H_A: \rho \neq 0$ at the 5 percent significance level?

3. A real estate agent, looking over past records, finds that ranch-style houses in the area come in only four sizes. The selling price for each house (in 1975 dollars) is:

Selling Price (in $1,000) by Aggregate Floor Space			
1,000 Square Feet	1,100 Square Feet	1,200 Square Feet	1,400 Square Feet
26	31	34	37
29	29	33	35
30	28	31	36
27	30	30	38
28	32	32	34

(That is, the first house listed sold for \$26,000.) Let $Y =$ selling price (in thousands of 1975 dollars) and $X =$ square feet of aggregate floor space. Using this data, we could find that $\Sigma X = 23,500$; $\Sigma Y = 630$; $\Sigma XY = 749,000$; $\Sigma Y^2 = 20,060$; and $\Sigma X^2 = 28,050,000$.
a. Find the regression line.

b. Interpret the meaning of a and b.

c. Find the standard error of estimate.

d. Plot the scatter diagram and the regression line on the same graph. Also, plot the two lines representing $\bar{Y}_c \pm s_{Y|X}$.

e. Calculate s_Y. Is $s_{Y|X}$ greater than s_Y? Interpret the relation between s_Y and $s_{Y|X}$.

f. For the five observed values of Y when $X = 1{,}000$, calculate s_Y (or $s_{Y|X=1{,}000}$). Do the same for the other three values of X.

g. From the answers to part f, should we be concerned about the homoscedasticity assumption for this problem?

h. Test the hypothesis that $\beta = 0$ for $\alpha = 0.05$.

i. Find r^2.

4. Consider the data and results from Exercise 3. A developer builds several ranch-style houses, each with an aggregate floor space of 1,300 square feet.

a. Construct the 95 percent prediction interval for the mean selling price (in thousands of 1975 dollars) for these houses.

b. Construct the 95 percent prediction interval for the selling price of the next house the developer sells.

c. Construct the 95 percent prediction interval for the mean selling price of the next four houses the developer sells.

d. How do the relative housing demand and housing supply within this area affect your answers to parts a through c?

e. Another developer comes along and builds several ranch-style houses, each with 1,600 square feet of floor space. What problems or concerns would you have in using the model developed in Exercise 3?

5. Reconsider the data in Exercise 3. Consider each of the four different house sizes to represent a different population. Specifically, let $A = 1{,}000$ square feet, $B = 1{,}100$ square feet, etc.

a. At $\alpha = 0.05$, use the analysis-of-variance techniques from Chapter 8 to test the hypothesis $H_0: \mu_A = \mu_B = \mu_C = \mu_D$.

b. If you accept the null hypothesis in part a, what would that say about the regression coefficient b?

c. If you reject the null hypothesis in part a, what would that say about the regression coefficient b?

6. On the Fruit Flight, from the Peachtree to the Big Apple, there is always the question of how many meals to place on the airplane. A random sample from past records provided data on the number of meals requested, by number of reservations 2 hours prior to takeoff, as given below.

Number of Meals Requested by Number of Reservations

10 Reservations	20 Reservations	40 Reservations	50 Reservations
12	24	48	60
10	23	50	58
14	26	46	62
12	22	48	59
12	25	49	61
	23	47	60
	24	47	59
	25	48	61
		49	

From this data we would find that $N = 30$; $\Sigma Y = 1{,}164$; $\Sigma X = 970$; $\Sigma XY = 45{,}720$; $\Sigma Y^2 = 54{,}908$; and $\Sigma X^2 = 38{,}100$.

a. Find the regression line such that Y = number of meals requested and X = number of reservations 2 hours before takeoff.

b. How do you interpret a and b?

c. Calculate the standard error of estimate.

d. Test the hypothesis that $\beta = 0$ (choose $0.01 < \alpha < 0.10$).

e. Calculate the coefficient of determination and the coefficient of correlation.

f. Is r^2 significantly different from 0?

g. If there are 30 reservations 2 hours prior to the next flight, find the 95 percent prediction interval for the number of meals that will be requested.

h. If there are 60 reservations (forecasted) for the next 10 flights, find the 95 percent extrapolation interval for the mean number of meals per flight requested.

i. The Fruit Flight has a crew of 5 and a capacity of 115 passengers. Compare the strategy "stock the plane with 120 meals" with your answer to part g.

7. If $\bar{Y}_c = 10 + 2X$, $r^2 = 0.90$, and b is significantly different from 0, then comment on each of the following situations:

a. Y_i = bushels per acre of corn for the ith farm
X_i = tons per acre of fertilizer for the ith farm

b. Y_i = weekly sales for the ith grocery store
X_i = lines of advertising space per week in the local newspaper for the ith store

c. Y_i = number of 2-ton trucks sold last month by the ith dealer
X_i = price of a 2-ton truck for the ith dealer

d. Y_i = amount spent on food last month by the ith family
X_i = last month's income for the ith family

8. Comment on the following statements.

a. Correlation and causation are not the same thing.

b. Extrapolation intervals and prediction intervals are not the same thing.

c. Neither a low value for $s_{Y|X}$ nor a high value of r^2 will ensure excellent predictions from a regression model. [Compare a model of the hours per week viewing TV (Y) with family income (X) constructed with 1950 data with a similar model using 1970 data.]

d. At the 10 percent significance level, saying that r^2 is different from 0 is not the same thing as saying b is different from 0.

9. Regression analysis requires several calculations. One common format for reporting regression results (in a company report or a journal article) is:

$$\bar{Y}_c = a + bX \qquad r^2 \qquad s_{Y|X}^2 \qquad n$$

$$(s_a) \ (s_b) \qquad \bar{X} \qquad \text{Range of } X \text{ values}$$

X units specified

Y units specified

For example, a regression analysis of past cost data on production runs for American flag lapel pins might be reported as:

$$\bar{Y}_c = \$200 + (\$.70 \text{ per pin})X \qquad r^2 = 0.9025 \qquad s_{Y|X}^2 = 1.00 \qquad n = 50$$

$$(0.52) \quad (0.009999) \qquad \bar{X} = 6{,}000 \text{ pins per run} \qquad X \text{ range: } 2{,}000 \text{ to } 10{,}000$$

X_i = the number of pins produced by the ith production run

Y_i = total cost, in dollars, of a production run

a. (1) How many production runs were used to construct this model?
    (2) The fewest pins produced in a run were how many?
    (3) The most pins produced in a run were how many?
    (4) What was the total number of pins produced?
b. (1) What is the estimated mean of total variable cost?
    (2) Why might $200 be an unreliable estimate of fixed costs?
c. (1) The standard error of a is?
    (2) The standard error of b is?
d. Provide a 95 percent confidence interval estimate of the $\bar{Y}$ intercept (α).
e. At the 5 percent significance level, test $H_0: \beta = 0$, $H_A: \beta \neq 0$.
f. Use the decision rule from part e and

$$t = r \sqrt{\frac{n-2}{1-r^2}}$$

to test $H_0: \rho = 0$, $H_A: \rho \neq 0$ at the 5 percent significance level. (Note: $\sqrt{0.9025} = 0.95$.)

10. The "reporting of regression results" format presented in Exercise 9 allows testing hypotheses *about* the model, but how are inferences made *with* the model? If you note that

$$s_{Y|X}\sqrt{\frac{1}{n} + \frac{(X_j - \bar{X})^2}{\Sigma X^2 - n\bar{X}^2}} = \sqrt{s_{Y|X}^2\left(\frac{1}{n}\right) + s_b^2(X_j - \bar{X})^2}$$

then everything except X_j, the "new" value of X, is given in the reported regression results. Using the data in Exercise 9 and noting that $s_b^2 = 0.00009998$, find:

a. The 95 percent prediction interval for the mean value of Y for all production runs of $(X_j =) 5,000$ pins by recognizing that Equation 10.12 can be written as

$$(\bar{Y}_c)_j \pm t_{\alpha/2, n-2}\sqrt{s_{Y|X}^2(1/n) + s_b^2(X_j - \bar{X})^2}$$

b. The 95 percent prediction interval for the next value of Y for a production run of 5,000 pins by recognizing that Equation 10.14 can be written as

$$(\bar{Y}_c)_j \pm t_{\alpha/2, n-2}\sqrt{s_{Y|X}^2(1 + 1/n) + s_b^2(X_j - \bar{X})^2}$$

c. The 95 percent prediction interval for the mean value of Y for the next two $(m = 2)$ production runs of 5,000 pins by recognizing that Equation 10.16 can be written as

$$(\bar{Y}_c)_j \pm t_{\alpha/2, n-2}\sqrt{s_{Y|X}^2(1/m + 1/n) + s_b^2(X_j - \bar{X})^2}$$

11. A national fast-food chain has built a regression model using Y = average daily sales for an establishment and X = average number of vehicles passing the establishment daily. The results are:

$$\bar{Y}_c = \$500 + (\$1 \text{ per vehicle})X \qquad r^2 = 0.92 \qquad s_{Y|X}^2 = 4 \qquad n = 100$$
$$(8.3) \quad (0.06245) \qquad\qquad \bar{X} = 1,220 \qquad X \text{ range: } 1,000 \text{ to } 1,500$$

a. For a 1 percent significance level, is b different from 0?
b. If the streets in front of these establishments were impassable for several days, would you expect these establishments to average $500 per day in sales?

c. How do you interpret b?
d. Using the expressions from Exercise 10:
   (1) Provide a 99 percent prediction interval of the mean average daily sales for all establishments with 1,300 vehicles per day passing in front.
   (2) A potential location for a new establishment in their chain has an average daily traffic flow of 1,300 vehicles. What is the 99 percent prediction interval of the average daily sales for this potential site?
   (3) Five potential sites each have an average daily traffic flow of 1,300 vehicles. Construct a 99 percent prediction interval for the mean of their average daily sales.

CASE QUESTIONS

1. Consider the Massachusetts sample of 40 renters. Let $Y = V06$ (i.e., coded values for the number of rooms in the housing unit) and $X = V23$ (i.e., number of bedrooms). The resulting regression results are:

$$Y = 1.268 + 1.027X \qquad r^2 = 0.617$$
$$s_b = 0.13124 \qquad s_{Y|X} = 0.66404$$

a. How would you interpret these results?
b. Note that "number of rooms" $= V06 + 1$ for $V06 < 8$. If you add one to each value of V06, how will that alter the regression results?

2. Consider the Massachusetts sample. Let V06, the number of rooms, be a measure of the amount of housing space purchased by a family. Recall that V32 is family income. Let $Y = V06$ and $X = V32$.
a. Find the regression line $Y = a + bX$.
b. Find r^2.
c. At the 5 percent level of significance, test $H_0: \beta = 0$.
d. Discuss the significance of this statistical model.
e. For all Massachusetts renters with $10,000 income in 1969, construct the 95 percent prediction interval for the average number of rooms they rented.
f. For nine Massachusetts renters, each with $10,000 income, construct the 95 percent prediction interval for the average number of rooms they rented.
g. For a particular renter with $10,000 income, construct the 95 percent prediction interval for the number of rooms that person rented.

11
MULTIVARIATE LINEAR AND NONLINEAR REGRESSION

There are two basic extensions of the simple linear regression model: (1) a linear model with more than one independent variable and (2) a nonlinear regression model with one independent variable. Other combinations are possible and are employed occasionally. The purpose of this chapter is not to develop a multitude of these models in great detail; our purpose is to review the basic concepts used in their construction and to comment on their powers and limitations.

11.1 THE GENERAL MULTIVARIATE LINEAR REGRESSION MODEL

As you might suspect, there are many modeling problems where you reason that the dependent variable Y responds to several independent variables. Let $X_{.j}$ be the jth variable and X_{ij} be the ith observation of the jth variable. For J independent variables, the general linear model is

$$Y_i = \alpha + \beta_1 X_{i1} + \beta_2 X_{i2} + \cdots + \beta_j X_{ij} + \cdots + \beta_J X_{iJ} + \xi_i \qquad (11.1)$$

where Y_i = the ith observation of the dependent variable
$\quad \alpha = Y$ intercept
$\quad \beta_j$ = parameter associated with $X_{.j}$, partial regression coefficient
$\quad \xi_i$ = error term, where the ξ_i are independent and normally distributed with $E(\xi) = 0$ and $\text{Var}(\xi) = \sigma^2$

The expected value of Y yields

$$E(Y) = \mu_{Y|X_1 X_2 \cdots X_J} = \alpha + \beta_1 X_{.1} + \beta_2 X_{.2} + \cdots + \beta_J X_{.J} \qquad (11.2)$$

In microeconomic theory, we frequently argue that the demand for a product is a function of its price and of consumer income. In terms of our model, we might say

$$Y = \alpha + \beta_1 X_1 + \beta_2 X_2 \qquad (11.3)$$

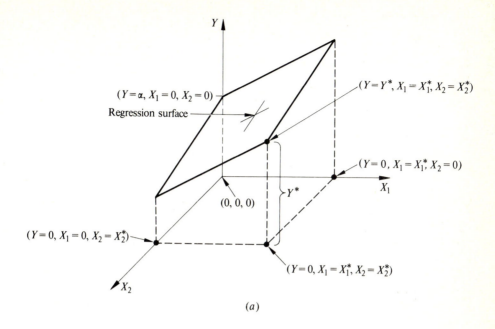

$(Y = \alpha, X_1 = 0, X_2 = 0)$

Regression surface

$(Y = Y^*, X_1 = X_1^*, X_2 = X_2^*)$

$(Y = 0, X_1 = X_1^*, X_2 = 0)$

X_1

$(0, 0, 0)$

Y^*

$(Y = 0, X_1 = 0, X_2 = X_2^*)$

$(Y = 0, X_1 = X_1^*, X_2 = X_2^*)$

X_2

(a)

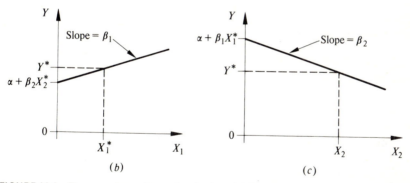

Y

Slope = β_1

Y^*

$\alpha + \beta_2 X_2^*$

0

X_1^*

X_1

(b)

Y

$\alpha + \beta_1 X_1^*$

Y^*

Slope = β_2

0

X_2

X_2

(c)

FIGURE 11.1   The regression surface and projections. (a) Three-dimensional view; (b) Projection on YX_1 plane when $X_2 = X_2^*$; (c) Projection on YX_2 plane when $X_1 = X_1^*$.

where X_1 = consumer income and X_2 = the price. As price increases, we expect the quantity demanded to decrease; thus, β_2 would be negative. For a normal good, as income increases, we expect the quantity demanded to increase; thus, β_1 would be positive.

Instead of a straight line, Equation 11.3 maps out a plane (or flat surface), as shown in Figure 11.1a. In Figure 11.1b, when X_2 is held constant at the value X_2^*, the sliced plane is projected onto the YX_1 plane. Since we assumed $\beta_1 > 0$, the projected line exhibits a positive slope. With X_1 held constant at the value of X_1^*, Figure 11.1c shows the sliced regression plane projected onto the YX_2 plane; the effect of β_2 being negative is demonstrated. As drawn in Figure 11.1a, the regression surface slants down toward you while slanting up to the right.

The Three-Variable Linear Regression Model

The three-variable linear regression model is a special case of the general model, Equations 11.1 and 11.2. A model of the form

$$\mu_{Y|X_1X_2} = \alpha + \beta_1 X_{.1} + \beta_2 X_{.2} \tag{11.4}$$

can be graphically represented (e.g., Figure 11.1), and it is sufficiently complex to demonstrate the power and limitations of the general linear model.

As with the simple linear model, we are searching for an estimate of the parameters in Equation 11.4:

$$\bar{Y}_c = a + bX_{.1} + cX_{.2} \tag{11.5}$$

Since we are dealing with only two independent variables, we can reduce confusion in the notation by letting $X_{.1} = X$ and $X_{.2} = W$. Then Equation 11.5 becomes

$$\bar{Y}_c = a + bX + cW \tag{11.6}$$

With the technique of least squares, we want the values of a, b, and c to be such that

$$G = \Sigma(Y - \bar{Y}_c)^2 = \Sigma(Y - a - bX - cW)^2$$

is a minimum. The resulting *normal equations*[1] are

$$an + b\Sigma X + c\Sigma W = \Sigma Y \tag{11.7a}$$

$$a\Sigma X + b\Sigma X^2 + c\Sigma(XW) = \Sigma(YX) \tag{11.7b}$$

$$a\Sigma W + b\Sigma(XW) + c\Sigma W^2 = \Sigma(YW) \tag{11.7c}$$

---

[1] First, the adjective "normal" has nothing to do with Gauss' or the normal distribution. Second, the normal equations are found by taking the first partial derivatives of G with respect to a, b, and c and setting those partials equal to 0:

$$\partial G/\partial a = -2\Sigma(Y - a - bX - cW) = 0$$

$$\partial G/\partial b = -2\Sigma[(Y - a - bX - cW)(X)] = 0$$

$$\partial G/\partial c = -2\Sigma[(Y - a - bX - cW)(W)] = 0$$

Finally, we can solve a system of linear equations (for a, b, and c) with matrix algebra and specifically with Cramer's rule. If A is nonsingular, then the solution to $Ax = d$ is $x = A^{-1}d$. In this case,

$$A = \begin{bmatrix} n & \Sigma X & \Sigma W \\ \Sigma X & \Sigma X^2 & \Sigma XW \\ \Sigma W & \Sigma XW & \Sigma W^2 \end{bmatrix} \qquad x = \begin{bmatrix} a \\ b \\ c \end{bmatrix} \qquad \text{and} \qquad d = \begin{bmatrix} \Sigma Y \\ \Sigma YX \\ \Sigma YW \end{bmatrix}$$

Solving the set of simultaneous equations given by Equations 11.7 for $a, b,$ and c yields[2]

$$a = \bar{a}/\Delta \qquad \text{(11.8a)}$$

$$b = \bar{b}/\Delta \qquad \text{(11.8b)}$$

$$c = \bar{c}/\Delta \qquad \text{(11.8c)}$$

where

$$\Delta = n[\Sigma X^2 \Sigma W^2 - (\Sigma X W)^2] - \Sigma X (\Sigma X \Sigma W^2 - \Sigma W \Sigma X W)$$
$$+ \Sigma W (\Sigma X \Sigma X W - \Sigma W \Sigma X^2) \qquad \text{(11.9a)}$$

$$\bar{a} = \Sigma Y[\Sigma X^2 \Sigma W^2 - (\Sigma X W)^2] - \Sigma X (\Sigma Y X \Sigma W^2 - \Sigma Y W \Sigma X W)$$
$$+ \Sigma W (\Sigma Y X \Sigma X W - \Sigma Y W \Sigma X^2) \qquad \text{(11.9b)}$$

$$\bar{b} = n(\Sigma Y X \Sigma W^2 - \Sigma Y W \Sigma X W) - \Sigma Y (\Sigma X \Sigma W^2 - \Sigma W \Sigma X W)$$
$$+ \Sigma W (\Sigma X \Sigma YW - \Sigma W \Sigma YX) \qquad \text{(11.9c)}$$

$$\bar{c} = n(\Sigma X^2 \Sigma YW - \Sigma X W \Sigma YX) - \Sigma X (\Sigma X \Sigma YW - \Sigma W \Sigma YX)$$
$$+ \Sigma Y (\Sigma X \Sigma X W - \Sigma W \Sigma X^2) \qquad \text{(11.9d)}$$

Of course, if $\Delta = 0$, then we have no unique solution—an unusual case which we shall return to later. The complicated appearance of Equations 11.9a through d has some redeeming features. Any "scare factor" associated with reading each equation is primarily a function of the equation's length. Each term (for example, ΣX^2, $\Sigma X W$, ΣW^2, etc.) is an easily identified arithmetic operation on the sample data; the "complicated" appearance arises because there are several algebraic manipulations to perform with these terms.

A "multitude of simple algebraic manipulations" suggests the use of a computer. Most multivariate regression models are constructed by substituting capital (i.e., the computer) for labor. Of course, there is some benefit in manually calculating Equations 11.9a through d for a few examples; however, the marginal benefit of manual calculation decreases rapidly as the number of examples increases.

Although Equations 11.9a through d may not enhance our interpretation of a, b, and c, they certainly should not detract from our understanding of those estimates. In Chapter 10, the least-squares estimates for a and b provided a *line* that minimized the variation of the data points from that *line*. That same best-fit concept is inherent in Equations 11.8 and 11.9. The least-squares estimate of a, b, and c provides a *plane* or "regression surface," given by $\bar{Y}_c = a + bX + cW$, which minimizes the variation between the data and that plane. If Equations 11.8 and 11.9 appear more complicated, that is directly related to the fact that our model, Equation 11.6, is more complicated.

---

[2] A useful check is $a = \bar{Y} - b\bar{X} - c\bar{W}$ from Equation 11.7a.

The *sample standard error of the estimate* is defined as

$$s_{Y|XW} = \sqrt{\frac{\Sigma(Y - \bar{Y}_c)^2}{n - 3}}$$

(11.10)

The concept is, of course, the same as that represented by Equation 10.9 for the regression line. The subscript $Y|XW$ simply denotes the fact that we are measuring the variation in Y values *given* a particular value of X and a particular value of W. The value of $s_{Y|XW}$ is assumed to be constant regardless of which known value of X (or W) is used: the homoscedasticity assumption. By estimating α, β_1, and β_2 with a, b, and c, the number of degrees of freedom has been reduced to $(n - 3)$—hence, the denominator of Equation 11.10.

A pragmatic equivalent to the definition given in Equation 11.10 is

$$s_{Y|XW} = \sqrt{\frac{\Sigma Y^2 - a\Sigma Y - b\Sigma XY - c\Sigma WY}{n - 3}}$$

(11.11)

Example 11.1 The Relationship between Beef Consumption, the Price of Beef, and the Price of Pork

In this hypothetical example, let Y = annual beef consumption in billions of pounds, X = a retail price index for beef, and W = a retail price index for pork. (For a rather dated example using real-world data, and including income as an independent variable, see H. Schultz, *The Theory and Measurement of Demand*, University of Chicago Press, Chicago, Ill., 1938, pp. 582ff.) From economic theory, we might expect the regression model, $\bar{Y}_c = a + bX + cW$, to exhibit the following characteristics: (1) b should be negative (i.e., the response to an increase in X, the price of beef, should be a decrease in Y, beef consumption, if everything else is held constant) and (2) c should be positive (i.e., beef and pork are generally considered to be "substitutes"). Consider the data for the last 5 years as shown in Table 11.1.

TABLE 11.1   DATA AND PRELIMINARY CALCULATIONS

Data			Preliminary Calculations					
Beef Consumption Y	Beef Price X	Pork Price W	YX	YW	XW	Y^2	X^2	W^2
6	1	1	6	6	1	36	1	1
5	1	2	5	10	2	25	1	4
4	2	1	8	4	2	16	4	1
3	2	2	6	6	4	9	4	4
7	4	9	28	63	36	49	16	81
25	10	15	53	89	45	135	26	91

Also, $n = 5$, $\bar{Y} = {}^{25}\!/_5 = 5$, $\bar{X} = {}^{10}\!/_5 = 2$, and $\bar{W} = {}^{15}\!/_5 = 3$.

In order to estimate a, b, and c, let us start with Equation 11.9a:

$$\Delta = n[\Sigma X^2 \Sigma W^2 - (\Sigma X W)^2] - \Sigma X(\Sigma X \Sigma W^2 - \Sigma W \Sigma X W) + \Sigma W(\Sigma X \Sigma X W - \Sigma W \Sigma X^2)$$

$$= (5)[(26)(91) - (45)^2] - (10)[(10)(91) - (15)(45)] + (15)[(10)(45) - (15)(26)]$$

$$= (5)(341) - (10)(235) + (15)(60) = +255$$

From Equation 11.9b:

$$\bar{a} = \Sigma Y[\Sigma X^2 \Sigma W^2 - (\Sigma X W)^2] - \Sigma X(\Sigma Y X \Sigma W^2 - \Sigma Y W \Sigma X W) + \Sigma W(\Sigma Y X \Sigma X W - \Sigma Y W \Sigma X^2)$$

$$= (25)[(26)(91) - (45)^2] - (10)[(53)(91) - (89)(45)] + (15)[(53)(45) - (89)(26)]$$

$$= (25)(341) - (10)(818) + (15)(71) = +1{,}410$$

From Equation 11.9c:

$$\bar{b} = n(\Sigma Y X \Sigma W^2 - \Sigma Y W \Sigma X W) - \Sigma Y(\Sigma X \Sigma W^2 - \Sigma W \Sigma X W) + \Sigma W(\Sigma X \Sigma Y W - \Sigma W \Sigma Y X)$$

$$= (5)[(53)(91) - (89)(45)] - (25)[(10)(91) - (15)(45)] + (15)[(10)(89) - (15)(53)]$$

$$= (5)(818) - (25)(235) + (15)(95) = -360$$

From Equation 11.9d:

$$\bar{c} = n(\Sigma X^2 \Sigma Y W - \Sigma X W \Sigma Y X) - \Sigma X(\Sigma X \Sigma Y W - \Sigma W \Sigma Y X) + \Sigma Y(\Sigma X \Sigma X W - \Sigma W \Sigma X^2)$$

$$= (5)[(26)(89) - (45)(53)] - (10)[(10)(89) - (15)(53)] + (25)[(10)(45) - (15)(26)]$$

$$= (5)(-71) - (10)(95) + (25)(60) = +195$$

Next, from Equations 11.8 a through c:

$$a = \bar{a}/\Delta = 1{,}410/255 = +5.5294$$

$$b = \bar{b}/\Delta = -360/255 = -1.4118$$

$$c = \bar{c}/\Delta = 195/255 = +0.7647$$

As a check,

$$a = \bar{Y} - b\bar{X} - c\bar{W} = 5 - (-1.4118)(2) - (0.7647)(3)$$

$$= 5.5295 \qquad \text{(difference due to rounding error)}$$

Thus, the estimated regression surface is

$$\bar{Y}_c = 5.5294 - (1.4118)X + (0.7647)W$$

From Equation 11.11, the sample standard error of estimate is

$$s_{Y|XW} = \sqrt{\frac{\Sigma Y^2 - a\Sigma Y - b\Sigma XY - c\Sigma WY}{n-3}}$$

$$= \sqrt{\frac{135 - (5.5294)(25) - (-1.4118)(53) - (0.7647)(89)}{5-3}}$$

$$= \sqrt{3.5321/2} = \sqrt{1.766}$$

$$= 1.33 \text{ billion pounds of beef}$$

For reasons which will be explained in Chapter 13, this regression model is called the "regression relationship" between beef consumption, beef price, and pork price. It is *not* called a "demand relationship." Using real-world data, it is entirely possible to find $b > 0$ and $c < 0$.

The Linear Regression Model for $(J + 1)$ Variables[3]

Consider J independent variables $(X_1, X_2, \ldots, X_J)$ and the dependent variable Y. For the model given by Equation 11.2, using a as an estimate of α and b_j as an estimate of β_j, the normal equations have the following form:

$$na \quad + b_1\Sigma X_1 \quad + b_2\Sigma X_2 \quad + b_3\Sigma X_3 \quad + \cdots + b_J\Sigma X_J \quad = \Sigma Y$$

$$a\Sigma X_1 + b_1\Sigma X_1{}^2 \quad + b_2\Sigma X_1 X_2 + b_3\Sigma X_1 X_3 + \cdots + b_J\Sigma X_1 X_J = \Sigma Y X_1$$

$$a\Sigma X_2 + b_1\Sigma X_1 X_2 + b_2\Sigma X_2{}^2 \quad + b_3\Sigma X_2 X_3 + \cdots + b_J\Sigma X_2 X_J = \Sigma Y X_2$$

$$a\Sigma X_3 + b_1\Sigma X_1 X_3 + b_2\Sigma X_2 X_3 + b_3\Sigma X_3{}^2 \quad + \cdots + b_J\Sigma X_3 X_J = \Sigma Y X_3$$

$$\cdots\cdots\cdots\cdots\cdots\cdots\cdots\cdots\cdots\cdots\cdots\cdots\cdots\cdots\cdots\cdots\cdots\cdots$$

$$a\Sigma X_J + b_1\Sigma X_1 X_J + b_2\Sigma X_2 X_J + b_3\Sigma X_3 X_J + \cdots + b_J\Sigma X_J{}^2 \quad = \Sigma Y X_J$$

There are $(J + 1)$ unknowns (that is, a and the J b's), and there are $(J + 1)$ equations. With the use of matrix algebra, this set of equations could be written as

$$AZ = d$$

---

[3] Can be omitted without loss of continuity.

where

$$A = \begin{bmatrix} n & \Sigma X_1 & \Sigma X_2 & \cdots & \Sigma X_J \\ \Sigma X_1 & \Sigma X_1{}^2 & \Sigma X_1 X_2 & \cdots & \Sigma X_1 X_J \\ \Sigma X_2 & \Sigma X_1 X_2 & \Sigma X_2{}^2 & \cdots & \Sigma X_2 X_J \\ \cdots\cdots\cdots\cdots\cdots\cdots\cdots\cdots\cdots\cdots\cdots\cdots \\ \Sigma X_J & \Sigma X_1 X_J & \Sigma X_2 X_J & \cdots & \Sigma X_J{}^2 \end{bmatrix}$$

$$Z = \begin{bmatrix} a \\ b_1 \\ b_2 \\ \vdots \\ b_J \end{bmatrix} \quad \text{and} \quad d = \begin{bmatrix} \Sigma Y \\ \Sigma Y X_1 \\ \Sigma Y X_2 \\ \vdots \\ \Sigma Y X_J \end{bmatrix}$$

The square matrix A is $(J + 1) \times (J + 1)$. If $|A| \neq 0$, then we can solve for the least-squares estimates by

$$Z = A^{-1} d$$

The sample standard error of estimate is

$$s_{Y|X_1 X_2 \cdots X_J} = \sqrt{\frac{\Sigma(Y - \bar{Y}_c)^2}{n - (J + 1)}}$$

We will leave the exact nature of the general linear regression model for more advanced texts. Even without any knowledge of matrix algebra, the general nature of the normal equations, the problem of solving a set of simultaneous linear equations, and the definition of the standard error of estimate should be remotely familiar. From the discussion of the three-variable model and this outline, you should be able to deduce the general nature of the "beast," even if you do not know its eating habits, its migratory pattern, or its mating behavior.

11.2 MULTIPLE CORRELATION

The *degree* of association between Y and the J independent variables is but one correlation concept when we consider the general linear model. We might be concerned with the degree of association between Y and one particular independent variable, X_j. But, what do we do with the other $(J - 1)$ independent variables? The *ceteris paribus* assumption, often used in economics, offers one solution: we simply hold the "other" $(J - 1)$ independent variables *constant* and look at the degree of association between Y and X_j.

Both correlation concepts will be discussed in the context of the previous section; the three-variable model will be treated in some detail while the general model will receive only a brief survey.

The Three-Variable Model

In the context of section 10.9, we will define the *multiple coefficient of determination,* $R^2_{Y|XW}$, by using the concepts "total variation," "explained variation," and "unexplained variation." Given specific values of X and W, say X_i and W_i, the regression equation yields

$$(\bar{Y}_c)_i = a + bX_i + cW_i$$

The total deviation is $Y_i - \bar{Y}$, the difference between an observed value of Y and the mean of all the observed values. The unexplained deviation is $Y_i - (\bar{Y}_c)_i$. As in Chapter 10, the proportion of explained variation is 1 minus the proportion of unexplained variation, or

$$R^2_{Y|XW} = 1 - \frac{\Sigma[Y_i - (\bar{Y}_c)_i]^2}{\Sigma(Y_i - \bar{Y})^2} \tag{11.12}$$

The sample multiple coefficient of determination can also be calculated from

$$R^2_{Y|XW} = 1 - \frac{s^2_{Y|XW}(n-3)}{s_Y{}^2(n-1)} \tag{11.13a}$$

or

$$R^2_{Y|XW} = \frac{b(\Sigma XY - n \cdot \bar{X} \cdot \bar{Y}) + c(\Sigma WY - n \cdot \bar{W} \cdot \bar{Y})}{\Sigma Y^2 - n\bar{Y}^2} \tag{11.13b}$$

For example, with the data from example 11.1, Equation 11.13b yields

$$R^2_{Y|XW} = \frac{(-1.4118)[53 - (5)(2)(5)] + (0.7647)[89 - (5)(3)(5)]}{(135) - (5)(5)^2}$$

$$= \frac{-4.2354 + 10.7058}{10} = 0.6470$$

That is, 64.7 percent of the variation of Y about its mean can be described by using the regression surface $\bar{Y}_c = 5.5294 - 1.4118X + 0.7647W$. The 35.3 percent of the total variation in beef consumption, unexplained by prices of beef and pork, could be attributed to such factors as fowl or fish prices and consumer income. Or, as we shall see in section 11.5, perhaps a logically derived nonlinear model would be more descriptive.

In Equations 11.12 and 11.13 the uppercase R^2 symbolized the multiple coefficient of determination; the lowercase r is reserved for the degree of association between just *two* variables.

Consider the degree of association between only beef consumption and the price of beef. With Equation 10.17 we use the subscript YX to denote which two variables are being measured, and we find that

$$1 - r_{YX}^2 = \text{proportion of variation in } Y \text{ } unexplained \text{ by } X$$

Presumably, introducing the independent variable W (e.g., pork price) will reduce the *unexplained* variation in Y. That is,

$$1 - r_{YX}^2 > 1 - R_{Y|XW}^2$$

Let us use that inequality as a ratio:

$$\frac{1 - R_{Y|XW}^2}{1 - r_{YX}^2} = \frac{\text{proportion of unexplained } Y \text{ variation using both } X \text{ and } W}{\text{proportion of unexplained } Y \text{ variation using only } X}$$

This ratio is the proportion of unexplained Y variation using X that remains after we also consider W. Or,

$$r_{YW|X}^2 = 1 - \frac{1 - R_{Y|XW}^2}{1 - r_{YX}^2} \qquad (11.14)$$

is the proportion of Y variation left unexplained by X that can be described by W. The *partial correlation coefficient*, $r_{YW|X}$, is a measure of the association between Y and W when we have already considered the influence of X.

In like manner, the partial correlation coefficient for Y and X, given W, is

$$r_{YX|W} = \sqrt{1 - \frac{1 - R_{Y|XW}^2}{1 - r_{YW}^2}} \qquad (11.15)$$

For example, from Equation 10.19, the coefficient of determination for Y and (only) X is

$$r_{YX}^2 = \frac{(\Sigma XY - n\bar{X}\bar{Y})^2}{(\Sigma X^2 - n\bar{X}^2)(\Sigma Y^2 - n\bar{Y}^2)}$$

Using the data from example 11.1,

$$r_{YX}{}^2 = \frac{[53 - (5)(2)(5)]^2}{[26 - (5)(2)^2][135 - (5)(5)^2]} = \frac{9}{(6)(10)} = 0.1500$$

and

$$r_{YW}{}^2 = \frac{(\Sigma WY - n\bar{W}\bar{Y})^2}{(\Sigma W^2 - n\bar{W}^2)(\Sigma Y^2 - n\bar{Y}^2)} = \frac{[89 - (5)(3)(5)]^2}{[91 - (5)(3)^2][135 - (5)(5)^2]}$$

$$= \frac{196}{(46)(10)} = 0.4261$$

From Equation 11.14,

$$r^2_{YX|W} = 1 - \frac{1 - 0.6470}{1 - 0.1500} = 1 - 0.4153 = 0.5847$$

By just considering X, 85 percent of the Y variation was unexplained. Of that 85 percent of unexplained Y variation, W accounts for 58.47 percent.

Now, from Equation 11.15,

$$r^2_{YX|W} = 1 - \frac{1 - 0.6470}{1 - 0.4261} = 1 - 0.6151 = 0.3849$$

The regression line for Y given W would leave 57.39 percent of the Y variation unexplained; of that 57.39 percent of unexplained Y variation, X will describe 38.49 percent.

Thus, in terms of explaining variation in Y, W appears relatively more important than X.

Correlation Measures for $(J + 1)$ Variables

Once again consider J independent variables, denoted $X_1, X_2, \ldots, X_J$, and the dependent variable Y. The least-squares regression model would be

$$\bar{Y}_c = a + b_1 X_1 + b_2 X_2 + b_3 X_3 + \cdots + b_j X_j + \cdots + b_J X_J$$

For a given value of *each* independent variable, a particular value of $\bar{Y}_c$ can be calculated from the regression model and compared with the observed value Y.

The general expression for the *multiple coefficient of determination* is

$$R^2_{Y|X_1 Y_2 \cdots X_J} = 1 - \frac{\Sigma(Y - \bar{Y}_c)^2}{\Sigma(Y - \bar{Y})^2} \tag{11.16}$$

where $R^2_{Y|X_1 X_2 \cdots X_J}$ is more frequently expressed as $R^2_{Y|12 \cdots J}$. An alternative form of calculation is given by

$$
R^2_{Y|123 \cdots J} = \frac{\sum\limits_{j=1}^{J} \left[b_j \left(\sum\limits_{i=1}^{n} X_{ij} Y_i - n\bar{X}_j \bar{Y} \right) \right]}{\sum\limits_{i=1}^{n} Y_i^2 - n\bar{Y}^2} \tag{11.17}
$$

$$
= \frac{b_1(\Sigma X_1 Y - n\bar{X}_1 \bar{Y}) + b_2(\Sigma X_2 Y - n\bar{X}_2 \bar{Y}) + \cdots + b_J(\Sigma X_J Y - n\bar{X}_J \bar{Y})}{\Sigma Y^2 - n\bar{Y}^2}
$$

By letting $X_1 = X$ and $X_2 = W$ and $J = 2$, the result of Equation 11.17 is given by Equation 11.13*b*.

The general expression for the *partial coefficient of determination* is

$$
r^2_{YX_k|\text{all } X_j \text{ except } X_k} = 1 - \frac{1 - R^2_{Y|\text{all } X_j}}{1 - R^2_{Y|\text{all } X_j \text{ except } X_k}} \tag{11.18}
$$

where $R^2_{Y|\text{all } X_j}$ is given by Equation 11.17, and $R^2_{Y|\text{all } X_j \text{ except } X_k}$ can be calculated from Equation 11.17 *only after a new regression model that excludes X_k but includes all other X_j is constructed.* The abundance of calculation required dictates the use of the computer.

Why would anyone want to do (or have performed) all those calculations just to get partial coefficients of determination? To pay such a price, there must be considerable value in the information obtained. The value originates in the interpretation of the partial coefficient of determination. The partial coefficient of determination is a *measure of the net benefit* of including X_k as an independent variable in the regression model. After all the other variables are considered, the partial coefficient is a relative measure of how well X_j describes otherwise unexplained variation in Y.

The contribution or explanatory quality of each independent variable is somewhat camouflaged by the other independent variables included in a particular general linear regression model. The partial coefficient of determination attempts to measure the relative importance of just one independent variable. Thus, if $r^2_{Y1|\text{all other } X_j}$ is equal to 0.90, we probably want to keep X_1 in the model; however, if $r^2_{Y2|\text{all other } X_j}$ is 0.10, then the contributions of X_2 is dubious.

Technical Note: Beta Coefficients

The novice who walks into the local computer center and submits a set of data for multiple regression analysis is likely to receive a printout with, among other things, a set of numbers under the heading of "beta coefficients." The "beta coefficient" is *not* the same thing as the use of the Greek letter β in the *population* regression model used in both this chapter and Chapter 10.

The beta coefficient is another measure of the relative importance of each independent variable. Although it is mathematically different from the partial coefficient of determination, the beta coefficient can be interpreted in essentially the same way.

Start with the regression line $\bar{Y}_c = a + bX + cW$ and recognize that Y, X, and W could all be measured in different units (for example, Y in billions of pounds of beef, X in cents per pound of beef, and W in dollars per pound of pork). Also, a will have the same units as Y; b will have units of Y per unit of X; and c will be in units of Y per unit of W. The next step is to *standardize* the regression equation; that is, we want to make all terms "unitless." The regression equation is standardized by using the standard deviation of each variable:

$$\frac{\bar{Y}_c}{s_Y} = \left[\frac{a}{s_Y}\right] + \left(b\frac{s_X}{s_Y}\right)\frac{X}{s_X} + \left(c\frac{s_W}{s_Y}\right)\frac{W}{s_W} \qquad (11.19)$$

Since a is in Y units, a/s_Y is unitless; since b is in Y units per unit of X, then bs_X/s_Y is unitless; and so forth.

The expressions in parentheses in Equation 11.19 are called "beta coefficients." We could rewrite the standardized equations as

$$\bar{Y}_c/s_Y = \beta_0^* + \beta_1^*(X/s_X) + \beta_2^*(W/s_W) \qquad (11.20)$$

where $\beta_0^* = a/s_Y$, $\beta_1^* = bs_X/s_Y$, and $\beta_2^* = cs_W/s_Y$.

Using the data in example 11.1, the three standard deviations are $s_Y = \sqrt{10/4} = 1.5811$, $s_X = \sqrt{6/4} = 1.2247$, and $s_W = \sqrt{46/4} = 3.3912$. Thus, the beta coefficients are

$$\beta_0^* = a/s_Y = 5.5294/1.5811 = 3.50$$
$$\beta_1^* = bs_X/s_Y = (-1.4118)(1.2247)/1.5811 = -1.09$$
$$\beta_2^* = cs_W/s_Y = (0.7647)(3.3912)/1.5811 = +1.64$$

Substituting the appropriate values in Equation 11.19 yields

$$(\bar{Y}_c/1.58) = 3.50 - 1.09(X/1.22) + 1.64(W/3.39)$$

This equation can be interpreted as follows:

1. Ignore $\beta_0^* = 3.50$.
2. For β_1^*, a "one standard deviation change in X" is associated with "a 1.09 standard deviation change in Y." (Ignore the algebraic sign.)
3. For β_2^*, a "one standard deviation change in W" is associated with "a 1.64 standard deviation change in Y."

4. Hence, variation in W (price of pork) is relatively more important in describing variation in Y (beef consumption) than variation in X (price of beef).

Let us compare the beta coefficients with the partial coefficients of determination:

Independent Variable	Beta Coefficient	Partial Coefficient of Determination
X	$(-)1.09$	0.3849
W	1.64	0.5847

Both measures provide the same rank-ordering; both measures have a higher value for W. Both indicate that W is relatively more important in describing the variation in Y. Although this rank-ordering may be typical, the beta coefficients and the partial coefficients of determination do not always provide the same rank-ordering. While the beta coefficients are easier to calculate, they are not as easy to interpret (i.e., in the context of the marginal contribution of each independent variable).

11.3 PREDICTION INTERVALS

For the simple linear model $\bar{Y}_c = a + bX$, we investigated the interval prediction for (1) the mean value of all possible Y values given a particular value of X, (2) the next value of Y given an X value, and (3) given an X value, the mean of the next m values of Y. With the three-variable regression model $\bar{Y}_c = a + bX + cW$, we will investigate the same three prediction intervals. The inclusion of one more independent variable (W) causes the calculations to look more involved; however, the basic strategy is the same as for the simple linear regression model. The prediction intervals for the general linear regression model for J independent variables can be conveniently expressed only with the use of matrix algebra; we reserve that discussion for more advanced texts.

Assume that a three-variable regression plane $\bar{Y}_c = a + bX + cW$ and the standard error of estimate $s_{Y|XW}$ have been determined. Consider a " new " value of X, say X_0, where X_0 is in the range of X values employed in the construction of the regression plane; also, consider a "new" value of W, say W_0, which is within the range of W values used to construct the regression plane. (If *either* X_0 or W_0 is outside that range of original values, then we have *extrapolation intervals* rather than prediction intervals. In section 11.5, we further restrict the range of values for X_0 and W_0 to avoid " hidden extrapolation.")

The Prediction Interval for $\mu_{Y|XW}$ Given "New" Values X_0 and W_0

For X_0 and W_0, assume that we are interested in predicting the "expected value of Y," or "the mean of all possible Y values." The point estimate is

$$\bar{Y}_0 = a + bX_0 + cW_0$$

The $(1 - \alpha)$ confidence interval is given by

$$\bar{Y}_0 \pm t_{\alpha/2,\, n-3} s_{Y|XW} \sqrt{H} \tag{11.21}$$

where H uses the expression Δ, as defined in Equation 11.9a, and is given by

$$
\begin{aligned}
H = (1/\Delta)\{ & [\Sigma X^2 \Sigma W^2 - (\Sigma X W)^2] + X_0{}^2[n\Sigma W^2 - (\Sigma W)^2] \\
& + 2X_0(\Sigma W \Sigma X W - \Sigma X \Sigma W^2) + 2X_0 W_0(\Sigma X \Sigma W - n\Sigma X W) \\
& + 2W_0(\Sigma X \Sigma X W - \Sigma W \Sigma X^2) + W_0{}^2[n\Sigma X^2 - (\Sigma X)^2] \}
\end{aligned}
\tag{11.22}
$$

For a general linear regression model, Equation 11.21 is valid; however, H becomes a much more lengthy expression.

The Prediction Interval for Y_{next} Given "New" Values of X_0 and W_0

The point prediction of the next value of Y, given X_0 and W_0, is still $\bar{Y}_0 = a + bX_0 + cW_0$. The $(1 - \alpha)$ prediction interval for the next value of Y is

$$\bar{Y}_0 \pm t_{\alpha/2,\, n-3} s_{Y|XW} \sqrt{1 + H} \tag{11.23}$$

where all the terms are defined as above.

The Prediction Interval for Y_m Given "New" Values of X_0 and W_0

For the "new" values X_0 and W_0, we want the mean of the next m values of Y. The point prediction is as given above, and the $(1 - \alpha)$ prediction interval is given by

$$\bar{Y}_0 \pm t_{\alpha/2,\, n-3} s_{Y|XW} \sqrt{1/m + H} \tag{11.24}$$

Example 11.2 Prediction Intervals from Example 11.1

From example 11.1, we have

$$\bar{Y}_c = 5.5294 - (1.4118)X + (0.7647)W$$

$$s_{Y|XW} = 1.33 \qquad \Delta = 255$$

and the data from Table 11.1 where $Y =$ beef consumption in billions of pounds, $X =$ price of beef, and $W =$ price of pork.

(a) Assume that, for next year, the composite price of beef is known to be $X_0 = 2$ and the composite price of pork is known to be $W_0 = 2$. The point estimate of the mean of all possible beef consumption values, $\bar{Y}$, is given by

$$\bar{Y}_0 = 5.5294 - (1.4118)(2) + (0.7647)(2) = 4.2352$$

Now consider the 95 percent confidence interval for $\bar{Y}$. The appropriate t value for $n - 3 = 5 - 3 = 2$ degrees of freedom is 4.303. In order to employ Equation 11.21, the value of H must be determined from Equation 11.22. The preliminary calculations from Table 11.1, $\Delta = 255$, $X_0 = 2$, and $W_0 = 2$, are all that is required to calculate H:

$$
\begin{aligned}
H &= (1/255)\{[(26)(91) - (45)^2] + (2)^2[(5)(91) - (15)^2] \\
&\quad + 2(2)[(15)(45) - (10)(91)] + 2(2)(2)[(10)(15) - (5)(45)] \\
&\quad + 2(2)[(10)(45) - (15)(26)] + (2)^2[(5)(26) - (10)^2]\} \\
&= (1/255)(341 + 920 - 940 - 600 + 240 + 120) \\
&= 81/255 = 0.3176
\end{aligned}
$$

From Equation 11.21, the 95 percent prediction interval for the average beef consumption in all years where $W_0 = 2 = X_0$ is

$$4.2352 \pm (4.303)(1.33)\sqrt{0.3176}$$

$$4.235 \pm 3.205$$

$$1.03 \leq \bar{Y} \leq 7.44 \text{ billion pounds of beef}$$

(b) Assume that, for next year, $X_0 = 2$ and $W_0 = 2$ and we want a 95 percent confidence interval for next year's beef consumption, that is, a particular value of Y called Y_{next}. From the above calculations, Equation 11.23 yields

$$4.235 \pm (4.303)(1.33)\sqrt{1 + 0.3176}$$

$$4.235 \pm 6.570$$

$$-2.335 \leq Y_{next} \leq 10.805$$

which does not make economic sense. Thus, we might reply

$$0 \leq Y_{next} \leq 10.805 \text{ billion pounds}$$

This interval is too large to be meaningful, but remember there were only $(n =)$ five observations. A larger sample size would at least reduce $t_{0.95, n-3}$; hopefully, the larger sample would otherwise reduce the prediction interval.

(c) Assume that for the next 3 years $X_0 = 2$ and $W_0 = 2$ and that we want the average annual beef consumption for the next 3 years. From Equation 11.24, the 95 percent prediction interval for $\bar{Y}_{m=3}$ is

$$4.235 \pm (4.303)(1.33)\sqrt{0.3333 + 0.3176}$$

$$4.235 \pm 4.617$$

$$-0.38 \leq \bar{Y}_m \leq 8.85$$

Of course, this interval exhibits the same problems as part b.

11.4 TESTS ABOUT THE THREE-VARIABLE LINEAR MODEL

In the three-variable regression plane $\bar{Y}_c = a + bX + cW$, we have three *regression coefficients*: a, b, and c. These three regression coefficients are estimates of their respective population parameters in the model $\mu_{Y|XW} = \alpha + \beta_1 X + \beta_2 W$. We wish to establish the tests of hypothesis for each.

Two-Tail Test of Hypothesis for $\alpha = \alpha_0$

Although it is a rarity, we may want to test

$$H_0 : \alpha = \alpha_0$$

$$H_A : \alpha \neq \alpha_0$$

at some level of significance. Let t = test statistic and the *decision rule* is:

If $-t_{\alpha/2,\,n-3} \leq t \leq t_{\alpha/2,\,n-3}$	accept H_0
Otherwise	reject H_0

The test statistic is given by

$$t = \frac{a - \alpha_0}{s_a}$$

where $\qquad s_a = s_{Y|XW}\sqrt{(1/\Delta)[\Sigma X^2 \Sigma W^2 - (\Sigma XW)^2]}$

and Δ is given by Equation 11.9a.

Two-Tail Test of Hypothesis for $\beta_1 = \beta_1^*$

For this test, the most common application is $\beta_1 = 0$; however, we shall treat the general statement that β_1 has some particular value, β_1^*. The *decision rule* is the same as that given above for testing α. The test statistic, however, is

$$t = \frac{b - \beta_1^*}{s_b}$$

where

$$s_b = s_{Y|XW}\sqrt{(1/\Delta)[n\Sigma W^2 - (\Sigma W)^2]}$$

and Δ is from Equation 11.9a.

Two-Tail Test of Hypothesis for $\beta_2 = \beta_2^*$

Again, for this test the most common application is $\beta_2 = 0$. The decision rule for $\alpha = \alpha_0$ serves here as well. The test statistic is given by

$$t = \frac{c - \beta_2^*}{s_c}$$

where

$$s_c = s_{Y|XW}\sqrt{(1/\Delta)[n\Sigma X^2 - (\Sigma X)^2]}$$

and Δ, from Equation 11.9a, monotonously reappears.

Example 11.3   Tests of Hypothesis about β_1 and β_2

(a)   Consider the hypothesis: H_0: $\beta_1 = 0, H_A$: $\beta_1 \neq 0$. At the 5 percent significance level, and for $n - 3 = 2$ degrees of freedom, $t_{\alpha/2, n-3} = 12.71$. The decision rule is:

If $-12.71 \leq t \leq +12.71$ accept H_0

Otherwise   reject H_0

In order to calculate $t = (b - 0)/s_b$, we must calculate s_b. From example 11.1, $s_{Y|XW} = 1.33$, $\Delta = 255$, $n = 5$, $\Sigma W^2 = 91$, and $\Sigma W = 15$.

$$s_b = s_{Y|XW}\sqrt{(1/\Delta)[n\Sigma W^2 - (\Sigma W)^2]}$$
$$= (1.33)\sqrt{(1/255)[(5)(91) - (15)^2]}$$
$$= (1.33)\sqrt{230/255} = (1.33)(0.9497)$$
$$= 1.26$$

Then, $t = b/s_b = -1.4118/1.26 = -1.12$. Since $t = -1.12$ is in the interval ± 12.71, we *cannot* reject the hypothesis that $\beta_1 = 0$. But, letting $\beta_1 = 0$ means that the variable X (i.e., beef prices) should be deleted from the model.

(b) For the hypotheses $H_0: \beta_2 = 0$, $H_A: \beta_2 \neq 0$, the 5 percent significance level would require the same decision rule as for part a.

In order to calculate $t = (c - 0)/s_c$, first we calculate s_c. From example 11.1, the additional information we require is $\Sigma X = 10$ and $\Sigma X^2 = 26$.

$$s_c = s_{Y|XW}\sqrt{(1/\Delta)[n\Sigma X^2 - (\Sigma X)^2]}$$
$$= (1.33)\sqrt{(1/255)[(5)(26) - (10)^2]}$$
$$= (1.33)\sqrt{30/255} = (1.33)(0.343)$$
$$= 0.456$$

Then, $t = c/s_c = 0.7647/0.456 = 1.677$. Since $t = 1.677$ is within the interval ± 12.71, we *cannot* reject the null hypothesis that $\beta_2 = 0$. This result would suggest deleting the variable W (pork prices) from the model.

(c) As a result of parts a and b, we have no regression model! Does this mean that beef prices and pork prices do not influence beef consumption? Not necessarily. In order to keep the calculations relatively simple in example 11.1, only five observations were used. Since $n = 5$, there are only $n - 3 = 2$ degrees of freedom. If a linear relationship exists between Y, X, and W, a much larger sample size is required to specify the nature of that relationship.

Note that prediction intervals can be constructed (see example 11.2) without first performing these tests of hypothesis. If the hypotheses that $\beta_1 = 0$ and $\beta_2 = 0$ cannot be rejected, then the projections in example 11.2 (while arithmetically correct) are meaningless. After estimating the regression model, *your first step should always be the test of hypothesis to see if β_j is significantly different from zero.* The following technical note is the first step in that process.

Technical Note: F Test for Regression Relation

Frequently, an "F-VALUE" statement will appear on a computer printout. The F test is designed to test the hypothesis:

$$H_0: \beta_1 = \beta_2 = \cdots = \beta_J = 0$$
$$H_A: \text{Not all } \beta_j = 0$$

We shall call the F value provided by the computer F^*. The decision rule is:

$$\text{If } F^* \leq F_{\alpha, J-1, n-J} \quad \text{accept } H_0$$
$$\text{If } F^* > F_{\alpha, J-1, n-J} \quad \text{reject } H_0$$

The critical value of F, from Table VIII, has $J - 1$ and $n - J$ degrees of freedom. (*Note:* $F^* = [\text{SSR}/(J - 1)]/[\text{SSE}/(n - J)]$ where SSR is the sum of squares from regression, $\Sigma(\bar{Y}_c - \bar{Y})^2$, and SSE is the error sum of squares, $\Sigma(Y - \bar{Y}_c)^2$.)

Rejecting the null hypothesis does *not* say that *all* $\beta_j \neq 0$; it says that *some* $\beta_j \neq 0$.

11.5 PROBLEMS WITH USING THE MULTIVARIATE REGRESSION MODEL

The multivariate regression model is a powerful tool for investigating problems with many variables. However, the statistical principles, the mathematical rules, and the logic inherent in this model are numerous enough occasionally to cause an investigator (without deceit) to misuse or misinterpret the model. The following discussion items serve as a partial "check list" for altering (1) the structure, (2) the use, or (3) the interpretation of a multivariate regression model.

"Hidden" Extrapolation

Given some "new" values for X and W, say X_0 and W_0, section 11.3 discussed the techniques by which prediction intervals could be constructed. In Figure 10.15, we suggested the dangers of extrapolating the simple linear regression line beyond "the limits of the data for X." When we have two (or more) independent variables, is there a similar concept for "the limits of the data for the independent variables"? Yes, but the concept is somewhat more devious.

We could find the maximum and minimum values of each independent variable. If the "new" value for just one independent variable were outside that range, then we would certainly be dealing with an extrapolation rather than a prediction. For example, in example 11.1, the range for the beef price index (X) was 1 to 4, and the range for the pork price index (W) was 1 to 9. Either $X_0 = 5$, $W_0 = 5$ or $X_0 = 2$, $W_0 = 10$ would represent a point outside "the limits of the data set for X and W." For these values of X_0 and W_0, we would have to extrapolate or extend the model beyond its data base to use the "prediction" techniques of section 11.3.

The maximum and minimum values of each independent variable are necessary conditions for defining "the limits of the data set for the independent variables," *but they are not sufficient.* Consider the point $X_0 = 3 = W_0$ for the problem in example 11.1. The value 3 is between the maximum and minimum values for both X and W *individually*, but the problem is that we have to consider X and W *jointly*.

Consider Figure 11.2, where the data on X and W is plotted. Considering just the maximum and minimum values of X and W would map out a rectangle (as shown) for the limits of the data set. However, if we place an envelope around those points (X, W) actually observed, then we end up with just the shaded area representing the limits of the data set for the independent variables.

The shaded arrowhead-shaped area is considerably smaller than the rectangle.

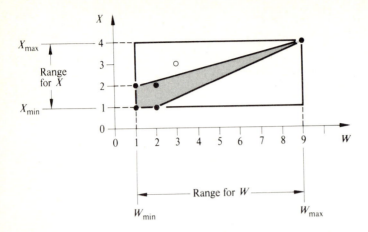

FIGURE 11.2 The problem of "Hidden extrapolation."

Note that the point $X_0 = 3$, $W_0 = 3$ is not in the shaded area. Points X_0, W_0 that are in the rectangle but not in the shaded area lead to *hidden extrapolation*. When points such as $X_0 = 3$, $W_0 = 3$ are used with the techniques of section 11.3, the result is an extrapolation interval. The fact that the resulting interval is *not* a prediction interval is not easily deciphered; the "extrapolation" nature of the interval is "hidden" because, for example, the "new" point is within the rectangle in Figure 11.2 but not in the shaded area. The limits of the data set are given by the shaded area.

If there are only two independent variables in a linear regression model, then the graphical technique demonstrated in Figure 11.2 is sufficient to detect hidden extrapolation. If the model includes more than two independent variables, detection of hidden extrapolations requires techniques beyond the scope of this discussion.

Dummy (or Shift) Variables

Thus far, the discussion of the regression model has assumed that all the independent variables are quantitative. Frequently a model builder will be faced with a list of potential "explanatory variables," which includes some nonquantitative variables. For example, in the length-of-cable problem from Chapter 10, the manufacturer of the cable might occasionally ship a metal spool rather than a wooden spool. While the weight of a spool is a quantitative variable, we might want to consider it as a qualitative variable.

Recall that Y was the length of cable remaining on a partially used spool and W was the gross weight of cable plus spool. Let us introduce a new variable X_i, where

$$X_i = \begin{cases} 0 & \text{if the spool is made of wood} \\ 1 & \text{if the spool is made of metal} \end{cases}$$

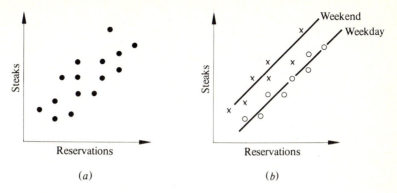

FIGURE 11.3   Scatter diagram for dummy-variable effect.

Then the regression model would be

$$\bar{Y}_c = a + bX_i + cW_i \qquad \text{(look familiar?)}$$

If the jth spool is wood, then $\bar{Y}_c = a + cW_j$; if it is of metal, then $\bar{Y}_c = (a + b) + cW_j$, which shifts the line parallel to itself by b units.

The scatter diagram will frequently suggest the importance of a qualitative variable. For example, consider the restaurant manager who wants to predict the number of steaks that will be requested that evening from the number of reservations. From past data the manager might plot a scatter diagram similar to Figure 11.3a. The restaurant manager also recognizes that proportionately more "nonreservation" customers request steak on the weekends than on weekdays. Thus, the scatter diagram is replotted using X's to represent a weekend and 0's to represent weekdays. Figure 11.3b suggests that a dummy variable should be used:

$$X_i = \begin{cases} 0 & \text{if weekday} \\ 1 & \text{if weekend} \end{cases}$$

Equations 11.8 and 11.9 can be used to calculate a, b, and c. In business and economics models, such variables as sex, marital status, occupation, and education are frequently useful. As long as mutually exclusive categories can be established, any classification scheme can be used as a dummy variable. However, Equations 11.8 and 11.9 are appropriate only if the dummy variable effects a parallel shift in the regression line.

For example, the restaurant manager might try to predict the drain on the wine cellar in a manner similar to the steak-forecasting problem. Suspecting that the customers' propensity to have wine with a meal is greater on the weekends, the manager plots Figure 11.4. Note that not only will the Y intercept be different, but the slope of the implied regression line is greater for the weekend data.

FIGURE 11.4 Dummy variable which affects both the slope and the Y intercept.

Consider the model

$$\bar{Y}_c = a + bX_i + cW_i + d(X_iW_i) \qquad (11.25)$$

where W_i = the number of reservations
$X_i = 0$ if i is a weekday
$X_i = 1$ if i is a weekend

For a weekday, the model is

$$\bar{Y}_c = a + cW_i$$

For a weekend the model is

$$\bar{Y}_c = (a + b) + (c + d)W_i$$

But, the coefficients in Equation 11.25 cannot be calculated from Equations 11.8 and 11.9. If we let $Z_i = X_iW_i$, then we have three independent variables. We would have to return to the general linear model in section 11.1 and, for $J = 3$, solve for the least-squares estimates of a, b, c, and d.

Lagged Variables

While surveying the list of potential independent variables, the model builder might reason that the value of X in some past time period is likely to be more influential on the value of Y for the current time period than is the value of X for the current time period. Consider a firm that specializes in automobile exhaust-system repairs. Let Y_t be the number of customers seeking exhaust-system repair in year t. We could let X_t be the number of new cars sold during year t.

For the sake of argument, let us assume that the exhaust system on a new car functions properly for exactly 2 years before repair is required. Thus, the model builder might conclude that the demand for repair service is a function of how many new cars were sold 2 years ago. That is,

$$\bar{Y}_t = a + bX_{t-2}$$

You still use paired values of Y and X, but Y_i is for year t and X_i is for year $(t - 2)$. One result for this model is that you should get a higher r^2 (coefficient of determination) than you would from $\bar{Y}_t = a + bX_t$. Also, the prediction interval for Y_{next} should be narrower, and it is relatively easy to predict the demand for repair service next year (all you need is last year's new-car sales).

Well, if this is such a good deal, why wouldn't

$$\bar{Y}_t = a + bX_{t-2} + cX_{t-3} + dX_{t-4} + eX_{t-5} \qquad \textbf{(11.26)}$$

be better? It might be, but there are several points to consider.

First, it might be preferable to define X_{t-k} to be the number of "cars on the road" which are k years old. If an automobile produced in year $t - k$ was unsafe at any speed, then there may not be many left after, say, 3 years. The demand for exhaust repair is a function of the number of drivable cars, not the number that were produced in that model year.

Second, are X_{t-2} and X_{t-3} independent of each other? We assumed that all "independent variables" are independent of one another in formulating the multivariate regression model. We shall return to this problem later.

Finally, the more independent variables we include in the model, the fewer the degrees of freedom—a result which can affect the model in several ways.

Equation 11.26 is a special case of a class of problems involving *distributed lags*. One of the problems with distributive lags is how long the lag period should be. Usually, theory is of little assistance in answering that question; the question is resolved statistically.

Even a general discussion of lagged variables quickly breaches the limit of this text. The usefulness of lagged variables should be clear.

Homoscedasticity

Now we investigate the importance of the assumptions employed in the general linear model. We assumed that the variance of the error term was constant regardless of the values of the independent variables. Reconsider the restaurant manager's problems. If we had many evenings when prior reservations were made for 20 people, for 30 people, etc., then for each value of X we would have many observations of Y. Figure 11.5 indicates that the homoscedastic assumption is valid for steaks; but for

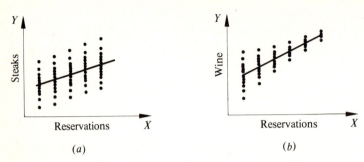

FIGURE 11.5 Homoscedasticity versus heteroscedasticity. (*a*) Homoscedastic; (*b*) Heteroscedastic.

wine, the variance in the error term seems to decrease as the number of reservations increases.

Of course, if you have data similar to that denoted in Figure 11.5, it is relatively easy to determine whether heteroscedasticity is a problem. If you have only one value of Y for each value of X, then it is much more difficult to test for heteroscedasticity.

The "cure" for heteroscedasticity includes recognizing the nature of the relationship between the independent variable and the variance of the error term and then transforming the data to accomplish homoscedasticity. For example, in Figure 11.5*b*, where X = number of reservations, if we let $W_i = 1/X_i^2$, then (Y_i, W_i) would produce a scatter diagram similar to Figure 11.5*a*, and the regression line $\bar{Y}_c = a + bW_i$ would satisfy the homoscedastic assumption.

Autocorrelation

The regression model also assumed that the error terms are independent of one another. Primarily we are concerned with successive values of ξ_i. Consider Figure 11.6 where we have a perfect curvilinear relation between steaks and reservations. A

FIGURE 11.6 Autocorrelation. (*a*) Scatter diagram; (*b*) Error terms are not random.

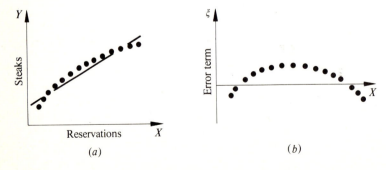

linear regression model with this data would yield $\xi_X = f(\xi_{X-1})$; that is, the error term for one value of X is, in part, determined by the error term for the next lower value of X. The error terms are not independent of one another; autocorrelation exists.

To test for autocorrelation, we could test the hypothesis that the residuals, $\xi_i = Y_i - (\bar{Y}_e)_i$, are random. Tests for randomness are discussed in Chapter 9.

If autocorrelation exists, it presents a serious problem. Our tests of significance (see sections 10.7 and 11.4) are affected, and their results are dubious. Autocorrelation may be remedied by data transformation (e.g., using first differences) or, as in Figure 11.6, by respecifying the expression for the model.

Multicollinearity

In the multivariate regression model, we assumed that the independent variables were independent of one another. Consider the model $Y = a + bX + cW$. If X and W are independent of one another, then the coefficient of determination, r^2, for X and W should be close to 0. But, let us assume that there is a perfect relation between X and W: $X = A + BW$. Multicollinearity is present, but its resolution is relatively easy. We redefine the model for Y:

Either
$$Y = a + b(A + BW) + cW = (a + bA) + (bB + c)W$$

or
$$Y = a + bX + c\left(-\frac{A}{B} + \frac{1}{B}X\right) = \left(a - \frac{Ac}{B}\right) + \left(b + \frac{c}{B}\right)X$$

The more common, and more troublesome, case is that in which the coefficient of determination between X and W is close to but not equal to 1. Multicollinearity is troublesome because the estimates of the regression coefficients a, b, and c are not very reliable.

For example, recall in example 11.3 that we could not reject the hypotheses that $\beta_1 = 0$ and $\beta_2 = 0$. Could there be a strong relationship between X and W (i.e., beef and pork prices)? The values of X and W are plotted in Figure 11.2, and from example 11.1 we have: $\Sigma X = 10$, $\Sigma W = 15$, $\Sigma XW = 45$, $\Sigma X^2 = 26$, $EW^2 = 91$, and $n = 5$. Using Equation 10.20,

$$r^2 = \frac{[n\Sigma XW - (\Sigma X)(\Sigma W)]^2}{[n\Sigma X^2 - (\Sigma X)^2][n\Sigma W^2 - (\Sigma W)^2]} = \frac{[5(45) - 10(15)]^2}{[5(26) - (10)^2][5(91) - (15)^2]}$$

$$= \frac{5,625}{6,900} = 0.815$$

or
$$r > 0.9$$

Indeed there is a strong relationship from this data. If for a larger-size sample this relationship still exists, then the results of example 11.3 might not change. Thus, we will have to specify the model in such a way that we exclude W or eliminate the multicollinearity.

Several computer programs provide a printout of a "correlation matrix" as a part of the calculations for the multivariate regression model. If the correlation between two independent variables is high, multicollinearity will influence all the results. A "correlation matrix" is not a perfect device for detecting multicollinearity, but it is a useful starting point.

Summary

The potential value and difficulties of constructing, interpreting, and using a general linear regression model are only outlined by this discussion. A more complete, albeit more rigorous, treatment may be found in: M. Dutta, *Econometric Methods*, South-Western Publishing Company, Incorporated, Cincinnati, 1975; H. J. Kelejian and W. E. Oates, *Introduction to Econometrics*, Harper & Row, Publishers, Incorporated, New York, 1974; and J. Johnston, *Econometric Methods*, 2d ed., McGraw-Hill Book Company, New York, 1972. Business students may find J. Neter and W. Wasserman, *Applied Linear Statistical Models*, Richard D. Irwin, Inc., Homewood, Ill., 1974, more useful.

11.6 NONLINEAR REGRESSION

In Chapters 10 and 11 we have occasionally made reference to nonlinear relations. Economic theory refers to a U-shaped average-cost curve (for example, $Y = a + bX + cX^2$) and the Cobb-Douglas production function (for example, $Y = AX^BW^C$), while businesspersons often refer to a constant rate of growth (for example, $Y = AB^X$).

In many cases, when either theory or the scatter diagram suggest a nonlinear representation, the general linear model can be used. The key to the use of least-squares estimates is the condition that *the model must be linear in parameters*. That is,

$$Y = \alpha + \alpha\beta X$$

$$Y = \alpha + \alpha^2 X$$

$$Y = \beta^2 + \beta X$$

are linear in variables but not linear in parameters; least-squares estimates could not be used for these models. We will look at several cases where the model is nonlinear in variables but linear in parameters. These cases require some form of *data transformation* in order to use the linear model.

Polynomial Regression

Consider the second-degree polynomial as a model of the population:

$$Y_i = \alpha + \beta_1 X_i + \beta_2 X_i^2 + \xi_i \qquad \text{(11.27)}$$

If we use the data transformation $W_i = X_i^2$, then Equation 11.27 becomes

$$Y_i = \alpha + \beta_1 X_i + \beta_2 W_i + \xi_i$$

which is the three-variable model discussed in section 11.1.

Extending this process to higher-degree polynomials,

$$Y_i = \alpha + \beta_i X_i + \beta_2 X_i^2 + \beta_3 X_i^3 + \cdots + \beta_J X_i^J$$

can be rewritten as

$$Y_i = \alpha + \beta_1 W_{1,i} + \beta_2 W_{2,i} + \beta_3 W_{3,i} + \cdots + \beta_J W_{J,i}$$

where $W_{j,i} = X_i^j$, and we have a general linear model.

Example 11.4   Second-Degree Polynomial Regression Curve

Consider the following paired values of Y and X, and let $W = X^2$.

Y_i	X_i	(X_i^2) W_i	YX	YW	XW	Y^2	X^2	W^2
2	0	0	0	0	0	4	0	0
6	1	1	6	6	1	36	1	1
12	2	4	24	48	8	144	4	16
20	3	9	60	180	27	400	9	81
40	6	14	90	234	36	584	14	98

From Equation 11.9a:

$$\Delta = n[\Sigma X^2 \Sigma W^2 - (\Sigma X W)^2] - \Sigma X(\Sigma X \Sigma W^2 - \Sigma W \Sigma X W) + \Sigma W(\Sigma X \Sigma X W - \Sigma W \Sigma X^2)$$

$$= 4[14(98) - (36)^2] - 6[6(98) - 14(36)] + 14[6(36) - 14(14)]$$

$$= 304 - 504 + 280 = 80$$

From Equation 11.9*b*:

$$\bar{a} = \Sigma Y[\Sigma X^2 \Sigma W^2 - (\Sigma X W)^2] - \Sigma X(\Sigma Y X \Sigma W^2 - \Sigma Y W \Sigma X W)$$
$$+ \Sigma W(\Sigma Y X \Sigma X W - \Sigma Y W \Sigma X^2)$$
$$= 40[14(98) - (36)^2] - 6[90(98) - 234(36)] + 14[90(36) - 234(14)]$$
$$= 3,040 - 2,376 - 504 = 160$$

From Equation 11.9*c*:

$$\bar{b} = n(\Sigma Y X \Sigma W^2 - \Sigma Y W \Sigma X W) - \Sigma Y(\Sigma X \Sigma W^2 - \Sigma W \Sigma X W)$$
$$+ \Sigma W(\Sigma X \Sigma Y W - \Sigma W \Sigma Y X)$$
$$= 4[90(98) - 234(36)] - 40[6(98) - 14(36)] + 14[6(234) - 14(90)]$$
$$= 1,584 - 3,360 + 2,016 = 240$$

From Equation 11.9*d*:

$$\bar{c} = n(\Sigma X^2 \Sigma Y W - \Sigma X W \Sigma Y X) - \Sigma X(\Sigma X \Sigma Y W - \Sigma W \Sigma Y X)$$
$$+ \Sigma Y(\Sigma X \Sigma X W - \Sigma W \Sigma X^2)$$
$$= 4[14(234) - 36(90)] - 6[6(234) - 14(90)] + 40[6(36) - 14(14)]$$
$$= 144 - 864 + 800 = 80$$

From Equations 11.8*a* through *c*:

$$a = \bar{a}/\Delta = 160/80 = 2$$
$$b = \bar{b}/\Delta = 240/80 = 3$$
$$c = \bar{c}/\Delta = 80/80 = 1$$

Thus, our regression model is

$$\bar{Y}_c = 2 + 3X_i + (1)W_i$$

or

$$\bar{Y}_c = 2 + 3X_i + X_i^2$$

Inverse Relations

The theoretical model

$$y = a + b(1/x)$$

can be transformed into a linear regression model

$$\bar{Y}_c = a + bW$$

by letting $W_i = 1/X_i$. By the same process, $1/y = a + bx$ can become $\bar{Y}_c = a + bX$ by letting $Y_i = 1/y_i$.

Exponential Relations

An exponential growth model $y = AB^x$ can be estimated by the method of least squares by using a logarithmic transformation. Taking logarithms of both sides yields $\log (y) = \log (A) + x \log (B)$. We can use the model

$$\bar{Y}_c = a + bX_i$$

where $Y_i = \log (y)$, $a = \log (A)$, and $b = \log (B)$. Taking antilogarithms of the results will get us back to the original model. Note that, in the context of interest rates or a constant rate of growth, $B - 1 = \text{rate}$.

In the economic theory of production, the Cobb-Douglas production function is frequently mentioned. This production function can be stated as $y = AK^B L^C$ where

$$y = \text{output}$$

$$K = \text{capital}$$

$$L = \text{labor}$$

and A, B, and C are constants.

Taking logarithms of both sides gives

$$\log (y) = \log (A) + B \log (K) + C \log (L)$$

We could use the linear regression model

$$\bar{Y}_c = a + bX_i + cW_i$$

where $Y_i = \log (y)$, $X_i = \log (K)$, $W_i = \log (L)$, $a = \log (A)$, $b = B$, and $c = C$. Before using the worksheet table in example 11.1, we would take the logarithms of y, K, and L and perform all calculations on those logarithms.

Summary

When either theory or a scatter diagram suggests the use of a nonlinear model, we have several models at our disposal which can be transformed into linear models. The least-squares techniques of the previous sections are valid only if the transformed model is linear in parameters. Of course, the flexibility and problems of the linear regression model that were discussed in section 11.5 still apply.

EXERCISES

1. Consider the following data:

Y	X	W
23	3	6
3	7	2
19	5	7
15	3	2
4	8	4
7	5	1
11	7	6
22	2	4

a. Set up the worksheet for a multiple regression equation (similar to Table 11.1).
b. Find the regression line $\bar{Y}_c = a + bX$ and calculate r^2 for only Y and X.
c. Find the regression line $\bar{Y}_c = a + bW$ and calculate r^2 for only Y and W.
d. Find the regression line $\bar{X}_c = a + bW$ and calculate r^2 for only X and W. (If X and W are used as independent variables, should we be concerned about multicollinearity? That is, is there a strong linear relationship between X and W?)
e. Find the regression equation $\bar{Y}_c = a + bX + cW$.
f. Calculate $s_{Y|XW}^2$.
g. Calculate $R_{Y|XW}^2$.
h. Compare your answers in parts b and c, simple linear regression lines, with your answers for parts e and g for the regression plane.
i. From your answers to parts b, c, and g, calculate the partial correlation coefficients. Which of the independent variables contributes the most to explaining variation in Y?
j. Which of the following points would represent hidden extrapolation?
    (1) $X_0 = 6$, $W_0 = 4$
    (2) $X_0 = 8$, $W_0 = 3$
    (3) $X_0 = 3$, $W_0 = 7$

2. Consider the following data:

Y	X	W
6	6	1
8	4	3
10	2	5
9	3	4
7	5	1

Repeat parts a through i, from Exercise 1, for this data.

3. Consider recent data for housing market transactions in Autoless City.

House Number i	Selling Price ($1,000) Y_i	Living Space (100 Square Feet) X_i	Nearest Bus Stop (Blocks) W_i
1	41	13	3
2	32	10	3
3	24	8	4
4	44	14	4
5	42	14	4
6	36	12	5
7	35	10	0
8	40	12	1
9	29	10	6
10	26	8	4
Total	349	111	34

We would find that: $\Sigma Y^2 = 12,619$, $\Sigma X^2 = 1,277$, $\Sigma W^2 = 144$, $\Sigma YX = 4,009$, $\Sigma YW = 1,157$, $\Sigma XW = 377$, $\Delta = 12,750$, $\bar{a} = 63,750$, $\bar{b} = 38,250$, and $\bar{c} = -12,750$.

a. Find the regression expression $\hat{Y}_c = a + bX + cW$.
b. Calculate the standard error of estimate and $R^2_{Y|XW}$.
c. At the 5 percent significance level, test the hypotheses that b and c (separately) differ from 0.
d. Interpret the values of a, b, and c.
e. For all houses two blocks from the nearest bus stop and with 1,000 square feet of living space, find the 95 percent prediction interval for the mean selling price.
f. A homeowner, two blocks from the nearest bus stop, wants to sell a 1,000-square-foot house. Provide a 95 percent prediction interval for the selling price.
g. A real estate agent has four houses for sale. All four are two blocks from the nearest bus stop, and all four have the same living space, 1,000 square feet. Find the 95 percent prediction interval for the mean selling price of these four houses.
h. Assume the real estate agent in part g receives 6 percent of the selling price in commission. Find the 95 percent prediction interval for the agent's earnings from selling all four houses.
i. If you could build a 1,000-square-foot house for $40 per square foot on an empty lot (which costs $6,000) that is two blocks from the nearest but stop, should you build it or wait until a similar house comes up for sale?

4. Reconsider Exercise 3.
a. Would you classify the data as time-series or cross-sectional data?
b. Is there a multicollinearity problem? (Find r^2 for X and W. Is the correlation between these two independent variables high or low?)
c. For $X_0 = 10$, $W_0 = 2$ is there hidden extrapolation? (Are parts f through h of Exercise 3 extrapolations or predictions?)

5. Consider Skunk Lake Park which is 50 miles from a large metropolitan area but not within 300 miles of anything else. The park superintendent feels that the major factors influencing the revenues are the price of gasoline in and the per capita income of the metropolitan area. The following data (all values are in 1967 dollars) is collected:

Summer of	Park Revenue ($1,000) Y_i	Average Price of Gasoline ($ per Gallon) X_i	Per Capita Income ($1,000 per Person) W_i
1966	290	0.25	8
1967	200	0.25	10
1968	250	0.30	11
1969	490	0.40	10
1970	410	0.40	12
1971	360	0.35	11
1972	300	0.30	10
1973	150	0.25	11
1974	200	0.30	12
1975	100	0.25	12
Total	2,750	3.05	107

From this data we could calculate: $\Sigma Y^2 = 886,900$; $\Sigma X^2 = 0.9625$; $\Sigma W^2 = 1,159$; $\Sigma YX = 896$; $\Sigma YW = 29,100$; $\Sigma XW = 32.8$; $\Delta = 4.275$; $a = 159.596$; $b = 2,013.684$; $c = -46.614$.

a. Interpret the values of a, b, and c for this exercise.
b. Find the standard error of estimate and $R^2_{Y|XW}$.
c. Test the hypothesis that b (and then c) is not significantly different from 0.
d. The park superintendent forecasts that next summer the average gasoline price will be 30 cents per gallon and the per capita income will be $8,000. Find the 95 percent prediction interval for park revenue.
e. Is part d a case of hidden extrapolation?

6. Reconsider Exercise 5.
a. Is there a multicollinearity problem?
b. Is there a problem with autocorrelation for X? [Hint: From Equation 9.4 use the von Neumann ratio to test

$$H_0: \text{Data is random (no autocorrelation)}$$

$$H_A: \text{Data is not random (autocorrelation)}$$

It may help to know that $\Sigma(X_i - X_{i-1})^2 = 0.0250$.]
c. Is there a problem of autocorrelation with W? [Hint: $\Sigma(W_i - W_{i-1})^2 = 14$.]
d. Is there a problem of autocorrelation with Y? [Hint: $\Sigma(Y_i - Y_{i-1})^2 = 115,700$.]
e. Consider the following table of residuals $(Y_i - \bar{Y})$ from the regression model in Exercise 5.

Summer of	Y Observed Y_i	Y Estimated $\bar{Y}_{xw}$	Residual or Error (ξ_i) $Y_i - \bar{Y}_{xw}$
1966	290	290.105	−0.105
1967	200	196.878	3.122
1968	250	250.948	−0.948
1969	490	498.930	−8.930
1970	410	405.702	4.298
1971	360	351.632	8.368
1972	300	297.562	2.438
1973	150	150.264	−0.264
1974	200	204.334	−4.334
1975	100	103.650	−3.650

Is there an autocorrelation problem with this model? [Use the von Neumann ratio with the added information that $\Sigma(\xi_i - \xi_{i-1})^2 = 341.734$ and $\Sigma(\xi_i - \bar{\xi})^2 = 217.061$.]

7. A farmer has two types of feed, feed K and feed L, and is interested in how different blends of these two feeds influence the weight gain of the hogs. For one month this farmer conducts an experiment in which a different blend is given to each of eight hogs, and the weight gain in each hog is recorded. The results are given below.

Hog Number	Weight Gain (Pounds) G_i	Pounds per Feeding of	
		Feed K K_i	Feed L L_i
1	40	1	1
2	49	1	2
3	61	2	1
4	75	2	2
5	56	1	3
6	84	2	3
7	95	3	2
8	77	3	1

The farmer wants to estimate a Cobb-Douglas production function of the form $G_i = A \cdot K_i^B \cdot L_i^C$. Taking the logarithm of both sides yields $\log(G_i) = \log(A) + B \cdot \log(K_i) + C \cdot \log(L_i)$.

a. Let $Y_i = \log(G_i)$, $a = \log(A)$, $b = B$, $X_i = \log(K_i)$, $c = C$, and $W_i = \log(L_i)$. Transform the above data to find Y_i, X_i, and W_i.

b. Construct the worksheet similar to Table 11.1.

c. Find the regression equation $\bar{Y}_c = a + bX + cW$.

d. Take the antilogarithms to find A, K, and L, and write the equation for the estimated production function.

e. Reconsider the original data. Start with the blend $K = 1$, $L = 1$ and observe the weight gain. Now consider twice as much of each feed, $K = 2$, $L = 2$. Does the weight gain double? Does the production function, your answer to part d, exhibit the same property?

8. Consider the following cost data:

Total Cost per Day ($100) Y_i	Production per Day (Units of Output) X_i
10	1
17	2
26	3
37	4
50	5

a. Let $W_i = X_i{}^2$ and set up a worksheet similar to Table 11.1.
b. Calculate a, b, and c, and state the regression expression.
c. From the data in part a, find the regression line $\bar{Y} = a + bX$.
d. How does the model in part b compare with the model in part c? For both models, find $\bar{Y}_c$ when $X_0 = 10$.

CASE QUESTIONS

1. In Chapter 10, case question 2 tried to relate number of rooms rented and income. For the Massachusetts sample, let $Y = $ V06, $X = $ V32, and $W = $ V11 (gross rent).
a. Find the regression plane $\bar{Y}_c = a + bX + cW$.
b. Find R^2.
c. Test the hypothesis that each coefficient is significantly different from zero.
d. How does this model compare with the model, $\bar{Y}_c = a + bX$, from case question 2 in Chapter 10?

2. Consider the Oregon sample. What affects the rent a landlord can obtain? Market conditions. But, we do not have sufficient data to consider all market conditions. When you look for a rental unit, what things do you consider? Location? (Sorry, no data.) Amount of space? (V06 and V23 might serve as imperfect substitutes for "square feet of living space.") How new is the unit? (V20 gives us some information on that.)
a. Let us investigate the model,

$$\bar{Y}_c = a + bX + cW$$

where $Y = $ monthly rent $= $ V11
$X = $ number of rooms $= $ V06
$W = 1$ if V20 $= 0$, 1, or 2
$W = 0$ if V20 $= $ anything but 1 or 2
Basically, $W = 1$ if the housing unit is less than 10 years old and $W = 0$ if the housing unit is more than 10 years old. W is called a "dummy variable."
(1) Find the regression plane.
(2) Find R^2.
(3) Test the hypothesis that the estimated parameters are significantly different from zero.
(4) Graph the regression line when $W = 0$. On the same graph, draw the regression line when $W = 1$.
(5) Interpret these results.
b. Repeat part a except let $X = $ number of bedrooms $= $ V23. Which model explains more of the variation in rent? Can we infer anything about the behavior of renters from these two models?

TIME-SERIES ANALYSIS

12.1 TIME-SERIES COMPONENTS

The familiar graphical representation of a time series is the use of time as the independent variable and some other variable (sales, GNP, housing starts) as the dependent variable. As a table of data, a time series is the value of the dependent variable and the point (or interval) in time when it occurred.

In contrast to time-series analysis, an alternative set of procedures is called *cross-sectional analysis*. For example, one way of investigating household expenditure on prepared foods (e.g., TV dinners, etc.) would be to investigate the gross retail sales per week over a period of several years; the resulting data could be used for a time-series analysis. In contrast, at one point in time, a survey of families could provide weekly expenditures on prepared food by income level of the family; the resulting data could be used for a cross-sectional analysis. Of course, hybrids of both approaches are possible, but they also represent larger data-collection costs. The techniques of regression analysis and analysis of variance are common in cross-sectional analysis; we have to develop the techniques for time-series analysis.

The use of time as an independent variable has some strange effects. We shall mention a few of them here and save the rest for the end of the chapter. First, the origin is arbitrarily chosen. Many economics variables, which could be considered for time-series analysis, are recorded by year. Using the Gregorian calendar, we may be interested in the data from 1950 to 1975. But, we can easily use 0 to represent 1950, 1 for 1951, 2 for 1952, and so forth. Frequently, we will employ this method of renumbering years in order to simplify the numbers we have to work with.

A time sequence of data may refer to either a *stock* (dependent) variable or a *flow* (dependent) variable. A stock variable is measured at a point in time; a flow variable is measured over some time interval.

Consider a water tank or reservoir into which water either flows or is pumped. The height of the water level can be measured at any point in time; it is a stock variable. The water flows from the reservoir to a system of pipes. The water pressure at a nozzle connected to this pipe system depends on the height of the water level (if only gravity is used). Thus, at 6 P.M. every day we may want two readings from this reservoir: the height of the water level (a stock variable that affects water pressure) and how much water has left the reservoir in the last 24 hours (a flow variable).

Inventory is an important stock variable in the operation of a firm. The number of picture tubes on hand at a TV repair shop and the number of unprepared steaks in a restaurant are inventory examples where measurement is made at a particular point in time. Sales is a flow variable; sales are measured over some specific interval of time.

To a student, the amount of note paper used each month is a flow variable; the amount of note paper available at any point in time is a stock variable.

Whether the dependent variable is a stock or a flow variable, the reason for using time as the independent variable is the hope that observing the past will yield clues to the future changes in the dependent variable. Time-series data and time-series analysis are frequently a useful starting point in dealing with a business problem or an economic concern. Time-series analysis and index number construction are techniques which are unique to business and economics analysis.

It is traditional to view time-series data as the net result of four components: the trend T, the cycle or cycles C, the seasonal variation S, and the irregular variation I. In fact, these components are precisely defined by the way in which the investigator chooses to measure or identify them. That is, "how each component is calculated" defines each component. This attribute of time-series analysis exhibits "good news" and "bad news." The good news is that there is flexibility in how the components can be employed and interpreted. The bad news is the credibility of the components as separable and identifiable entities. After we see what these techniques are, we will reintroduce this issue.

The general interpretation of the four components is presented in Panel 12.1. Consider the logic of each component separately, and then we shall discuss how they might be merged to describe a time sequence of real-world data.

The trend component can be visualized by plotting a time series with time on the horizontal axis. Then stand way back from this time-series chart and observe the general *long-run* growth or decline in the data. Strictly speaking, the trend is an attempt to describe that general long-run pattern with a smooth, fairly simple mathematical expression (e.g., a straight line). In a sense, the trend is an attempt to measure the regular movement of the data over the "longest" time period applicable for a particular problem.

The cyclical component has two attributes: (1) it considers time intervals somewhat shorter than that implied by the trend, and (2) it is usually characterized by a wavelike shape. The so-called "business cycles" appearing in economic time-series data (for example, GNP over time) are examples. Unlike the nice sinusoidal curves shown in Panel 12.1, cycles in economic data are not very regular in their amplitude or period. That is, the severity of a "boom" period and a "bust" period may change through time (i.e., the height of each peak may change—the amplitude varies). Also, the trough-peak-trough sequence (which measures the "period" of a cycle) may take 10 years, then 8 years, then 9 years, etc., as we move through time.

The seasonal component is usually considered to be those changes that regularly occur within the time frame of 1 calendar year and that repeat themselves year after year. For example, we would not expect the retail sales of Christmas trees in the United States to be relatively constant during the year; however, year after year the

---

Panel 12.1   Components of a Time-Series Model

Trend
The *trend* is the general tendency the data exhibits over a long period of time. It is the general path that the data maps through time. Either a straight line or a curve may be used to represent the growth or decline of the time series.

Cycle
The cyclical component represents those recurring variations that make large swings through time. The "period" of a cycle must be more than 1 year in duration.

Seasonal
Either by custom or by climate, there are patterns that are associated with the seasons of the year; these patterns repeat themselves every year. This pattern may take any shape.

Irregular
Everything not accounted for in the other three components is lumped into this one; however, this component should exhibit a random series.

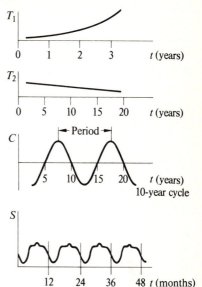

---

monthly data on retail sales of Christmas trees would exhibit a large "blip" in December. The seasonal component is an attempt to identify those variations that are regularly associated with the seasons of the year. In geographic regions where climatic conditions do not change much during the year, seasonal variations may still occur because of custom (e.g., the sales of Easter bunnies in Hawaii).

Since the seasonal component attempts to identify regular changes that occur within the period of 1 year, the cycle is considered to be fairly regular changes that occur with periods of greater than 1 year.

The irregular component is essentially all changes not falling into one of the other components. Except for "obvious but unpredictable influences" the irregular component of an economic time-series should exhibit changes that are very small relative to the changes represented by the cyclical and seasonal components. Further, the "irregular" component should exhibit a random pattern. The "obvious but unpredictable influences" could refer to natural disasters (e.g., earthquakes, tornadoes, floods, etc.), human-induced irregularities (e.g., embargoes, strikes, a presidential assassination, etc.), or other influences which might not consistently fit either classification (e.g., epidemics).

Our strategy for investigating time-series analysis will focus on these four components. First, we shall treat several ways of combining these four components into a model. We will start with the components, and built a hypothetical model from them; this is a pedagogical device to clarify how the structure of the model influences the calculational techniques used to identify each component. Finally, we will investigate some of the techniques used to identify the components from time-series data.

12.2 VARIOUS TIME-SERIES MODELS

The Additive Time-Series Model

One way of merging the four components of a time series is to simply add all four components for each point (or interval) in time. If we let Y_t be the actual or observed value of the dependent variable in time period t, then

$$Y_t = T_t + C_t + S_t + I_t \qquad (12.1)$$

represents the additive model.

If we let the seasonal component represent that regular pattern of change in Y that occurs *within* each year, and if we have annual data for Y (that is, t is measured in 1-year units), then the seasonal component would not be included in the model.

Consider the 4 years of monthly data presented in Panels 12.2 a and b. The trend is a simple straight line with positive slope; the trend component contributes an additional amount each month. The cyclical component is a smooth wavelike function that repeats itself every 4 years. The seasonal component is a straight line that starts low at the beginning of each year, increases as the year ages, and ends high. This simple seasonal pattern repeats itself each year. The irregular component follows a random pattern.

This data is fabricated, but let us assume it is for monthly sales of Christmas cards. Further, we might hypothesize that the trend is a function of income per family only and that the seasonal component is a function of buying habits and the calendar location of Christmas Day.

In order to get the gross sales per month (Y_t), we add the four components for each month: $Y = T + C + S + I$. The top diagram in Panel 12.2 shows the resulting gross sales per month.

There are several characteristics of the additive model. First, note the units on Y (e.g., thousands of dollars) must be the units on T, S, C, and I. That is, if we are going to add the values from each of the four components to obtain Y, then all four components must be measured in the same units.

Second, in order to add the values from each component, the components should be independent of one another. If they were not independent of one another, then a singular influence (e.g., family income) could affect two components simultaneously. Thus, this model implies that any "other variables" that are influencing the trend do not influence any of the other three components. The same must be true for the other three components.

Finally, the seasonal component has a special characteristic. Over the period of 1 year, the values of the seasonal component should total 0. Choose any year of monthly data listed at the bottom of Panel 12.2. The seasonal component has values of -33, -27, -21, ..., 21, 27, and 33, which, of course, totals 0. This means that January averages 33 (thousand dollars) worse than the average for the year, February 27 worse, etc. If the seasonal component represents seasonal influences that occur over the period of 1 year, then the average annual influence of the seasonal component should be 0.

The Multiplicative Time-Series Model

Another way of merging the four components of a time series is to multiply them for each time period. Letting Y_t be the actual or observed value of the dependent variable in time period t, then

$$Y_t = T_t \cdot C_t \cdot S_t \cdot I_t \qquad (12.2)$$

represents the multiplicative time-series model.

Consider the 4 years of monthly (hypothetical) data in Panels 12.3a and b. Again, the trend is a straight line with positive slope (but it is a different trend than the one in Panels 12.2a and b). The cyclical component is a wavelike function that repeats itself every 4 years. The simple seasonal component is a straight line that starts low and ends high; the seasonal repeats itself every year. The irregular follows a random pattern.

Now if all four components were in dollar units, then Y would be in units of dollars to the fourth power. If Y, the dependent variable, has units of dollars, then only one of the four components may have dollar units and the other three components must be unitless. It is traditional to measure the trend in dollars and designate C, S, and I as unitless index numbers.

Another characteristic of the multiplicative time-series model is that the four components are assumed to influence one another (in contrast to the additive model); however, the four components are still assumed to arise from different causes (as in the additive model). If you argue that the same variable (e.g., income) affect both the trend and the cyclical components, then you cannot isolate the trend from the cycle. Thus, in order to isolate the four components of a time-series model, a necessary assumption is that the set of variables which affects one component must be different (mutually exclusive) from that set of variables which influences another component.

Panel 12.2a Additive Model (Years 1 and 2)

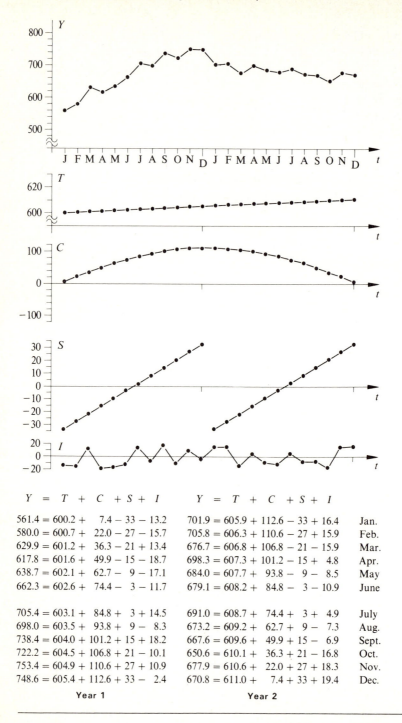

Y	$=$	T	$+$	C	$+ S +$	I		Y	$=$	T	$+$	C	$+ S +$	I	
561.4	=	600.2	+	7.4	− 33 −	13.2		701.9	=	605.9	+	112.6	− 33 +	16.4	Jan.
580.0	=	600.7	+	22.0	− 27 −	15.7		705.8	=	606.3	+	110.6	− 27 +	15.9	Feb.
629.9	=	601.2	+	36.3	− 21 +	13.4		676.7	=	606.8	+	106.8	− 21 −	15.9	Mar.
617.8	=	601.6	+	49.9	− 15 −	18.7		698.3	=	607.3	+	101.2	− 15 +	4.8	Apr.
638.7	=	602.1	+	62.7	− 9 −	17.1		684.0	=	607.7	+	93.8	− 9 −	8.5	May
662.3	=	602.6	+	74.4	− 3 −	11.7		679.1	=	608.2	+	84.8	− 3 −	10.9	June
705.4	=	603.1	+	84.8	+ 3 +	14.5		691.0	=	608.7	+	74.4	+ 3 +	4.9	July
698.0	=	603.5	+	93.8	+ 9 −	8.3		673.2	=	609.2	+	62.7	+ 9 −	7.3	Aug.
738.4	=	604.0	+	101.2	+ 15 +	18.2		667.6	=	609.6	+	49.9	+ 15 −	6.9	Sept.
722.2	=	604.5	+	106.8	+ 21 −	10.1		650.6	=	610.1	+	36.3	+ 21 −	16.8	Oct.
753.4	=	604.9	+	110.6	+ 27 +	10.9		677.9	=	610.6	+	22.0	+ 27 +	18.3	Nov.
748.6	=	605.4	+	112.6	+ 33 −	2.4		670.8	=	611.0	+	7.4	+ 33 +	19.4	Dec.

Year 1   **Year 2**

Panel 12.2b   Additive Model (Years 3 and 4)

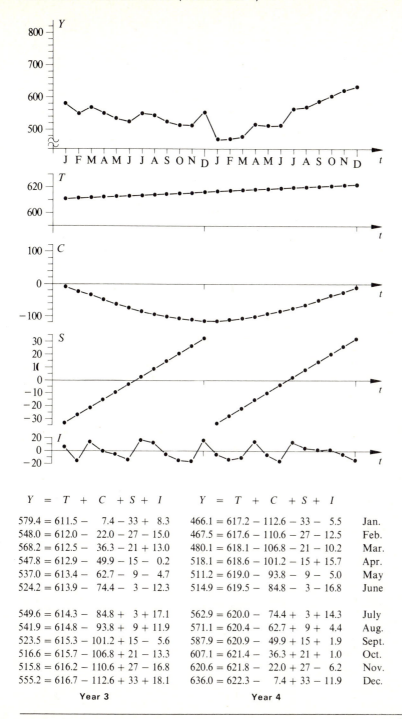

Y	$=$	T	$+$	C	$+ S +$	I		Y	$=$	T	$+$	C	$+ S +$	I	
579.4	=	611.5	−	7.4	− 33 +	8.3		466.1	=	617.2	−	112.6	− 33 −	5.5	Jan.
548.0	=	612.0	−	22.0	− 27 −	15.0		467.5	=	617.6	−	110.6	− 27 −	12.5	Feb.
568.2	=	612.5	−	36.3	− 21 +	13.0		480.1	=	618.1	−	106.8	− 21 −	10.2	Mar.
547.8	=	612.9	−	49.9	− 15 −	0.2		518.1	=	618.6	−	101.2	− 15 +	15.7	Apr.
537.0	=	613.4	−	62.7	− 9 −	4.7		511.2	=	619.0	−	93.8	− 9 −	5.0	May
524.2	=	613.9	−	74.4	− 3 −	12.3		514.9	=	619.5	−	84.8	− 3 −	16.8	June
549.6	=	614.3	−	84.8	+ 3 +	17.1		562.9	=	620.0	−	74.4	+ 3 +	14.3	July
541.9	=	614.8	−	93.8	+ 9 +	11.9		571.1	=	620.4	−	62.7	+ 9 +	4.4	Aug.
523.5	=	615.3	−	101.2	+ 15 −	5.6		587.9	=	620.9	−	49.9	+ 15 +	1.9	Sept.
516.6	=	615.7	−	106.8	+ 21 −	13.3		607.1	=	621.4	−	36.3	+ 21 +	1.0	Oct.
515.8	=	616.2	−	110.6	+ 27 −	16.8		620.6	=	621.8	−	22.0	+ 27 −	6.2	Nov.
555.2	=	616.7	−	112.6	+ 33 +	18.1		636.0	=	622.3	−	7.4	+ 33 −	11.9	Dec.

Year 3   **Year 4**

Panel 12.3a   Multiplicative Model (Years 1 and 2)

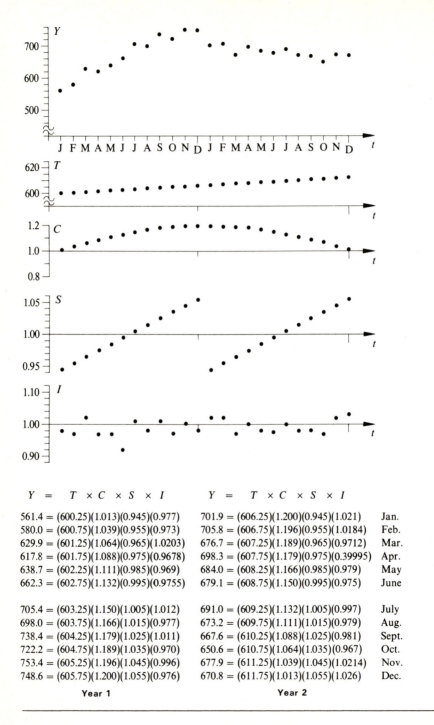

Y	$=$	T	$\times$	C	$\times$	S	$\times$	I		Y	$=$	T	$\times$	C	$\times$	S	$\times$	I	

$561.4 = (600.25)(1.013)(0.945)(0.977)$ $\qquad$ $701.9 = (606.25)(1.200)(0.945)(1.021)$ $\qquad$ Jan.
$580.0 = (600.75)(1.039)(0.955)(0.973)$ $\qquad$ $705.8 = (606.75)(1.196)(0.955)(1.0184)$ $\qquad$ Feb.
$629.9 = (601.25)(1.064)(0.965)(1.0203)$ $\qquad$ $676.7 = (607.25)(1.189)(0.965)(0.9712)$ $\qquad$ Mar.
$617.8 = (601.75)(1.088)(0.975)(0.9678)$ $\qquad$ $698.3 = (607.75)(1.179)(0.975)(0.39995)$ $\qquad$ Apr.
$638.7 = (602.25)(1.111)(0.985)(0.969)$ $\qquad$ $684.0 = (608.25)(1.166)(0.985)(0.979)$ $\qquad$ May
$662.3 = (602.75)(1.132)(0.995)(0.9755)$ $\qquad$ $679.1 = (608.75)(1.150)(0.995)(0.975)$ $\qquad$ June

$705.4 = (603.25)(1.150)(1.005)(1.012)$ $\qquad$ $691.0 = (609.25)(1.132)(1.005)(0.997)$ $\qquad$ July
$698.0 = (603.75)(1.166)(1.015)(0.977)$ $\qquad$ $673.2 = (609.75)(1.111)(1.015)(0.979)$ $\qquad$ Aug.
$738.4 = (604.25)(1.179)(1.025)(1.011)$ $\qquad$ $667.6 = (610.25)(1.088)(1.025)(0.981)$ $\qquad$ Sept.
$722.2 = (604.75)(1.189)(1.035)(0.970)$ $\qquad$ $650.6 = (610.75)(1.064)(1.035)(0.967)$ $\qquad$ Oct.
$753.4 = (605.25)(1.196)(1.045)(0.996)$ $\qquad$ $677.9 = (611.25)(1.039)(1.045)(1.0214)$ $\qquad$ Nov.
$748.6 = (605.75)(1.200)(1.055)(0.976)$ $\qquad$ $670.8 = (611.75)(1.013)(1.055)(1.026)$ $\qquad$ Dec.

Year 1 $\qquad\qquad\qquad\qquad\qquad\qquad$ Year 2

Panel 12.3b   Multiplicative Model (Years 3 and 4)

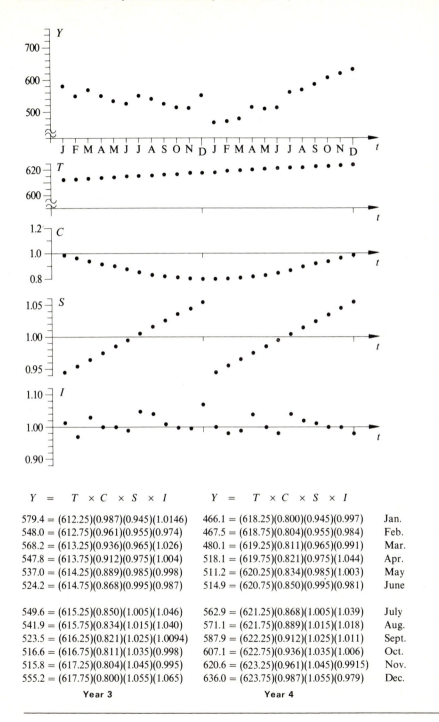

$$Y = T \times C \times S \times I \qquad Y = T \times C \times S \times I$$

$579.4 = (612.25)(0.987)(0.945)(1.0146)$	$466.1 = (618.25)(0.800)(0.945)(0.997)$ Jan.
$548.0 = (612.75)(0.961)(0.955)(0.974)$	$467.5 = (618.75)(0.804)(0.955)(0.984)$ Feb.
$568.2 = (613.25)(0.936)(0.965)(1.026)$	$480.1 = (619.25)(0.811)(0.965)(0.991)$ Mar.
$547.8 = (613.75)(0.912)(0.975)(1.004)$	$518.1 = (619.75)(0.821)(0.975)(1.044)$ Apr.
$537.0 = (614.25)(0.889)(0.985)(0.998)$	$511.2 = (620.25)(0.834)(0.985)(1.003)$ May
$524.2 = (614.75)(0.868)(0.995)(0.987)$	$514.9 = (620.75)(0.850)(0.995)(0.981)$ June
$549.6 = (615.25)(0.850)(1.005)(1.046)$	$562.9 = (621.25)(0.868)(1.005)(1.039)$ July
$541.9 = (615.75)(0.834)(1.015)(1.040)$	$571.1 = (621.75)(0.889)(1.015)(1.018)$ Aug.
$523.5 = (616.25)(0.821)(1.025)(1.0094)$	$587.9 = (622.25)(0.912)(1.025)(1.011)$ Sept.
$516.6 = (616.75)(0.811)(1.035)(0.998)$	$607.1 = (622.75)(0.936)(1.035)(1.006)$ Oct.
$515.8 = (617.25)(0.804)(1.045)(0.995)$	$620.6 = (623.25)(0.961)(1.045)(0.9915)$ Nov.
$555.2 = (617.75)(0.800)(1.055)(1.065)$	$636.0 = (623.75)(0.987)(1.055)(0.979)$ Dec.
Year 3	**Year 4**

Other Time-Series Models

You might want to build a combination of these two models. For example, you might reason that the trend does not influence the cycle and seasonal components, but the cycle and seasonal components do somehow interact. The resulting model might be

$$Y = T + CSI$$

Another possibility is that the trend and cycle influence one another, but they do not affect the seasonal component. Thus, the model might be

$$Y = TCI + S$$

Of course, depending upon the nature of the problem, other "arrangements" of the four components could be suggested. We will limit ourselves to the two basic models.

12.3 ISOLATION OF COMPONENTS FOR THE ADDITIVE TIME-SERIES MODEL ·

The basic statement of the additive model suggests the approaches generally used:

(Actual value of Y)

$$= \text{(trend value)} + \text{(cyclical value)} + \text{(seasonal value)} + \text{(irregular value)}$$

Assume that from the actual data of Y values we can decide upon an expression that describes the trend. By whatever means we use to develop the trend component, we could remove the influence of the trend by

$$Y - T = (T + C + S + I) - T = C + S + I$$

The values of $(Y - T)$ for each time period are frequently called "detrended" values. Then, if we can isolate the seasonal component, we can remove the seasonal influence by

$$(C + S + I) - S = C + I$$

Isolation of the cyclical component will leave only the irregular component.

Although the components are not always treated in this particular order, it is this "subtraction strategy" that characterizes the isolation of the various components. Following common practice, we shall start with the trend.

Trend

Consider the annual production data on crude oil produced by Aramco in Saudi Arabia from 1946 to 1967, which is graphed in Figure 12.1. Since the data is annual production, there is no seasonal component and the additive model is reduced to $Y = T + C + I$. There is a general upward trend, but precisely how do we express that component?

We could use a *straight-line* model of the trend; a straight line would imply a constant *amount* of increase in crude oil production each year. We could use an *exponential* model of the trend; an exponential might indicate a constant *percent* increase per year. Further, a third-order polynomial (for example, $T = a + bt + ct^2 + dt^3$) might be the best explanation, but do we have any hypothesis or theory that would support using a third-order polynomial as a model for the long-run growth of crude oil production in Saudi Arabia?

For the moment, let us stick to the simpler models. The straight-line model for the trend would be

$$T_t = a + bt \tag{12.3}$$

where T_t is the trend value for time period t. In the crude oil example, T is millions of barrels of crude oil and t is measured in years. The values of a and b have to be determined. The method of least squares developed in Chapter 10 is one way of calculating a and b.

For example, for the trend line $Y_t = a + bt$, the method of least squares yields

$$b = \frac{n\Sigma Yt - (\Sigma Y)(\Sigma t)}{n(\Sigma t^2) - (\Sigma t)^2} \tag{12.4}$$

and

$$a = \bar{Y} - b\bar{t} = (\Sigma Y)/n - b(\Sigma t/n) \tag{12.5}$$

FIGURE 12.1   Time series for Aramco crude oil production (1946–1967). (*Source*: See Table 12-1.)

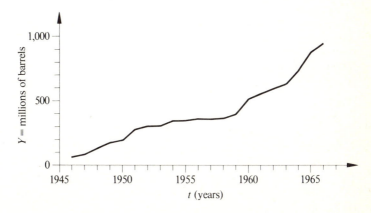

From Table 12.1, we find that

$$b = \frac{(22)(135,002) - (9,050)(253)}{(22)(3,795) - (253)^2} = \frac{680,394}{19,481} = 34.93$$

and $a = (9,050/22) - (34.93)(253/22) = 9.669$. Thus, $T_t = 9.7 + 34.93t$, where $t = 0$ at 1945, t is in years, and T_t is millions of barrels of crude oil.

TABLE 12.1   ARABIAN CRUDE OIL PRODUCTION (1946–1967)

Year (1)	Crude Oil Production (Millions of Barrels) Y (2)	Renumbered Years t (3)	$Y \cdot t$ (4)	t^2 (5)	Trend Values T (6)	Detrended Values $(Y - T)$ $= C + I$ (7)
1946	60	1	60	1	44.63	15.37
1947	90	2	180	4	79.56	10.44
1948	143	3	429	9	114.49	28.51
1949	174	4	696	16	149.42	24.58
1950	200	5	1,000	25	184.34	15.66
1951	278	6	1,668	36	219.27	58.73
1952	302	7	2,114	49	254.20	47.80
1953	308	8	2,464	64	289.12	18.88
1954	348	9	3,132	81	324.05	23.95
1955	352	10	3,520	100	358.97	−6.97
1956	361	11	3,971	121	393.90	−32.90
1957	362	12	4,344	144	428.83	−66.83
1958	370	13	4,810	169	463.75	−93.75
1959	400	14	5,600	196	498.68	−98.68
1960	456	15	6,840	225	533.60	−77.60
1961	508	16	8,128	256	568.53	−60.53
1962	555	17	9,435	289	603.46	−48.46
1963	595	18	10,710	324	638.38	−43.38
1964	628	19	11,932	361	673.31	−45.31
1965	739	20	14,780	400	708.23	30.77
1966	873	21	18,333	441	743.16	129.84
1967	948	22	20,856	484	778.06	169.91
Total	9,050	253	135,002	3,795	—	—

Source: *Aramco Handbook: Oil and the Middle East*, Arabian American Oil Company, Dhahran, Saudi Arabia, 1968, p. 135.

Now, we employ the "subtraction strategy." For each value of t we calculate a value of T_t in column 6 of Table 12.1. Note that

$$Y_t - T_t = (T_t + C_t + I_t) - T_t = C_t + I_t$$

Subtracting the trend value from the observed value yields a number which denotes the combined effects of the cyclical and irregular components. The values for $C + I$ are given in column 7 of Table 12.1 and are plotted as a solid line in Figure 12.2.

FIGURE 12.2   The $C + I$ data and the moving-average smoothing effect.

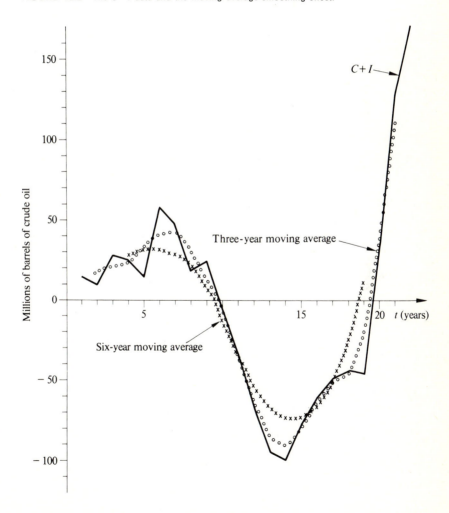

Cycle

In Figure 12.2, the solid line is rather jagged. Going back to Panels 12.2a and b, we see that adding a smooth cyclical function and the random nature of the irregular component could produce a result similar to that found in Figure 12.2. If we could eliminate the irregular component, we could identify the cyclical component.

How do we eliminate the irregular component? If we had a technique that would "smooth out" that jagged line, we would feel more comfortable about calling the smoothed curve a "cycle." We will investigate two smoothing techniques: the method of moving averages and exponential smoothing.

The method of moving averages If the irregular component is to be comprised of numbers that are essentially random in their sequential appearance in the time series, then they should average to 0. Let n be the number of time periods in our time series, and let m be some integer value less than n but greater than 1. If we take m consecutive values of I_t, the irregular component, then the average of those m values should be close to 0. That is, we will assume that

$$\frac{\left(\sum_{t=k}^{k+m} I_t\right)}{m} \simeq 0$$

If that assumption is true, then

$$\frac{1}{m} \sum_{t=k}^{k+m} Y_t = \frac{1}{m} \sum_{t=k}^{k+m} (T_t + C_t + I_t)$$
$$= (1/m)\Sigma T_t + (1/m)\Sigma C_t + (1/m)\Sigma I_t$$
$$= (1/m)\Sigma T_t + (1/m)\Sigma C_t + 0$$

If we can further assume that $(1/m)\Sigma T_t$ is represented by $T_{(2k+m)/2} = a + b(2k+m)/2$ (i.e., the least-squares regression line value for the middle of the time interval from k to $k+m$), then we could subtract the trend value from $(1/m)\Sigma Y$ and have a value representative of C for the time period $(2k+m)/2$. This is the general strategy employed to isolate the cyclical component with the method of moving averages.

Fortunately, the method of moving averages is easier to calculate than it is to describe in general terms. Consider Table 12.2 where a $(m =)$ 3-year moving average is used to smooth the Y values (the crude oil production in Saudi Arabia). In column 4, the 3-year moving total is the sum of the three Y values centered on time period t.

TABLE 12.2 3-YEAR MOVING AVERAGE

Time Period (Year) t (1)	Crude Oil Production ($\times 10^6$ Barrels) Y (2)	Trend Values T (3)	3-Year Moving Total (4)	3-Year Moving Average $(T + C)$ (5)	Cycle (5) − (3) $(T + C) - T$ $= C$ (6)
1	60	44.64	n.a.	n.a.	n.a.
2	90	79.57	293	97.67	18.10
3	143	114.49	407	135.67	21.18
4	174	149.42	517	172.33	22.91
5	200	184.34	652	217.33	32.99
6	278	219.27	780	260.00	40.73
7	302	254.20	888	296.00	41.80
8	308	289.12	958	319.33	30.21
9	348	324.05	1,008	336.00	11.95
10	352	358.97	1,061	353.67	− 5.30
11	361	393.90	1,075	358.33	− 35.57
12	362	428.83	1,093	364.33	− 64.50
13	370	463.75	1,132	377.33	− 86.42
14	400	498.68	1,226	408.67	− 90.01
15	456	533.60	1,364	454.67	− 78.93
16	508	568.53	1,519	506.33	− 62.20
17	555	603.46	1,658	552.67	− 50.79
18	595	638.38	1,778	592.67	− 45.71
19	628	673.31	1,962	654.00	− 19.31
20	739	708.23	2,240	746.67	38.44
21	873	743.16	2,560	853.33	110.17
22	948	778.06	n.a.	n.a.	n.a.

Thus,

$$\sum_{t=1}^{3} Y_t = Y_1 + Y_2 + Y_3 = 60 + 90 + 143 = 293$$

$$\sum_{t=2}^{4} Y_t = Y_2 + Y_3 + Y_4 = 90 + 143 + 174 = 407$$

$$\sum_{t=3}^{5} Y_t = Y_3 + Y_4 + Y_5 = 143 + 174 + 200 = 517$$

The 3-year moving average in column 5 is simply $(\frac{1}{3})\Sigma Y_t$ or

$$293/3 = 97.67 \qquad 407/3 = 135.67 \qquad 517/3 = 172.33$$

and so forth.

The moving total changes by dropping the "oldest value" and including the next "new" value in the time sequence; in this sense, it is a "moving" average. Column 5, the 3-year moving average, represents $T + C$. By subtracting the trend values in column 3 from the values in column 5, we obtain a value for C in column 6. The data from column 6 is plotted as a sequence of dots in Figure 12.2.

TABLE 12.3 6-YEAR MOVING AVERAGE

Time Period (Year) t (1)	Crude Oil Production ($\times 10^6$ Barrels) Y (2)	Trend Values T (3)	6-Year Moving Total (4)	6-Year Centered Moving Total (5)	6-Year Moving Average (6)	(6) − (3) Cycle $(T+C)-T$ $= C$ (7)
1	60	44.64	n.a.	n.a.	n.a.	n.a.
2	90	79.57	n.a.	n.a.	n.a.	n.a.
3	143	114.49	945	n.a.	n.a.	n.a.
4	174	149.42	1,187	1,066.0	177.67	28.25
5	200	184.34	1,405	1,296.0	216.00	31.66
6	278	219.27	1,610	1,507.5	251.25	31.98
7	302	254.20	1,788	1,699.0	283.17	28.97
8	308	289.12	1,949	1,868.5	311.42	22.30
9	348	324.05	2,033	1,991.0	331.83	7.78
10	352	358.97	2,101	2,067.0	344.50	− 14.47
11	361	393.90	2,193	2,147.0	357.83	− 36.07
12	362	428.83	2,301	2,247.0	374.50	− 54.33
13	370	463.75	2,457	2,379.0	396.50	− 67.25
14	400	498.68	2,651	2,554.0	425.67	− 73.01
15	456	533.60	2,884	2,767.5	461.25	− 72.35
16	508	568.53	3,142	3,013.0	502.17	− 66.36
17	555	603.46	3,481	3,311.5	551.92	− 51.54
18	595	638.38	3,898	3,689.5	614.92	− 23.46
19	628	673.31	4,338	4,118.0	686.33	13.02
20	739	708.23	n.a.	n.a.	n.a.	n.a.
21	873	643.16	n.a.	n.a.	n.a.	n.a.
22	948	778.06		n.a.	n.a.	n.a.

In Table 12.2, the number of time periods included in each "moving average" was an odd number; $m = 3$. When the number of time periods selected is even (for example, $m = 6$), an additional problem occurs. In Table 12.3 we construct a 6-year moving average for the same data on crude oil production.

Note that the first 6-year moving total is centered not on time period 3 or time period 4, but on the dividing line between them. The second 6-year moving total is centered between time periods 4 and 5. Thus, to get a moving total which is centered on $t = 4$, we take the arithmetic mean of the two consecutive 6-year moving totals:

$$\text{(Centered moving total for } t = 4) = (945 + 1{,}187)/2 = 1{,}066$$

$$\text{(Centered moving total for } t = 5) = (1{,}187 + 1{,}405)/2 = 1{,}296$$

Dividing the centered moving total by $m = 6$ gives the 6-year moving average presented in column 6 of Table 12.3.

Since the moving average has smoothed out the original data and (hopefully) removed the irregular component, subtracting the trend values in column 3 from the 6-year moving average in column 6 should leave column 7 representing the cyclical component for each time period t. The X's in Figure 12.2 represent the data from column 7; they indicate a rather smooth wavelike function.

There are two noteworthy characteristics of the simple method of moving averages as presented here. First, the larger m (the number of time periods included in the moving average), the more data points are lost at both ends of the time series. In Tables 12.2 and 12.3 this characteristic is noted by the number of n.a.'s (i.e., not available) listed; in Figure 12.2, there are no points plotted at either end of the time series.

Second, the amplitude or magnitude of the cyclical component is dampened or reduced by the method of moving averages. This characteristic is best visualized in Figure 12.2. The "peaks" and "troughs" of the solid line (that is, $Y - T$ before any smoothing) are greater or more severe than those for the C values when a 3-year moving average is used. Further, the height (and depth) of the swing in the 6-year moving average is less than that for the 3-year moving average. In general, the larger m, the more damping of the resulting cyclical component.

The higher m, the better the chance that irregular component will be "averaged out," the smoother the curve, and the greater the reduction in the amplitude of the cyclical values. The lower m, the less the chance of "averaging out" the irregular component; the curve may be "kinkier," but the amplitude is not reduced as much. You make your choice of m and take your chances.

Finally, the simple method of moving averages weighs equally all m values of Y in calculating the moving average for any time period t. (When m is even, the "centering process" weighs $m - 1$ values equally.) The method of exponential smoothing is a technique for placing more emphasis on the more current values.

Simple exponential smoothing   A relatively new method of smoothing time-series data is called "exponential smoothing."[1] Instead of weighing all values equally, as is done with moving averages, we may feel that the more current observations deserve a relatively higher weight. Let X be the dependent variable in a time series, and let $S(X)_t$ be the "exponentially smoothed value of X in time period t." Then, $S(X)_t$ is given by

$$S(X)_t = \alpha X_t + (1 - \alpha)S(X)_{t-1} \qquad (12.6)$$

where α is an arbitrarily chosen weight such that $0 < \alpha < 1$, and $S(X)_{t-1}$ is the exponentially smoothed value for the previous time period.

Note that $\alpha + (1 - \alpha) = 1$. Thus, α is a measure of the relative emphasis placed on the current value of X, and "what's left," or $(1 - \alpha)$, is the measure of relative emphasis placed on the exponentially weighted average of all previous values. If you want the "smoothed value of X," $S(X)_t$, to be very responsive to the current value of X, X_t, then you set α close to 1. If you want to give emphasis to past values, you set α close to 0. The value of α is called "the smoothing constant." The *lower* α, the less the influence of the irregular component on $S(X)_t$.

It is true that $S(X)_{t-1}$ assumes that there are an infinite number of previous values for X; however, after just a few periods, those ancient values of X receive so little weight that they are essentially equal to 0. For example, let us consider the smoothing constant $\alpha = 0.5$ and look what happens to the weight placed on the value of X as time progresses. X_1 occurred when $t = 1$. The weight implied[2] by Equation 12.6 for X in the next time period ($t = 2$) is 0.25. Table 12.4 shows the weight placed on X, in future time periods. The name "exponential smoothing" is derived from the exponential decrease in emphasis placed on past data in the time series. Thus, the value of $S(X)_t$ is a weighted average of all past values of X, where the weights get exponentially smaller as t gets smaller.

But how do we choose a value for α? We have already seen that the higher α, the less relevant are past values of X; the lower α, the more relevant are past X values. Also, we can draw an analogy with the number of time periods (m) used in a moving average. It can be shown that

$$\alpha = \frac{2}{m + 1} \qquad (12.7)$$

---

[1] There are several variations on this general technique. For an elementary but lucid discussion of several models, see Barry Shore, *Operations Management*, McGraw-Hill Book Company, New York, 1973, chapter 11. For a more complete and more advanced treatment, see the pioneering work by Robert G. Brown, *Smoothing, Forecasting and Prediction of Discrete Time Series*, Prentice-Hall, Inc., Englewood Cliffs, N.J., 1963.

[2] Equation 12.6 can be written as the infinite series

$$S(X)_t = \alpha X_n + \alpha(1 - \alpha)X_{n-1} + \alpha(1 - \alpha)^2 X_{n-2} + \alpha(1 - \alpha)^3 X_{n-3} + \cdots$$

$$= W_n X_n + W_{n-1} X_{n-1} + W_{n-2} X_{n-2} + W_{n-3} X_{n-3} + \cdots$$

and has the property that $\sum_{i=1}^{n} W_i = 1$.

TABLE 12.4 WEIGHTS FOR $\alpha = 0.05$

t	Proportion of X_1 in $S(X)_t$
1	0.5000
2	0.2500
3	0.1250
4	0.0625
5	0.0312
6	0.0156
7	0.0078
8	0.0039

Thus, if it is easier to think in terms of the number of time periods used in a moving average, then Table 12.5 will assist in choosing a value for the smoothing constant (α).

In Table 12.1, we calculated a least-squares trend line, $T = a + bt$, for crude oil production in Saudi Arabia and found $C + I$ by subtracting the trend value in each time period from Y_t. We saw in Figure 12.2 that these $(C + I)$ values produced a rather jagged "cyclical plus irregular" graph. Now, let

$$X_t = C + I$$

These "detrended" values of crude oil production are rewritten in Table 12.6.

We can attempt to eliminate the irregular component in these X values by exponential smoothing. First, let us choose a "smoothing constant." Let $\alpha = 0.5$, which corresponds to a 3-year moving average (see Table 12.5).

From Equation 12.6, we need a value of $S(X)_{t-1=0}$ in order to calculate $S(X)_{t-1}$. This is always a problem when using exponential smoothing; we lack the *initial condition*. In the crude-oil production problem, we could go back and get the

TABLE 12.5 EQUIVALENT VALUES OF m AND α

m	α
3	0.5000
4	0.4000
5	0.3333
6	0.2857
7	0.2500
8	0.2222
9	0.2000
10	0.1818

TABLE 12.6 EXPONENTIAL SMOOTHING WITH $\alpha = 0.5$

t	$C + I$ X_t	Weighted New Value $(0.5)X_t$	+	Weighted Smoothed Previous Values $(0.5)S(X)_{t-1}$	=	Smoothed Value $S(X)_t$ C	
						(18.1)	← Initial Condition
1	15.36	7.680		9.05		16.73	
2	10.43	5.215		8.36 (5)		13.58	
3	28.51	14.255		6.79		21.04	
4	24.58	12.290		10.52		22.81	
5	15.66	7.830		11.41		19.24	
6	58.73	29.365		9.62		38.98	
7	47.80	23.900		19.49		43.39	
8	18.88	9.440		21.70		31.14	
9	23.95	11.975		15.57		27.54	
10	−6.97	−3.485		13.77		10.29	
11	−32.90	−16.450		5.14		−11.31	
12	−66.83	−33.415		−5.65		−39.07	
13	−93.75	−46.875		−19.53		−66.41	
14	−98.68	−49.340		−33.20		−82.54	
15	−77.60	−38.800		−41.27		−80.07	
16	−60.53	−30.265		−40.04		−70.30	
17	−48.46	−24.230		−35.15		−59.38	
18	−43.38	−21.690		−29.69		−51.38	
19	−45.31	−22.655		−25.69		−48.34	
20	30.77	15.385		−24.17		−8.79	
21	129.84	64.920		−4.39		60.53	
22	169.91	84.955		30.26		115.22	

$$\text{3-year moving average} = \frac{15.36 + 10.43 + 28.51}{3} = \frac{54.3}{3} = 18.1 \leftarrow \text{Initial Condition}$$

pre-1946 production data, find the arithmetic mean of the pre-1946 data, and set it equal to $S(X)_{t-1}$. But, what if that data did not exist?

Since $\alpha = 0.5$ corresponds to $m = 3$, let us take the first three time periods and calculate the arithmetic mean

$$\bar{X} = (\tfrac{1}{3})(15.36 + 10.43 + 28.51) = 18.1$$

and set it equal to $S(X)_{t=0}$. What if 18.1 is a poor approximation of $S(X)_0$? It will give us a "bad start," but the exponentially decreasing weight placed upon $S(X)_0$ as t

increases will eliminate the influence of $S(X)_0$; the *process* of exponential smoothing will wash out the influence of the "bad start."

Thus, for $\alpha = 0.5$, the exponentially smoothed values of X are

$$\alpha X_t + (1 - \alpha)S(X)_{t-1} = S(X)_t$$

$t = 1$	$(0.5)(15.36) + (0.5)(18.1)$	$= 16.73$
$t = 2$	$(0.5)(10.43) + (0.5)(16.73)$	$= 13.58$
$t = 3$	$(0.5)(28.51) + (0.5)(13.58)$	$= 21.04$

TABLE 12.7 EXPONENTIAL SMOOTHING FOR $\alpha = \frac{1}{3}$

t	$C + I$ X_t	Weighted New Value $(\frac{1}{3})X_t$	+	Weighted Smoothed Previous Values $(\frac{2}{3})S(X)_{t-1}$	=	Smoothed Value $S(X)_t$ C	
						(28.72)	← Initial Condition
1	15.36	5.12		19.15		24.27	
2	10.43	3.48		16.18		19.66	
3	28.51	9.50		13.11		22.61	
4	24.58	8.19		15.07		23.26	
5	15.66	5.22		15.51		20.73	
6	58.73	19.58		13.82		33.40	
7	47.80	15.93		22.27		38.20	
8	18.88	6.29		25.46		31.75	
9	23.95	7.98		21.17		29.15	
10	−6.97	−2.32		19.43		17.11	
11	−32.90	−10.97		11.41		0.44	
12	−66.83	−22.28		0.29		−21.99	
13	−93.75	−31.25		−14.66		−45.91	
14	−98.68	−32.89		−30.61		−63.50	
15	−77.60	−25.87		−42.33		−68.20	
16	−60.53	−20.18		−45.47		−65.65	
17	−48.46	−16.15		−43.76		−59.91	
18	−43.38	−14.46		−39.94		−54.40	
19	−45.31	−15.10		−36.27		−51.37	
20	30.77	10.26		−34.25		−23.99	
21	129.84	43.28		−15.99		27.29	
22	169.91	56.64		18.19		74.83	

$$\text{An average} = \frac{15.36 + 10.43 + 28.51 + 24.58 + 15.66 + 58.73 + 47.80}{7} = 28.72$$

If the exponential smoothing eliminates the irregular component, then $S(X)_t = C_t$ in this example. The unsmoothed values of $C + I$ and the exponentially smoothed values, column 5 of Table 12.6, are graphed in Figure 12.3. The effect of choosing $\alpha = \frac{1}{3}$ (corresponding to $m = 5$) for this data is calculated in Table 12.7 and graphed in Figure 12.3.

As compared to the method of moving averages, the technique of exponential smoothing has several advantages. First, a practical advantage, the method of exponential smoothing does not require that you "store" or keep track of past data. In the

FIGURE 12.3 Exponentially smoothed data for $Y - T$.

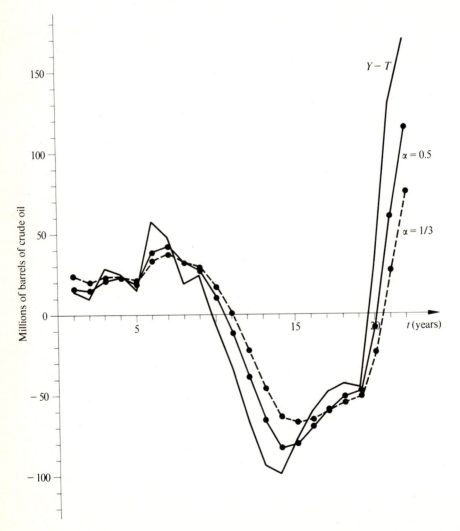

method of moving averages, you are continually dropping the oldest value (therefore, you must keep track of it) and adding a new value. In contrast, exponential smoothing requires you to know just $S(X)_{t-1}$, the value you last calculated.

Second, there is no "data loss" at the most recent end of the time series. Unlike Table 12.2, there are no "n.a's" in Table 12.6. The "bad start" associated with the initial conditions in exponential smoothing may be comparable to the "n.a's" for the oldest observations in calculating the moving average, but the most recent data is not plagued with "n.a."

Like the moving-average technique, exponential smoothing will also dampen cyclical swings in the raw data. Also, if the trend is not adequately accounted for, the method of exponential smoothing will pick it up in a "lag effect"; the $S(X)_t$ values will appear shifted to the left or right of the X_t values.

To summarize, when faced with *annual* time-series data for which an additive model appears relevant, the first step is to find an appropriate expression for the trend and then try to measure it (e.g., the method of least squares). Next, a smoothing technique, either moving averages or exponential smoothing, can be used in an attempt to eliminate the irregular component, which leaves the cyclical component as a residual.

Seasonal Component

For the additive model, the last component to consider is the seasonal component. If our time-series data has time periods of days, weeks, months, quarters, etc., then there may be an annually repeating regularity in that data. For ease of calculations only, we will consider only quarterly data. The procedures are the same if we divide the calendar year into 4 parts (quarters), 12 parts (months), 52 parts (weeks), or 365 parts (days).

First, let us consider a hypothetical case where there is no cycle and we have 6 years of quarterly data (see Table 12.8). The first step is to calculate a trend line from the values of Y and t: $T_t = 7.087 + (9.823)t$. If we have no cycle, then the model is $Y_t = T_t + S_t + I_t$. Detrending the Y values yields $Y_t - T_t = S_t + I_t$. Hence, for $t = 1$ to $t = 24$, the trend values T_t are calculated from the least-squares trend line and subtracted from Y_t.

The last column of Table 12.8 shows the values of $S + I$; Table 12.9 simply regroups this data. Now we consider each quarter separately. Consider the six "first-quarter" values:

$$(\tfrac{1}{6}) \sum_{t=1}^{6} (S_i + I_i) = (\tfrac{1}{6})\Sigma S_i + (\tfrac{1}{6})\Sigma I_i$$

but $(\tfrac{1}{6})\Sigma I_i$ should be close to 0 if the I values are random. Thus, the average of the six first-quarter values is $(23.16)/6 = 3.86$. The same calculation is performed for each quarter. The averaging process presumably eliminates the irregular component, and the average becomes a "representative value" for the seasonal fluctuation.

TABLE 12.8 DETRENDING HYPOTHETICAL TIME SERIES FOR ADDITIVE MODEL

	Time-Series Data			Regression Calculations		Detrending Y	
Year	Quarter	Time Period t	Actual Value Y_t	$Y_t \cdot t$	t^2	T_t	$S_t + I_t$
First	I	1	16	16	1	16.91	−0.91
	II	2	36	72	4	26.73	9.27
	III	3	28	84	9	36.56	−8.56
	IV	4	27	108	16	46.38	−19.38
Second	I	5	59	295	25	56.20	2.80
	II	6	84	504	36	66.03	17.97
	III	7	66	462	49	75.85	−9.85
	IV	8	75	600	64	85.67	−10.67
Third	I	9	112	1,008	81	95.49	16.51
	II	10	123	1,230	100	105.32	17.68
	III	11	114	1,254	121	115.14	−1.14
	IV	12	104	1,248	144	124.96	−20.96
Fourth	I	13	138	1,794	169	134.79	3.21
	II	14	159	2,226	196	144.61	14.39
	III	15	159	2,385	225	154.43	4.57
	IV	16	142	2,272	256	164.26	−22.26
Fifth	I	17	173	2,941	289	174.08	−1.08
	II	18	193	3,474	324	183.90	9.10
	III	19	194	3,686	361	193.72	0.28
	IV	20	197	3,940	400	203.55	−6.55
Sixth	I	21	216	4,536	441	213.37	2.63
	II	22	242	5,324	484	223.19	18.81
	III	23	240	5,520	529	233.02	6.98
	IV	24	220	5,280	576	242.84	−22.84
Total		300	3,117	50,259	4,900	—	—

$$b = \frac{n\Sigma Yt - (\Sigma Y)(\Sigma t)}{n\Sigma t^2 - (\Sigma t)^2} = \frac{(24)(50,259) - (3,117)(300)}{(24)(4,900) - (300)(300)} = \frac{271,116}{27,600} = 9.823$$

$$a = \bar{Y} - b\bar{t} = \frac{3,117}{24} - (9.823)\frac{300}{24} = 129.875 - 122.788 = 7.087$$

Thus, the trend line is $T_t = 7.087 + (9.823)t$.

TABLE 12.9 "AVERAGING" THE $S + I$ VALUES TO ELIMINATE I

	Year						Quarter	
Quarter	1	2	3	4	5	6	Total	Average
I	−0.91	2.80	16.51	3.21	−1.08	2.63	23.16	3.86
II	9.27	17.97	17.68	14.39	9.10	18.81	87.22	14.54
III	−8.56	−9.85	−1.14	4.57	0.28	6.98	−7.72	−1.29
IV	−19.38	−10.67	−20.96	−22.26	−6.55	−22.84	−102.66	−17.11
							Total	0

Our seasonal component for quarter I is 3.86; for quarter II, it is 14.54; and so forth. This technique does not ensure that the "total" in the last column of Table 12.9 will be zero. Pretend that the quarter II average is 13.54.

The seasonal component should add to 0 over the period of 1 year. Since

$$\sum_{Q=1}^{4} \bar{S}_Q = 3.86 + 13.54 + (-1.29) + (-17.11) = -1.00$$

we have to adjust our values of $\bar{S}_Q$ to make $\Sigma \bar{S} = 0$. If we add $\frac{1}{4}$ to each $\bar{S}_i$, then

$$\bar{S}_{Q=1} = \quad 3.86 + 0.25 = \quad 4.11$$

$$\bar{S}_{Q=2} = \quad 13.54 + 0.25 = \quad 13.79$$

$$\bar{S}_{Q=3} = \quad -1.29 + 0.25 = \quad -1.04$$

$$\bar{S}_{Q=4} = -17.11 + 0.25 = \quad -16.86$$

$$\Sigma \bar{S}_Q = \quad 0.00$$

This is a matter of adding one-fourth of the "excess" to each seasonal component.

Note that the values $Y_t - T_t$ depend upon which expression is used for a trend line. An exponential trend, rather than a straight-line trend, would yield different values for each $\bar{S}_Q$. Second, note that there are several ways to "average out" the irregular component. We could have used the median value (midway between 2.63 and 2.80 for the first quarter) instead of the arithmetic mean to determine each $\bar{S}_Q$. Or, considering only the detrended values for one quarter, we could exponentially smooth the detrended values. Finally, whatever process we use, some form of adjustment must be made to ensure that $\sum_{Q=1}^{4} \bar{S} = 0$.

By now it should be painfully clear that there is as much "art" as "science" to time-series analysis. The "subtraction strategy," the *choice* of the type of trend line that will be used, the *choice* of smoothing technique, the *choice* of adjustment procedure and averaging technique for the seasonal component all leave plenty of room for human judgment.

12.4 DETERMINATION OF COMPONENTS FOR THE MULTIPLICATIVE MODEL

In contrast to the additive model where we employed a "subtraction strategy," the multiplicative model employs a "division strategy" to separate the time-series components. In the additive model, a nonexistent component was represented by 0; that is, if there is no cycle, then $Y = T + C + S + I = T + 0 + S + I$, or $Y = T + S + I$. For the multiplicative model, a neutralized or nonexistent component is represented by a 1. If there is no cycle, then $Y = T(1)SI = TSI$. Except for the trend, which is measured in the same units as Y, we are interested in how the time-series components C, S, and I vary about 1.

Although the progression of the division strategy can vary, consider the following:

$$Y_t = T_t \times C_t \times S_t \times I_t \qquad \text{Observed data}$$

Assume we can develop a seasonal index. Then

$$\frac{Y_t}{S_t} = T_t \times C_t \times I_t \qquad \text{Deseasonalized data}$$

Assume we can develop a trend expression. Then

$$\frac{T_t \times C_t \times I_t}{T_t} = C_t \times I_t \qquad \text{Detrended and deseasonalized data}$$

Now, with some smoothing technique (i.e., moving averages or exponential smoothing) we can, hopefully, eliminate the irregular component and isolate the cycle. This process is sometimes called the "decomposition of a time series."

Seasonal Component

Frequently, business managers will feel that the seasonal fluctuations (due to custom and/or climate) are the slowest to change. This does *not* mean that there is not a drastic change in sales from December to January; what it does mean is that *there is always* a drastic change in sales from December to January. The pattern of change within a year changes very slowly, if at all, from year to year.

Put another way, if we are going to "summarize" the past, or if we are going to extrapolate the past into the future, of the four components of time-series analysis, many business persons have more confidence in the constancy of the seasonal pattern. Thus, we will start with the seasonal component.

First, consider several years of quarterly (or monthly or weekly) data. A four-quarter (or 12-month or 52-week) "centered" moving average will not include the seasonal component (because the average considers an entire year). Thus,

$$\frac{Y}{\text{Centered moving average}} = \frac{T \times C \times S \times I}{T \times C} = S \times I$$

Using the $S \times I$ data for the first quarter (or month or week) of each year, we could use some "average" to eliminate the irregular component, leaving a measure of the seasonal fluctuation for that quarter (or month or week). Using the data from Table 12.8, this process has been used to find an *index of seasonal fluctuation* in Tables 12.10 and 12.11.

TABLE 12.10 FOUR-QUARTER MOVING AVERAGE TO FIND $S \times I$

			Four-Quarter Moving			
Year (1)	Quarter (2)	Actual Value Y (3)	Total (4)	Centered Total (5)	Centered Average $T \times C$ (6)	(3)/(6) $Y/(T \times C)$ $= S \times I$ (7)
First	I	16		n.a.	n.a.	n.a.
	II	36	n.a.	n.a.	n.a.	n.a.
	III	28	107	128.5	32.125	0.8716
	IV	27	150	174.0	43.500	0.6207
			198			
Second	I	59	236	217.0	54.250	1.0876
	II	84	284	260.0	65.000	1.2923
	III	66	337	310.5	77.625	0.8502
	IV	75	376	356.5	89.125	0.8415
Third	I	112	424	400.0	100.000	1.1200
	II	123	453	438.5	109.625	1.1220
	III	114	479	466.0	116.500	0.9785
	IV	104	515	497.0	124.250	0.8370
Fourth	I	138	560	537.5	134.375	1.0270
	II	159	598	579.0	144.750	1.0984
	III	159	633	615.5	153.875	1.0333
	IV	142	667	650.0	162.500	0.8738
Fifth	I	173	702	684.5	171.125	1.0110
	II	193	757	729.5	182.375	1.0583
	III	194	800	778.5	194.625	0.9968
	IV	197	849	824.5	206.125	0.9557
Sixth	I	216	895	872.0	218.000	0.9908
	II	242	918	906.5	226.625	1.0678
	III	240	n.a.	n.a.	n.a.	n.a.
	IV	220		n.a.	n.a.	n.a.

TABLE 12.11 CALCULATION OF SEASONAL INDEX FROM $S \times I$ DATA

Year	Quarter I	Quarter II	Quarter III	Quarter IV	
First			0.8716	0.6207	
Second	1.0876	1.2923	0.8502	0.8415	
Third	1.1200	1.1220	0.9785	0.8370	
Fourth	1.0270	1.0984	1.0333	0.8738	
Fifth	1.0110	1.0583	0.9968	0.9557	
Sixth	0.9908	1.0678			
Total	5.2364	5.6388	4.7304	4.1287	Total ↓
Average	1.0473	1.1278	0.9461	0.8257	3.9469
Seasonal Index	1.0614	1.1430	0.9588	0.8368	4.0000

Column 7 of Table 12.10 is rearranged and placed in Table 12.11. If the average influence of the quarter average is to be 1.000, then the total of the four quarter averages should be 4.000; but, from Table 12.11, it is only 3.9469. In order to force the sum of the seasonal indices to be 4, we do the following:

$$\text{(Seasonal index)} = \frac{4.0000}{3.9469} \text{(quarter average)}$$

There are several variations on this theme. Instead of finding the arithmetic mean for the five values for each quarter, the median could be used for the quarter average. If there is reason to believe that the seasonal pattern is slowly changing, a linear regression line could be constructed for each quarter; that is, quarter average = $a + b(S \times I)$. If we feel that the more recent values are more relevant than the older values, we could exponentially smooth the $S \times I$ values, for each quarter, with a smoothing constant close to 1.

Whatever method is used to calculate the quarter average, there must still be some adjustment to ensure that the arithmetic mean of the seasonal indices is equal to 1. Also, the seasonal index is frequently given in percentages:

Quarter:	I	II	III	IV
Seasonal index:	106.14 percent	114.30 percent	95.88 percent	83.68 percent

The index can be interpreted as follows: (1) The first quarter is 6.14 percent better than average; (2) the second quarter is 14.3 percent better than average; (3) the third quarter is only 95.88 percent of the average; (4) the fourth quarter is only 83.68 percent of the average. A seasonal index is obviously useful information for a decision maker; it may be equally informative to know how the index was generated.

The Trend

The usual process is to plot the original data of a time series and observe the general tendency of the data. From that observation, you select a mathematical expression (a few are listed in Panel 12.4) and somehow (e.g., the method of least squares) measure the trend expression from the time-series data. But, should there be a logical justification for the mathematical expression chosen? The trend expression that best fits the data may have "best fit" as its only justification. These are questions which have long plagued time-series analysis. Without really answering them, we shall return to these questions in the next section.

Panel 12.4   A Sample of Possible Expressions for the Trend

Name	Mathematical Expresson	Selected Graphs
Exponential	$T = ab^t$ $\log (T) = \log (a)$ $+ (t) \log (b)$	
Nth-degree polynomial	$T = a + b_1 t + b_2 t^2$ $+ b_3 t^3 + \cdots$ $+ b_N t^N$	One case of $N = 3$
Gompertz	$T = ab^{c^t}$ $\log (T) = \log (a)$ $+ c^t \log (b)$	
Modified exponential	$T = a + bc^t$	
Pear-Reed or Logistic	$1/T = a + bc^t$	

TABLE 12.12 MOVING AVERAGE FOR CYCLICAL COMPONENT OF MULTIPLICATIVE MODEL

Year (1)	Quarter (2)	t (3)	Y (4)	Seasonal Index S (5)	Deseasonalized TCI (6)	Trend Values T (7)	Detrended Data CI (8)	5-Quarter Moving Total (9)	5-Quarter Moving Average C (10)	Irregular I (11)
1	I	1	16	1.0614	15.07	16.91	0.8915			
	II	2	36	1.1430	31.50	26.73	1.1783			
	III	3	28	0.9588	29.20	36.56	0.7988	4.5534	0.9107	0.8771
	IV	4	27	0.8368	32.27	46.38	0.6957	4.7749	0.9550	0.7285
2	I	5	59	1.0614	55.59	56.20	0.9891	4.5041	0.9008	1.0980
	II	6	84	1.1430	73.49	66.03	1.1130	4.7515	0.9503	1.1712
	III	7	66	0.9588	68.84	75.85	0.9075	5.1608	1.0322	0.8792
	IV	8	75	0.8368	89.63	85.67	1.0462	5.1935	1.0387	1.0072
3	I	9	112	1.0614	105.52	95.49	1.1050	5.1131	1.0226	1.0806
	II	10	123	1.1430	107.61	105.32	1.0218	5.2002	1.0400	0.9825
	III	11	114	0.9588	118.90	115.14	1.0326	5.1186	1.0237	1.0087
	IV	12	104	0.8368	124.28	124.96	0.9946	4.9755	0.9951	0.9995
4	I	13	138	1.0614	130.02	134.79	0.9646	5.0275	1.0055	0.9593
	II	14	159	1.1430	139.11	144.61	0.9619	5.0280	1.0056	0.9565
	III	15	159	0.9588	165.83	154.43	1.0738	4.9697	0.9939	1.0804
	IV	16	142	0.8368	169.69	164.26	1.0331	4.9233	0.9847	1.0491

Year (1)	Quarter (2)	t (3)	Y (4)	Seasonal Index S (5)	Deseasonalized TCI (6)	Trend Values T (7)	Detrended Data CI (8)	5-Quarter Moving Total (9)	5-Quarter Moving Average C (10)	Irregular I (11)
5	I	17	173	1.0614	162.99	174.08	0.9363	5.0059	1.0012	0.9352
	II	18	193	1.1430	168.85	183.90	0.9182	5.0887	1.0177	0.9022
	III	19	194	0.9588	202.34	193.72	1.0445	5.0094	1.0019	1.0425
	IV	20	197	0.8368	235.42	203.55	1.1566	5.0217	1.0043	1.1516
6	I	21	216	1.0614	203.50	213.37	0.9538	5.1777	1.0355	0.9211
	II	22	242	1.1430	211.72	223.19	0.9486	5.2158	1.0432	0.9093
	III	23	240	0.9588	250.31	233.02	1.0742			
	IV	24	220	0.8368	262.91	242.84	1.0826			

Seasonal Index from Table 12.8

Example: $\dfrac{Y}{S} = \dfrac{16}{1.0614} = 15.07$

Example: $\dfrac{TCI}{T} = \dfrac{15.07}{16.91} = 0.8915$

Table 12.5

Example: $\dfrac{CI}{C} = \dfrac{0.7988}{0.9107} = 0.8771 = I$

In the previous section, we used the data in Table 12.7 (or 12.10) to calculate a least-squares trend line of $T = 7.087 + (9.823)t$. We will use that trend line for the rest of this discussion.

The Cycle

We started this discussion of the multiplicative model by saying that we could deseasonalize the original data, detrend the deseasonalized data, and use the residual as a measure of $C \times I$. Using an averaging process to eliminate I would result in the cyclical component. With the assistance of Table 12.12, we will follow that "division strategy."

The previously calculated seasonal index for each quarter is recorded in column 5 of Table 12.12. The seasonal index is used to "deseasonalize" the original data (see column 6). Column 7, the values calculated from $T = 7.087 + (9.823)t$, is used to "detrend" the deseasonalized data. That is, the values for $Y/(TS) = CI$ are presented in column 8. A five-quarter moving average is used to smooth the CI values; the result is column 10, which is presumably the cycle. Although you might not often calculate the irregular component, that component is shown in column 11.

For any given time period t, the model is

$$Y_t = T_t \times C_t \times S_t \times I_t$$

or
$$\text{Col. 4} = (\text{col. 7}) \times (\text{col. 10}) \times (\text{col. 5}) \times (\text{col. 11})$$

For example, consider $t = 12$; the model says that

$$Y_{12} = (124.96)(0.9951)(0.8368)(0.9995) = 104.00$$

and from column 4 we have $Y_{12} = 104$. Of course, it has to be! The division strategy used to get all four components ensures that their product will get us right back to the actual value, Y_t.

Now consider Figure 12.4, where the results of Table 12.12 have been graphed. All four graphs have the same horizontal time scale, and C, S, and I have the same vertical scale. No, the titles have not been reversed on the "cycle" and "irregular" graphs! With the methods we used, there is no discernible cycle for this data. (Recall that in the previous section we used this data for the additive model and assumed that no cycle existed in it.)

Given these results, we would redefine our multiplicative model to be $Y = T \times S \times I$, and column 8 of Table 12.12 would be the "irregular index" for each time period.

To summarize, the procedure for decomposing the time-series data into a multiplicative model involves four steps:

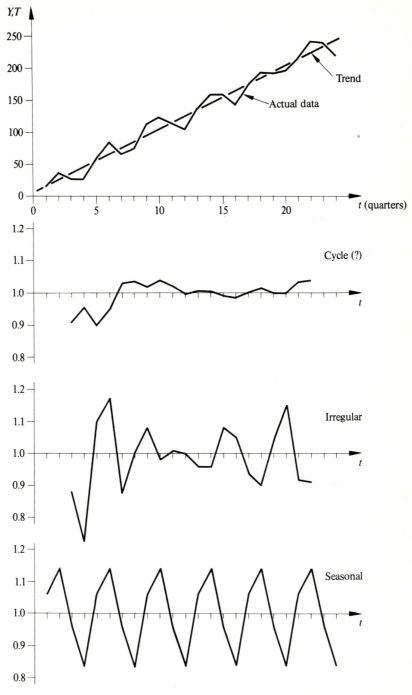

FIGURE 12.4   Time-series components from Table 12.12.

1. Find the seasonal index. Tables 12.10 and 12.11 outline one such method.
2. Choose an expression for the trend; calculate the trend expression; for each time period calculate a trend value. For the hypothetical data in Table 12.7, we chose a straight-line trend, $T_t = a + bt$; the values of a and b were determined by the method of least squares for the paired values of Y and t; for $t = 1$ to 24, T_t was calculated.
3. Deseasonalize and detrend the actual values; $Y/(T \times S) = C \times I$.
4. Use a smoothing technique (e.g., a moving average) to remove the irregular component.

A fifth step, calculating the "irregular index," $(CI)/C$, can be performed for completeness.

12.5 SUMMARY OF TIME-SERIES ANALYSIS

The previous four sections are intended as a survey of a collection of techniques that are employed in an attempt to separate four components: the trend, the seasonal, the cyclical, and the irregular. What have we observed?

First, there are a lot of "choices." Which type of time-series model should be used? An additive model? A multiplicative model? Some combination of the two? Should we look to the data for suggestions? Reconsider Panels 12.2a and b and 12.3a and b. Panels 12.2a and b took four components and created Y values for an additive model. Panels 12.3a and b took four components and created Y values for a multiplicative model. But carefully compare the Y values in both panels. They are identical! Obviously this data is fabricated, but faced with these Y values, what model would you choose? Either? Both? The uniqueness of an answer is illusive. There are not many national polls on the subject, but the author will offer the opinion that the multiplicative model is "more popular."

Another choice is the selection of technique for isolating a component. There is more than one way to determine the seasonal component. Which one is most appropriate? To a large degree, we have to rely on the investigator's judgment, understanding of the particular problem for which the data is used, and artistic or creative ability.

There is the choice of which mathematical expression should be used to represent the trend. For example, with the actual data for Aramco production of crude oil in Saudi Arabia, we used a straight line to represent the trend for crude oil production data. Does a third-degree polynomial represent the concept of "the trend" better than a straight line? Not necessarily.

Finally, whatever choices are made regarding the trend and the seasonal components, those "choices" will affect what we identify as the "cycle." The cyclical component is the subject of much debate. In Figure 12.3 a rather smooth cycle was observed when a straight-line trend was used for the oil production data. In Figure 12.5 the residuals (that is, $Y - T$) are plotted when the third-degree polynomial is used

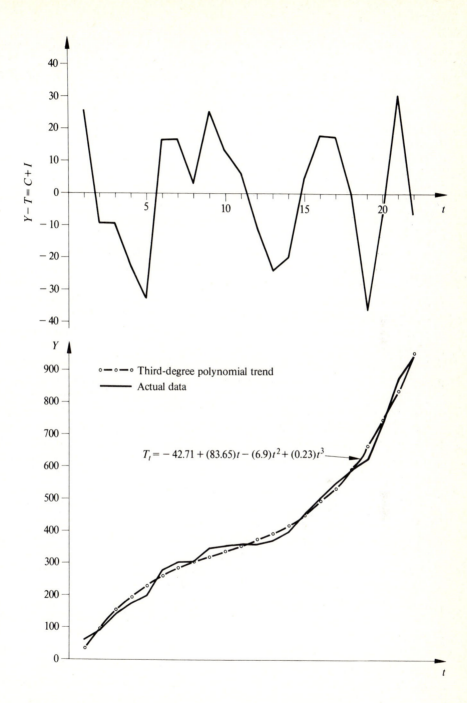

FIGURE 12.5 Nonlinear trend and residuals for Aramco crude-oil production.

for the trend. It should be obvious that when we smooth those residuals, not much of a "cycle" will appear.

Also, there is a disturbing phenomenon known as the "Slutzky-Yule effect." Basically, this effect says that a moving average of a series of random numbers may generate or create a wavelike result. Thus, in an attempt to smooth values of $(C + I)$ or $(C \times I)$, if I is a series of random numbers, the *method* of moving averages may create, rather than isolate, the cyclical component.

It should be clearly recognized that time-series analysis is an attempt to *describe* the historical movement of some variable through time. As an independent variable, "time" may or may not "cause" changes in Y. Certain biological variables (the weight of an ear of corn over the growing season) and certain physical processes (radioactive decay of isotopes), at one level of generality, may be thought of as changing because of the passage of time.

For many economic variables, "time" is not *the* (and possibly not even *a*) causal variable. Will GNP go up 6 percent *just because* 1 year has passed? Will the use of electric energy by American industry go up 2 percent *just because* 3 months have gone by? Of course not! GNP and electric energy demands are functions of other (and probably more important) variables. Up to this point, we have not necessarily assumed that time did cause changes in Y; so, why all the fuss? In the next section, we will use time series to "forecast" the future value of Y. When the variable "time" is used as a substitute for those "other variables," such forecasts should be christened with caution.

This argument is not universal among economic variables. Time, or some measure of time, may be a very useful substitute (or summary) variable for the combined effects of several variables. Does movie attendance increase just because it is Saturday night? For a complex array of reasons, Saturday night may be an institution for many Americans. "Just because it is Saturday night" may summarize much of the behavior that causes a change in an economic variable. Do retail sales of beer increase just because it is St. Patrick's Day? Do retail sales of wrapping paper go up just because it is December? Does the amount of red, white, and blue displayed increase just because it is July 4? Does the tide change just because the relative position of the moon changed? The answers may be the same, but the reasons for those answers differ. The last question is answered with reference to physical laws of gravity and motion. The first three are answered with reference to human behavior and American custom.

Therefore, while it might be very useful to use time as an explanatory variable for seasonal fluctuations in Y, it might be very dangerous to use time as an explanatory variable for the trend and cyclical variations in the same dependent variable Y. But, it is not just the seasonal component that offers benefit.

Much of this discussion might be interpreted as a demeaning critique of time-series analysis. If asked to summarize this section in two sentences, the reader might state: "With time-series analysis we have, at best, correctly identified the appropriate components and have well described the change in Y with the passage of time, without knowing why or how each component occurs. At worst, time-series analysis massages the data until it produces components which 'appear to describe' the movement of Y through time; we have a fabricated description of Y, and we still don't know why Y changes."

As a logical game, assume that statement is valid. From the viewpoint of forecasting some (near) future value of Y, does that statement demolish the usefulness of time-series analysis? Not necessarily. If your purpose is to provide some "ballpark" forecast of Y in the next time period (or in the next couple of time periods), time-series analysis may be sufficient.

Even if you want the most thoroughly researched forecast for some future value(s) of Y (which, by the way, does not necessarily make it more accurate), time-series analysis may play a very important role. First, the seasonal fluctuations may not be easily explained with "other variables." Second, as historical description, time-series data does provide some perspective. Third, time-series analysis is frequently a useful starting point; it may suggest other variables. The actual time-series data or one of the components may exhibit a striking similarity to some other time series.

No matter how well we understand the causes of changes in Y, if a forecast of some future value of Y is required, some form of extrapolation or extending into the future is necessary. Time-series analysis is one way of doing that, and understanding the weakness of these procedures should permit more informed interpretations of the forecast.

12.6 FORECASTING

The importance of a *forecast* for the value of some economic variable, in some future time period, is self-evident. In practice, business managers and economists use a variety of forecast models. Some are predominantly subjective and are frequently called *judgment forecasts*. Some forecast models are complicated mathematical models, a subset of which is called "econometric models" (see Chapter 13). Some firms use several different models as a way of getting several independent forecasts.

Briefly, forecasting with a time-series model is a matter of extending into the future the model's description of the past. Assume a model which exhibits no cyclical component. By assuming that whatever system of influences that led to the past trend will persist in the future, we can extend the trend into future time periods. If 6 years of quarterly data (that is, $t = 1$ to 24) provided the trend $T_t = a + bt$, then letting $t = 25$, 26, ... will provide forecasts for future values of the trend. In like manner, if we can assume that the seasonal fluctuations are stable and will continue to be stable, then we can adjust the quarterly trend forecast with the seasonal index.

The existence of a cyclical component presents even greater problems. Cycles frequently demonstrate variation in both their duration and their severity. For short-term (i.e., in the near future) forecasts, a 15-year cycle might not have much influence on a forecast that included only the trend and the seasonal adjustments. However, a 10-year forecast would find it difficult to ignore the possibility of a 15-year cyclical component. Hence, the frequently illusive cyclical component becomes important. Partially as a result, forecasts are usually accompanied by a set of "conditions." Instead of a detailed explanation of the forecast model used, the conditions are the assumptions about the future environment that the forecasters deemed most relevant in evaluating the subjective probability that the forecast would be realized.

Forecasting is a segment of almost any type of planning. Time-series analysis has been an important ingredient in the history of economic forecasting. Shall we extrapolate that experience into the future?

EXERCISES

1. Consider the following four times series separately:

Year	Quarter	Series W	X	Y	Z
1	I	71	29	48	11
	II	92	32	12	16
	III	4	35	15	18
	IV	45	38	45	16
2	I	15	41	48	11
	II	1	44	12	6
	III	38	47	15	4
	IV	66	50	45	6
3	I	54	53	48	11
	II	72	56	12	16
	III	18	59	15	18
	IV	74	62	45	16

Also, consider the following preliminary calculations:

Variable	Median	Sum of Squared Successive Differences	Sum of Squared Deviations from the Mean
W	49.5	19,635	10,263.6
X	45.5	99	1,287.0
Y	30.0	6,633	3,294.0
Z	13.5	149	272.9

For each time series:

a. From section 9.2, use the runs test to determine if the data is random.
b. From section 9.2, use the von Neumann ratio to test the randomness of each series?
c. Plot each series.
d. What component or components would you look for in each series?
e. How do the results from parts *a* and *b* relate to each component in a time series?

2. Consider the following data from Madame Easter's Hat Shoppe.

Year	Quarter	Sales ($)
1973	I	808
	II	2,040
	III	824
	IV	416
1974	I	840
	II	2,120
	III	856
	IV	432
1975	I	872
	II	2,200
	III	888
	IV	448

a. Plot the data. What components does it suggest?

Part A: The Additive Model

b. Find a trend line, with the method of least squares, such that $t = 1$ in the first quarter of 1973.
(*Hint:* $\Sigma Y = 12{,}744$, $\Sigma t = 78$, $\Sigma Y^2 = 18{,}375{,}500$, $\Sigma t^2 = 650$, $\Sigma Yt = 80{,}384$.)
c. Detrend the original data.
d. Using the method of moving averages with the data from part *c*, find the seasonal component.
e. Find the irregular component for each quarter.
f. With your answers from parts *b* and *d*, forecast sales for each quarter of 1976.

Part B: The Multiplicative Model

g. Use the trend line from part *b* to detrend the data.
h. From the results of part *g*, use the method of moving averages to find the seasonal component.
i. Find the irregular component for each quarter.
j. With your answers from parts *b* and *h*, forecast sales for each quarter of 1976.
k. Compare your answers from parts *f* and *j*.

3. Reconsider the data in Exercise 2. There appears to be very strong seasonal influences. So, let us try a slightly different approach.

Part A: The Additive Model ($Y = T + S + I$)

a. Use a four-quarter moving average on the original data to find the seasonal components.
b. Deseasonalize the original data.

c.   With the deseasonalized data, calculate a least-squares trend line with $t = 1$ at the first quarter of 1973.

d.   Use the results of parts *a* and *c* to forecast quarterly sales for 1976.

Part B: The Multiplicative Model ($Y = TSI$)

e.   Use the moving averages from part *a* to get the seasonal indices.

f.   Deseasonalize the original data.

g.   From the results of part *d* calculate a least-squares trend line with $t = 1$ for the first quarter of 1973.

h.   Forecast the quarterly sales for 1976 with this multiplicative model.

4.   Comment on the following statements:

a.   For the model $Y = T + CSI$, you must detrend the data before you attempt to deseasonalize the data.

b.   The model $Y = TSCI$ is appropriate for annual data.

5.   Use a multiplicative model for the number of barn fires per year in a farm state.

Year	Number of Barn Fires	Year	Number of Barn Fires
1	189	11	288
2	293	12	484
3	379	13	595
4	404	14	520
5	382	15	446
6	265	16	349
7	177	17	180
8	80	18	104
9	95	19	108
10	194	20	259

a.   Find the least-squares trend line. ($\Sigma Y = 5{,}791$; $\Sigma t = 210$; $\Sigma Y^2 = 2{,}113{,}269$; $\Sigma t^2 = 2{,}870$; $\Sigma Yt = 60{,}131$.)

b.   Use a 3-year moving average to smooth the original data.

c.   Take the ratio of the 3-year moving average to the trend values to find an index of the cyclical component.

d.   Detrend the original data.

e.   Use exponential smoothing ($\alpha = 0.5$) on the detrended data and determine the cyclical component.

f.   Does either part *c* or part *e* tend to indicate a cyclical component for barn fires? If so, what "period" does the cycle have?

6.   The ABC Corporation has constructed a time-series model for the number of sickdays per employee per month. The model has this form: $Y = TSI$.

Trend: $T = 0.50 + (0.0001)t$ where $t = 1$ refers to January 1970; t units are months; T units are sickdays per employee per month.

Seasonal:

Jan.	Feb.	Mar.	Apr.	May	June	July	Aug.	Sept.	Oct.	Nov.	Dec.
1.09	0.98	0.97	0.99	1.01	1.02	1.01	1.03	1.05	1.10	0.88	0.87

a. In which months are sickdays above average?

b. Forecast the number of sickdays per employee per month for January 1975. For April 1975.

c. Does the trend line add much to this model? Would the company be just as well off with $\bar{Y}$ and the seasonal component?

d. Without changing the units or the origin, assume the trend line to be $T = 0.5 + 0.1(t)$. Forecast the number of sickdays per employee for January and April 1975.

e. Do the answers to part d make much sense?

THE CONSTRUCTION
OF ECONOMIC MODELS

13.1 MODEL BUILDING

As any body of knowledge which becomes branded as a "discipline," economics faces a methodological problem of *how* the human investigators (usually referred to as "economists") do whatever it is they do. Such terms as "the scientific method," "deductive logic," and "inductive logic" frequently differentiate various methodological considerations. Even the most practical of economists occasionally indulges in (what at first glance may appear to be irrelevant, philosophical) questions of methodology. Although it is not our purpose to pursue the questions of methodology in economics, it should be obvious that most methodologies would require a comparison between "what is" (e.g., what is observed in the famous "real world") and a hypothesis, theory, or model.

One elementary economics text touts that "Einstein started with facts"[1] and proceeds to precede all introductory-level economic principles with a discussion of "facts"—usually with regard to the United States economy. These "facts" are tables of data, series of index numbers, rank-orderings, graphs of "paired observations" (i.e., consumption expenditures and disposable income), and so forth. A little reflection should reveal that these "facts" employed many of the "tools of statistical analysis" that we have endured in the previous chapters. Hopefully, those "facts" can now be viewed with a more jaundiced eye. But that is not the sole purpose, as valid as it may be, of statistical analysis in economics.

It is difficult, unless you define away the problem, to classify the "facts" as the chicken or the egg. But for the purpose of our discussion (we are defining away the problem), it does not make any difference. Whether the collection, manipulation, and presentation of economic data are used to formulate a hypothesis, theory, or model *or* whether data is used to test a non-data-derived hypothesis, theory, or model is irrelevant. The point is that data is collected, manipulated, presented, and interpreted in the development, refinement, and application of economic theory.

We have discussed the collection, summarization, and presentation of data. We have spent even more time developing techniques for testing hypotheses. The techniques of regression, correlation, and time-series analysis may be used to formulate or investigate economic theories and models. This seems to beg the question

---

[1] R. G. Lipsey and P. O. Steiner, *Economics*, Harper & Row, New York, 1972.

"What is the difference between hypothesis, theory, and model?" Abbreviating Webster's *Third International Dictionary* we find:

Hypothesis a proposition tentatively assumed in order to draw out its logical or empirical consequences and so test its accord with facts that are known or may be determined.

Theory imaginative contemplation of reality ... a working hypothesis given probability by experimental evidence or by factual or conceptual analysis but not conclusively established or accepted as a law.

Model a description, a collection of statistical data, or an analogy used to help visualize often in a simplified way something that cannot be directly observed ... a theoretical projection in detail of a possible system of human relationships.

Obviously a high degree of semantic selection was used in associating those particular definitions with those three words. A "model" who puts on a flesh-colored bra is photographed and pictured in the Sears Roebuck Catalogue may be "a theoretical projection in detail of a possible system of human relationships"; however, that is not exactly the meaning the economists intend to convey when they discuss a model. Thus, these three definitions are a subjective estimate of the most frequently used meanings in the literature concerning economics.

It may be useful to think of a hypothesis as a proposition that does not have widespread support (i.e., a large proportion of those humans who have considered the hypothesis do not believe it to be valid—the majority may be "reserving judgment"); a theory is a hypothesis which enjoys widespread support; and a model is a collection of theories and/or hypotheses. This "view" of the interrelationships among hypotheses, theories, and models may not enjoy "widespread support," but it does serve as a starting point for trying to determine what each particular author means by these three words.

What, then, are the ingredients of an "economic model"? Having an economic problem to consider is certainly a necessary (but not sufficient) condition. The most explicit possible statement of the problem is a prerequisite for efficient construction of an economic model. Next, you might collect the various hypotheses and theories that have attempted to deal with aspects of the (hopefully) well-defined problem. Or you may generate your own hypothesis (which usually includes a brew of logic, perception, and imagination). You may wish to specify particular relationships (using statistical analysis) to evaluate which variables and/or considerations appear to be influential. Those hypotheses and theories not dismissed in this process are then merged (again the inclusion of imagination and logic adds flavor) into a model. Depending upon your interest (i.e., description or prediction—if they are different), the resulting model should enable you to make some statement about the problem.

This summary of model construction is undeniably vague (some would even classify it as "denoting criminal neglect"), but the overview provides a useful reference for the rest of this chapter. We will consider a particular economic problem, further discuss strategy in developing an economic model, and investigate the process of evaluating the "accuracy," as opposed to "validity," of an economic model.

13.2 THE IDENTIFICATION PROBLEM

The generalization "economic model" discussed in the previous section has meaning, but relating it to your knowledge of economics may be somewhat "strained." Perhaps you have heard of the Brookings Institution's "Model of the Economy" with its dozens of equations and variables. The economic affairs of the United States are extremely complicated, and so is the Brookings model. But even the Brookings model is an abstraction. Not all possible variables are included; not all possible relationships are attempted.

Any model (economic or not) includes some variables and excludes others. *Ceteris paribus* (i.e., "other things being constant") is a familiar assumption in economics; however, we seldom, if ever, collect economic data in the controlled conditions of a laboratory experiment. The "other things" frequently forget not to change. So let us take a simple model and see how economic theory and statistical analysis embrace and inflate or strangle and inflict one another.

Consider the market for a product and some familiar concepts: supply and demand. Suppose we are interested in constructing a demand curve for a particular product. We usually define the *market demand curve* for a product as the maximum rate of purchase (i.e., units of product per unit of time) that buyers would indulge in at each price. That is, $Q_d = f(P)$. Further, we normally define the *market supply curve* as the maximum rate of output sellers will offer at each price. That is, $Q_s = g(P)$. The intersection of this supply curve and demand curve will yield the equilibrium price and output.

Consider a hypothetical food product: feep (*f*ish, b*ee*f, *p*ork). We would like to generate a "model" of the demand for feep. We might hypothesize the following:

$$Q_d = \alpha_d + \beta_d P + \xi_d \tag{13.1}$$

where Q_d = millions of pounds of feep purchased per year
   P = price per pound
   α_d, β_d are parameters and ξ_d is an error term

A substantial part of economic theory has been devoted to rationalizing a downward-sloping demand curve (remember marginal utility and indifference curves?). It is almost an automatic reaction—the lower the price, the larger the rate of purchase. Thus, we would expect $\alpha_d > 0$ and $\beta_d < 0$.

TABLE 13.1   MARKET EXCHANGE DATA FOR FEEP

Year	Q_d Quantity Sold (Millions of Pounds)	P Price (Dollars per Pound)
1962	16	8
1963	24	10
1964	32	14
1965	32	12

Table 13.1 gives us the amount of feep purchased and its price for a 4-year period.

To eliminate some initial problems concerning the reliability and derivation of this data, let us assume that on January 1 of each year a price of feep is established (which is not necessarily the same thing as the producers setting the price) and that market price does not change during the year. Further, let us assume that there was no inflation (or deflation) during the period 1962 to 1965 in the United States. Finally, let us assume that there is a remarkably precise reporting system concerning the quantity of feep purchased. You may also be interested in knowing that feep is a fantastically homogeneous product.

With this "laboratory-type data" and our "model" of the demand curve, it should be a simple process to estimate the parameters in Equation 13.1. Try it!

Do you get the following?

$$Q_d = -4.8 + (2.8)P \qquad (r^2 = 0.944) \tag{13.2}$$

What do our results imply? Feep is a Giffen-good? (For the uninitiated, a Giffen-good is a product or service that experiences an *increase* in the rate of units purchased when the price *increases*.) Is 94 percent of the variation explained by changes in price?

Feep may or may not be a Giffen-good, but the data presented in Table 13.1, *by itself*, will not permit us to come to that conclusion. This is the problem we wish to discuss. The high value of r does not indicate that we have included all important variables.

One view of Table 13.1 is the tautology that the amount buyers purchased is equal to the amount that sellers sold. Alternatively, the data is four paired observations on a price-quantity diagram where each paired observation represents the intersection of the supply curve and the demand curve. Either the supply curve or the demand curve or both could have "shifted" from one year to the next. These "shifts" could have been caused by changes in "other" variables affecting either supply or demand.

Our first mistake consisted of not correctly stating our model. The "model" suggested by economic theory is:

$$Q_d = f(P) \qquad \textit{ceteris paribus}$$

$$Q_s = g(P) \qquad \textit{ceteris parabis}$$

$$Q_d = Q_s \qquad \text{in equilibrium}$$

Moreover, let us assume these demand and supply relationships are linear within the relevant range:

$$Q_d = \alpha_d + \beta_d P \qquad \textit{ceteris paribus} \qquad \text{(13.3a)}$$

$$Q_s = \alpha_s + \beta_s P \qquad \textit{ceteris paribus} \qquad \text{(13.3b)}$$

$$Q_d = Q_s \qquad \text{in equilibrium} \qquad \text{(13.3c)}$$

Since we have already discussed the sign of the parameters in the demand equation, what would microeconomic theory suggest for the supply equation's[2] parameters?

We would expect that the offer price and the quantity offered would be directly proportional: the greater the price, the larger the quantity offered for sale. Thus, $\beta_s > 0$. When we discuss some "relevant range of output" where $Q_s \neq 0$, then α_s may be either positive or negative without violating any principles of economic theory.

For the moment, let us ignore sample errors. Without considering what the "other variables" are, let us consider their potential influence on the market equilibrium condition.

In Panel 13.1 we depict the situation where either no "other variables" affect supply and demand or there is no change in those "other variables" which does affect supply and/or demand (i.e., the *ceteris paribus* conditions hold). We could have 2, 20, or 2 million observations, and we would not be able to distinguish a supply curve from a demand curve. In this case we say that neither the supply curve nor the demand curve are identified; both curves are *underidentified*.

Next, assume that either there are no "other variables" influencing the supply curve or there are no changes in those "other variables" which do influence the supply curve (i.e., the *ceteris paribus* conditions hold only for supply). With regard to the demand curve, assume that there are influential "other variables" which change each time period (i.e., the *ceteris paribus* conditions are violated for demand). This situation is portrayed by Panel 13.2 where the paired observations lie along a station-

---

[2] To simplify the concept of a "market supply curve," we will assume the market meets the conditions of perfect competition; thus, to maintain a touch of reality, feep is an agricultural product rather than, for example, a motor vehicle.

Panel 13.1   No Change in the "Other Variables" in Either the Supply or Demand Functions

In every time period we would observe the same price and quantity. There is no way to estimate parameters in Equation 13.3.

$$P_0 = (\alpha_s - \alpha_d)/(\beta_d - \beta_s)$$

$$Q_0 = \alpha_s + \beta_s P_0 = \alpha_d + \beta_d P_0$$

$$Q_s = ?   \quad Q_d = ?$$

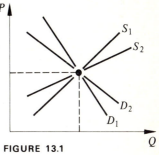

FIGURE 13.1

Panel 13.2   "Other Variables" Change *Only* in Demand Function

Supply curve is the same from one time period to the next and is determined. Demand curve shifts around and is not determined from P and Q alone.

$$Q_s = 10 + 2P$$

$$Q_d = ?$$

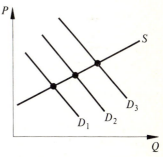

FIGURE 13.2

Panel 13.3   "Other Variables" Change *Only* in Supply Function

Supply curve shifts from one period to another and is not determined by P and Q alone. Demand curve is the same for each period and is determined.

$$Q_s = ?$$

$$Q_d = 10 + 2P   \quad \text{(a Giffen-good)}$$

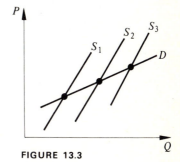

FIGURE 13.3

Panel 13.4 "Other Variables" Change in *Both* Supply and Demand Functions

Both the supply and demand curves shift from one time period to the next, and neither curve is determined from only P and Q data.

$$Q_s = ?$$

$$Q_d = ?$$

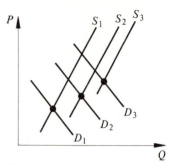

FIGURE 13.4

---

ary supply curve but represent only one point on each of several demand curves. Therefore, "the data" would enable us to estimate the parameters in the supply equation (Equation 13.3b), but we can say nothing about the demand curve (Equation 13.3a). In this case, the supply curve is said to be *identified* while the demand curve is *underidentified*.

Panel 13.3 presents a situation similar to Panel 13.2; the only difference is that the roles of the demand curve and the supply curve have been switched.

Finally, if there are "other variables" which independently influence supply and demand (but not both) and the *ceteris paribus* conditions are violated, then both the supply curve and the demand curve will shift around from period to period (see Panel 13.4). Neither the supply curve nor the demand curve is identified from only the price-quantity data.

Note that we have employed the economic model of supply and demand (Equations 13.3a, b, and c) to direct *how* statistical analysis (i.e., method of least squares for a simple linear regression equation) may or may not be used to estimate a particular demand curve. In order to decide which situation the price-quantity data for feep demonstrates, we need to (1) specify the "other variables" in our model and (2) obtain data for those "other variables." The situations described in Panels 13.1 to 13.4 represent what is known as *the identification problem when dealing with a model consisting of a system (more than one) of simultaneous equations.*

What variables other than price might influence the rate of purchase of a product or service? The level of income? Size of the population? Prices of "other" (substitutes and complements) commodities? Tastes of consumers? Remember that the demand curve, $Q_d = f(P)$ is defined *ceteris paribus* and that "shifts" in the demand curve are caused by changes in these "other variables." Thus, we might say that $Q_d = f(P, Y, H)$, where Y = level of income per period and H = average number of humans per period.

What factors other than price might influence the rate of production of a product or service? Reverting back to economic theory, under certain conditions the supply curve for a perfectly competitive industry is the "horizontal summation" (at a particular price how much each firm would offer) of each firm's supply curve, which is its marginal-cost curve above average variable cost. (If you didn't follow that and wish you could, then please refer to the "short-run supply curve" section of your elementary economics text. The remaining discussion should make sense even if you don't understand the derivation of the "short-run supply curve.") The marginal-cost curve is determined by (1) technology available to the firm and (2) the relative prices of the variable inputs used by the firm. Even if we simplify and say there is only one variable input, then Q_s will depend upon the price of the input. As you can see, even if we hypothesize a market supply curve (it generally requires the product market to be perfectly competitive), specifying "other" variables and the form (i.e., exponential, second-degree polynomial, etc.) of their appearance in the supply function is not always easy.

Before you give up hope, we can return to our feep example and suggest an alternative. Assume that the feep is a snooty gormandizer that field-grazes only on "whoats." "Whoats" depends primarily on the average annual rainfall. Therefore, the amount of feep supplied will depend, in part, upon the rainfall each year. With this dainty piece of information, we will rewrite our demand and supply functions as follows:

$$Q_d = \alpha_d + \beta_d P + \gamma_d Y + \theta_d H + \xi_d \qquad \text{Demand} \qquad \text{(13.4)}$$

$$Q_s = \alpha_s + \beta_s P + \phi_s R + \xi_s \qquad \text{Supply} \qquad \text{(13.5)}$$

$$Q_s = Q_d \qquad \text{Equilibrium} \qquad \text{(13.6)}$$

It is important to note that disposable income (Y) and population (H) are assumed to be independent of each other and neither have any influence on the output considerations of suppliers. In like manner, average annual rainfall (R) has no influence on the demand for feep.

We introduced Y, H, and R as "shift" variables because when their values change, they cause the supply (in the case of R) and demand (for Y and H) curves to "shift." These "shift" variables are more frequently called *exogenous* variables; that is, their values are not determined by the model, by Equations 13.4 and 13.5, but are "given" or "provided" from outside the model. Likewise, variables whose values are determined by the model (for example, Q and P) are called *endogenous*.

If the values of Y and H are essentially constant for all time periods under consideration, but the average annual rainfall varies a great deal from one year to the next, then we have a supply curve shifting around "mapping out" an essentially stationary demand curve. See Panel 13.3. This would appear to solve our original problem: construct the demand curve for feep.

Just Identified Demand Function

If Panels 13.1 through 13.4 were four candidates for the title "Most Similar to Mr. Real World," then Panel 13.4 would receive at least a plurality. Panel 13.4 presents two cases of interest. In both cases, we will assume that average annual rainfall (R) demonstrates variability over time.

For our first case, assume that the population (H in Equation 13.4) does not vary from time period to time period (ZPG has been attained); however, the disposable income per time period (Y in Equation 13.4) does vary. (Statistically, the null hypothesis $\theta_d = 0$ could not be rejected under this assumption.) Rewriting Equation 13.4 to account for this assumption, our model becomes:

$$Q_d = \alpha_d + \beta_d P + \gamma_d Y + \xi_d \qquad \text{Demand} \qquad (13.7)$$

$$Q_s = \alpha_s + \beta_s P + \phi_s R + \xi_s \qquad \text{Supply} \qquad (13.5)$$

$$Q_d = Q_s \qquad \text{Equilibrium} \qquad (13.6)$$

This model or pair of simultaneous equations is called *just identified*.[3] That is, the supply equation contains one exogenous (or shift) variable R, which is not found in the demand equation; therefore, the demand equation is (just) identified. In like manner, Y is not found in the supply equation, and the parameters in Equation 13.5 can be estimated. In order to clarify this concept of "just identified," we need to introduce two other concepts of model building: "structural" and "reduced-form" equations.

The demand and supply functions in Equations 13.6 and 13.7 are relationships suggested by economic theory. Both are known as *structural equations;* they represent the theoretical structure of the hypothesized model. Please note, however, that both contain two endogenous variables (Q and P) and one exogenous variable (Y for demand and R for supply). Given the values of Y and R for any particular time period, the theoretical model says that P and Q are *jointly* determined by solving this pair of simultaneous equations; P and Q are obtained from the intersection of supply and demand.

That is, in any given year we interpret the data to say that there was some price where the quantity demanded was equal to the quantity supplied. By setting Equation 13.5 equal to Equation 13.7 we can solve for that price, or

$$P = (\alpha_s - \alpha_d)/(\beta_d - \beta_s) - [\gamma_d/(\beta_d - \beta_s)]Y + [\phi_s/(\beta_d - \beta_s)]R$$
$$+ (\xi_s - \xi_d)/(\beta_d - \beta_s) \qquad (13.8)$$

or
$$P = \pi_1 - \pi_2 Y + \pi_3 R + \pi_4 \qquad (13.9)$$

---

[3] "Just" is employed here as an adverb to indicate "exactly" and does not denote "fair," "right," or "legal."

where $\pi_1 = (\alpha_s - \alpha_d)/(\beta_d - \beta_s)$, $\pi_2 = [\gamma_d/(\beta_d - \beta_s)]$, etc. In like manner, the quantity supplied and demanded at this price is

$$Q_s = Q_d = Q = [(\alpha_s \beta_d - \alpha_d \beta_s)/(\beta_d - \beta_s)] - [\gamma_d \beta_s/(\beta_d - \beta_s)]Y$$
$$+ [\phi_s \beta_d/(\beta_d - \beta_s)]R + [(\xi_s \beta_d - \xi_d \beta_s)/(\beta_d - \beta_s)]$$
$$(13.10)$$

or
$$Q = \psi_1 - \psi_2 Y + \psi_3 R + \psi_4 \qquad (13.11)$$

where $\psi_1 = [(\alpha_s \beta_d - \alpha_d \beta_s)/(\beta_d - \beta_s)]$, and so forth.

Equations 13.9 and 13.11 are called the *reduced form* of the structural equations given in Equations 13.5 and 13.7. Why should we go to all this algebraic effort? Note that the left side of each reduced-form equation is a single endogenous variable, and the right side consists of constants, (only) exogenous variables, and an error term.

Now we may use the method of least squares to get consistent and unbiased estimators of the coefficients in the reduced-form equations (that is, π_1, π_2, π_3, ψ_1, ψ_2, ψ_3). In the case where each equation in the model (i.e., Equations 13.5 and 13.7) are *just identified*, we can use these coefficients from the reduced-form equations to determine the consistent (but not unbiased) estimators of the structural equations. To see how this is accomplished please note that

$$\beta_s = \frac{\psi_2}{\pi_2} = \frac{\gamma_d \beta_s/(\beta_d - \beta_s)}{\gamma_d/(\beta_d - \beta_s)} \qquad (13.12)$$

and
$$\beta_d = \frac{\psi_3}{\pi_3} = \frac{\phi_s \beta_d/(\beta_d - \beta_s)}{\phi_s/(\beta_d - \beta_s)} \qquad (13.13)$$

With that information, we may proceed to find

$$\gamma_d = \pi_2(\beta_d - \beta_s) = \frac{\gamma_d(\beta_d - \beta_s)}{(\beta_d - \beta_s)} \qquad (13.14)$$

and
$$\phi_s = \pi_3(\beta_d - \beta_s) = \frac{\phi_s(\beta_d - \beta_s)}{(\beta_d - \beta_s)} \qquad (13.15)$$

Finally,
$$\alpha_d = \psi_1 - \pi_1 \beta_d \qquad (13.16)$$

and
$$\alpha_s = \psi_1 - \pi_1 \beta_s \qquad (13.17)$$

Appropriately, this technique is called the *method of indirect least squares*. Equations 13.12 to 13.17 yield estimates of all the parameters in the structural equations (e.g., supply and demand) from the estimates of the parameters in the reduced-form equations.

Example 13.1 Method of Indirect Least Squares for Feep Data

Let our variables be represented by the following (also see Table 13.2):

$$Y = \text{disposable income per time period}$$

$$Q = \text{quantity of feep purchased per time period}$$

$$P = \text{price of feep during the time period}$$

$$R = \text{average rainfall in feep-producing areas during time period}$$

TABLE 13.2 CALCULATIONS FOR FEEP DATA

P	Q	Y	R	Y^2	R^2	RY	YP	RP	YQ	RQ
8	16	6	4	36	16	24	48	32	96	64
10	24	9	6	81	36	54	90	60	216	144
14	32	13	2	169	4	26	182	28	416	64
12	32	12	8	144	64	96	144	96	384	256
44	104	40	20	430	120	200	464	216	1,112	528

From the reduced-form equations given by Equation 13.9 and 13.11, we can see that $P = \pi_1 - \pi_2 Y + \pi_3 R + \pi_4$ and $Q = \psi_1 - \psi_2 Y + \psi_3 R + \psi_4$ are the two expressions we wish to start with. By taking the expected value of both equations and noting that if $E(\xi_s) = E(\xi_d) = 0$, then $E(\pi_4) = E(\psi_4) = 0$, we obtain

$$\hat{P} = \hat{\pi}_1 - \hat{\pi}_2 Y + \hat{\pi}_3 R \quad \text{and} \quad \hat{Q} = \hat{\psi}_1 - \hat{\psi}_2 Y + \hat{\psi}_3 R$$

The least-squares estimates[4] of the parameters in the reduced-form equation for $\hat{P}$ are given by:

$$\Delta = [(\Sigma Y)^2 - n\Sigma Y^2][n\Sigma R^2 - (\Sigma R)^2] + (\Sigma Y \Sigma R - n\Sigma YR)^2$$

$$= [(40)^2 - (4)(430)][4(120) - (20)^2] + [(40)(20) - 4(200)]^2$$

$$= -9,600$$

---

[4] For the equation $Y = a + bX + cZ$, the *normal equations* are:

$$na + (\Sigma X)b + (\Sigma Z)c = \Sigma Y$$

$$(\Sigma X)a + (\Sigma X^2)b + (\Sigma XZ)c = \Sigma YX$$

$$(\Sigma Z)a + (\Sigma XZ)b + (\Sigma Z^2)c = \Sigma YZ$$

(1) Solve the first equation for a and substitute the result into the second and third equations; (2) solve the second and third equations for b and c; (3) employ the symbol substitution $a = \hat{\pi}_1$ or $\hat{\psi}_1$, $b = -\hat{\pi}_2$ or $-\hat{\psi}_2, \ldots, Y = P$ or Q, etc.

$$\Delta_1 = (n\Sigma PY - \Sigma P\Sigma Y)[n\Sigma R^2 - (\Sigma R)^2] - (n\Sigma PR - \Sigma P\Sigma R)(n\Sigma YR - \Sigma R\Sigma Y)$$

$$= [4(464) - (44)(40)][4(120) - (20)^2] - [4(216) - 44(20)][4(200) - 20(40)]$$

$$= 7,680$$

$$\Delta_2 = [(\Sigma Y)^2 - n\Sigma Y^2](n\Sigma PR - \Sigma P\Sigma R) - (\Sigma Y\Sigma R - n\Sigma YR)(n\Sigma PY - \Sigma P\Sigma Y)$$

$$= [(40)^2 - 4(430)][4(216) - 44(20)] - [40(20) - 4(200)][4(464) - 44(40)]$$

$$= 1,920$$

$$\hat{\pi}_2 = \Delta_1/\Delta = (7,680)/(-9,600) = -0.8$$

$$\hat{\pi}_3 = \Delta_2/\Delta = (1,920)/(-9,600) = -0.2$$

$$\hat{\pi}_1 = \bar{P} + \bar{Y}\hat{\pi}_2 - \bar{R}\hat{\pi}_3 = 11 + 10(-0.8) - 5(-0.2) = 4$$

The least-squares estimates of the parameters in the reduced-form equation for $\hat{Q}$ are given by:

$$\Delta = -9,600$$

$$\Delta_3 = (n\Sigma QY - \Sigma Q\Sigma Y)[n\Sigma R^2 - (\Sigma R)^2] - (n\Sigma QR - \Sigma Q\Sigma R)(n\Sigma YR - \Sigma R\Sigma Y)$$

$$= [4(1,112) - 104(40)][4(120) - (20)^2] - [4(528) - 104(20)][4(200) - 20(40)]$$

$$= 23,040$$

$$\Delta_4 = [(\Sigma Y)^2 - n\Sigma Y^2](n\Sigma QR - \Sigma Q\Sigma R) - (\Sigma Y\Sigma R - n\Sigma YR)(n\Sigma QY - \Sigma Q\Sigma Y)$$

$$= [(40)^2 - 4(430)][4(528) - 104(20)] - [40(20) - 4(200)][4(1,112) - 104(40)]$$

$$= -3,840$$

$$\hat{\psi}_2 = \Delta_3/\Delta = (23,040)/(-9,600) = -2.4$$

$$\hat{\psi}_3 = \Delta_4/\Delta = (-3,840)/(-9,600) = 0.4$$

$$\hat{\psi}_1 = \bar{Q} + \bar{Y}\hat{\psi}_2 - \bar{R}\hat{\psi}_3 = 26 + 10(-2.4) - 5(0.4) = 0$$

We may now return to the estimates of the parameters in the structural equations by using Equations 13.12 through 13.17:

$$\hat{\beta}_s = \hat{\psi}_2/\hat{\pi}_2 = (-2.4)/(-0.8) = 3$$

$$\hat{\beta}_d = \hat{\psi}_3/\hat{\pi}_3 = (0.4)/(-0.2) = -2$$

$$\hat{\gamma}_d = \hat{\pi}_2(\hat{\beta}_d - \hat{\beta}_s) = (-0.8)(-2 - 3) = 4$$

$$\hat{\phi}_s = \hat{\pi}_3(\hat{\beta}_d - \hat{\beta}_s) = (-0.2)(-5) = 1$$

$$\hat{\alpha}_d = \hat{\psi}_1 - \hat{\pi}_1\hat{\beta}_d = 0 - 4(-2) = 8$$

$$\hat{\alpha}_s = \hat{\psi}_1 - \hat{\pi}_1\hat{\beta}_s = 0 - 4(3) = -12$$

Inserting these results into the structural equations yields

$$\hat{Q} = 8 - 2P + 4Y \qquad \text{Demand}$$

$$\hat{Q} = -12 + 3P + R \qquad \text{Supply}$$

Please note from the demand curve that feep is not a Giffen-good; quantity decreases as price increases.

Since there is perfect correlation, you may generate the data for this problem by solving for P and Q, given the values of Y and R, the above demand and supply equations.

Overidentification

Returning to the supply and demand relationships given by Equations 13.4 and 13.5, consider a market where there have been significant population increases during the relevant time periods; H is not constant. Let us construct the reduced-form equations and consider the indirect least-squares technique. By setting Equation 13.4 equal to Equation 13.5 and solving for P (try it, you'll like it), we obtain

$$P = [1/(\beta_d - \beta_s)][(\alpha_s - \alpha_d) - \gamma_d Y - \theta_d H + \phi_s R + (\xi_s - \xi_d)] \qquad \text{(13.18)}$$

or
$$P = \pi'_1 - \pi'_2 Y - \pi'_3 H + \pi'_4 R + \pi'_5 \qquad \text{(13.19)}$$

where $\pi'_1 = (\alpha_s - \alpha_d)/(\beta_d - \beta_s)$, $\pi'_2 = \gamma_d/(\beta_d - \beta_s)$, etc. Rewriting both Equations 13.4 and 13.5 in the form of $P = f(Q, \ldots)$, setting the rewritten equations equal to one another, and solving for Q yield

$$Q = [1/(\beta_d - \beta_s)][(\beta_d \alpha_s - \beta_s \alpha_d) - (\beta_s \gamma_d)Y - (\beta_s \theta_d)H$$
$$+ (\beta_d \phi_s)R + (\beta_d \xi_s - \beta_s \xi_d)] \qquad \text{(13.20)}$$

or
$$Q = \psi'_1 - \psi'_2 Y - \psi'_3 H + \psi'_4 R + \psi'_5 \qquad \text{(13.21)}$$

Thus, Equations 13.19 and 13.21 represent the reduced-form equations (check the placement of the endogenous and exogenous variables) for the structural relationships given by Equations 13.4 and 13.5.

Again, let us translate the π's and ψ's of the reduced-form equations into the parameters of the structural equations (that is, α, β, etc.). Carefully recognize that *two* estimates of β_s are now available; that is,

$$\beta_s = \frac{\psi'_2}{\pi'_2} = \frac{\beta_s \gamma_d/(\beta_d - \beta_s)}{\gamma_d/(\beta_d - \beta_s)} \qquad \text{(13.22)}$$

and
$$\beta_s = \frac{\psi'_3}{\pi'_3} = \frac{\beta_s \theta_d/(\beta_d - \beta_s)}{\theta_d/(\beta_d - \beta_s)} \qquad \text{(13.23)}$$

Does this simply mean that $\psi'_2/\pi'_2 = \psi'_3/\pi'_3$? Algebraically, yes; however, in terms of using the method of least squares to estimate the parameters (i.e., the π's and ψ's) of the reduced-form equations, no! The ratio of the two estimators[5] $\hat{\psi}'_2$ and $\hat{\pi}'_2$ will frequently not be equal to the ratio of $\hat{\psi}'_3$ and $\hat{\pi}'_3$. Thus, *we frequently have two estimates for* β_s—*this is the problem presented by overidentification.*

Two reactions immediately come to mind. First, if we are interested in estimating the demand function, who cares if there are two estimates for β_s, a parameter in the supply function? It is true that there is only one estimate for β_d:

$$\beta_d = \frac{\psi'_4}{\pi'_4} \qquad (13.24)$$

Consequently, there is only one value estimated for α_d:

$$\alpha_d = \psi'_1 - \pi'_1\beta_d \qquad (13.25)$$

However, consider the remaining parameters in Equation 13.4:

$$\gamma_d = \pi'_2(\beta_d - \beta_s) \qquad (13.26)$$

and $$\theta_d = \pi'_3(\beta_d - \beta_s) \qquad (13.27)$$

Since both of these parameters depend upon β_s, there will be two estimates for each parameter associated with an exogenous variable in the demand equation. The demand equation is affected!

The second reaction to consider is: Why not change the model? If we drop population or income from the demand equation, then the model's equations are just identified. Why not? Although it is uncommon for any scientist to use the word "never," it is the only word appropriate for this situation. The *validity* of the model does not depend upon identification. If we develop a theoretical model whose structural equations lead to overidentification, the relevant question is: How do we obtain consistent estimates of the parameters? Thus, it is not our structural equations that require alteration: it is our method of estimation that demands attention! The indirect least-squares method does not yield consistent estimates of the parameters in the structural equation when the system is overidentified.

There are other methods for estimating these parameters; one of the more popular is the method of *two-stage least squares*. The details of the two-stage least-

---

[5] Recall from Chapter 10 that, except in the case of perfect correlation, when you regress X on Y you will obtain an equation algebraically different from the one where Y was regressed on X. Intuitively, this should help in understanding why the two ratios will not always be equal. It is equally important to recognize that while the analogy of regressing X on Y or Y on X is similar in effect, it is different in form from the "dual estimate" problem considered here.

squares and other methods are beyond the scope of this book (they belong, by tradition at least, in an econometrics course).

Results of the Identification Problem

Let us summarize briefly. If the model is underidentified, then it will be *impossible*, regardless of sample size, to estimate the parameters for at least one of the model's structural equations. Consider Equations 13.3*a*, *b*, and *c*. Neither the supply nor the demand equation can be estimated! If we add income (Y) to the demand equation (Equation 13.1), then the supply equation is identified, but the demand equation is not. Instead, if we include rain (R) in the supply function, Equation 13.3, and leave Equation 13.1 unchanged, the demand equation cannot be estimated.

If a model yields underidentified the equation of particular interest to our problem, then our only alternative is to develop a logical hypothesis (or hypotheses) that makes the model identifiable. We must return to the proverbial "drawing boards."

If the structural equations of a model are just identified, all equations may be estimated by the method of indirect least squares. Reconsider Equations 13.5 and 13.7 where only income (Y) appears as an exogenous variable in the demand equation and average annual rainfall (R) appears in the supply equation. The parameters of both the supply and demand equations are estimated by the method of indirect least squares. (See Equations 13.12 through 13.17.)

In an overidentified model, all the structural equations can be estimated by method of indirect least squares; however, these estimates will not be consistent, and the technique will produce a multiplicity of estimates for some parameters. (See Equations 13.22 and 13.23.) In order to obtain consistent estimators for an overidentified model, some other method (e.g., two-stage least squares) of estimation is required.[6]

Summary

The identification problem logically precedes the problems of statistical estimators discussed in Chapter 7. The state of identifiability (i.e., underidentified, just identified, or overidentified) can be investigated without any empirical data.

The state of identification does not really apply to a model; it denotes the measurability of an equation within the model. Using structural Equations 13.4 and 13.5, the demand equation is just identified and the supply equation is overidentified; we do *not* say that the model is partially overidentified. The state of identification of a particular equation does, of course, depend upon the model in which it exists. It is the simultaneous existence of several factors that determines whether a coal mine is

---

[6] There are alternative techniques such as the full-information maximum-likelihood method.

likely to (1) cave in, (2) explode, or (3) be safe for human occupancy. Analogously, the identification of a particular equation is a function of its "environment" or model.

While you could argue that statistical analysis is a necessary method of evaluating (or testing) economic hypotheses, theories, or models, this section should make it clearer that to correctly apply statistical tools, a generous understanding of economic theory (and your explicit assumptions) is required. In the construction of a demand curve (e.g., for feep), economic theory (e.g., supply and demand) and statistical tools (e.g., linear regression) are not just two distinct steps which yield an "economic model." Instead, economic theory and statistical tools are two components in a feedback loop, which must be viewed carefully in order to produce a viable economic model.

13.3 ACCURACY ANALYSIS

In the previous section we investigated the interconnected problems of economic theory and statistical methods. We *surveyed* the problems of constructing models and did not treat any of these problems in depth. At best, we only suggested solution techniques for each of these problems. The detailed investigation of such problems and the listing of possible solutions are generally regarded as the topic called *econometrics*.

Gerhard Tintner[7] suggests:

> Econometrics is a method now widely used in economic research. It consists in the application of modern statistical procedures to theoretical models, which have been formulated in mathematical terms. These methods are of interest in connection with the verification of economic laws and also are potentially useful for economic policy.

This text is not a discussion of econometrics; however, it is intended as an introduction to modern statistical procedures, which is a necessary ingredient to bake the "econometrics" cake. Before you take the next step, attempt an understanding of econometrics, you may well want to use or evaluate the use of an already constructed econometric model. In part, you can participate in this exercise without any more knowledge of econometrics than you currently possess.

For example, suppose you wish to forecast the migration into (or out of) your particular region of the United States. Further, assume that you have two different but equally acceptable (relative to your problem) models. One factor you might consider is the accuracy of each model. That is, if you have migration data for several past periods of time, how accurately would each model have predicted those population movements?

---

[7] Gerhard Tintner, *Econometrics*, Science Editions, John Wiley & Sons, Inc., New York, 1952, p. vii.

If one model yielded forecasts that were consistently closer to the actual values than the other model, the choice of which model is "more accurate" is trivial. It is not unrealistic to assume that many situations would produce a roughly equal number of "closer" or "more accurate" predictions by each model. What criteria do we employ, then, to "pass judgment" on the relative accuracy of these two models? We shall consider two such "measures of accuracy."

Theil's Inequality Coefficient

Henri Theil has considered several measures for "accuracy analysis." We will only consider the *U inequality coefficient* which Theil presents in *Applied Economic Forecasting* (North-Holland, Amsterdam, 1966). Theil's analysis is based on the concept of a "loss function," which fortunately is not necessary for our "first look" at accuracy analysis. Further, we will only consider point estimates.

13.3(a)   The Prediction-Realization Diagram

First, we need a way to visualize the accuracy of a model's predictions. Theil's "prediction-realization diagram" is an uncommonly simple device for observing the "accuracy" of a set of forecasts. The paired observations we need to consider are "percent change predicted" by the model and "percent change realized" (or "observed") from the real-world data. Denote F_i as the percent change forecasted and A_i as the actual percent change in whatever variable is under consideration. Also, we will use time-series data; thus, i refers to a particular time period.

TABLE 13.3   HYPOTHETICAL DATA ON PERCENT CHANGE IN GNP

i (Time Period, Quarter)	F_i (Percent Change Forecasted)	A_i (Actual Percent Change)	Point Designation
1	2	4	G
2	-1	2	H
3	-3	-3	J
4	0	2	K
5	3	-1	L
6	-2	0	M
7	-4	-3	N
8	-1	-4	P
9	3	1	Q
10	4	4	R

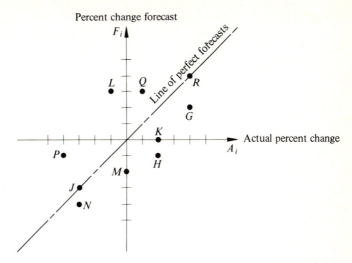

FIGURE 13.5   A plot of the forecasted versus actual percent change from Table 13.3.

The hypothetical data presented in Table 13.3 assumes that the actual GNP for the last quarter was known before the GNP forecast for the next quarter is generated. Further, let us assume that the actual GNP for each quarter is known with certainty; there are no recording errors or estimation errors in the A_i values.

Each paired value (F_i, A_i) in Table 13.3 has been plotted in Figure 13.5.[8] For example, in the first time period (3 months), the predicted change in GNP was a 2 percent increase; for that time period, a 4 percent increase was observed (point G). If a 1 percent change was represented by the same distance on both the A and the F axes, then the 45° line represents the *line of perfect forecasts* (LPF). Note that in period $(i =)$ 3, the 3 percent decrease in GNP was accurately predicted. In time period 10, the observed 4 percent increase was accurately predicted. *Thus, the closer the points in Figure 13.5 are to the LPF, the more accurate the model.*

Although points that lie off the "line of perfect forecasts" represent "errors in predicting change," they may represent different kinds of error. For one reason or another we may view one kind of error as more important than other. These "errors" can be classified as turning-point errors and non-turning-point errors.

---

[8] Theil points out that it is tempting to plot F_i on the horizontal axis. If the "actual values" are composed of a nonstochastic component *plus* a stochastic factor and the "forecasting model" attempts to account for only the nonstochastic component, then (analogous to the linear regression model) it is tempting to place the "first known" point estimate F on the horizontal axis and the "later known" actual value (which includes the stochastic component) on the vertical axis. However, this alternative presentation would make it more difficult to generate Figure 13.6, which has some pedagogical advantages. H. Theil, *Applied Economic Forecasting*, North-Holland, Amsterdam, 1966, p. 20.

13.3(a.1) Turning-Point Errors

The *direction of change* was not predicted correctly. Either an increase was predicted (see point *L*) and a decrease was observed, or a decrease was predicted (see point *H*) and an increase was experienced. In evaluating your (subjective?) model for forecasting price changes in your favorite common stock, you may well place a great deal more importance on turning-point errors (but probably less importance on point *H* than on point *L*) than on other types of error.

13.3(a.2) Direction of Change Correctly Predicted but Error in Predicting Magnitude of Change

Underestimation refers to the error of predicting less change than was observed. Consider point *G*. A 2 percent increase was forecasted, and a 4 percent increase was realized. The direction of change was correctly predicted; however, the absolute value, or magnitude, of the change was underestimated. A quarterly GNP increase of 2 percent may call for mild warnings about inflationary problems whereas a 4 percent increase, correctly predicted, would have invoked anti-inflationary government action. Note that point *P* represents a similar case. The decrease was correctly predicted, but the decrease was more drastic than the prediction indicated. In this case, $|F_8| = 1 < 4 = |A_8|$.

Overestimation occurs when the direction of change is correctly predicted but the absolute value of the predicted value is greater than the absolute value of the observed change; then we say the model overestimated the change. Points *N* and *Q* illustrate overestimation.

It is quite reasonable for you to ask: But if the quarterly GNP were forecasted to decrease by 4 percent, would not the government take steps which would normally lead to an observed change in GNP to be something like 3 percent? Assuming that governments can react that quickly, of course that action could occur; however, that is *not* the situation which F_i and A_i are designed to represent. The model producing the forecasts (the F_i's) assumes, implicitly or explicitly, certain behavioral trends and institutional arrangements. If the actions of the government alter these trends or arrangements, then we may have a bias in the model's predictions. Not only should it be expected, but Figure 13.5 should assist in detecting that bias.

Another relevant consideration is the exogenous variables within the predictive model. If the GNP prediction model has the rate of increase in the money supply as an exogenous variable, and that variable is influential within the time period, then the relevant F_i is generated by the new value of the exogenous variable. "Announcement effects" have been discussed elsewhere and cannot be eliminated entirely in the accuracy analysis of models where human behavior is included.

Return to Figure 13.5. If we rotate (clockwise) the diagram 45°, we have "line of perfect forecasts" serving as the horizontal axis (Figure 13.6). Theil calls this rotated diagram the *prediction-realization diagram*. The sets of points representing the above-

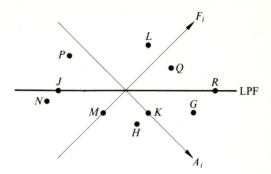

FIGURE 13.6 Prediction-realization diagram.

mentioned "errors" are noted on Figure 13.6. The purpose of rotating Figure 13.5 by 45° was not just to write in those "errors" already defined.

Now we want to introduce a different scheme for classifying the "errors" and superimpose this new scheme on the old. This new scheme for classifying errors can be deduced by solving the riddle: What do points P, L, and Q have in common? First, they all lie above the line of perfect forecasts in the prediction-realization diagram. But that is just the pedagogical reason for constructing the prediction-realization diagram in this fashion. What is the significance of a point being above the LPF in Figure 13.5?

Note that for points P, L, and Q, the following relationship exists: $F_i > A_i$. The *value* of the forecasted variable (*not* the percent change in that variable) is always greater than the actual value of the variable. If in time period 13 the GNP was 100 and the forecasted GNP for period 14 is 98, then the percent change is $F_{14} = -2$ percent. If the actual GNP for period 14 is 96, then $A_{14} = -4$ percent. The direction of change was correctly predicted. The magnitude of the change was underestimated (that is, $|F| < |A|$), but the *level of GNP was overestimated* (that is, $F_i > A_i$).

Thus, all points above the LPF in Figure 13.6 represent overestimation of the level of GNP. In like manner, *all points below LPF in Figure 13.6 represent underestimation of the level of GNP.* Therefore, the prediction-realization diagram permits us to visualize, in one diagram, a model's predictive accuracy; where the model is in "error" we can detect: (1) overestimation or underestimation of the variable itself, (2) the ability to predict direction of change in the variable, and (3) the ability to forecast the magnitude of change.

Further, this diagram has built in another forecasting model with which you can compare. Consider the *no-change forecast model.* That is, you take the actual GNP this quarter and say that that is your estimate for next quarter's GNP. Reflect a moment. Have you ever used the "no-change forecast model"? Some people even use it for forecasting GNP. Reconsider Figure 13.5. All possible results from the no-change forecast model lie along the A axis, where $F_i = 0$. The only point where this model could be correct is at the origin, where the LPF cuts the A axis (where there is no change observed).

The no-change forecast model is cheap to construct; you don't need any knowledge of statistical methods or economic theory. Thus, any other model that requires knowledge, time, and money to construct should be more accurate than the no-change model. Let us translate this argument of the prediction-realization diagram in Figure 13.6. In general, *a model should generate a set of points that are closer to the LPF than is the A axis*. It is in this context that there is a "built-in" model for comparison.

Now let us move on to algebraic measures of "accuracy." We mentioned that we could completely describe the normal (or Gauss') distribution with only two parameters: the mean and standard deviation. Can we summarize the accuracy of predictions from a particular model with one number?

13.3(b)   The Root-Mean-Square (RMS) Prediction Error

Henri Theil did not develop the RMS measure of predictive error; it is a measure that, for some time, has been popular in the physical sciences. The "mean-square prediction error" for n pairs of prediction-observation is

$$\text{MS} = (1/n)\Sigma(F_i - A_i)^2 \tag{13.28}$$

The "error" is $(F_i - A_i)$, the difference between the numerical value of the prediction and the actual numerical value. Obviously $\Sigma(F_i - A_i) = 0$ does not ensure that $F_i = A_i$ for all i. Recall that $\Sigma(X_i - \bar{X}) = 0$ even where $X_i \neq \bar{X}$ for all i. Thus, $\Sigma(F_i - A_i)^2$ yields a measure of aggregate error in the n predictions, and MS is an average "squared error per prediction."

Just as the standard deviation permits us to work with the same units as the mean, so the root-mean-square

$$\text{RMS} = \sqrt{(1/n)\Sigma(F_i - A_i)^2} \tag{13.29}$$

permits us to discuss "error" in the same units as the predicted variable. (See Table 13.4.) For example, from our previous example:

$$\text{RMS} = \sqrt{(1/n)\Sigma(F_i - A_i)^2}$$
$$= \sqrt{(\tfrac{1}{10})(51)} = \sqrt{5.1}$$
$$= 2.26 \text{ percent per quarter}$$

TABLE 13.4 CALCULATION OF
ROOT-MEAN-SQUARE ERROR

F_i	A_i	$F_i - A_i$	$(F_i - A_i)^2$
2	4	−2	4
−1	2	−3	9
−3	−3	0	0
0	2	−2	4
3	−1	4	16
−2	0	−2	4
−4	−3	−1	1
−1	−4	3	9
3	1	2	4
4	4	0	0
			51

The RMS has some merit.[9] Our purpose here, however, is to introduce one traditional measure of "accuracy" and the building blocks for constructing Theil's inequality coefficient.

13.3(c)   Theil's Inequality Coefficient U

In discussing the prediction-realization diagram we used the built-in comparative model of no-change forecast. Theil's inequality coefficient is essentially the "accuracy" of any given model relative to the "accuracy" of the no-change forecast model. Consider the no-change forecast model. Since the next period's forecast is this period's observed value, then the percentage change is *zero*. That is, $F_i = 0$ for all i. Thus, for the no-change forecast model

$$\text{RMS} = \sqrt{(1/n)\Sigma A_i^{\,2}} \qquad (13.30)$$

Theil's inequality coefficient (U) is simply the ratio of the RMS for a particular model to the RMS of the no-change forecast model.

$$U = \frac{\sqrt{(1/n)\Sigma(F_i - A_i)^2}}{\sqrt{(1/n)\Sigma(A_i)^2}}$$

or
$$U = \sqrt{[\Sigma(F_i - A_i)^2]/(\Sigma A_i^{\,2})} \qquad (13.31)$$

---

[9] If you have no information about the model generating the predictions, then the RMS can be used as a rough estimate of the "standard error of estimate" or the standard deviation of the stochastic variable within the model. Theil states that if ". . . the prediction errors can be regarded as, say, independent random variables with zero mean [our example has $(1/n)\Sigma(F_i - A_i) = 0.1$] and a certain RMS value, then we can use this result to formulate probability statements about future predictions even if the forecaster himself refrains from doing so." (Henri Theil, *Applied Economic Forecasting*, North-Holland, Amsterdam, 1966, p. 27. Bracketed material not in original.)

TABLE 13.5 INTERPRETATION OF RELATIVE VALUES OF U

$U = 0$	Model yields *perfect predictions.*
$0 < U < 1$	Model yields *more* accurate predictions than would the no-change forecast model.
$U = 1$	The *same* accuracy could have been obtained from the no-change forecast model.
$U > 1$	Model yields *less* accurate predictions than would the no-change forecast model.

Now let us interpret the meaning of U. If all the predictions were perfect, then $F_i = A_i$ for all i and $U = 0$. If $U = 1$, then the model's predictions are no more (or no less) accurate than the no-change forecast model. We may consider the no-change forecast model to be naive (and usually cheap to construct), but if $U = 1$, whatever other model was used is no more accurate. (See Table 13.5.)

Some judgments are in order. Except for the extremely desirable but unusual case of perfect predictions, the predictive accuracy of economic models is essentially a relative concept. As a standard for comparison, the no-change forecast model is as good as any. Theil's U does not require any more data than the RMS, but it does provide a method of comparison within one measure of accuracy. Thus, in a sense, Theil's U is preferred to the RMS.

For example, return to our hypothetical GNP quarterly forecast (which contained every possible type of error plus two perfect forecasts).

$$U = \sqrt{[\Sigma(F_i - A_i)^2]/(\Sigma A_i^2)} = \sqrt{51/76} = \sqrt{0.671} = 0.819$$

That is, our hypothetical model produced an RMS prediction error that is 82 percent of the RMS prediction error that would have been produced by a no-change forecast model. (By eliminating the two perfect results, you would find that $U = 1$, which is purely coincidental.) The model is more accurate than the no-change forecast model.

13.3(d) Theil's "First Set of Inequality Proportions"[10]

The evaluation of a model's accuracy (or error) can be carried one step further. That is, we wish to investigate the question: Which particular kinds of error contribute to the inaccuracy of the model's predictions? We start by observing that for any given set of A_i, the value of U will be uniquely determined by the numerator of U. Consider the numerator of U^2 (or the MS given by Equation 13.28), which can be written as

$$(1/n)\Sigma(F_i - A_i)^2 = (\bar{F} - \bar{A})^2 + (s_F - s_A)^2 + 2(1 - r)s_F s_A \qquad \textbf{(13.32)}$$

---

[10] "First set" implies the existence of a second set, which is presented in *Applied Economic Forecasting*, pp. 33–36. They are equally useful and should be referenced by any serious participant in "accuracy analysis." The second set places primary emphasis on considerations *after* you have eliminated all systematic error in your prediction; that is, $(F_i - A_i)$ represents a random variable with 0 mean but nonzero variance.

where the means are $\bar{F} = (\Sigma F_i)/n$ and $\bar{A} = (\Sigma A_i)/n$, the standard deviations are $s_F = \sqrt{(1/n)\Sigma(F_i - \bar{F})^2}$ and $s_A = \sqrt{(1/n)\Sigma(A_i - \bar{A})^2}$, and the correlation coefficient is $r = \Sigma[(F_i - \bar{F})(A_i - \bar{A})]/(ns_F s_A)$. (You should check the algebraic validity of the identity given by Equation 13.32.) If we divide both sides of Equation 13.32 by the left side of that equation, we find

$$1 = \frac{(\bar{F} - \bar{A})^2}{(1/n)\Sigma(F - A)^2} + \frac{(s_F - s_A)^2}{(1/n)\Sigma(F - A)^2} + \frac{2(1 - r)s_F s_A}{(1/n)\Sigma(F - A)^2}$$

$$= U^m + U^s + U^c \tag{13.33}$$

where U^m is called the *bias proportion,* U^s the *variance proportion,* and U^c the *covariance proportion.*[11]

Hang on, all this algebraic manipulation has some useful results. Consider the bias proportion U^m. For any given number of time periods (n), U^m is equal to 0 if and only if the average predicted change ($\bar{F}$) is equal to the average actual change ($\bar{A}$). As long as n is not very small, we would expect U^m to be small; a small proportion of the model's predictive error originates with error in the average change. If U^m is not small, then you would suspect that it could be reduced by altering the model (e.g., reinvestigating the trend and seasonal components). Thus, U^m should be small. If it is not, then our initial subjective opinion is that you should be able to reduce U^m, and thus U, by reinvestigating the predictive model itself. The bias proportion U^m is a serious but presumably correctable component of "predictive error."

The variance proportion U^s is 0 if and only if the variances of predicted and actual changes are equal. But what if $s_F \neq s_A$? What should we expect when dealing with predictions from economic models? First, if there is a very small variation in the actual changes, $s_A \simeq 0$, then we should expect to generate a fairly accurate and simple predictive model. That is, if we observe that quarterly values of GNP increase at 1 percent per quarter over several consecutive years, then we have just stated our predictive model. In this case U^m and U^s should be very close to 0! (By the way, explaining why there is a constant percentage increase every quarter may be difficult; consider the identification problem analogous to Panel 13.1.)

Most economic variables are interesting because the standard deviation of observed changes is neither equal to, nor close to, 0. Thus, we would expect $s_A > 0$. What about the standard deviation of predicted changes, s_F? Let us assume that the actual values are generated by systematic changes plus random changes and that the predicted changes are generated by accounting for some of the systematic changes. Hence, we would expect that s_A would be larger than s_F, and by accounting for more of the systematic changes we could reduce the difference between s_A and s_F—reduce U^s. Although this need not always be the case, if U^s is large, because $s_A > s_F$, we

---

[11] Theil uses the U superscript notation for the proportions of mean-square error. Equation 13.33 is algebraically identical to dividing Equation 13.32 by the RMS for A (to get U^2) and then dividing the result expression by U^2.

might try including "other" variables in our model (hopefully with due respect to autocorrelation, etc.).

What if the standard deviation of predicted changes is greater than the standard deviation of observed changes? Have we made our predictive model too complicated? Possibly, but if simplifying the predictive model results in increasing U^m and U^c more than it reduces U^s, you have reduced the "accuracy" of your predictions.

In summary, U^s may not be as easy to correct for as was U^m, the bias inequality. We would expect that $s_A > 0$ and $s_A > s_F$. In any case, a large proportion of the error originating in U^s suggests some possible courses of action. Reducing U^s while increasing U is not usually a desirable exercise.

The third component, the covariance proportion (U^c), decreases as the correlation (r) between predicted and observed changes approaches 1. Recall the prediction-realization diagram in Figure 13.6. If $r = 1$, then $U^c = 0$, and all the points in a prediction-realization diagram would lie on a straight line. However, that straight line would coincide with the "line of perfect forecasts" if and only if U^m and U^s were also equal to 0.

Theil relates that "... because we can never hope that forecasters will be able to predict such that their points (F_i, A_i) are all located on a straight line," we should expect that $U^c > 0$. In comparison with U^m and U^s, there is not much we can do about U^c. The forecaster is relatively "helpless" in dealing with U^c.

Finally, what can we deduce by considering the "inequality proportions"? First, be sure to recognize that U^m, U^s, and U^c just represent percentages of the "total error," U, that can be attributed to the means, the variances, and the covariance respectively. If model A gives $U = 0.50$ and model B has $U = 0.80$, then model A, in its current form, is more accurate in prediction than model B. If model C has $U = 0.52$, and we wish to choose model A or model C, then we would want to investigate the "inequality proportions." Consider the following:

Model	U	U^m	U^s	U^c
A	0.50	0.01	0.23	0.76
C	0.52	0.20	0.25	0.55

Model A is "refined to the limit"; it does not appear that much of the error in model A can be reduced. Model C, however, has quite large U^m, a presumably correctable component of error. Thus reinvestigation and refinement of model C have the possibility of increasing the predictive accuracy. The fact that $U^m = 0.20$ does not prove anything, but it does give us some guidance in deciding which model has the greatest potential for the improvement of predictive accuracy.

Summary

Theil's U provides a measure of accuracy. More important, however, are the "model alterations" suggested by the inequality proportions. Further, the continuous feedback nature of economic model building should be more evident.

Example 13.2   The Accuracy of Forecasting the United States Money Multiplier
Albert E. Burger[12] is concerned with forecasting a United States money multiplier. He
constructs a model and uses that model to forecast the value of the multiplier for each
month. Let us look at only the 12 forecasts for 1971, as shown in Table 13.6.
From the data in Table 13.6 we find:

$$\Sigma(F_i - A_i)^2 = 1.9061 \qquad \Sigma(A_i)^2 = 11.217 \qquad n = 12$$

$$\bar{F} = (1/n)\Sigma F_i = 0.0655 \qquad \bar{A} = (1/n)\Sigma A_i = -0.08375$$

$$s_F = \sqrt{(1/n)[\Sigma(F_i - \bar{F})^2]} = 0.9109 \qquad s_A = \sqrt{(1/n)[\Sigma(A_i - \bar{A})^2]} = 0.9633$$

$$r = \Sigma[(F_i - \bar{F})(A_i - \bar{A})]/(n s_F s_A) = 0.9237$$

Therefore,

$$U = \sqrt{[\Sigma(F_i - A_i)^2]/(\Sigma A_i^{\ 2})} = \sqrt{(1.9061)/(11.217)} = \sqrt{0.16993} = 0.413$$

which indicates that Burger's model for forecasting the monthly value of the United States

TABLE 13.6   BURGER'S FORECASTS AND ACTUAL VALUES FOR
UNITED STATES MONEY MULTIPLIER

Month (1971)	Forecast Multiplier	Actual Multiplier	(F) Percent Change Forecast	(A) Percent Change Actual	$(F - A)^2$
		(2.744)			
January	2.765	2.741	0.765	−0.109	0.7639
February	2.687	2.690	−1.970	−1.861	0.0119
March	2.696	2.705	0.223	0.558	0.1122
April	2.730	2.732	0.924	0.998	0.0055
May	2.690	2.679	−1.537	−1.940	0.1624
June	2.705	2.718	0.971	1.456	0.2352
July	2.722	2.714	0.147	−0.147	0.0864
August	2.699	2.703	−0.553	−0.405	0.0219
September	2.707	2.697	0.148	−0.222	0.1369
October	2.711	2.699	0.519	0.074	0.1980
November	2.710	2.700	0.408	0.037	0.1376
December	2.720	2.715	0.741	0.556	0.0342
Total	32.542	32.493	0.786	−1.005	1.9061

[12] Albert E. Burger, "Money Stock Control," *Federal Reserve Bank of St. Louis Review*, Vol. 54,
No. 10, October 1972, pp. 10–16.

money multiplier demonstrated only 41.3 percent of the mean-square error that would have been produced by the no-change forecast model.

Now, we calculate the inequality coefficients:

$$U^m = [(\bar{F} - \bar{A})^2]/\{(1/n)[\Sigma(F_i - A_i)^2]\} = 0.140$$

$$U^s = [(s_F - s_A)^2]/\{(1/n)[\Sigma(F_i - A_i)^2]\} = 0.017$$

$$U^c = [2(1 - r)s_F s_A]/\{(1/n)[\Sigma(F_i - A_i)^2]\} = 0.843$$

The bias inequality coefficient, U^m, accounts for 14 percent of the forecasting error in Burger's model. Theil implies that this "bias inequality" component of forecasting error has the best chance of being corrected or reduced. However, we have never stated what Burger's model is! We have used the *results* of Burger's model to evaluate its forecast accuracy. Essentially, Burger's model is a regression equation which employs a moving average. Thus, it *may* be possible to reduce the mean-square error by using a 4-month instead of a 3-month moving average, or by some such reinvestigation within the model.

Also note that about 85 percent of the forecasting error is in the term U^c. Thus, we would expect that very little could be done to reduce most of the (already low?) forecasting error in Burger's model.

EXERCISES

1. Discuss the identification problem associated with each of the following models (or systems of equations).
a. $Y_1 = \alpha_1 + \beta_1 X_1; \quad Y_2 = \alpha_2 + \beta_2 X_1; \quad Y_1 = Y_2$
b. $Q_1 = \alpha_1 + \beta_1 X_1 + \gamma_1 X_2; \quad Q_2 = \alpha_2 + \beta_2 X_1 + \gamma_2 X_3; \quad Q_1 = Q_2$
c. $Q_1 = \alpha_1 + \beta_1 X + \gamma_1 Y; \quad Q_2 = \alpha_2 + \beta_2 X + \gamma_2 Y; \quad Q_1 = Q_2$
d. $Y_1 = \alpha_1 + \beta_1 P + \beta_2 H; \quad Y_2 = \alpha_2 + \beta_3 P + \beta_4 K + \beta_5 L; \quad Y_1 = Y_2$

2. Explain in your own words the meaning of:
a. underidentified
b. overidentified
c. just identified
d. structural equations
e. reduced-form equations
f. indirect least-squares estimation

3. For each model in Exercise 1, which variables would be:
a. endogenous?
b. exogenous?

4. Consider the structural equations

$$Q_d = \alpha_d + \beta_d P + \gamma_d Y + \xi_d$$

$$Q_s = \alpha_s + \beta_s P + \phi_s R + \xi_s$$

$$Q_d = Q_s$$

the following data

Q	P	Y	R		
11	1	1	15	$\Sigma Y^2 = 70$	$\Sigma YQ = 224$
7	7	2	7	$\Sigma R^2 = 742$	$\Sigma RQ = 748$
13	5	4	12	$\Sigma RY = 203$	$\Sigma Q^2 = 780$
21	3	7	18	$\Sigma YP = 56$	$\Sigma P^2 = 84$
				$\Sigma RP = 178$	
52	16	14	52		

and the reduced-form equations

$$\hat{P} = {}^{31}\!/_3 + \tfrac{2}{3}Y - \tfrac{2}{3}R$$
$$\hat{Q} = -\tfrac{1}{3} + \tfrac{4}{3}Y + \tfrac{2}{3}R$$

a. Find the indirect least-squares estimate of the structural equations.
b. Find the regression line $Q = a + bP$, the resulting r^2, and compare that regression line with your answer to part a.
c. Find the regression plane $Q = a + bP + cY$ and the resulting $R^2_{Q|PY}$, and compare that result with your answer to part a.
d. Assume the model is for the demand and supply of feep. How would you interpret the coefficients in part a?

5. Consider the following structural equations:

$$Q_d = \alpha_d + \beta_d P + \gamma_d Y + \theta_d H + \xi_d$$
$$Q_s = \alpha_s + \beta_s P + \phi_s R + \xi_s$$
$$Q_d = Q_s$$

and the following data:

Q	P	Y	R	H		
13	3	1	4	11	$\Sigma Y^2 = 50$	$\Sigma RP = 148$
18	2	2	12	11	$\Sigma R^2 = 480$	$\Sigma YQ = 295$
20	4	3	8	11	$\Sigma RY = 148$	$\Sigma RQ = 924$
31	5	6	16	11	$\Sigma YP = 49$	
82	14	12	40	44		

a. From the original data, how does the model degenerate to a just identified model?
b. Estimate the reduced-form equations.
c. Estimate the structural equations.
d. Interpret your results from part c assuming the variables are defined as they were in example 13.1.
e. For $Y = 2$, $R = 10$, and $H = 11$, what quantity and price of feep would you predict?
f. For $Y = 4$, $R = 6$, and $H = 7$, what quantity and price of feep would you predict?

6. Consider the following data:

Period i	Model Forecasted Y Value to Be: Y_f	Actual Y Value Was: Y_a	Point Designation
		(26.00)	
1	28	27.04	A
2	32	32.00	B
3	30	32.32	C
4	36	35.00	D
5	26	25.00	E
6	28	28.00	F

a. Find the percent change forecasted (F_i) and the actual percent change (A_i). Note:

$$F_i = \left(\frac{Y_{f, i} - Y_{a, i-1}}{Y_{a, i-1}} \right) 100 \quad \text{and} \quad A_i = \left(\frac{Y_{a, i} - Y_{a, i-1}}{Y_{a, i-1}} \right) 100$$

b. Plot (F_i, A_i) on a graph similar to Figure 13.5.
c. Identify the type of error associated with each point.
d. Find the root-mean-square error.
e. Find Theil's U.
f. Find Theil's first set of inequality proportions (U^m, U^s, and U^c).

7. You are interested in forecasting the number of dishwashers sold each year. Consider the following data:

Year	t	Thousands of Dishwashers Sold Y	Disposable United States Income (Billions of Dollars) X
1950	0	230	206.9
1951	1	260	226.6
1952	2	175	238.3
1953	3	180	252.6
1954	4	215	257.4
1955	5	295	275.3
1956	6	400	293.2
1957	7	390	308.5
1958	8	425	318.8
1959	9	547	337.3

Source: *Y: Electrical Merchandising; X: Survey of Current Business.*

Research group A builds a least-squares trend line from this data:

$$T = 152.89 + (35.29)t$$

where T = thousands of dishwashers sold per year, $t = 0$ in 1950, and the t units are years.

Research group B builds a least-squares regression line with this data:

$$Y = -382.21 + 2.56(X)$$

where Y = thousands of dishwashers sold per year and X = billions of dollars of annual disposable income.

Consider the following data from the same sources:

Year	t	Y	X
1960	10	555	350.0
1961	11	620	364.4
1962	12	720	385.3
1963	13	880	404.6

Part A: The Time-Series Model

a. Use the time-series model from research group A to forecast the number of dishwashers sold for 1960, 1961, 1962, and 1963.
b. Find the percent change forecasted and the actual percent change. (See note in Exercise 6, part *a*.)
c. Find Theil's U.

Part B: The Regression Model

d. Use the model from research group B and the X values for 1960 to 1963 (given above) to forecast the number of dishwashers sold for each of those 4 years.
e. Find F_i and A_i.
f. Find Theil's U.

Part C: Comparison

g. Using U as the only measure for comparison, which model is more accurate in forecasting the four Y values?
h. Is either model more accurate than the no-change forecast model?
i. What changes would you suggest to improve both models?

CASE QUESTIONS

1. Reconsider case question 1 in Chapter 11. Do we have an identification problem? Should we consider both supply and demand for rental units?

2. Reconsider case question 2 in Chapter 11. Do we have an identification problem with either of the two models discussed in that question?

14

MODERN DECISION THEORY

In this chapter we wish to merge some concepts of probability, some of the statistical methods previously investigated, and a touch of economics to concoct what is sometimes called "modern decision theory." The economic aspects enter early. We want to describe the consequences of our actions either in monetary terms (dollars) or in measures of satisfaction (utilities). We want to include the cost of sample information in our decision process; we want to assess the value of information to our decisions. To evaluate the relative merit of each decision, we must have some goal. Finally, we will try to incorporate "the experience" of the decision maker in the decision-making techniques with the use of subjective probabilities.

First, we must narrow the field of decision problems to *statistical decision problems*. In order to qualify as a statistical decision problem, the situation must be such that:

1. More than one decision or action is possible.
2. More than one event could occur.
3. For each event that could occur, the result of each possible decision can be represented by a number.

14.1 MEASURING THE CONSEQUENCES OF YOUR ACTION

For any situation that can classify as a statistical decision problem, we have to measure the extent of the situation. First, we construct a list of possible events, or states of nature (SON). The listed states of nature must be mutually exclusive and collectively exhaustive.

Second, a list of mutually exclusive decisions must be constructed. Finally, we must place a value on each "outcome"—that is, some numerical measure of the result from a particular decision when a specific event or state of nature occurs. The "values" can be easily recorded in a *payoff table*. The values may be either profit, costs, or utilities.

Accounting Profit

Assume that you are the manager of a budding young New York City rock group who are eager to play whenever and wherever they can. This weekend you have scheduled a Friday night concert in Atlanta and a Saturday night concert in Boston.

If the concert is performed, the group receives $1,200 per concert (plus expenses); however, the promoters in Atlanta and Boston each want an $800 bond to cover their costs in case the rock group fails to show. Thus, if both concerts are performed, the weekend profit is $2,400; if neither concert is performed, the weekend profit is −$1,600 (a loss).

As manager, you recognize that transportation foul-ups, accidents, etc., could prevent the group from performing. You approach an insurance agent who offers you the following deal: (1) For either the New York City–Atlanta or the Atlanta–Boston trip, if the group is unable to perform, the agent will pay $700 and the cost of this insurance is $200. (2) For a weekend package that covers both trips: if neither concert is performed, the agent will pay $1,500; if the Boston concert is not performed, the agent will pay $800; and the cost of this insurance is $300.

Assume that whatever prevents the Atlanta concert will also prevent the Boston concert. The possible states of nature (or events) are: (1) neither concert will be performed; (2) only the Atlanta concert will be performed; (3) both concerts will be performed.

Your decision alternatives are: (1) don't insure anything; (2) insure only the Atlanta concert; (3) insure only the Boston concert; (4) insure both concerts. From the above information, a payoff table showing the profit for each decision and each state of nature is shown in Table 14.1.

For example, for SON_2 (i.e., only the Atlanta concert is performed) and D_1 (i.e., neither concert is insured), the profit is the receipt from performing the Atlanta concert minus the amount paid to the Boston promoters for not performing there; profit = $1,200 − $800 = $400. For D_2 and SON_1,

$$\text{(Net payment to Atlanta promoters)} + \text{(insurance premium)} + \text{(payment to Boston promoters)} = -(\$800 - \$700) - (\$200) - (\$800) = -\$1,100$$

TABLE 14.1   ROCK GROUP PAYOFF TABLE (PROFITS)

	Decisions			
States of Nature (SON)	D_1 Insure Neither	D_2 Insure Only Atlanta	D_3 Insure Only Boston	D_4 Insure Both
1. Neither concert is performed.	− $1,600	− $1,100	− $1,100	− $400
2. Only Atlanta concert is performed.	$400	$200	$900	$900
3. Both concerts are performed.	$2,400	$2,200	$2,200	$2,100

For D_3 and SON_2,

(Profit from Atlanta) − (net payment to Boston) − (insurance premium)
  = \$1,200 − (\$800 − \$700) − (\$200) = \$900

This payoff table specifies the outcomes (profits) for each decision depending upon which state of nature occurs.

Utilities

Instead of measuring the payoffs in dollars of profit, we may want to measure the payoffs in terms of level of satisfaction. For example, \$2,400 is 6 times as much profit as \$400, but will the rock group get 6 times as much satisfaction from \$2,400 profit? A net loss of \$1,600 is 4 times larger than a net loss of \$400, but how does the rock group view it? A \$400 loss for the weekend may be viewed as a minor setback (postpone buying a more powerful amplifier) while a \$1,600 loss may be drastic (pawn some equipment).

Consider Figure 14.1 where the relation between dollar profit and the rock group's attitude toward various levels of profit is diagrammed. Using this diagram, we could convert the payoff table in Table 14.1 to a payoff table like Table 14.2.

The conceptual key to converting the profits to utilities is the relation described in Figure 14.1. Discerning that relation is at least difficult. We shall continue the process of looking at alternative decisions primarily with the use of payoff tables in profit values.

The purpose of introducing utilities is twofold: (1) The utility concept reflects a behavioral aspect of decision making that is not totally foreign to our experience. It ties into the marginal-utility concept in economic theory. It presents a conceptual (if not a practical) mechanism for expressing the fact that the magnitude of a loss (or negative profit) may have personal impact out of proportion with the dollar values attached to each payoff. (2) A more important purpose is to demonstrate that the payoff for each outcome must be measured in units that correspond to the decision maker's goal. If the decision maker's goal is to maximize profits, then the profit payoffs in Table 14.1 are appropriate. If the decision maker's goal is to maximize satisfaction, then the utility-type payoffs in Table 14.2 are appropriate.

This introduction to several aspects of statistical decision theory will treat only dollar profits because the step of converting profits into utilities can be eliminated without destroying the *process* of identifying an optimal decision. This pedagogical simplification is not intended to imply either that a decision maker *should* attempt to maximize profits or that most decision makers *do* attempt to maximize profits. What decision makers do attempt to optimize is a testable, but difficult to test, hypothesis. What business decision makers *should* attempt to optimize is not an irrelevant question; however, its only influence on the payoffs is their units of measurement.

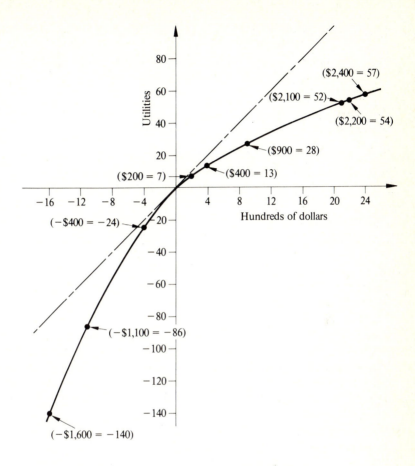

($2,400 = 57)

($2,100 = 52)

($2,200 = 54)

($900 = 28)

($400 = 13)

($200 = 7)

(−$400 = −24)

(−$1,100 = −86)

(−$1,600 = −140)

Utilities

Hundreds of dollars

FIGURE 14.1   Relation between profit and utility for rock group.

TABLE 14.2   ROCK GROUP PAYOFF TABLE
(UTILITIES)

States of Nature	Decisions			
	D_1	D_2	D_3	D_4
SON_1	−140	−86	−86	−24
SON_2	13	7	28	28
SON_3	57	54	54	52

14.2 CRITERIA FOR CHOOSING A DECISION

By what standard do we judge one decision to be better than another? There are several strategies that one could employ, and we shall introduce a sample of these decision criteria.

The Minimax Criterion

The minimax strategy is based upon an attempt to avoid the worst outcomes. For each decision, determine the worst thing that could happen. From this list of "worst outcomes," choose the best outcome. The decision associated with that outcome is called the *minimax decision.*

How does the word "minimax" describe this strategy? Start with the end of mini*max*. For each decision, find the outcome representing the *max*imum *loss*, or the worst thing that could happen. From only this set of "maximum loss" outcomes, choose the *mini*mum loss. That decision will *mini*mize the *max*imum *loss;* hence, the minimax decision is the traditional abbreviation for this strategy.

Reconsider Table 14.1. Table 14.3 lists the worst outcome for each decision and the minimax decision.

The Maximax Criterion

The maximax strategy is based on an attempt to obtain the maximum possible payoff. Find the outcome with the highest profit, and choose the decision associated with that outcome. From inspection of Table 14.1, the maximum profit ($2,400) is associated with D_1; the maximax decision is to insure neither concert.

For a payoff table with a multitude of outcomes (i.e., either many states of nature or many decisions, or both), you can determine the maximum payoff for each decision (or each state of nature), and from that subset you can choose the maximum profit. Table 14.4 demonstrates this.

TABLE 14.3 MINIMAX DECISION FOR ROCK GROUP

State of Nature	Decisions			
	D_1	D_2	D_3	D_4
SON_1	−1,600	−1,100	−1,100	−400
SON_2	400	200	900	900
SON_3	2,400	2,200	2,200	2,100
Worst outcome	−1,600	−1,100	−1,100	−400

←Maximum loss for each decision

−$400 is the minimum loss

Minimax decision is D_4 (insure both concerts)

TABLE 14.4 MAXIMAX DECISION FOR ROCK GROUP

State of Nature	Decisions				Maximum Profit for Each SON
	D_1	D_2	D_3	D_4	
SON_1	$-1,600$	$-1,100$	$-1,100$	-400	-400
SON_2	400	200	900	900	900
SON_3	2,400	2,200	2,200	2,100	2,400←
Maximum profit for each decision	2,400	2,200	2,200	2,100	

$\llcorner$The maximax decision is D_1, to insure neither concert.⎤

Although we are uncertain as to which state of nature will occur, neither the minimax strategy nor the maximax strategy considers the relative likelihood that each state of nature will occur. We shall now consider a strategy which employs the $P(SON_i)$ in evaluating the merit of each decision.

The Optimal Expected Value Criterion

Assume that the probability that both concerts will be performed is 0.64; the probability that only the Atlanta concert will be performed is 0.16; the probability that neither concert is performed is 0.20. For each decision, we could calculate the expected value:

$$E(D_j) = \sum_{i=1}^{n} [V_{ij} \cdot P(SON_i)] \qquad (14.1)$$

where V_{ij} = the payoff from decision j if state of nature i occurs and $P(SON_i)$ = the probability that state of nature i will occur.

Table 14.5 serves as the worksheet for calculating the expected value for each decision.

TABLE 14.5 EXPECTED VALUE OF DECISIONS FOR ROCK GROUP

Probability $P(SON_i)$	State of Nature	Decision				Expected Value Calculations for $P(SON_i)V_{ij}$			
		D_1	D_2	D_3	D_4	D_1	D_2	D_3	D_4
0.20	Neither	$-1,600$	$-1,100$	$-1,100$	-400	-320	-220	-220	-80
0.16	Only A	400	200	900	900	64	32	144	144
0.64	Both	2,400	2,200	2,200	2,100	1,536	1,408	1,408	1,344
1.00			Expected values →			1,280	1,220	1,332	1,408

Maximum expected profit ⎤

Since the expected value for D_4 (insure both concerts),

$$E(D_4) = 0.20(-400) + 0.16(900) + 0.64(2,100) = \$1,408$$

is greater than the expected value for the other three decisions, D_4 is the optimal decision using the expected value criterion.

The probability for each state of nature could be based on past experience or purely on the manager's hunch. Frequently, past experience is extremely limited, and subjective probabilities are the only type available.

Both as a brief review and for a slightly different interpretation of the expected value criterion, consider example 14.1.

Example 14.1 A Repair-Now versus Repair-Later Decision

Assume that you are a manufacturer of room humidifiers. You purchase the humidity controlled off-on switch from another firm. Every day you assemble four humidifiers. Before assembly, it costs \$5 per switch to test for proper functioning, and it costs \$10 per unit to repair a defective switch. After assembly (and shipment) it costs \$20 per unit to replace a defective switch.

Since four humidifiers are to be assembled each day, the five states of nature are "no defectives," "one defective switch," ..., "four defectives." Consider only two decisions: D_1: Test all four switches and repair all defectives; D_2: Do not test any of the four switches but replace all defective switches on returned humidifiers. Finally, assume that all humidifiers with defective switches are returned.

The payoff table is constructed with *costs* (not profits) as follows:

1. To inspect the four switches it costs (\$5 per switch) (4 switches) = \$20. Since it costs \$10 per switch to repair defectives, for D_1 the costs are:

   Payoff (cost in dollars) = \$20 + (\$10 per defective) (number of defectives)

2. For D_2, the costs are:

   Payoff (cost in dollars) = (\$20 per returned defective switch)(number of defectives)

Recall that costs, not profits, are the measure of payoffs in the table. Thus, we are interested in minimizing cost to reach an optimal decision.

For D_1, the worst thing that could happen is that we experience a cost of \$60 per day; for D_2, \$80 per day. Thus, the *minimax* decision is D_1.

For D_1 the best thing that could happen is a cost of only \$20 per day; for D_2, \$0 per day. Thus, the *maximax* decision is D_2.

Now consider the probabilities given in Table 14.6. Based upon past experience, the manager's judgment is that on 40 percent of the workdays there will be no defectives in the batch of four switches selected; 20 percent of the time, one defective; 20 percent of the time, two defectives; and so forth.

TABLE 14.6 PAYOFF TABLE (COSTS) FOR HUMIDIFIER SWITCH PROBLEM

Probability $P(SON_i)$	State of Nature (Number of Defectives) SON_i	Decisions		Expected Value Calculations for	
		Test All D_1	Test None D_2	D_1 $P(SON) \cdot V_1$	D_2 $P(SON) \cdot V_2$
0.4	0	$20	$ 0	$8	$0
0.2	1	30	20	6	4
0.2	2	40	40	8	8
0.1	3	50	60	5	6
0.1	4	60	80	6	8
1.0	Total	–	–	$33	$26
	Maximum loss	$60	$80		
	Maximum gain	$20	$ 0		

If all four switches were tested and any defectives repaired (D_1), then 40 percent of the workdays the cost would be $20 per day; 20 percent of the workdays, $30 per day; and so forth. This would average to $33 per day for D_1.

If the manager did not test the switches but replaced them later, then 40 percent of the workdays the cost would be $0 per day; 20 percent of the workdays the cost would be $20 per day; and so forth. For D_2, the average cost would be $26 per day.

Since $E(D_2) = \$26 < E(D_1) = \$33$, then D_2 is the optimal decision using the expected value criterion. By not checking the switches before assembly, the manager can expect to save $33 - 26 = \$7$ per day in the cost of producing these humidifiers.

14.3 PUTTING A LIMIT ON INFORMATION COSTS

Usually the acquisition of information costs something. To the decision maker, the value of information is determined by how much the information (reduces expected costs or) increases expected profits. It is reasonable to assume that the value of the information should exceed the cost of obtaining it. Our first step in investigating the cost and value of further information about a decision problem is to determine the maximum amount the decision maker should pay for further information.

The first concept we need is the *expected value with certain prediction*. (This concept has various titles which include "expected payoff with perfect information," "expected monetary value with perfect prediction," etc.) The strategy here is to assume that we know which state of nature will occur, choose the decision with the optimal payoff for that state of nature, and find the expected value of those optimal payoffs. For each state of nature we consider only the optimal payoff; then we weigh

TABLE 14.7 EXPECTED VALUE WITH CERTAIN PREDICTION FOR SWITCHES

$P(\text{SON}_i)$	Number of Defectives SON_i	Decisions — Test All D_1	Decisions — Test None D_2	Optimal (Least Cost) Payoff for Each SON_i $(V_{\text{opt}} \mid \text{SON}_i)$	Calculation $(V_{\text{opt}} \mid \text{SON}_i) \cdot P(\text{SON}_i)$
0.4	0	$20	$ 0	$ 0	$0
0.2	1	30	20	20	4
0.2	2	40	40	40	8
0.1	3	50	60	50	5
0.1	4	60	80	60	6
—					—
1.0					$E(\text{CP}) = \$23$

each optimal payoff by the probability that that state of nature will occur. Symbolically,

$$E(\text{CP}) = \sum_{i=1}^{n} [(V_{\text{opt}} \mid \text{SON}_i) \cdot P(\text{SON}_i)] \qquad (14.2)$$

In Table 14.7 the use of Equation 14.2 is demonstrated for example 14.1.

The expected value with certain prediction can be interpreted as follows: Each morning a clairvoyant calls you and says, "Of the four switches you use today, X of them will be defective." If the reported value of X is 0, then you would choose D_2 (do not test the switches). If the reported value of X is 1, but you do not know which one of the four switches is defective,[1] then we assume that you would choose D_2. If the reported value of X is 2, then either decision can be chosen. If $X = 3$, D_1; if $X = 4$, D_1.

The probability that there are 0 defectives, or $P(\text{SON}_1) = 0.4$, can be interpreted as the relative frequency with which the clairvoyant reports "zero defectives this morning." Over a long period of time, the manager would spend $23 per day when there was certain prediction regarding the number of defectives for each day.

In Table 14.6 we calculated that "without any information," the best the manager could do is $26 per day in costs. With certain prediction the manager would spend $23 per day. Thus, the manager is willing to pay *less than $3 per day* to the clairvoyant for perfect information. At $3 per day the manager is no better off with

---

[1] The information is "perfect" in the sense that it tells you exactly *how many* switches are defective. The "perfect" information is incomplete in the sense that it does not tell you which switch(s) is (are) defective. If you knew that only one switch out of the four were defective and if the first switch you tested were defective, would you bother to test the remaining three switches? Probably not. However, the action we called D_1 was to test all four switches. We could, of course, define D_3 as "randomly select one switch and test it; if it is defective, repair it." But the set of decisions the decision maker wishes to consider does not have to be exhaustive.

the perfect information ($3 per day + $23 per day) than with no information and decision D_2 (that is, $E(D_2) = \$26$ per day). Paying the clairvoyant more than $3 per day for perfect information leaves the manager worse off than with D_2.

Using this strategy, the maximum the decision maker is willing to pay for perfect information (PI), $3 per day, is called the *expected value of perfect information*. In general,

$$E(\text{PI}) = [E(D_j)]_{\text{opt}} - E(\text{CP}) \qquad \text{(for costs only)} \qquad \textbf{(14.3)}$$

where $[E(D_j)]_{\text{opt}}$ is the expected value of the optimal decision when the optimal expected value criterion is used, and $E(\text{CP})$ is the expected value of certain prediction (CP). The expected value of certain prediction is a hypothetical measure, as calculated with Equation 14.2, which provides a "standard" or "benchmark" against which we can measure the value of additional information.

If the payoff table is constructed in terms of profits rather than costs, then the expected value of perfect information is

$$E(\text{PI}) = E(\text{CP}) - [E(D_j)]_{\text{opt}} \qquad \text{(for profits)} \qquad \textbf{(14.4)}$$

The expected value of perfect information can be 0, but it cannot be negative.

As the maximum the decision maker would pay for perfect information, the expected value of perfect information is a useful "first-step" tool in evaluating the value of further information.

14.4 BAYES' RULE REVISITED

We now want to investigate the influence of sample information, not perfect information, on a decision problem. The first step includes how we go about adjusting the $P(\text{SON}_i)$, the probability that each state of nature will occur, when we have the results of sample information. If we had sample information, how would that change things?

Reconsider the humidifier producer and the on-off switch portion of the assembly problem. A defective or faulty switch will be designated by F, and a nondefective switch will be designated by F'. Also, let us introduce a new decision strategy. For a particular day's assembly, four switches are brought from the stock room, and one of those four is randomly selected and tested. That one switch is either faulty or not faulty. Immediately, given either result, the values of $P(\text{SON}_i)$ are affected.

If that one switch is defective, then $P(\text{SON}_1) \neq 0.4$, but $P(\text{SON}_1) = 0$; that is, the probability that "there are no defective switches in the lot of four switches" is 0, and not 0.40. If that one switch is not defective, then $P(\text{SON}_5) \neq 0.1$, but $P(\text{SON}_5) = 0$; that is, the probability that "all four switches are defective" is 0, not 0.10. But how are the other values of $P(\text{SON}_i)$ affected by the sample results?

To answer that question requires the use of Bayes' rule, as presented in Chapter 2. Recall that

$$P(B_1 \mid A) = \frac{P(B_1)P(A \mid B_1)}{P(A)}$$

$$= \frac{P(B_1)P(A \mid B_1)}{\sum_i [P(B_i) \cdot P(A \mid B_i)]}$$

In terms of our current problem, we want to find $P(SON_i \mid F)$ and $P(SON_i \mid F')$ for each state of nature.

Starting with the population of four switches and randomly selecting one of them, we could use the principle of insufficient reason to calculate the following:

$P(F \mid SON_1) = P$ (selecting a defective switch given none are defective) $= \frac{0}{4}$

$P(F \mid SON_2) = P$ (selecting a defective switch given one is defective) $= \frac{1}{4}$

$P(F \mid SON_3) = P$ (selecting a defective switch given two are defective) $= \frac{2}{4}$

$P(F \mid SON_4) = P$ (selecting a defective switch given three are defective) $= \frac{3}{4}$

$P(F \mid SON_5) = P$ (selecting a defective switch given all are defective) $= \frac{4}{4}$

In like manner, we could calculate $P(F' \mid SON_1) = \frac{4}{4}$, $P(F' \mid SON_2) = \frac{3}{4}$, $P(F' \mid SON_3) = \frac{2}{4}$, $P(F' \mid SON_4) = \frac{1}{4}$, and $P(F' \mid SON_5) = \frac{0}{4}$.

Using the laws of multiplication and addition of probabilities,

$P(F) = P$ (selecting a faulty switch)

$\quad = P[(F \cap SON_1) \cup (F \cap SON_2) \cup (F \cap SON_3) \cup \cdots \cup (F \cap SON_5)]$

$\quad = P(SON_1)P(F \mid SON_1) + P(SON_2)P(F \mid SON_2) + \cdots + P(SON_5)P(F \mid SON_5)$

Using the values of $P(SON_i)$ as given in Table 14.7, we can calculate: $P(F) = (0.4)(\frac{0}{4}) + (0.2)(\frac{1}{4}) + (0.2)(\frac{2}{4}) + (0.1)(\frac{3}{4}) + (0.1)(\frac{4}{4}) = 0.325$. In like manner, the probability of selecting a nonfaulty switch is:

$$P(F') = \sum_i [P(SON_i) \cdot P(F' \mid SON_i)]$$

$$= (0.4)(\frac{4}{4}) + (0.2)(\frac{3}{4}) + (0.2)(\frac{2}{4}) + (0.1)(\frac{1}{4}) + (0.1)(\frac{0}{4}) = 0.675$$

or $P(F') = 1 - P(F) = 1 - 0.325 = 0.675$.

TABLE 14.8 BAYES' RULE CALCULATIONS FOR ALTERING $P(\text{SON}_i)$ GIVEN SAMPLE INFORMATION

	Sample Result: The Switch Is Defective		
A Priori Probabilities $P(\text{SON}_i)$	**State of Nature (Number of Defectives)**	**Calculations** $\dfrac{P(\text{SON}_i)P(F\mid\text{SON}_i)}{P(F)} = P(\text{SON}_i\mid F)$	**Posterior Probabilities**
0.4	0	$\dfrac{(0.4)(0/4)}{0.325} =$	0.0000
0.2	1	$\dfrac{(0.2)(1/4)}{0.325} =$	0.1538
0.2	2	$\dfrac{(0.2)(2/4)}{0.325} =$	0.3077
0.1	3	$\dfrac{(0.1)(3/4)}{0.325} =$	0.2308
0.1	4	$\dfrac{(0.1)(4/4)}{0.325} =$	0.3077
1.0		Total	1.0000

	Sample Result: The Switch Is Not Defective		
A Priori Probabilities $P(\text{SON}_i)$	**State of Nature (Number of Defectives)**	**Calculations** $\dfrac{P(\text{SON}_i)P(F'\mid\text{SON}_i)}{P(F')} = P(\text{SON}_i\mid F')$	**Posterior Probabilities**
0.4	0	$\dfrac{(0.4)(4/4)}{0.675}$	0.5926
0.2	1	$\dfrac{(0.2)(3/4)}{0.675}$	0.2222
0.2	2	$\dfrac{(0.2)(2/4)}{0.675}$	0.1481
0.1	3	$\dfrac{(0.1)(1/4)}{0.675}$	0.0370
0.1	4	$\dfrac{(0.1)(0/4)}{0.675}$	0.0000
1.0		Total	0.9999

Employing Bayes' rule, we can now calculate

$$P(\text{SON}_i|F) = \frac{P(\text{SON}_i)P(F|\text{SON}_i)}{P(F)}$$

and

$$P(\text{SON}_i|F') = \frac{P(\text{SON}_i)P(F'|\text{SON}_i)}{P(F')}$$

These calculations are presented in Table 14.8.

Figure 14.2 diagrams the results calculated in Table 14.8. There is a noticeable change in the probabilities associated with each state of nature, depending upon the results of testing one switch.

The original probability assessments for each state of nature, $P(\text{SON}_i)$, are called *a priori probabilities*, and they may be subjective assessments. The conditional

FIGURE 14.2 The effect of sample information on $P(\text{SON}_i)$. (*a*) No sample information; (*b*) Sample of size 1: defective switch; (*c*) Sample of size 1: nondefective switch.

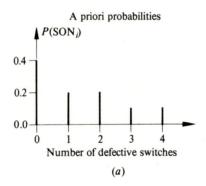

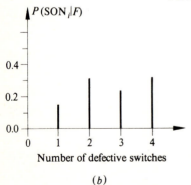

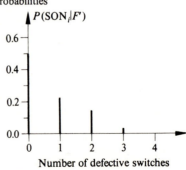

probabilities calculated from Bayes' rule, $P(SON_i|F)$, are called *posterior probabilities*. Bayes' rule is a logical mechanism for merging the a priori probabilities with the results of the sample to produce a new assessment for the probability associated with each state of nature.

   We have seen how sample information could alter the probability associated with each state of nature, but how do we use these posterior probabilities?

14.5   INCLUDING SAMPLE RESULTS IN THE DECISION PROCESS

The humidifier manufacturer started with two alternative courses of action: D_1: test all switches and D_2: test none of the switches. We have noted that by testing just one randomly selected switch, the probabilities associated with each state of nature are altered. In testing the one switch, there are two possible results: the switch is faulty; the switch is not faulty.

   Depending upon the sample result, the decision maker has several possible decisions. These decisions would include:

1.   Test no more switches.
     a.   If the tested switch was faulty, repair and install it.
     b.   If the tested switch was not faulty, install it.
2.   Test no more switches.
     a.   If the tested switch was faulty, throw it away and select another from the store room, but do not test it.
     b.   If the tested switch was not faulty, install it.
3.   Test one more switch.
4.   Test the remaining three switches and repair all faulty switches.

   To prevent the excessive cluttering of the process involved, let us assume that the decision maker only considers options 1 and 4. Figure 14.3 is a *decision-tree* representation of this new decision problem.

Backward Induction

Chronologically first is the choice among D_1 (test all), D_2 (test none), and D_3 (test one). In Table 14.6 we found $E(D_1) = \$33$ and $E(D_2) = \$26$, but what is $E(D_3)$? Obviously, the expected value of "testing one randomly selected switch from the population of four switches" will depend upon the result of that one tested switch and how the decision maker reacts to that result (i.e., test no more or test the rest).

   The first *procedural* step is to start with the end branches of the decision tree and work back to $E(D_1)$, $E(D_2)$, and $E(D_3)$. This process is sometimes referred to as *backward induction*. We start with the payoffs associated with each set of outcomes. Multistage decision problems require starting with the possible outcomes.

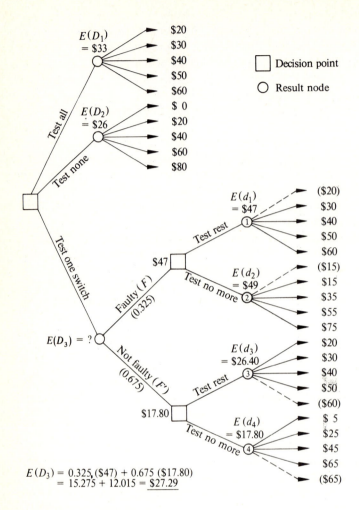

FIGURE 14.3 Decision tree for humidifier manufacturer.

Our first procedural step is to start with the payoffs. From example 14.1, the payoffs for D_1 and D_2 are reproduced from Table 14.6.

Now consider the branch that indicates "test one switch, it is faulty (and it is repaired), and we decide to test the remaining three switches." This is essentially the same as testing all the switches (D_1); thus, the payoffs for this branch of the decision tree are simply a restatement of those for D_1. The $20 payoff is in parentheses because it could not possibly occur. That is, if we tested one switch and it was faulty, then we cannot possibly have no faulty switches in our population of four switches.

Next, consider the branch that indicates "test one switch, it is faulty (and it is repaired), and we decide not to test any more switches from that batch." Recall from example 14.1 that it costs $5 per switch to test, $10 per switch to repair a switch before it is installed, and $20 per switch to replace a faulty switch after it has been installed. Since we have found one defective switch, then out of the original four switches there cannot be no faulty switches. The $15 in parentheses is used to maintain consistency with D_1 and D_2, but it is not a possible outcome.

It cost $5 to test the one switch and $10 to repair it. If that was the only faulty switch among the four, then the cost will be $15. If there is one other faulty switch among the original four, then it will cost an additional $20 to replace it after it is installed; if there are two faulty switches, it will cost $15 + $20 = $35 for that outcome. If there are three faulty switches, it will cost $15 to test and repair the sample switch and $20 each for the two faulty switches that are detected after installation. Hence, a total of $55 for the state of nature "three defective switches." In like manner, if all four switches are defective, the cost will be $15 + 3($20) = $75.

The third multistage decision branch reflects "test one, it is not faulty, and the remaining three are tested." Again, this branch is the same as testing all four except that we now know that not all four are defective.

Finally, consider the branch representing "test one, it is not defective, we test no more switches before installation." If none of the four are defective, then our only cost was testing the one sample switch, which cost $5. If there is one faulty switch in the batch of four, but it was not the one tested, then we pay $5 for testing a nonfaulty switch and $20 for replacing the faulty switch after installation. If there are two faulty switches, the cost is $5 + 2($20) = $45, and so forth.

The decision tree in Figure 14.3 has designated the nodal points d_1, d_2, d_3, and d_4. The next step is to use the payoffs in Figure 14.3 with the posterior probabilities from Table 14.8 to calculate the expected value for d_1, d_2, d_3, and d_4. Table 14.9 presents those expected value calculations.

When one switch is tested and found to be defective, the decision maker must choose either to test the remaining three switches or to test none of the remaining switches. Using the optional expected value criterion, the choice must be to "test the rest." Why? Because $E(d_1) = \$47 < E(d_2) = \$49$, and since we are dealing with costs, the lower value is "optimal." The expected cost of $47 is then entered above the "decision box" in Figure 14.3, and the branch which yields $E(d_2) = \$49$ is ignored in the rest of the decision process.

By the same reasoning, when one switch is tested and found to function properly, the remaining three switches (and the one tested) would be installed without further testing. Since $E(d_4) = \$17.80 < E(d_3) = \$26.40$, the node represented by d_3 in Figure 14.3 suffers benign neglect.

Finally, we are in a position to calculate $E(D_3)$, the expected value of testing just one switch. Previously, we found the probability of randomly selecting a faulty switch to be $P(F) = 0.325$ and the probability of selecting a nonfaulty switch to be $P(F') = 0.675$. If the switch is faulty, the optimal (least) expected cost is $47; if the switch is not faulty, the optimal expected value is $17.80. Hence,

$$E(D_3) = 0.325(\$47) + 0.675(\$17.80) = \$27.29$$

TABLE 14.9 EXPECTED VALUE CALCULATIONS

Sample Result: The Switch Is Faulty

Posterior Probabilities $P(\text{SON}_i|F)$ (1)	State of Nature SON_i (2)	Payoff for d_1 (Costs) (3)	Calculations (1) × (3) (4)	Payoff for d_2 (Costs) (5)	Calculations (1) × (5) (6)
0.00	0	(20)	0.00	(15)	0.00
0.15	1	30	4.50	15	2.25
0.31	2	40	12.40	35	10.85
0.23	3	50	11.50	55	12.65
0.31	4	60	18.60	75	23.25
1.00			$E(d_1) = 47.00$		$E(d_2) = 49.00$

Sample Result: The Switch Is Not Faulty

Posterior Probabilities $P(\text{SON}_i|F')$ (1)	State of Nature SON_i (2)	Payoff for d_3 (Costs) (3)	Calculations (1) × (3) (4)	Payoff for d_4 (Costs) (5)	Calculations (1) × (5) (6)
0.59	0	20	11.80	5	2.95
0.22	1	30	6.60	25	5.50
0.15	2	40	6.00	45	6.75
0.04	3	50	2.00	65	2.60
0.00	4	(60)	0.00	(65)	0.00
1.00			$E(d_3) = \$26.40$		$E(d_4) = \$17.80$

As shown in Figure 14.3, we have three decisions to choose from:

D_1: Test all four switches, $E(D_1) = \$33$
D_2: Test none of the four switches, $E(D_2) = \$26$
D_3: Test one randomly selected switch and follow contingency plan depending upon result, $E(D_3) = \$27.29$

Following the optimal expected value criterion, D_2 (test none of the four switches), still offers the least-cost decision strategy. With D_2 the humidifier producer can expect only \$26 per day in costs associated with the switch.

There are several aspects of this problem that we need to reconsider. First, in the calculation of $E(D_1)$ and $E(D_2)$, we directly considered the payoffs from each outcome and their associated probabilities while the calculation of $E(D_3)$ was more indirect in considering the payoffs and probabilities associated with each outcome. The decision designated D_3 represents multiple decisions. Suboptimal decisions (that is, d_2 and d_3) were eliminated.

Second, the calculation of $E(D_1)$, $E(D_2)$, and $E(D_3)$ did not require conducting any physical experiment. This procedure is frequently called *preposterior*: the process allows conceptual treatment of the problem before you have to make a decision.

Third, the cost of sample information (or the cost of testing a switch) was given. The payoffs associated with d_1, d_2, d_3, and d_4 incorporated the $5-per-switch cost of testing. Frequently, decision problems will not have that information, and the question is: What is the maximum amount the decision maker should pay for sample (imperfect) information?

14.6 EXPECTED VALUE OF SAMPLE INFORMATION

An example will serve to demonstrate the basic process for evaluating the expected value of sample information.

Example 14.2   A Souvenir Tennis Program

Ms. Robbie Biggs has the exclusive rights to produce "the souvenir pamphlet" for a challenge tennis match. Two months before the match Ms. Biggs had 5,000 copies of the souvenir program printed at a cost of $5,200.

The program includes the rules for the match, score sheet, and some tennis history, but the program is primarily a biography of the players. The $2 per copy sale price was boldly printed on the cover. Three weeks before the match, a general sports magazine prints a biographical sketch of the players that essentially duplicates the material in the souvenir program. Ms. Biggs is discouraged and reassesses the probability associated with the levels of sales for the program.

Ms. Alacrity offers to pay Ms. Biggs $6,200 for the souvenir programs and the exclusive rights to distribute them. Mr. Fondant offers to pay Ms. Biggs $6,100 and 20 percent of any profit. All agree that the program will sell for $2 per copy. Ms. Biggs is faced with the following decisions:

D_1: Sell rights and printed copies to Ms. Alacrity
D_2: Sell rights and printed copies to Mr. Fondant
D_3: Keep rights and sell them herself

TABLE 14.10   PAYOFF TABLE ($ PROFIT) FOR MS. BIGGS

Ms. Biggs' Probability Assessment $P(SON_i)$	Number of Copies Demanded SON_i	Decisions		
		Sell to Alacrity D_1	Sell to Fondant D_2	Sell Them Herself D_3
0.50	2,000	1,000	900	−1,200
0.20	3,000	1,000	900	800
0.20	4,000	1,000	1,280	2.800
0.10	5,000	1,000	1,680	4,800

Ms. Biggs considers only four possible states of nature: 2,000 copies, 3,000 copies, 4,000 copies, or 5,000 copies will be sold. If she takes Ms. Alacrity's offer, Ms. Biggs will receive $6,200 - $5,200 = $1,000 profit regardless of the number of souvenir programs sold.

If Ms. Biggs sells the programs herself, her profit function is:

$$\text{Profit} = (\$2 \text{ per copy})(\text{number of copies sold}) - \$5,200$$

For the four states of nature, the profits are shown under D_3 in Table 14.10.

For 2,000 copies sold, there is a loss. If Mr. Fondant's offer is accepted, Ms. Biggs will receive $6,100 - $5,200 = $900 profit for that state of nature. If Mr. Fondant makes a profit, then Ms. Biggs will receive

$$\$900 + (0.20)[(\$2)(\text{number of copies sold}) - \$6,100]$$

or $900 plus 20 percent of Mr. Fondant's profit. If only 3,000 copies are sold, then Mr. Fondant's revenue is $6,000, but his costs were $6,100. If 4,000 copies are sold, then Mr. Fondant's profit is $8,000 - $6,100 = $1,900, of which 20 percent is $380; thus, Ms. Biggs would receive $900 + $380 = $1,280 in profit.

From Table 14.10, the minimax decision is D_1 and the maximax decision is D_3. From the expected-value calculations in Table 14.11, the optimal expected value is $1,054 (recall that optimal is "maximum" in this problem because the payoffs are "profits").

TABLE 14.11 EXPECTED VALUE CALCULATIONS

$P(\text{SON}_i)$	D_1	D_2	D_3	(1) × (2)	(1) × (3)	(1) × (4)	$P(\text{SON}) \cdot (V_{opt}|\text{SON})$
(1)	(2)	(3)	(4)	(5)	(6)	(7)	(8)
0.50	1,000	900	−1,200	500	450	−600	500
0.20	1,000	900	800	200	180	160	200
0.20	1,000	1,280	2,800	200	256	560	560
0.10	1,000	1,680	4,800	100	168	480	480
				1,000	1,054	600	1,740
				$E(D_1)$	$E(D_2)$	$E(D_3)$	$E(\text{CP})$

Since the expected value of certain prediction is $1,740, then the expected value of perfect information is

$$E(\text{PI}) = 1,740 - 1,054 = \$686$$

Now, let us introduce another wrinkle in Ms. Biggs' decision problem. Most of the ticket sales for the match were transacted by mail. A marketing research firm offers to randomly select a sample of n persons from this population, find out how many of those n people will buy a souvenir program, and report the results to Ms. Biggs. How much should she pay for that sample information? Certainly not more than \$686, the expected value of perfect information. Perhaps we can obtain a better evaluation of the worth of imperfect information.

First, let us make the following assumptions:

1. There will be 10,000 people attending the tennis match.
2. The proportion of attendees who will buy a souvenir program is represented by π.
3. The proportion of mail-order ticket recipients who will buy a souvenir program is also equal to π.
4. Ms. Biggs' assessment of the states of nature is correct. (That is, exactly 2,000, 3,000, 4,000, or 5,000 programs will be sold.)
5. A sample of size $(n =)$ 4 will be selected from the mail-order ticket recipients.
6. Although the sample will be selected without replacement, the sample size is so small relative to the population size that binomial probabilities can be used to calculate the probability of the various sample results.

Second, let us define D_4 as the "sample of size 4 plan." That is, the marketing research firm will select a random sample of size 4 and report to Ms. Biggs that either zero, one, two, three, or four of those people will buy a souvenir program. Based upon what they report, Ms. Biggs will choose the decision (i.e., sell rights to Ms. Alacrity, sell to Mr. Fondant, or keep the rights) with the highest expected value. To keep track of all this, consider the decision tree in Figure 14.4.

Finally, let $_k d_j =$ decision D_j given the sample result that k people will buy the souvenir program. Then, the expected value for $_k d_j$ is

$$E(_k d_j) = \sum_{i=1}^{4} [V_i \cdot P(\text{SON}_i | k)]$$

where $P(\text{SON}_i | k)$ is the posterior probability. For example, the probability that 5,000 copies will be sold (SON_4) given that $(k = 2)$ two of the four people sampled will buy the souvenir program is $P(\text{SON}_4 | k = 2)$. To obtain the posterior probabilities (via Bayes' rule), we need the conditional probabilities: $P(k | \text{SON}_i)$. Using the assumptions stated above, we can use the binomial probabilities from Appendix Table III for the conditional probabilities.

How? If 2,000 people will buy the program, which represents SON_1, then $\pi = 2,000/10,000 = 0.20$. Since $n = 4$,

$$P(k | \text{SON}_1) = P(k | \pi = 0.20, n = 4)$$

In general, $P(k | \text{SON}_i) = P(k | \pi = \text{SON}_i/10,000, n = 4)$. Table 14.12 presents the appropriate binomial probabilities from Appendix Table III.

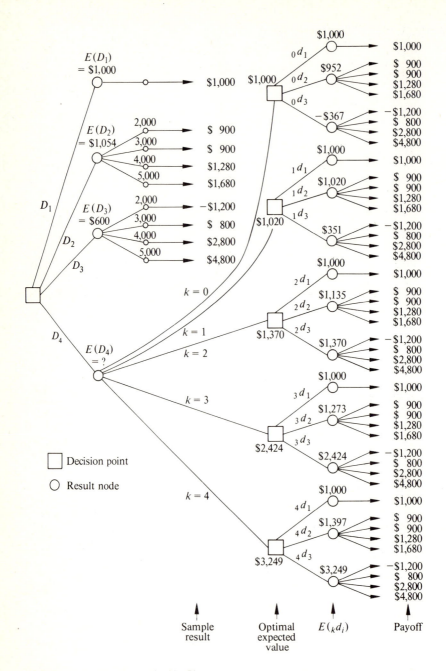

FIGURE 14.4 Decision tree for Ms. Biggs.

TABLE 14.12 CONDITIONAL (BINOMIAL) PROBABILITIES

Number of People in Sample Who Will Buy k	$P(k\|\pi = SON_i/10,000,\ n = 4)$			
	SON_1 $\pi = 2,000/10,000$ $= 0.2$	SON_2 $\pi = 3,000/10,000$ $= 0.3$	SON_3 $\pi = 4,000/10,000$ $= 0.4$	SON_4 $\pi = 5,000/10,000$ $= 0.5$
0	0.4096	0.2401	0.1296	0.0625
1	0.4096	0.4116	0.3456	0.2500
2	0.1536	0.2646	0.3456	0.3750
3	0.0256	0.0756	0.1536	0.2500
4	0.0016	0.0081	0.0256	0.0625
	1.0000	1.0000	1.0000	1.0000

Consider the sample result $k = 0$; none of the four mail-order ticket recipients will buy a souvenir program. That sample could have been selected from a population where 5,000 (or half the attendees) will buy the program. From Table 14.12, we find that

$$P(k = 0 | SON_4 \text{ or } 5,000 \text{ copies sold}) = 0.0625$$

For $k = 0$, we need to use these conditional probabilities, $P(k = 0 | SON_i)$, Ms. Biggs' assessment of which state of nature will occur (or the a priori probabilities), $P(SON_i)$, and Bayes' rule to obtain the posterior probabilities.

$$P(SON_i \mid k = 0) = \frac{P(k = 0 | SON_i)}{P(k = 0)}$$

Table 14.13 serves as the worksheet for (1) calculating the posterior probabilities and (2) calculating the expected value for the decisions to sell to Mr. Fondant or keep the rights. Regardless of the sample result and regardless of which state of nature occurs, the decision to sell to Ms. Alacrity always yields an expected value of $1,000.

Finally, we are able to calculate $E(D_4)$, the expected value of "the sample of size 4 plan." For a given sample result (for example, $k = 0$), we consider *only* the decision with the optimal expected value [for example, $E(_0 d_1) = \$1,000 > E(_0 d_2) = \$951.78 > E(_0 d_3) = -\$367$]. From Table 14.13 we also obtain $P(k)$, the probability of getting each sample result. We combine all this information to calculate $E(D_4)$, as shown in Table 14.14.

The expected value for the sample plan, $E(D_4) = \$1,244.99$, is greater than the expected values for the other three alternatives. But wait! How much did the sampling cost? (How much did the marketing research firm charge?)

TABLE 14.13 POSTERIOR PROBABILITY AND EXPECTED-VALUE CALCULATIONS

A Priori Probability $P(SON_i)$ Table 14.10 (1)	Conditional Probability $P(k \mid SON_i)$ Table 14.12 (2)	Joint Probability $P(k \cap SON_i)$ (1) × (2) (3)	Posterior Probability $P(SON_i \mid k)$ (3) ÷ $P(k)$ (4)	Payoff from D_2 Table 14.10 (5)	$E(_k d_j)$ Calcu- lation (4) × (5) (6)	Payoff from D_3 Table 14.10 (7)	$E(_k d_j)$ Calcu- lation (4) × (7) (8)
$k = 0$							
0.50	0.4096	0.2048	0.7186	900	646.74	−1,200	−862.32
0.20	0.2401	0.0480	0.1684	900	151.56	800	134.72
0.20	0.1296	0.0259	0.0909	1,280	116.35	2,800	254.52
0.10	0.0625	0.0063	0.0221	1,680	37.13	4,800	106.08
	$P(k = 0) = 0.2850$		1.0000		$E(_0 d_2) = \$951.78$		$E(_0 d_3) = -367.00$
$k = 1$							
0.50	0.4096	0.2048	0.5372	900	483.48	−1,200	−644.64
0.20	0.4116	0.0823	0.2159	900	194.31	800	172.72
0.20	0.3456	0.0691	0.1813	1,280	232.06	2,800	507.64
0.10	0.2500	0.0250	0.0656	1,680	110.21	4,800	314.88
	$P(k = 1) = 0.3812$		1.0000		$E(_1 d_2) = \$1,020.06$		$E(_1 d_3) = 350.60$
$k = 2$							
0.50	0.1536	0.0768	0.3250	900	292.50	−1,200	−390.00
0.20	0.2646	0.0529	0.2239	900	201.51	800	179.12
0.20	0.3456	0.0691	0.2924	1,280	374.27	2,800	818.72
0.10	0.3750	0.0375	0.1587	1,680	266.62	4,800	761.76
	$P(k = 2) = 0.2363$		1.0000		$E(_2 d_2) = \$1,134.90$		$E(_2 d_3) = 1,369.60$
$k = 3$							
0.50	0.0256	0.0128	0.1531	900	137.79	−1,200	−183.72
0.20	0.0756	0.0151	0.1806	900	162.54	800	144.48
0.20	0.1536	0.0307	0.3672	1,280	470.02	2,800	1,028.16
0.10	0.2500	0.0250	0.2990	1,680	502.32	4,800	1,435.20
	$P(k = 3) = 0.0836$		0.9999		$E(_3 d_2) = \$1,272.67$		$E(_3 d_3) = 2,424.12$
$k = 4$							
0.50	0.0016	0.0008	0.0580	900	52.20	−1,200	−69.60
0.20	0.0081	0.0016	0.1159	900	104.31	800	92.72
0.20	0.0256	0.0051	0.3696	1,280	473.09	2,800	1,034.88
0.10	0.0625	0.0063	0.4565	1,680	766.92	4,800	2,191.20
	$P(k = 4) = 0.0138$		1.0000		$E(_4 d_2) = \$1,396.52$		$E(_4 d_3) = 3,249.20$

TABLE 14.14 EXPECTED VALUE OF SAMPLE PLAN (EXCLUDING SAMPLING COSTS)

Sample Result k (1)	Probability of Sample Result (Table 14.13) $P(k)$ (2)	Optimal Expected Value Given Sample Result (Figure 14.4) $[E(_k d_j)]_{max}$ (3)	Expected Value Calculation (2) × (3) (4)
0	0.2850	$1,000.00 (d_1)	$ 285.00
1	0.3812	1,020.06 (d_2)	388.85
2	0.2363	1,369.60 (d_3)	323.64
3	0.0836	2,424.12 (d_3)	202.66
4	0.0138	3,249.20 (d_3)	44.84
	0.9999		$1,244.99 = E(D_4)$

Without sampling information, the optimal expected value is associated with selling to Mr. Fondant, $E(D_2) = \$1,054$. Thus, the most Ms. Biggs would be willing to pay for the sample information is

$$\left(\begin{array}{c} \text{Maximum value} \\ \text{of sample information} \end{array}\right) = \left(\begin{array}{c} \text{optimal expected value} \\ \textit{with} \text{ sample information} \end{array}\right)$$

$$- \left(\begin{array}{c} \text{optimal expected value} \\ \textit{without} \text{ sample information} \end{array}\right)$$

$$= \ \$1,244.99 - \$1,054.00 = \$190.99$$

If the cost of sampling is less than \$190.99, then the optimal expected value criterion would suggest the following decision rule. Engage the marketing research firm to obtain the sample information. If they report

$k = 0$	then sell rights to Ms. Alacrity
$k = 1$	then sell rights to Mr. Fondant
$k = 2, 3, \text{ or } 4$	then Ms. Biggs should keep the rights

If the cost of sampling is more than \$190.99 for a sample of 4, the optimal expected value criterion would suggest that Ms. Biggs sell the rights to Mr. Fondant.

Note that this analysis is only specific for a sample of size 4. Consider $n = 5$. For $n = 5$, we would start with the binomial probabilities for $n = 5$ and reconstruct Table 14.12. Then for each sample result (that is, $k = 0, 1, \ldots, 5$), Table 14.13 would be recalculated. As in Table 14.14, we would use the optimal expected value for each sample result and $P(k)$ to find $E(D_5)$.

In general, the expected value for "a plan using a sample of size n" will vary with the size of the sample, or $E(D_n) = f(n)$. Also, the cost of sampling usually varies with the size of sample, or $C_n = g(n)$. Let $E(D_j^*)$ represent the optimal expected value of those decisions which do not use sampling information. Then the *net expected gain from sampling* (NEGS) is

$$\text{NEGS} = E(D_n) - C_n - E(D_j^*)$$

which varies with n, the size of the sample.

If, for all values of n, NEGS is less than 0, then the optimal expected value criterion yields D_j^* as the optimal decision (i.e., do not sample).

If NEGS is greater than 0 for one or more values of n, then the optimal expected value criterion suggests that you should use sample information and use a sample size that provides the largest value of NEGS. Hence, the procedures discussed above not only assess the expected value of sample information but also provide a basis for deciding what size sample should be employed.

SUMMARY

This concludes the introduction to Bayesian decision analysis. The essential power of this logic, or technique, should be obvious. Most of the material in the preceding chapters can be classified as "classical statistics." The techniques in this chapter share some of that material, but there are differences which have fired controversy. That controversy has produced both heat and light. For those who progress beyond the scope of this book, do not ignore the light of Bayesian analysis.

In retrospect, the logic of (1) identifying what could happen (the states of nature), (2) deciding what alternatives the decision maker has (the decisions), and (3) measuring the payoff for each decision, given a state of nature, in units that are most appropriate to the goals of the decision maker, seems almost trivial. But implementing that logic forces the decision maker to be very explicit about the problem. And, as we have seen, the payoff table is only the beginning.

Assigning a value to the probability of occurrence for each state of nature may require the use of subjective probabilities. There are techniques (not covered here) for checking the decision maker's "degree of belief."

Bayes' rule is a device for revising those initial probability assessments in the light of sample information. Finally, the expected value of information (either perfect or imperfect) provides a conceptual mechanism for evaluating sample information *before* that information is obtained.

Remember, this is only an introduction—an overview and an enticement. For those who seek it, there is more.

EXERCISES

1. You have developed and obtained a patent on a new superduper coffee maker. The coffee maker is primarily useful to those who prefer a high chicory content in their coffee. The demand for this coffee maker is concentrated in an area of Louisiana where there are 4,000 families. The coffee maker will be purchased by only these families, and you are planning to sell it for $20.

Your "how to" alternatives can be summarized as follows:

D_1: Shelve the patent and do not produce the crazy thing.

D_2: Produce it yourself. Your fixed costs will be $3,000, and the variable costs will be $10 per unit.

D_3: Supreme Steamfitters Incorporated has offered to produce the coffee maker for you. They will charge you $5,000 plus $8 per unit.

At the price of $20, your demand situation can be described as follows:

Number of Families Who Will Purchase the Coffee Maker SON_i	$P(SON_i)$
0	0.3
400	0.3
800	0.2
1,200	0.2

Finally, you have to make your "how to" produce decision (and incur the fixed costs) before you find out exactly what the demand is; however, the decision concerning "how many" coffee makers to produce can be made after the exact nature of the demand is known.

a. Construct the payoff table in profits.
b. What is the optimal decision using:
  (1) the minimax criterion?
  (2) the maximax criterion?
  (3) the Bayesian (or optimal expected value) criterion?
c. What is the expected value of perfect information, and what does it indicate?

2. In Exercise 1, you reason that if you sell the coffee maker for $15, the demand situation changes to:

SON_i	$P(SON_i)$
0	0.1
400	0.3
800	0.4
1,200	0.2

Assume that there is no change in the production cost situation.
a. Construct the payoff table in profits.
b. Find the optimal decision using:
  (1) the minimax criterion.
  (2) the maximax criterion.
  (3) the Bayesian criterion.

c. Find the expected value of perfect information.
d. For the Bayesian criterion:
   (1) Does the "how to" produce decision change because of the lower price?
   (2) Which price would you charge and why?

3. You are building a house in the country, and you have to drill a well. Well driller A charges $1,000, regardless of depth to be drilled; well driller B charges $300 for every 20-foot section of pipe used in the job. After talking to people in the area, you arrive at the following probabilities:

Drilling Depth to Find Water SON_i	$P(SON_i)$
0– 20 feet	0.0
20– 40 feet	0.1
40– 60 feet	0.4
60– 80 feet	0.3
80–100 feet	0.2
Over 100 feet	0.0

a. Construct the payoff table in dollars of cost.
b. Find the optimal decision using:
   (1) the minimax criterion.
   (2) the maximax criterion.
   (3) the Bayesian criterion.
c. Find the expected value of perfect information.

4. Reconsider Exercise 3. For a $25 fee, a state hydrologist will come to the site and give you an estimate as to what depth you will find water. The hydrologist is not always correct, but the success rate (of the decisions) could be summarized as follows:

TABLE VALUES $= P(H_j|SON_i)$

	Hydrologist's Estimate				
SON_i	H_1 0–20 feet	H_2 20–40 feet	H_3 40–60 feet	H_4 60–80 feet	H_5 80–100 feet
0– 20 feet	0.9	0.1	0	0	0
20– 40 feet	0.1	0.8	0.1	0	0
40– 60 feet	0	0.1	0.8	0.1	0
60– 80 feet	0	0	0.1	0.8	0.1
80–100 feet	0	0	0	0.1	0.9

Further, you are certain to hit water before 100 feet. Let H_i = the hydrologist's estimate.
a. Find $P(H_j)$ for each j.
b. Find each $P(SON_i|H_j)$ using Bayes' rule.
c. For each possible estimate by the hydrologist, what is the optimal decision?
d. Let D_3 = obtaining the hydrologist's estimate. What is $E(D_3)$?
e. What is the net expected gain from sampling (the hydrologist's estimate)?

5. Reconsider Exercise 1. Let $\pi_i =$ the proportion of families that will buy the coffee maker (e.g., market share $= \pi_i$). Assume that you take a sample of size 2 from the relevant population in a manner which approximates a Bernoulli process.
a. List all possible sample outcomes.
b. For each state of nature find the probability of observing each sample result.
c. Find the probability of getting each sample result, regardless of which state of nature it came from.
d. Given each sample result, find the probability of having each state of nature.
e. For each sample result, find the optimal decision using the Bayesian criterion.
f. Find the expected value of taking this sample of size 2.
g. What is the maximum you would pay for the sample of size 2?
h. If the sampling cost is $10 per family sampled, what is the net expected gain from selecting a sample of size 2?

6. Repeat Exercise 5 for the situation given in Exercise 2.

Appendix A

ANSWERS TO SELECTED EXERCISES

CHAPTER 1

4. **a.** $2(21 + 27 + 24) = (2)(3)(7 + 9 + 8) = 144$
    b. $24 + \Sigma Y_i$
    c. $48 + 3\Sigma Y_i$
    d. 22

CHAPTER 2

1. **a.** $\{1, 4\}$
    b. $\{1, 2, 3, 4, 5, 9, 16, 25\}$
    c. $\{1, 3, 5\}$
    d. $\{1, 2, 3, 4, 5, 7, 9\}$
    e. $\{7, 9\}$
    f. $\{9, 16, 25\}$
    g. $\{0, 1, 2, 3, 4, 5, 6, 7, 8, 9\}$
    h. $\{\phi\}$
    i. $\{1, 2, 3, 4, 5, 7, 9, 16, 25\}$
    j. $\{1\}$
    k. $\{0, 1, 2, 4, 6, 8, 9\}$
    l. $\{1, 4, 9\}$

3. **a.** They are pairwise mutually exclusive because no two pairs of events, A_i and A_j, contain the same (forecasted) unemployment rate. They are "collectively exhaustive" because all possible forecasts are contained in those five events.
    b. (1) The forecasted unemployment was 8 to 10 percent while the actual unemployment was 6 to 8 percent. (2) Unemployment was forecasted to be 4 percent or more, but less than 8 percent. (3) A correct forecast. (4) Actual unemployment was higher than forecasted.

5. **a.** HH, HT, TH, TT
    b. $\frac{1}{4}$
    c. $\frac{2}{4} = 0.50$
    d. $\frac{3}{4} = 0.75$

7. **a.** 40

   b. $^{38}/_{40} = 0.95$

   c. $^{17}/_{40} = 0.425$

   d. $^{38}/_{40} = 0.95$

9. **a.** Since $0.6 + 0.2 + 0.1 < 1$, they cannot be collectively exhaustive.

   b. Since $P(A \cup B) = 0.5$, then $P(A \cap B)$ must be 0; A and B are mutually exclusive. Since $P(A \cup C) = 0.6$, then $P(A \cap C)$ must be 0; A and C are mutually exclusive. $P(B \cap C) = 0$ was given; B and C are mutually exclusive. Thus, A, B, and C are pairwise mutually exclusive. Since $0.1 + 0.4 + 0.5 = 1$, they are collectively exhaustive.

   c. By the same logic as part b, A, B, and C are pairwise mutually exclusive and collectively they exhaust all possible outcomes.

   d. Since $3(0.35) > 1$, then A, B, and C may or may not be collectively exhaustive, but they cannot be mutually exclusive.

11. First, we must find the composition of this crew. *Circled* numbers were given; numbers in *squares* are implied from "Everyone speaks French except . . . ."

MALES (M)

		F	F*	Total
	B_1	[5]	0	(5)
A_1	B_2	4	(6)	(10)
	B_3	2	(1)	(3)
	B_4	[10]	0	(10)
A_2	B_5	[5]	0	(5)
Total		26	7	33

FEMALES (M*)

		F	F*	Total
	B_1	3	(2)	(5)
A_1	B_2	[4]	0	(4)
	B_3	2	(1)	(3)
	B_4	[10]	0	(10)
A_2	B_5	0	0	0
Total		19	3	22

Experiment A. **a.** (1) Yes, they are mutually exclusive because no runner speaks English. (2) Yes, all secretaries speak French. (3) No, three females do not speak French.

   b. No. For example, $P(M) = {}^{33}/_{55} = 0.6$ while $P(M \mid B_2) = 10/(10 + 4) = 0.71$; in order to be independent, $P(M)$ must equal $P(M \mid B_2)$.

   c. Yes.

$$P(M) = 0.6 = P(M \mid A_1) = \frac{(5 + 10 + 3)}{(5 + 10 + 3) + (5 + 4 + 3)} = \frac{18}{30}$$

$$= P(M \mid A_2) = \frac{(10 + 5)}{(10 + 5) + (10 + 0)} = \frac{15}{25} = 0.6$$

Experiment B. **d.** $\dfrac{(3 + 4 + 2)}{(5 + 10 + 3) + (5 + 4 + 3)} = \dfrac{9}{30} = 0.30$

	e.	$24/30 = 0.80$
Experiment C.	**f.**	$10/25 = 0.40$
	g.	$20/25 = 0.80$
Experiment D.	**h.**	$20/30 = 0.67$
	i.	$12/30 = 0.40$
Experiment E.	**j.**	$(8/30)(7/29) = 0.06$
	k.	$(9/30)(8/29) = 0.08$

13. **a.** 0.38

b. $$\frac{(0.08)(0.5)}{(0.08)(0.5) + (0.15)(0.1) + (0.05)(0.1) + (0.3)(0.3)} = 0.27$$

CHAPTER 3

1. **a.** (X, Y)

1,1	2,1	3,1	4,1	5,1	6,1
1,2	2,2	3,2	4,2	5,2	6,2
1,3	2,3	3,3	4,3	5,3	6,3
1,4	2,4	3,4	4,4	5,4	6,4
1,5	2,5	3,5	4,5	5,5	6,5
1,6	2,6	3,6	4,6	5,6	6,6

There are 36 possible outcomes.

b. $P(X = 1) = P(X = 2) = \cdots = P(X = 6) = 1/6.$

c. $P(Y = 1) = P(Y = 2) = \cdots = P(Y = 6) = 1/6.$

d.

Z	$P(Z)$	
2	$1/36$	$\leftarrow P(\{X = 1\} \cap \{Y = 1\}) = P(\text{"snake eyes"}) = P(X = 1)P(Y = 1)$
3	$2/36$	$\leftarrow P((\{X = 1\} \cap \{Y = 2\}) \cup (\{X = 2\} \cap \{Y = 1\})) = P(X = 1)P(Y = 2)$
		$\qquad + P(X = 2)P(Y = 1)$
4	$3/36$	
5	$4/36$	
6	$5/36$	
7	$6/36$	
8	$5/36$	
9	$4/36$	
10	$3/36$	
11	$2/36$	
12	$1/36$	
	$36/36 = 1$	

e. The value of Z occurs by chance; Z is a random variable. In a finite range of possible Z values, say between 0 and 20, there are only a finite number of values (i.e., the *eleven* values 2, 3, ..., 12) that Z can possibly assume; Z is a *discrete* random variable.

f. Both conditions are met; the answer to part *d* is a probability distribution.

g. Values of $D = X - Y$

		X							D	$P(D)$
		1	**2**	**3**	**4**	**5**	**6**			
	1	0	1	2	3	4	5		-5	$1/36$
	2	-1	0	1	2	3	4		-4	$2/36$
Y	**3**	-2	-1	0	1	2	3		-3	$3/36$
	4	-3	-2	-1	0	1	2		-2	$4/36$
	5	-4	-3	-2	-1	0	1		-1	$5/36$
	6	-5	-4	-3	-2	-1	0		0	$6/36$
									1	$5/36$
									2	$4/36$
									3	$3/36$
									4	$2/36$
									5	$1/36$
										$36/36$

h. Yes. $P(D) \geq 0$ for all values of D and $\sum\limits_{D=-5}^{5} P(D) = 1$.

i. $E(X) = 3.5 = E(Y)$, $E(Z) = 7$, $E(D) = 0$

j. $Var(X) = (\frac{1}{6})(1 - 3.5)^2 + (\frac{1}{6})(2 - 3.5)^2 + \cdots + (\frac{1}{6})(6 - 3.5)^2$
$= (\frac{1}{6})(17.5) = 2.9167$
$Var(Y) = Var(X)$
$Var(Z) = (\frac{1}{36})(2 - 7)^2 + (\frac{2}{36})(3 - 7)^2 + \cdots + (\frac{1}{36})(12 - 7)^2 = \frac{210}{36}$
$= 5.8333 = Var(D)$

k. $E(X + Y) = E(X) + E(Y) = 3.5 + 3.5 = 7 = E(Z)$; it checks.
$E(X - Y) = E(X) - E(Y) = 3.5 - 3.5 = 0 = E(D)$; it checks.

l. $Var(X + Y) = Var(X) + Var(Y) = \dfrac{17.5}{6} + \dfrac{17.5}{6} = \dfrac{35}{6} = 5.8333$

$$= Var(Z)$$

$$Var(X - Y) = Var(X) + Var(Y) = 5.8333 = Var(D)$$

5. a. Base $= \frac{1}{3}$, height $= 3X = 3(\frac{1}{3}) = 1$. One half of the base times height equals $\frac{1}{6}$.

b. $P(\tfrac{1}{3} \leq X \leq 2) = \tfrac{1}{2}(\tfrac{5}{3})[\tfrac{6}{5} - (\tfrac{3}{5})(\tfrac{1}{3})] = \tfrac{5}{6}$

Note: To be a probability density function, the area under the density function must equal 1. Also, note that

$$P(-\infty \leq X \leq +\infty) = P(-\infty \leq X \leq 0) + P(0 \leq X \leq \tfrac{1}{3})$$
$$+ P(\tfrac{1}{3} \leq X \leq 2) + P(2 \leq X \leq +\infty)$$
$$= 0 + \tfrac{1}{6} + \tfrac{5}{6} + 0 = 1$$

c. $P(X > 1) = \tfrac{1}{2}(1)(\tfrac{6}{5} - \tfrac{3}{5}) = \tfrac{3}{10}$

d. $P(0 \leq X \leq \tfrac{2}{3}) = 1 - P(\tfrac{2}{3} \leq X \leq 2)$
$$= 1 - \tfrac{1}{2}(\tfrac{4}{3})[\tfrac{6}{5} - (\tfrac{3}{5})(\tfrac{2}{3})] = 1 - \tfrac{8}{15} = \tfrac{7}{15}$$

7. a.

r	0	1	2	3	4	5	6	7
$P(r)$	0.0000	0.0000	0.0002	0.0026	0.0230	0.1240	0.3720	0.4783

$P(r)$ values are from Appendix Table III, for $\pi = 0.9$ and $n = 7$.

b. $P(r = 7) = 0.4783$

c. $P(r \geq 5) = P(r = 5) + P(r = 6) + P(r = 7) = 0.9743$

9. For parts *a* through *c*, let a "success" be a fish which is above the legal size limit, and let r equal the number of successes when $\pi = 0.40$ and $n = 5$. For parts *d* and *e*, let a "success" be a fish which is below the legal size limit, and let q equal the number of successes when $\pi = 0.60$ and $n = 5$. From Appendix Table III:

r	0	1	2	3	4	5	
$P(r)$	0.0778	0.2592	0.3456	0.2304	0.0768	0.0102	$P(q)$
	5	4	3	2	1	0	q

a. $P(r = 5) = 0.0102$

b. $P(r = 0) = 0.0778$

c. $P(r \geq 1) = 1 - P(r = 0) = 1.0000 - 0.0778 = 0.9222$
d. $P(q \geq 2) = 0.2304 + 0.3456 + 0.2592 + 0.0778 = 0.9130 = P(r \leq 3)$
e. $P(q \geq 3) = 0.3456 + 0.2592 + 0.0778 = 0.6826 = P(r \leq 2)$

11.   a. $P(X \geq 100) = P\left(z \geq \dfrac{100 - 103}{4}\right) = P(z \geq -0.75)$
$$= P(-0.75 \leq z \leq 0) + P(z > 0)$$
$$= 0.2734 + 0.5000 = 0.7734$$

b. $P(X < 100) = 1 - P(X \geq 100) = 1 - 0.7734 = 0.2266 \ \{= P(z < -0.75)\}$

c. $P(X > 104) = P\left(z \geq \dfrac{104 - 103}{4}\right) = P(z \geq +0.25) = 0.4013$

d. $P(X < 104) = 1 - P(X \geq 104) = 1 - P(X = 104) - P(X > 104)$
$$= 1 - 0 - 0.4013 = 0.5987$$

e. $P(100 \leq X \leq 104) = P\left(\dfrac{100 - 103}{4} \leq z \leq \dfrac{104 - 103}{4}\right)$
$$= P(-0.75 \leq z \leq +0.25)$$
$$= P(-0.75 \leq z \leq 0) + P(0 \leq z \leq 0.25)$$
$$= 0.2734 + 0.0987 = 0.3721$$

f. $1 - P(100 \leq X \leq 104) = 1 - 0.3721 = 0.6279$

g. $P(96.42 \leq X \leq 109.58) = P\left(\dfrac{96.42 - 103}{4} \leq z \leq \dfrac{109.58 - 103}{4}\right)$
$$= P(-1.645 \leq z \leq +1.645)$$
$$= P(-1.645 \leq z \leq 0) + P(0 \leq z \leq 1.645)$$
$$= 0.4500 + 0.4500 = 0.9000$$

h. $1 - P(96.42 \leq X \leq 109.58) = 1 - 0.9000 = 0.1000$

15.   *Experiment A.* $\mu_X = 200, \ \sigma_X = 10$

a. $P(X \geq 200) = P\left(z \geq \dfrac{200 - 200}{10}\right) = P(z \geq 0) = 0.5000$

b.   1 percent of 4,240 is $(0.01)(4,240) = 42.4$ million gallons.

$$P(X \leq 42.4) = P\left(z \leq \dfrac{42.4 - 200}{10}\right) = P(z \leq -15.76) \simeq 0$$

c.   A fire warning will be issued if, at 8 A.M., the water remaining is $4,240 - (0.1)(4,240) = 3,816$ million gallons. At 7 A.M., we have $4,240 - (0.05)(4,240) = 4,028$ million gallons (due to the "5 percent below capacity" statement). Therefore, if the reservoir's outflow is $4{,}028 - 3{,}816 = 212$ million gallons or more between 7 and 8 A.M., then a fire warning will be issued.

$$P(X \geq 212) = P\left(z \geq \frac{212 - 200}{10}\right) = P(z \geq 1.20) = 0.1151$$

d. $P(180 \leq X \leq 210) = P\left(\dfrac{180 - 200}{10} \leq z \leq \dfrac{210 - 200}{10}\right) = P(-2 \leq z \leq +1)$

$$= P(-2 \leq z \leq 0) + P(0 \leq z \leq 1)$$
$$= 0.4772 + 0.3413 = 0.8185$$

Experiment B. $\mu_Y = 100$, $\sigma_Y = 30$

e. $P(Y \leq 10) = P\left(z \leq \dfrac{10 - 100}{30}\right) = P(z \leq -3) = 0.0013$

f. $P(X \geq 222) = P\left(z \geq \dfrac{222 - 200}{10}\right) = P(z \geq 2.20) = 0.0139$

Then $P(\{Y \leq 10\} \cap \{X \geq 222\}) = P(Y \leq 10)P(X \geq 222)$
$$= (0.0013)(0.0139) = 0.00002$$

CHAPTER 4

1. a.

	X	For Part *b* Method 1		For Part *b* Method 2
		$(X - \bar{X})$	$(X - \bar{X})^2$	X^2
	1	-3	9	1
	3	-1	1	9
Median →	5	1	1	25
	5	1	1	25
	6	2	4	36
	20	0	16	96

Mean = $^{20}\!/_5 = 4$ machines
Median = 5 machines
Mode = 5 machines

b. *Method 1.* $\text{Var}(X) = s_X{}^2 = \dfrac{\Sigma(X - \bar{X})^2}{n - 1} = \dfrac{16}{5 - 1} = 4 \text{ (machines)}^2$

Sample standard deviation $= s_X = \sqrt{4} = 2$ machines

Method 2. $s_X{}^2 = \dfrac{n\Sigma X^2 - (\Sigma X)^2}{n(n - 1)} = \dfrac{(5)(96) - (20)^2}{(5)(4)} = \dfrac{480 - 400}{20} = 4$

$s_X = \sqrt{4} = 2$

   c. There is one mode and the mean, $\bar{X} = 4$, is less than the median, which is 5. Thus, the distribution (small as it is) is skewed to the left or negatively skewed. The tail goes to the smaller values of X.

   d. Range $= 6 - 1 = 5$.

2. a. Mean $= \mu = {}^{40}\!/_{10} = 4$ washing machines

      Median $=$ average of two middle values in array $= 4.5$

      Mode $= 5$

   b. Summary of calculations: $N = 10$, $\Sigma X = 40$, $\Sigma(X - \bar{X})^2 = 42$, $\Sigma X^2 = 202$.

      Method 1.   Population variance of $X = \sigma_X{}^2 = \dfrac{\Sigma(X - \mu)^2}{N} = \dfrac{42}{10} = 4.2$

                Population standard deviation $= \sigma_X = \sqrt{4.2} = 2.049$

      Method 2. $\sigma_X{}^2 = \dfrac{N\Sigma X^2 - (\Sigma X)^2}{N^2} = \dfrac{(10)(202) - (40)^2}{100} = \dfrac{2{,}020 - 1{,}600}{100} = 4.2$

                $\sigma_X = \sqrt{4.2} = 2.049$

   c. Range $= 7 - 0 = 7$.

   d. If $\Sigma X/10 = 3.8$, then the total number of washing machines sold last week must be 38. At \$100 profit per machine, profit $=$ \$3,800 last week.

6. a. Array:

\$0.39	2.49	3.85	5.94	7.94
0.50	3.05	3.95	5.97	9.18
1.02	3.74	4.80	6.86	9.89
1.47	3.78	Median → 5.09	7.16	9.92
1.87	3.79	5.46	7.74	10.94
2.30	3.83	5.62	7.92	11.12

      Mean $= \mu = \dfrac{\Sigma X}{N} = \dfrac{157.58}{30} = \$5.25$

      Median $= \dfrac{4.80 + 5.09}{2} = \$4.945$

   b. Since $\sigma_X = \$3.04$, then $(1.5)\sigma = 1.5(3.04) = \$4.56$ and $\mu_X \pm (1.5)\sigma_X = \$5.25 \pm \$4.56 = \$0.69$ to \$9.81. There are two values below \$0.69 (that is, 39¢ and 50¢) and there are four values above \$9.81. Thus, there are $30 - 6 = 24$ values, or ${}^{24}\!/_{30}(100) = 80$ percent of the values in the interval $\mu \pm (1.5)\sigma$.

   c. Chebyshev's inequality says that at least $1 - 1/K^2 = 1 - 1/(\tfrac{3}{2})^2 = 1 - \tfrac{4}{9} = \tfrac{5}{9} = 0.55556$ (or 55.6 percent) of the observations will lie in the interval $\mu \pm (\tfrac{3}{2})\sigma$. Since we observed 80 percent of the observations in that interval, Chebyshev was correct; at least 55.6 percent of the observations were in the interval.

Part B

d.

Class Intervals	Frequency f	Midpoints M
$ 0.00 to under $ 2.00	5	1.00
$ 2.00 to under $ 4.00	9	3.00
$ 4.00 to under $ 6.00	6	5.00
$ 6.00 to under $ 8.00	5	7.00
$ 8.00 to under $10.00	3	9.00
$10.00 to under $12.00	2	11.00
Total	30	—

e.

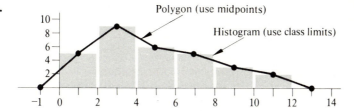

Polygon (use midpoints)

Histogram (use class limits)

f.

f_j	M_j	$f_j M_j$	Method 1 $M_j - \mu$	$(M_j - \mu)^2$	$f_j(M_j - \mu)^2$	Method 2 M_j^2	$f_j M_j^2$
5	1	5	−3.87	14.9769	74.8845	1	5
9	3	27	−1.87	3.4969	31.4721	9	81
6	5	30	0.13	0.0169	0.1014	25	150
5	7	35	2.13	4.5369	22.6845	49	245
3	9	27	4.13	17.0569	51.1707	81	243
2	11	22	6.13	37.5769	75.1538	121	242
30		146			255.4670		966

Mean: $\mu_x = \Sigma fM / \Sigma f = {}^{146}\!/_{30} = \$4.87$. This approximation represents a $[(5.25 - 4.87)100]/5.25 = 7.24$ percent error from the actual mean, \$5.25. This error is attributable to the fact that the ΣX within each class is not equal to $f_j M_j$ for each class.

Median: The median from grouped data requires us to locate the class interval that includes the median. We are looking for the $N/2 = {}^{30}\!/_2 =$ fifteenth item in the frequency distribution. There are 14 (that is, $5 + 9$)

observations in the range from \$0.00 to \$3.99; hence, the first observation in the \$4.00 to \$5.99 interval is the median. To approximate that value:

$$\text{Median} = L + \left(\frac{n/2 - F}{f_{med}}\right)c = \$4.00 + \left(\frac{15 - 14}{6}\right)(2) = \$4.333$$

From part a, the raw data gave the true median as \$4.945. The \$0.612 difference, or the $0.612(100)/4.945 = 12.4$ percent error, in approximating the median occurred because the six observations in the interval from \$4.00 to \$5.99 were not, in fact, evenly distributed within that interval.

g.   The modal class is the one with the highest frequency. The highest frequency is 9; thus, the \$2.00 to \$3.99 class interval is the modal class. This distribution is unimodal; it has only one peak. From part e and from the observation that the mean is greater than the median, \$4.87 > \$4.33, the distribution is skewed to the right or positively skewed (i.e., mean minus median is positive).

h.   From the worksheet in part f:

$$\text{Method 1.} \quad \sigma^2 = \frac{\Sigma f(M - \mu)^2}{N} = \frac{255.467}{30} = 8.5156; \ \sigma = \sqrt{8.5156} = \$2.92$$

$$\text{Method 2.} \quad \sigma^2 = \frac{N\Sigma f M^2 - (\Sigma f M)^2}{N^2} = \frac{(30)(966) - (146)^2}{(30)(30)}$$

$$= \frac{28,980 - 21,316}{900} = 8.52$$

The approximated standard deviation, \$2.92, is \$0.12 lower than the standard deviation calculated from the raw data, which was given in part b as \$3.04. The variance and standard deviation are measures of "squared deviations from the mean." If the approximated mean is in error, then you should expect the approximations of σ^2 and σ to contain some error.

8. a.   LPI(71) = 100.00,   LPI(73) = 145.32
    b.   PPI(71) = 100.00,   PPI(73) = 144.71
    c.   WARP(71) = 100.00,   WARP(73) = 145.32
    d.   WARP is always equal to LPI. Parts a and c show a 45.32 percent price increase, while part b shows a 44.71 percent price increase.

CHAPTER 5

1.   Although they may look the same, the widths of pages in this book vary ever so slightly. Thus, if not all pages were measured, some sampling error could be introduced when you state "the average width of pages." Using your little finger as a measuring device will introduce nonsampling error to the extent that

Length between joints $\neq$ 1 inch

Nonsampling error is more likely.

4. a. All possible samples of size 20 must have had the same chance of being selected.
    b. Sampling with replacement, there are $(10,000)^{20}$ possible samples. If all are equally likely, then the probability is $(1/10,000)^{20}$.
    c. Same as b.
    d. It tells us the relevant population from which one item (i.e., one collection of 20 observations) will be selected.

6. a.

$\bar{X}$	$5.00	$7.50	$10.00
$P(\bar{X})$	0.64	0.32	0.04

    c. Yes
    d.

$\bar{X}$	$5.00	$7.50	$10.00
$P(\bar{X})$	$\frac{56}{90}$	$\frac{32}{90}$	$\frac{2}{90}$

    e. $E(\bar{X}) = \$6,\quad \mathrm{Var}(\bar{X}) = \frac{16}{9}\ \$^2.$
    f. $E(X) = E(\bar{X}) = \$6,\quad \mathrm{Var}(\bar{X}) = \frac{\sigma^2}{n}\frac{N-n}{N-1} = \frac{4}{2}\frac{(10-2)}{(10-1)} = \frac{16}{9}$

7. a. 7.76 to 12.24 milligrams $(z = 2.24)$.
    b. Yes. No; it is possible to say that either (1) 2.5 percent weigh less than 8.04 milligrams or (2) 5 percent weigh less than 8.355 milligrams.
    c. Yes.
    d. Yes.
    e. 9.776 to 10.224 milligrams.
10. 0.6950
11. a. 0.0488
    b. 0.3124
    c. 0.2352
    d. 0.1160
16. $P(r_A > r_B) = 0.1496$

CHAPTER 6

1. a. $\frac{528}{16} \pm (1.96)(0.1) = 32.80$ to 33.20 gallons per keg.
    b. $100(33 \pm 0.196) = 3,280$ to 3,320 gallons.
    c. If we took all possible samples of size 16 and found $\bar{X}$ for each sample, then 95 percent of those sample means would produce an interval, $\bar{X} \pm 0.196$, which would include the population mean μ. Ninety-five percent of the time the procedure will yield the desired results. The population mean is either in

the interval 32.8 to 33.2 or it is not; the value of μ is a constant. It is the procedure (of constructing an interval estimate for μ) in which we have "95 percent confidence."

   d. 32.74 to 33.26 gallons.

   e. 32.84 to 33.16 gallons.

   f. 0.05, 0.025.

5. **a.** $\bar{X} = 29.5$, $s = \sqrt{0.875} = 0.94$

   b. $t = 2.306$, the interval is 28.78 to 30.22°F.

   c. This question asks for a slightly different use of statistical inference. One approach would be to assume that $\mu = 32°F$ and find the probability of observing the reported sample results. That is,

$$P(\bar{X} \leq 29.5 \,|\, \mu = 32) = P\left(t \leq \frac{29.5 - 32}{0.94/3}\right) = P(t \leq -7.98)$$

For 8 degrees of freedom, Appendix Table V does not provide a t value as large as 7.98. Thus, the best we can do is say that $P(t \geq +7.98) < 0.0005$. There is a chance, but it is not very likely that $\mu = 32°F$ (or more) given these sample results. (*Note:* The next chapter investigates this form of statistical inference.)

   d. Without proper ventilation, the freezer may not be as efficient. Depending upon the variety of environments in which the nine freezers are tested, the mean interior temperature may vary and the standard deviation might increase.

7. **a.** Regardless of population shape, the sample size is large enough to approximate the distribution of $\bar{X}$ with the normal distribution.

$$120 \pm (1.645)\frac{66}{\sqrt{1089}} = 116.71 \text{ to } 123.29 \text{ gallons}$$

   b. There are many reasons why this interval estimate is not the net increase in gasoline consumption per family during August. For example, if they were not on vacation they would have used some gasoline anyway.

9. **a.** Chebyshev. 285 to 315 feet.

   b. The 88.89 percent represents the minimum proportion of intervals that would include the population mean.

13. **a.** 0.783 to 0.817.

   b. $49.95 to $50.05

   c. $39.13 to $40.87

15. 56 to 64 percent favor park expansion.

17. **a.** 9,604

   b. 2,401

CHAPTER 7

1. a.

r	$\bar{p}$	$P(r) = P(\bar{p})$
0	0.0	0.3487
1	0.1	0.3874
2	0.2	0.1937
3	0.3	0.0574
4	0.4	0.0112
5	0.5	0.0015
6	0.6	0.0001
7	0.7	0.0000
8	0.8	0.0000
9	0.9	0.0000
10	1.0	0.0000

b.

r	$\bar{p}$	$P(r) = P(\bar{p})$
0	0.0	0.0282
1	0.1	0.1211
2	0.2	0.2335
3	0.3	0.2668
4	0.4	0.2001
5	0.5	0.1029
6	0.6	0.0368
7	0.7	0.0090
8	0.8	0.0014
9	0.9	0.0001
10	1.0	0.0000

c. $\alpha = P(\text{Type I error}) = P(\text{Reject } H_0 \,|\, H_0 \text{ is true})$
$= P(\bar{p} \geq 0.2 \,|\, \pi = 0.10,\ n = 10) = 1 - P(\bar{p} = 0) - P(\bar{p} = 0.1)$
$= 1 - 0.3487 - 0.3874 = 0.2639$
$\beta = P(\text{Type II error}) = P(\text{Accept } H_0 \,|\, H_0 \text{ is false})$
$= P(\bar{p} < 0.2 \,|\, \pi = 0.30,\ n = 10) = P(\bar{p} = 0) + P(\bar{p} = 0.1)$
$= 0.0282 + 0.1211 = 0.1493$

d. $\alpha = P(\bar{p} \geq 0.1 \,|\, \pi = 0.10,\ n = 10) = 1 - 0.3487 = 0.6513$
$\beta = P(\bar{p} < 0.1 \,|\, \pi = 0.30,\ n = 10) = 0.0282$

e. $\alpha = 0.0702$
$\beta = 0.3828$

f. For a constant sample size ($n = 10$), as α gets smaller, β becomes larger.

3. **a.** $H_0: \mu = 32$; $H_A: \mu \neq 32$
   b. $H_0: \mu \geq 20$; $H_A: \mu < 20$
   c. $H_0: \pi \leq 0.50$; $H_A: \pi > 0.50$, where π = proportion who buy.
   d. $H_0: \mu = 98.6$; $H_A: \mu \neq 98.6$
   e. $H_0: \pi \geq 0.80$; $H_A: \pi < 0.80$, where π = proportion with two or more sets.
   f. $H_0: \mu \leq 2000$; $H_A: \mu > 2000$

5. **a.** Type I: Deciding that the car does not average 32 mpg when it does.
   Type II: Deciding that the car does average 32 mpg when it does not.
   c. Type I: Deciding that more than half the customers purchase something when, in fact, half or less buy something.
   Type II: Deciding that half or fewer purchase something when, in fact, more than half buy.
   e. Type I: Deciding incorrectly that less than 80 percent have two or more sets.
   Type II: Deciding incorrectly that 80 percent or more have two or more sets.

7. **a.** The sample size is large. The CLT says that the sampling distribution of $\bar{X}$ can be approximated with the normal distribution; z can be used. For a two-tail test with $\alpha = 0.05$, $z_{\alpha/2} = 1.96$. FPC $= 1$ since $n/N < 0.10$.
   Method 1. Since the critical values of $\bar{X}$ are $\mu_0 \pm z_{\alpha/2}\sigma/\sqrt{n}$, or $13 \pm (1.96)(2)/\sqrt{100} = 13 \pm 0.392$, then the decision rule is: If $12.608 \leq \bar{X} \leq 13.392$, accept H_0; otherwise, reject H_0. Since $\bar{X} = 12.5 < 12.608$, reject H_0.
   Method 2. Decision rule: If $-1.96 \leq z \leq +1.96$, accept H_0; otherwise, reject H_0. Since

   $$z = \frac{\bar{X} - \mu_0}{\sigma/\sqrt{n}} = \frac{12.5 - 13.0}{2/\sqrt{100}} = \frac{-0.50}{0.20} = -2.5$$

   reject H_0.
   c. Since n is large, use z. For a left-tail test with $\alpha = 0.10$, $z = -1.28$, FPC $= 1$, and $\bar{X} = \Sigma X/n = {}^{810}/_{81} = 10$

   $$s^2 = \frac{n\Sigma X^2 - (\Sigma X)^2}{n(n-1)} = \frac{81(8,180) - (810)^2}{81(80)} = \frac{6,480}{6,480} = 1$$

   Method 1. $\mu_0 - z_\alpha s/\sqrt{n} = 10.12 - (1.28)(1)/9 = 9.98$
   If $\bar{X} \leq 9.98$, reject H_0; if $\bar{X} > 9.98$, accept H_0. Since $\bar{X} = 10 > 9.98$, accept H_0.
   Method 2. If $z \leq -1.28$, reject H_0; if $z > -1.28$, accept H_0. Since $z = (10 - 10.12)/\frac{1}{9} = -1.08 > -1.28$, accept H_0.

e. Use the normal distribution as an approximation of the binomial.

$$\mu_{\bar{p}} = \pi = 0.80 \qquad \sigma_{\bar{p}} = \sqrt{\frac{\pi(1-\pi)}{n}} = \sqrt{\frac{(0.8)(0.2)}{400}} = 0.02$$

For a right-tail test with $\alpha = 0.05$, $z_\alpha = 1.645$.
Method 1. $\bar{p}_c = \mu_{\bar{p}} + z_\alpha \sigma_{\bar{p}} = 0.80 + (1.645)(0.02) = 0.8329$
Decision rule: If $\bar{p} \geq 0.8329$, reject H_0; otherwise, accept H_0. Since $\bar{p} = 0.84 > 0.8329$, reject H_0.
Method 2. If $z \geq 1.645$, reject H_0; otherwise, accept H_0. Since $z = (0.84 - 0.80)/0.02 = 2 > 1.645$, reject H_0.

11. a. If $2.55 \leq \bar{X} \leq 3.45$, accept H_0; otherwise, reject H_0.
    b. $\beta = P(2.55 \leq \bar{X} \leq 3.45 \mid \mu = 3.5) = (3.45 - 3.00)(1) = 0.45$
    c. $\beta = 0.50$
    d. $\beta = 0$

13. If he is certain that 3,600 inches of string were cut into 200 shoestrings, then $\mu = \Sigma X/N = 3,600/200 = 18$ inches per shoestring. He knows the population mean; there is no need to test the hypothesis about μ if you know its value.

15. $H_0: \mu_X - \mu_Y \geq 0$; $H_A: \mu_X - \mu_Y < 0$ (The newspaper's claim is in H_A.) For $\alpha = 0.05$, $z_\alpha = 1.645$.

$$S_d{}^2 = \frac{S_X{}^2}{n_X} + \frac{S_Y{}^2}{n_Y} = \frac{(800)^2}{100} + \frac{(1,200)^2}{400} = 6,400 + 3,600 = 10,000$$

and

$$S_d = \sqrt{10,000} = \$100$$

Decision rule: If $\bar{X} - \bar{Y} \leq -z_\alpha S_d = -(1.645)(\$100) = -\$164.50$, reject H_0; otherwise, accept H_0. Since $\bar{X} - \bar{Y} = \$7,270 - \$7,410 = -\$150$, we can not reject the null. The newspaper's own survey results do not support their claim.

CHAPTER 8

1. a. $E(X) = 6$
    b. $\pi = 0.5$
    e. H_0: Data is binomially distributed. H_A: The data is not. Decision rule: If $\chi^2 > 11.070$, reject H_0; if $\chi^2 \leq 11.070$, accept H_0. Calculated $\chi^2 = 20.41$.

3. **a.** H_0: A_i and B_j are independent. H_A: They are not independent. Decision rule: If $\chi^2 > 21.026$, reject H_0; otherwise, accept H_0. Calculated $\chi^2 = 51.665$.

5. **a.** We can not use this data for a test of independence. In problem 3, we have an estimate of the proportion of customers who purchase in each division. Because a random sample was selected within each division, we do not have that estimate in problem 5. In this problem we treat each division as a separate population and we are interested in the proportion of items that fall within each classification, A_i. This problem is called a test of homogeneity; in general, it asks: "How alike are the various populations?" The procedure for conducting a test of homogeneity is exactly the same as for the test of independence. The only differences are (1) the null and alternative hypotheses and (2) the interpretation of the results.
    b. If $\chi^2 > 11.070$, reject H_0; otherwise, accept H_0.
    c. Calculated $\chi^2 = 65.67$.

7. H_0: $\sigma^2 \geq 1,600$ hours; H_A: $\sigma^2 < 1,600$ hours; $\chi^2_{0.95,50} = 34.764$. Decision rule: If $s^2 < (1,600/50)(34.764) = 1,112.45$, reject H_0; if $s^2 \geq 1,112.45$, accept H_0. Since $s^2 = 1,225$, the new alloy appears to reduce the variance of bulb life.

9. **a.** Decision rule: If $19.902 \leq \bar{X} \leq 20.098$, accept H_0; otherwise, reject H_0.
    b. (1) The population must be normally distributed. You could use a goodness-of-fit test on both your winter data and this summer sample.
    (2) H_0: $\sigma^2 = 1$; H_A: $\sigma^2 \neq 1$. Since $n = 400$, use the normal-curve approximation. $\chi_L{}^2 = 399 - (1.96)\sqrt{798} = 343.63$ and

    $$\chi_U{}^2 = 399 + (1.96)(28.25) = 454.37$$

    $$\left(\frac{\sigma_0{}^2}{n-1}\right)\chi_L{}^2 = \frac{1}{399}(343.63) = 0.861$$

    and

    $$\left(\frac{\sigma_0{}^2}{n-1}\right)\chi_U{}^2 = \frac{1}{399}(454.37) = 1.139$$

    Decision rule: If $0.861 \leq s^2 \leq 1.139$, accept H_0; otherwise, reject H_0. Since $s^2 = (1.2)^2 = 1.44$, reject H_0. Would you now retest part a?

13. **a.** H_0: $\mu_{\text{Small}} = \mu_{\text{Medium}} = \mu_{\text{Large}}$; H_A: They are not all equal. $F(0.05, v_b = 2, v_w = 25) = 3.38$
    Decision rule: If $F > 3.38$, reject H_0; if $F \leq 3.38$, accept H_0. SSW = 46, MSSW = 1.84, SSB = 1,155, MSSB = 577.5
    b. Scale economies should be investigated. Size of plant does seem to affect ATC.

CHAPTER 9

1. **a.** Median $= 4.5$, $R = 11$, decision rule: If $7 < R < 15$, accept H_0; otherwise, reject H_0.
   b. Decision rule: If $1.3680 < \text{VNR} < 2.8425$, accept H_0; otherwise, reject H_0. $\text{VNR} = {}^{318}\!/_{19}/7.61$.
3. **a.** $\text{VNR} = 4 > 3.1151$, reject H_0. Look for trend.
   b. $\text{VNR} = 1.5 < 1.55$, reject H_0. Look for trend.
   c. $1.896 < \text{VNR} = 2.02 < 2.104$, accept H_0. Time series is random.
5. **a.** 0.5, 0.5
   b. (1) 0.25, (2) 0.25, (3) 0.50
   c. (1) $(\frac{1}{2})^3 = 0.125$
      (2) $(\frac{1}{2})^3 = 0.125$
      (3) $1 - 0.125 - 0.125 = 0.75$
7. For $1 - \alpha = 0.95$, $z = 1.96$. From Equation 9.8,

$$r' = \tfrac{1}{2}(n - z\sqrt{n}) = \tfrac{1}{2}[50 - (1.96)\sqrt{50}] = 18.07$$

Let $r = 18$, which makes $1 - \alpha$ slightly more than 0.95. Placing the data in an array, we find that $X_{18} = \$0.77$ and $X_{50+1-18} = X_{33} = \$1.21$. Thus, the $95(+)$ percent confidence interval for the median is $\$0.77$ to $\$1.21$.

9. $H_0: \phi \geq \$9,500$; $H_A: \phi < \$9,500$ (mayor's claim)
Let a success be a family income that is above $\$9,500$.

$$\tfrac{1}{2}(n + z\sqrt{n}) = \tfrac{1}{2}(100 + 2.58\sqrt{100}) = 62.9$$

Decision rule: If the number of families with incomes above $\$9,500$ is greater than 62.9, reject the null hypothesis; if the number of families is 69.2 or fewer, accept the null. Since the number of families with more than $\$9,500$ income is 60, we accept H_0; this test does not support the mayor's claim.

CHAPTER 10

1. **b.** For case A:

$$b = \frac{n\Sigma XY - \Sigma X\Sigma Y}{n\Sigma X^2 - (\Sigma X)^2} = \frac{4(68) - 12(20)}{4(110) - 400} = \frac{32}{40} = 0.80$$

$$a = \frac{\Sigma Y - b\Sigma X}{n} = \frac{12 - 0.8(20)}{4} = -1$$

$\bar{Y}_c = -1 + (0.8)X$
For case B: $\bar{Y}_c = 1.90 + (0.4)X$
For case C: $\bar{Y}_c = 4.5 - (0.5)X$

c. For case A: $s_{Y|X} = \sqrt{\dfrac{\Sigma Y^2 - a\Sigma Y - b\Sigma XY}{n-2}}$

$$= \sqrt{\dfrac{44 - (-1)(12) - 0.8(68)}{2}}$$

$$= \sqrt{0.8000} = 0.894$$

Case B: $\sqrt{6.90} = 2.627$

Case C: 0

d.

Case	A	B	C
r^2	0.800	0.188	1.000

e. $H_0: \beta = 0, H_A: \beta \neq 0$. For $\alpha = 0.05$, $n - 2 = 2$ degrees of freedom, and a two-tail test, $t = \pm 4.303$. Decision rule: If $-4.303 \le t \le +4.303$, accept H_0; otherwise, reject H_0.

For case A: Since $\bar{X} = \Sigma X/n = {}^{20}\!/_4 = 5$, then

$$t = \frac{b}{s_b} = \frac{b\sqrt{\Sigma X^2 - n\bar{X}^2}}{s_{Y|X}} = \frac{(0.8)\sqrt{110 - (4)(5)^2}}{0.894} = 2.830$$

Accept H_0; β is not significantly different from zero.

Thus, $H_0: \rho = 0$ would also be accepted.

Case B: $t = 0.681$

Case C: $t = -\infty$

f. Using Equation 10.12, for case A,

$$[-1 + (0.8)(4)] \pm (4.303)(0.894)\sqrt{\frac{1}{4} + \frac{(4-5)^2}{110 - 4(25)}}$$

$$= 2.2 \pm (3.847)\sqrt{0.3500} = 2.2 \pm 2.28, \text{ or from } -0.08 \text{ to } 4.48$$

Case B: 3.5 ± 5.65

Case C: 2.5 (no interval possible)

g. Case A: 2.2 ± 2.98

Case B: 3.5 ± 7.99

h. Case A: 2.2 ± 4.47

Case B: 3.5 ± 12.64

k. To obtain $\bar{X}_c = c + dY$, use $d = \dfrac{n\Sigma XY - \Sigma X \Sigma Y}{n\Sigma Y^2 - (\Sigma Y)^2}$ and $c = \bar{X} - d\bar{Y}$.

Case A: $d = \dfrac{32}{4(44) - 144} = 1$ and $c = 5 - (1)3 = 2$

$\bar{X}_c = 2 + Y$ while solving $\bar{Y}_c = -1 + (0.8)X$ yields

$X = 1.25 + 1.25\bar{Y}_c$.

Case B: $\bar{X}_c = 2.36 + (0.47)Y$ which is not the same as
$X = -4.75 + 2.5\bar{Y}_c$.

Case C: $\bar{X}_c = 9 - 2Y$ which is the same as $X = 9 - 2\bar{Y}_c$ because $r^2 = 1$.

3. **a.** $\bar{Y}_c = 8 + (0.02)X$, where Y units = thousands of dollars and X units are square feet.

b. The Y intercept, $a = 8$, does not mean that a house with no space would sell for \$8,000. It is tempting to interpret the Y intercept as an estimate of the average value of land upon which these houses sit. However, the range of X values used (i.e., 1,000 to 1,400 square feet) to construct this regression line does not include $X = 0$. Using the \$8,000 value for a land value is the same thing as extrapolating the regression line back to the origin. The slope coefficient, $b = 0.02$ thousands of dollars per square foot (or \$20 per square foot), is a measure of the "addition to market value for an additional square foot of living space."

c. $s_{Y|X} = \sqrt{2.2222} = 1.491$ thousands of dollars.

d.

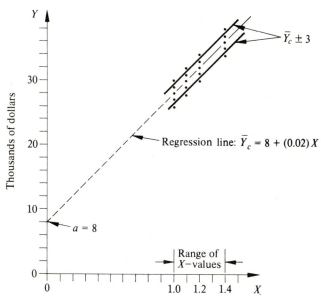

Regression line: $\bar{Y}_c = 8 + (0.02)X$

Thousands of square feet

e. $s_Y = \sqrt{11.3158} = 3.364$. $s_Y > s_{Y|X}$; knowing the relationship between Y and X reduces the variation in Y.

f. $\bar{Y}_{X=1,000} = \dfrac{26 + 29 + 30 + 27 + 28}{5} = 28$

$s_{Y|X=1,000} = \sqrt{\dfrac{(26-28)^2 + (29-28)^2 + \cdots}{4}} = \sqrt{2.5} = 1.581$

g. No; they are all the same.

h. $H_0: \beta = 0; H_A: \beta \neq 0$. For $\alpha = 0.05$, $20 - 2 = 18$ degrees of freedom, and for a two-tail test, $t = \pm 2.101$. The calculated value of t is 8.87.

i. $r^2 = 0.814$, $r = 0.90$.

CHAPTER 11

1. a.

$\Sigma Y = 104$	$\Sigma Y^2 = 1{,}794$	$\Sigma XW = 160$
$\Sigma X = 40$	$\Sigma X^2 = 234$	$\Sigma YW = 484$
$\Sigma W = 32$	$\Sigma W^2 = 162$	$\Sigma YX = 418$

   b. $\overline{Y}_c = 28 - 3X$

   $r_{YX}^2 = {}^{306}\!/_{442} = 0.6923$

   c. $\overline{Y}_c = 5 + 2W$

   $r_{YW}^2 = 0.3077$

   d. $\overline{X}_c = 5 + (0)W$

   $r_{XW}^2 = 0$

   e. $\overline{Y}_c = 20 - 3X + 2W$

   f. 0

   g. $R^2 = 1$

   h. This is an ideal case. From part g, we can see that all the variation in Y is explained by the variation in X and W. While X, by itself, explains 69 percent of the variation in Y, W explains the rest. From part d, there is no linear relation between the independent variables X and W. This means that the marginal contribution of each independent variable can be measured with or without the other independent variable. That is, in parts b and e, the slope coefficient for X is minus 3 whether W is included or excluded. In the same way, observe that the slope coefficient for W is plus 2 in part c and in part e.

3. a. $\overline{Y}_c = 5 + 3X - W$

   b. $s_{Y|XW} = \sqrt{0.5714} = 0.756$; $R^2 = 0.9909$

CHAPTER 12

1. a. Decision rule: If $4 \leq R \leq 10$, accept H_0; otherwise, reject H_0.

Variable	R	Decision
W	5	Accept H_0
X	2	Reject H_0
Y	7	Accept H_0
Z	4	Accept H_0

b. Decision rule ($\alpha = 0.10$): If $1.23 < \text{VNR} < 3.13$, accept H_0; otherwise, reject H_0.

Variable	VNR	Decision
W	2.09	Accept H_0
X	0.08	Reject H_0
Y	2.20	Accept H_0
Z	0.60	Reject H_0

d. W: ? X: T Y: S Z: C

e. Both the runs test and the VNR indicate that the sequence of X values is not random; a rather discernible trend is evident in the plot of X. The Y values have a distinct seasonal pattern but no trend. The "no trend" conclusion is consistent with parts a and b; however, parts a and b indicate "random sequence" whereas the Y values plot a rather specific pattern. Because a complete cycle is not included in the Z values, the VNR indicates we might look for a trend component; the plot does not suggest a trend component.

3. a.   **b.**

t	Y	MT	CMT	(T) CMA	(Y − T) S + I	Y − S
1	808					1,034.44
2	2,040					958.44
3	824	4,088	4,104	1,026	−202	1,035.94
4	416	4,120	4,160	1,040	−624	1,059.19
5	840	4,200	4,216	1,054	−214	1,066.44
6	2,120	4,232	4,240	1,060	1,060	1,038.44
7	856	4,248	4,264	1,066	−210	1,067.94
8	432	4,280	4,330	1,082.5	−650.5	1,075.19
9	872	4,380	4,396	1,099	−227	1,098.44
10	2,220	4,412	4,420	1,105	1,115	1,138.44
11	888	4,428				1,099.94
12	448					1,091.19

	Quarter					
	1	2	3	4		
1973			−202	−624		
1974	−214	1,060	−210	−650.5		
1975	−227	1,115				
Total	−441	2,175	−412	−1,274.5	Total	Correction
Average	−220.5	1,087.5	−206	−637.25	23.75	−23.75/4 = −5.94
S	−226.44	1,081.56	−211.94	−643.19	−0.01	

c. $T = 998.50 + 10.025t$ where $t = 1 \rightarrow$ first quarter, 1973

d.

Quarter t	T $998.5 + 10.025t$	S	Forecast
13	1,128.82	−226.44	902
14	1,138.85	1,081.56	2,220
15	1,148.88	−211.94	937
16	1,158.90	−643.19	516

CHAPTER 14

1. a.

SON$_i$	D_1	D_2	D_3
0	0	−3,000	−5,000
400	0	1,000	−200
800	0	5,000	4,600
1,200	0	9,000	9,400

b. (1) D_1
 (2) D_3
 (3) D_2

c. $1,180; it is the maximum we would pay for additional information.

5. a. Let r = number who say they will buy the coffee maker. Then, $r = 0, 1, 2.$

b. From the binomial tables:

π for SON$_i$	$r = 0$	$r = 1$	$r = 2$
$0/4{,}000 = 0.00$	1.00	0.00	0.00
$400/4{,}000 = 0.10$	0.81	0.18	0.01
$800/4{,}000 = 0.20$	0.64	0.32	0.04
$1{,}200/4{,}000 = 0.30$	0.49	0.42	0.09

c.
$$P(r = 0) = P(\{r = 0\} \cap \{SON_1\}) + P(\{r = 0\} \cap \{SON_2\}) + \cdots$$
$$= P(r = 0 \mid SON_1)P(SON_1) + P(r = 0 \mid SON_2)P(SON_2) + \cdots$$
$$= (1.00)(0.3) + (0.81)(0.3) + (0.64)(0.2) + (0.49)(0.2)$$
$$= 0.769$$
$$P(r = 1) = 0.202 \qquad \text{and} \qquad P(r = 2) = 0.029$$

d. Bayes' rule:

$$P(SON_1 | r = 0) = \frac{P(r = 0 | SON_1)P(SON_1)}{P(r = 0)} = \frac{(1.00)(0.3)}{0.769} = 0.390$$

$$P(SON_2 | r = 0) = \frac{(0.81)(0.3)}{0.769} = 0.316$$

$$P(SON_3 | r = 0) = \frac{(0.64)(0.2)}{0.769} = 0.166$$

$$P(SON_4 | r = 0) = \frac{(0.49)(0.2)}{0.769} = 0.127$$

$P(SON_1 | r = 1) = 0.000$	$P(SON_1 | r = 2) = 0.000$
$P(SON_2 | r = 1) = 0.267$	$P(SON_2 | r = 2) = 0.103$
$P(SON_3 | r = 1) = 0.317$	$P(SON_3 | r = 2) = 0.276$
$P(SON_4 | r = 1) = 0.416$	$P(SON_4 | r = 2) = 0.621$

e. If $r = 0$:

SON_i	$P(SON_i | r = 0)$	D_1		D_2		D_3	
1	0.390	0	0	−3,000	−1,170	−5,000	−1,950.0
2	0.316	0	0	1,000	316	−200	−63.2
3	0.166	0	0	5,000	830	4,600	763.6
4	0.127	0	0	9,000	1,143	9,400	1,193.8
			$E(D_1 | r = 0) = 0$		$E(D_3 | r = 0) = 1,119$		$E(D_2 | r = 0) = -55.8$

Max: $E(D_2 | r = 1) = \$5,594$

Max: $E(D_3 | r = 2) = \$7,083$

f. $(0.769)(1,119) + (0.202)(5,594) + (0.029)(7,083) = \$2,196$

g. $2,196 - 1,700 = \$496$

h. $476

Appendix B

CASE DATA

INTRODUCTION

The following data is drawn from the 1970 Census of Population and Housing. Since there are several such data banks, the one used here is referred to as "State Public Use Samples, the 5 Percent Sample." For each housing unit, there are more than 50 variables (i.e., different questions) on the "household record" and more than 50 variables for each person within that household. Of course, there are thousands of housing units in the data bank.

To give the reader an idea of what a large data bank might contain, without presenting hundreds of pages of data, we have selected 3 states, 42 variables (from the "housing unit" and "head-of-household" records), and a total of 170 housing units. Given a specific problem for investigation, you would only use those variables that might be relevant. Since the primary purpose of this case is to introduce a real-world data bank, not all the 42 chosen variables will appear relevant. The three states (Massachusetts, Illinois, and Oregon) were chosen for their latitudinal commonality and their longitudinal difference.

For the Illinois sample, the first 100 housing units that appear on the computer tape are reproduced here.

For the Oregon and Massachusetts samples, the household's response to variable number 18 was used to segregate the data. Variable number 18 (or $V18$) refers to the "payment of fuel." If the value of that variable is 3, then (1) the housing unit is owner occupied, or (2) it is a vacant unit, or (3) the unit is rented without cash rent, or (4) the unit is on a place of 10 acres or more (i.e., implies "farm"), or (5) the unit is defined as "group quarters" (e.g., a fraternity house). If $V18 \neq 3$, then we have a nonfarm, nonvacant, rented housing unit. The first 30 housing units in the Oregon file and the first 40 units in the Massachusetts file that have $V18 \neq 3$ are reproduced here.

Since the households are placed on the computer tape in a scrambled sequence (to preserve confidentiality), and since that order has been preserved here, do not infer anything about the order of the households. For example, in the Illinois sample, one "city center" household may be followed by another "city center" household, yet neither household may live in Chicago and the two households may be more than 100 miles apart.

At the same time it is not completely safe to assume that the 100 Illinois households reported here represent a simple random sample. The scrambling may not have been random. While it is possible to select a random sample from each state (i.e., the computer tapes are indexed for that purpose), there is some value in observing the data set as it is placed in the file.

In summary, the Illinois sample represents the first 100 housing units in a file while the Massachusetts and Oregon samples represent the first 30 or 40 households that meet a particular requirement: they are nonfarm, nonvacant, non-owner-occupied housing units.

DESCRIPTION OF VARIABLES

Variable Number	Item	Code	Code Description	
01	Household serial number	0–100	Illinois	Number used to
		0–30	Oregon	identify each
		0–40	Massachusetts	household
02	Urban/Rural	0	Not available (NA)	
		1	Urban, inside SMSA, inside central city	
		2	Urban, inside SMSA, outside central city	
		3	Urban, outside SMSA	
		4	Rural, inside SMSA	
		5	Rural, outside SMSA	
03	Total number of persons at unit	XX	Number of persons living in this housing unit; 0 if vacant	
04	Number of housing units at this address	0	One apartment or living quarters	
		1	Two apartments or living quarters	
		2	Three apartments or living quarters	
		:	:	
		8	Nine apartments or living quarters	
		9	Ten or more apartments, etc.	
		10	Mobile home or trailer	
		11	NA (group quarters)	
05	Complete kitchen facilities?	0	Yes, for this household only	
		1	Yes, but also used by another household	
		2	No	
		3	NA (group quarters)	
06	Number of rooms in this housing unit	0	One room	
		1	Two rooms	
		2	Three rooms	
		:	:	
		7	Eight rooms	
		8	Nine or more rooms	
		9	NA (or group quarters)	

DESCRIPTION OF VARIABLES—continued

Variable Number	Item	Code	Code Description
07	Plumbing facilities	0	With all plumbing facilities
		1	Without plumbing facilities (lacking hot water only)
		2	Without plumbing facilities (lacking other plumbing facilities)
		3	NA (group quarters, vacant, seasonal, or migratory units)
08	Tenure	0	Owned or being bought
		1	Cooperative or condominium
		2	Rented for cash rent
		3	No cash rent (e.g., minister's home)
		4	NA (vacant or group quarters)
09	Commercial use	0	With commercial establishment or medical office on a place of less than 10 acres
		1	On a place of 10 acres or more
		2	On a place of less than 10 acres with no commercial use
		3	NA (multiple units, mobile home, trailer, or group quarters)
10	Value of property for specified units	0	Under $5,000
		1	$ 5,000–$ 7,499
		2	$ 7,500–$ 9,999
		3	$10,000–$12,499
		4	$12,500–$14,999
		5	$15,000–$17,499
		6	$17,500–$19,999
		7	$20,000–$24,999
		8	$25,000–$34,999
		9	$35,000–$49,999
		10	$50,000 or more
		11	NA (vacant units not for sale and anything where $V09 \neq 2$, as given above)
11	Gross monthly rent	001–999	Dollar amount. "Gross monthly rent" is defined as "contract rent" (or what the renter agrees to pay) plus utilities.
		000	NA (owner-occupied, rented without cash rent, vacant but not for rent, one-family houses on places of 10 acres or more, or group quarters)
12	Payment of electricity	0	Yes
		1	No, included in rent
		2	No, not used
		3	NA (same as $V11 = 000$)

DESCRIPTION OF VARIABLES—continued

Variable Number	Item	Code	Code Description
13	Average monthly cost of electricity	XX 0	Dollar amount NA ($V12 \neq 0$)
14	Payment of gas	0 1 2 3	Yes No, included in rent No, not used NA (same as $V11 = 000$)
15	Average monthly cost of gas	XX 0	Dollar amount NA ($V14 \neq 0$)
16	Payment of water	0 1 2	Yes No, included in rent or no charge NA (same as $V11 = 000$)
17	Yearly cost of water	XXX 0	Dollar amount NA ($V16 \neq 0$)
18	Payment of fuel	0 1 2 3	Yes No, included in rent No, fuel not used NA (same as $V11 = 000$)
19	Yearly cost of fuel	XXX 0	Dollar amount NA ($V18 \neq 0$)
20	Year structure built	0 1 2 3 4 5 6	1969 or 1970 1965–1968 1960–1964 1950–1959 1940–1949 1939 or earlier NA (group quarters)
21	Sales of farm products	0 1 2 3 4 5 6	Less than \$50 (or none) \$ 50–\$ 249 \$ 250–\$2,499 \$2,500–\$4,999 \$5,000–\$9,999 \$10,000 or more NA (on a city lot, vacant, group quarters)
22	House heating fuel	0 1 2 3 4	Gas from underground pipes Bottled, tank, or LP gas Electricity Fuel oil, kerosene, etc. Coal or coke

DESCRIPTION OF VARIABLES—continued

Variable Number	Item	Code	Code Description
		5	Wood
		6	Other fuel
		7	No fuel used
		8	NA (group quarters or vacant)
23	Number of bedrooms	0	No bedroom
		1	One bedroom
		⋮	⋮
		4	Four bedrooms
		5	Five or more bedrooms
		6	NA (group quarters)
24	Second home	0	Yes
		1	No
		2	NA (vacant or group quarters)
25	Race of head of household	0	White
		1	Negro
		2	Indian (American)
		3	Japanese
		4	Chinese
		5	Filipino
		6	Hawaiian
		7	Korean
		8	Other
		9	NA (vacant or group quarters)
26	Age of head of household	XX	Years of age (from 14 to 99)
		00	100 or more
		01	NA (vacant or group quarters)
27	Number of persons under 18	X	0 to 8 persons under 18 in household
		9	9 or more persons under 18
28	Number of related children under 18	X	0 to 8 related children under 18
		9	9 or more
29	Number of related children under 6	X	0 to 8 related children under age 6
		9	9 or more
30	Number of persons 65 and over	X	0 to 8 persons age 65 or older
		9	9 or more
31	Persons per room	0	0.50 or less
		1	0.51–0.75
		2	0.76–1.00

DESCRIPTION OF VARIABLES—continued

Variable Number	Item	Code	Code Description
		3	1.01–1.50
		4	1.51–2.00
		5	2.01 or more
		6	NA (vacant unit or group quarters)
32	Total income of family or primary individual	999	No income
		0	$1 to $99 last year
		XXX	1–499; multiply XXX by $100 (e.g., 374 = $37,400 per year)
		500	$50,000 or more
		−00	$1 to $99 loss that year
		−XX	−1 to −98; multiply by $100
		−99	Loss of $9,900 or more
		998	NA (vacant or group quarters)
33	Gross rent as a percent of family income	0	Less than 10 percent
		1	10 to 14 percent
		2	15 to 19 percent
		3	20 to 24 percent
		4	25 to 34 percent
		5	35 percent or more
		6	Not computed (0 or negative income, or no cash rent, but otherwise in the rent universe)
		7	NA ($V11 = 000$, which means "not in the rent universe")
34	Sex of head of household	0	Male
		1	Female
35	Marital status	0	Married, spouse present
		1	Married, spouse absent
		2	Widowed
		3	Divorced
		4	Separated
		5	Never married
		6	NA (persons under 14)
36	Highest grade attended (not highest grade completed)	0	Never attended school
		1	Nursery school
		2	Kindergarten
		3	Elementary 1
		4	Elementary 2
		5	Elementary 3
		6	Elementary 4
		7	Elementary 5
		8	Elementary 6

DESCRIPTION OF VARIABLES—continued

Variable Number	Item	Code	Code Description
		9	Elementary 7
		10	Elementary 8
		11	Elementary 9
		12	High school 10
		13	High school 11
		14	High school 12
		15	College 1
		16	College 2
		17	College 3
		18	College 4
		19	College 5
		20	College 6 or more
		21	NA
37	Children ever born to head of household	XX	0 to 11 children ever born
		12	12 or more children ever born
		13	NA (males, females under 14)
38	Hours worked last week by head of household	0	1–14 hours
		1	15–29 hours
		2	30–34 hours
		3	35–39 hours
		4	40 hours
		5	41–48 hours
		6	49–59 hours
		7	60 or more hours
		8	NA (persons not at work)
39	Industry of current employment (For more detail, see 1970 Industrial Code List for 1970 Census) Head of household	17– 29	Agriculture, forestry, and fishing
		47– 58	Mining
		67– 78	Construction
		107–267	Manufacturing (durable goods)
		268–399	Manufacturing (nondurable goods)
		407–429	Transportation
		447–449	Communications
		467–479	Utilities and sanitary services
		507–599	Wholesale trade
		607–699	Retail trade
		707–719	Finance, insurance, and real estate
		727–767	Business and repair services
		769–799	Personal services
		807–817	Entertainment and recreational services
		828–899	Professional services
		907–947	Public administration
		0	NA (never worked, in armed forces, etc.)

DESCRIPTION OF VARIABLES—continued

Variable Number	Item	Code	Code Description
40	Current occupation of head of household (For more detail, see 1970 Occupation Code List)	1–196	Professional, technical, and kindred workers
		201–246	Managers and administrators, except farm
		260–296	Sales workers
		301–396	Clerical and kindred workers
		401–586	Artisans and kindred workers
		601–696	Operatives, except transport
		701–726	Transport equipment operatives
		740–796	Laborers, except farm
		801–806	Farmers and farm managers
		821–846	Farm laborers and farm foremen
		901–976	Service workers, except private household
		980–986	Private household workers
		0	NA
41	Class of worker (head of household)	0	Employee of private company, etc.
		1	Federal employee (includes armed forces)
		2	State government employee
		3	Local government employee
		4	Self-employed; own business, not incorporated
		5	Self-employed; own business, incorporated
		6	Working without pay
		7	NA
42	Ratio of family income to poverty cutoff		Below poverty cutoff
		0	Less than 0.50
		1	0.50–0.74
		2	0.75–0.99
			Above poverty cutoff
		3	1.00–1.24
		4	1.25–1.49
		5	1.50–1.99
		6	2.00–2.99
		7	3.00 or more
		8	NA

ILLINOIS SAMPLE ($n = 100$)

								Variable Number													
01	02	03	04	05	06	07	08	09	10	11	12	13	14	15	16	17	18	19	20	21	
1	1	2	5	0	2	0	2	3	11	117	0	7	1	0	1	0	1	0	5	6	
2	2	5	0	0	5	0	0	2	7	0	3	0	3	0	2	0	3	0	4	6	
3	2	4	0	0	5	0	0	2	8	0	3	0	3	0	2	0	3	0	3	6	
4	2	3	0	0	4	0	0	2	8	0	3	0	3	0	2	0	3	0	3	6	
5	1	2	2	0	5	0	0	3	11	0	3	0	3	0	2	0	3	0	5	6	
6	1	2	9	0	2	0	2	3	11	162	0	9	0	2	0	12	2	0	2	6	
7	5	2	0	0	3	0	0	1	11	0	3	0	3	0	2	0	3	0	5	0	
8	2	2	2	0	3	0	2	3	11	128	0	8	1	0	1	0	1	0	5	6	
9	1	6	0	0	5	0	0	2	8	0	3	0	3	0	2	0	3	0	2	6	
10	1	5	8	0	4	0	2	3	11	110	0	12	0	18	1	0	1	0	5	6	
11	1	2	0	0	5	0	0	2	8	0	3	0	3	0	2	0	3	0	5	6	
12	2	5	0	0	4	0	0	2	9	0	3	0	3	0	2	0	3	0	3	6	
13	2	4	0	0	5	0	0	2	9	0	3	0	3	0	2	0	3	0	3	6	
14	1	4	5	0	3	0	2	3	11	110	0	12	0	3	1	0	1	0	5	6	
15	2	5	0	0	5	0	0	2	7	0	3	0	3	0	2	0	3	0	1	6	
16	2	3	10	0	3	0	0	3	11	0	3	0	3	0	2	0	3	0	1	6	
17	2	5	0	0	5	0	0	2	7	0	3	0	3	0	2	0	3	0	3	6	
18	2	2	0	0	5	0	0	2	7	0	3	0	3	0	2	0	3	0	3	6	
19	2	3	0	0	4	0	0	2	4	0	3	0	3	0	2	0	3	0	5	6	
20	2	2	9	0	2	0	2	3	11	135	0	10	1	0	1	0	2	0	1	6	
21	1	7	0	0	3	0	0	2	4	0	3	0	3	0	2	0	3	0	3	6	
22	1	8	0	0	5	0	2	2	11	153	0	18	0	55	1	0	2	0	5	6	
23	1	3	8	0	3	0	2	3	11	79	0	10	0	4	1	0	2	0	5	6	
24	1	2	1	0	2	0	2	3	11	109	0	5	1	0	1	0	2	0	5	6	
25	1	2	2	0	4	0	2	3	11	95	0	10	0	10	1	0	2	0	5	6	
26	1	2	4	0	3	0	2	3	11	162	0	8	0	4	1	0	1	0	3	6	
27	2	3	9	0	2	0	2	3	11	191	0	13	1	0	1	0	2	0	1	6	
28	3	2	0	0	5	0	0	2	0	0	3	0	3	0	2	0	3	0	5	6	
29	1	2	0	0	5	0	0	2	7	0	3	0	3	0	2	0	3	0	5	6	
30	1	2	0	0	3	0	3	2	11	0	3	0	3	0	2	0	3	0	4	6	
31	2	2	0	0	6	0	0	2	1	0	3	0	3	0	2	0	3	0	5	6	
32	2	3	0	0	5	0	2	2	11	114	0	2	0	2	0	120	2	0	4	6	
33	2	6	0	0	6	0	0	2	8	0	3	0	3	0	2	0	3	0	2	6	
34	1	3	3	0	3	2	2	3	11	81	0	12	0	19	1	0	2	0	5	6	
35	1	2	0	0	4	0	0	2	8	0	3	0	3	0	2	0	3	0	3	6	
36	2	4	0	0	6	0	0	2	10	0	3	0	3	0	2	0	3	0	3	6	
37	1	2	2	0	2	0	2	3	11	116	1	0	1	0	1	0	2	0	5	6	
38	1	5	0	0	5	0	2	3	11	115	0	15	1	0	1	0	1	0	5	6	
39	2	6	0	0	4	0	0	2	6	0	3	0	3	0	2	0	3	0	3	6	
40	2	2	0	0	6	0	0	2	7	0	3	0	3	0	2	0	3	0	3	6	

ILLINOIS SAMPLE ($n = 100$)—continued

Variable Number

22	23	24	25	26	27	28	29	30	31	32	33	34	35	36	37	38	39	40	41	42
4	1	1	0	75	0	0	0	2	0	28	5	0	0	10	13	8	0	0	7	4
0	3	1	1	43	2	2	0	0	1	181	7	0	0	14	13	5	418	715	0	7
0	3	1	0	51	2	2	0	0	0	447	7	0	0	16	13	6	607	284	0	7
0	3	1	0	48	1	1	0	0	0	130	7	0	0	12	13	2	417	715	0	7
0	3	1	1	65	0	0	0	1	0	27	7	0	0	10	13	8	747	962	0	3
2	1	1	0	70	0	0	0	2	0	126	2	0	0	12	13	4	259	323	0	7
1	2	1	0	72	0	0	0	2	0	22	7	0	0	10	13	8	699	726	0	2
0	1	1	0	62	0	0	0	1	0	166	0	0	0	12	13	8	937	964	3	7
0	3	1	0	38	4	4	1	0	0	171	7	0	0	14	13	4	717	192	0	7
0	3	1	0	40	3	3	0	0	0	55	3	0	0	8	13	2	669	910	0	4
0	2	0	0	70	0	0	0	2	0	121	7	0	0	14	13	8	0	0	7	7
0	3	1	0	39	3	3	2	0	0	500	7	0	0	16	13	4	738	183	4	7
0	3	1	0	51	1	1	0	1	1	168	7	0	0	16	13	5	847	83	0	7
4	2	1	1	32	3	3	0	0	0	999	6	1	5	12	3	8	0	0	7	0
0	3	1	1	25	1	1	0	0	1	156	7	0	5	14	13	5	467	433	0	7
0	2	1	0	27	1	1	1	0	0	90	7	0	0	10	13	4	417	694	0	7
0	3	1	0	48	2	2	0	0	0	500	7	0	0	14	13	6	779	285	0	7
0	3	1	0	32	0	0	0	0	0	236	7	0	0	16	13	4	69	11	0	7
0	2	1	0	45	1	1	0	0	0	83	7	0	0	10	13	4	628	762	0	6
0	1	1	0	21	0	0	0	0	0	86	2	0	0	15	13	4	407	325	0	7
0	2	1	0	53	3	3	2	0	1	198	7	0	0	14	13	4	408	473	0	7
0	3	1	0	34	6	6	1	0	0	161	1	0	0	12	13	5	527	631	0	6
7	1	1	8	41	1	1	1	0	1	51	2	1	4	6	0	5	259	602	0	5
0	1	1	0	24	0	0	0	0	0	154	0	0	0	14	13	3	889	343	0	7
0	2	1	0	37	0	0	0	1	1	103	1	0	5	15	13	4	298	14	0	7
0	2	1	0	71	0	0	0	2	0	89	3	0	0	10	13	8	587	643	0	7
0	1	1	0	38	1	1	1	0	0	181	1	0	0	18	13	6	748	260	0	7
0	2	1	0	60	0	0	0	0	1	76	7	1	2	15	0	8	0	0	7	7
0	2	1	0	43	1	1	0	0	0	35	7	1	2	14	1	1	857	912	3	4
0	2	1	0	27	1	1	0	0	0	17	6	1	4	12	1	4	259	690	0	1
0	2	1	0	57	0	0	0	0	0	92	7	0	0	5	13	1	67	415	0	7
0	3	1	0	26	1	1	1	0	0	16	5	0	0	18	13	6	707	202	0	1
0	3	1	0	38	4	4	2	0	0	100	7	0	0	18	13	5	889	1	0	6
0	2	1	0	42	0	0	0	0	0	141	0	0	3	10	13	4	139	706	0	7
0	2	1	0	60	0	0	0	0	0	218	7	0	0	14	13	4	358	381	0	7
0	5	1	0	57	1	1	0	0	0	326	7	0	0	18	13	6	297	245	0	7
4	1	1	0	24	0	0	0	0	0	48	4	0	0	10	13	6	197	481	0	5
3	3	1	0	32	2	2	0	1	1	160	0	0	0	9	13	4	758	470	0	7
0	3	1	0	30	4	4	4	0	0	120	7	0	0	19	13	2	857	144	0	6
3	3	1	0	43	0	0	0	0	0	204	7	0	0	14	13	4	387	281	0	7

ILLINOIS SAMPLE ($n = 100$)—continued

Variable Number

01	02	03	04	05	06	07	08	09	10	11	12	13	14	15	16	17	18	19	20	21
41	1	2	0	0	5	0	0	2	3	0	3	0	3	0	2	0	3	0	5	6
42	1	4	1	0	4	0	2	3	11	98	0	9	0	19	1	0	2	0	5	6
43	1	2	1	0	2	0	2	3	11	125	1	0	1	0	1	0	2	0	5	6
44	1	3	0	0	3	0	0	2	7	0	3	0	3	0	2	0	3	0	3	6
45	1	7	1	0	3	0	0	3	11	0	3	0	3	0	2	0	3	0	5	0
46	3	10	0	0	4	0	0	2	3	0	3	0	3	0	2	0	3	0	5	6
47	5	4	0	0	3	0	3	1	11	0	3	0	3	0	2	0	3	0	2	5
48	5	4	0	0	4	0	2	1	11	0	3	0	3	0	2	0	3	0	5	2
49	3	3	0	0	5	0	0	2	2	0	3	0	3	0	2	0	3	0	3	6
50	2	2	0	0	4	0	0	2	4	0	3	0	3	0	2	0	3	0	5	6
51	2	6	0	0	5	0	0	2	8	0	3	0	3	0	2	0	3	0	3	6
52	2	2	2	0	2	0	2	3	11	132	0	9	0	3	1	0	2	0	5	6
53	1	3	0	0	5	0	3	2	11	0	3	0	3	0	2	0	3	0	5	6
54	1	3	1	0	4	0	0	3	11	0	3	0	3	0	2	0	3	0	5	6
55	1	2	0	0	5	0	0	2	8	0	3	0	3	0	2	0	3	0	3	6
56	1	2	5	0	4	0	2	3	11	170	0	15	0	5	1	0	2	0	5	6
57	2	2	1	0	3	0	2	3	11	85	0	5	1	0	1	0	1	0	5	6
58	1	5	0	0	5	0	2	3	11	94	0	11	0	13	1	0	1	0	5	6
59	1	2	0	0	5	0	0	2	3	0	3	0	3	0	2	0	3	0	5	6
60	1	5	0	0	7	0	0	2	5	0	3	0	3	0	2	0	3	0	5	6
61	1	3	0	0	4	0	0	2	8	0	3	0	3	0	2	0	3	0	5	6
62	2	2	3	0	3	0	2	3	11	190	0	15	1	0	1	0	2	0	2	6
63	2	5	0	0	5	0	0	2	7	0	3	0	3	0	2	0	3	0	4	6
64	1	3	9	0	2	0	2	3	11	151	0	10	1	0	1	0	1	0	3	6
65	1	2	9	0	4	0	2	3	11	265	0	10	1	0	1	0	1	0	2	6
66	5	5	0	0	8	0	0	1	11	0	3	0	3	0	2	0	3	0	5	4
67	2	3	0	0	4	0	2	2	11	185	0	12	0	12	0	12	0	60	4	6
68	2	6	0	0	6	0	0	2	6	0	3	0	3	0	2	0	3	0	4	6
69	1	2	1	0	4	0	0	3	11	0	3	0	3	0	2	0	3	0	5	6
70	1	3	0	0	5	0	0	2	7	0	3	0	3	0	2	0	3	0	2	6
71	1	4	0	0	5	0	0	2	5	0	3	0	3	0	2	0	3	0	3	6
72	2	5	0	0	6	0	0	2	8	0	3	0	3	0	2	0	3	0	1	6
73	2	2	0	0	6	0	0	2	6	0	3	0	3	0	2	0	3	0	5	6
74	1	2	9	0	1	0	2	3	11	202	0	12	1	0	1	0	1	0	0	6
75	2	4	0	0	5	0	0	2	8	0	3	0	3	0	2	0	3	0	1	6
76	2	2	0	0	6	0	0	2	9	0	3	0	3	0	2	0	3	0	1	6
77	1	5	0	0	5	0	0	2	6	0	3	0	3	0	2	0	3	0	3	6
78	1	3	5	0	3	2	2	3	11	112	1	0	0	7	1	0	1	0	5	6
79	2	2	0	0	5	0	0	2	8	0	3	0	3	0	2	0	3	0	5	6
80	1	5	0	0	5	0	0	2	6	0	3	0	3	0	2	0	3	0	2	6

ILLINOIS SAMPLE ($n = 100$)—continued

											Variable Number									
22	23	24	25	26	27	28	29	30	31	32	33	34	35	36	37	38	39	40	41	42
0	3	1	0	39	0	0	0	0	0	131	7	0	0	14	13	4	69	510	0	7
0	3	1	0	42	0	0	0	0	1	183	0	0	0	12	13	5	69	510	0	7
0	1	1	0	30	1	1	1	0	0	999	6	1	5	14	1	4	239	305	0	0
0	2	1	0	23	1	1	1	0	0	107	7	0	0	14	13	5	167	441	0	7
0	3	1	0	40	4	4	0	0	0	85	7	0	0	8	13	4	187	692	0	4
0	3	1	0	39	8	8	2	0	0	113	7	0	1	12	13	4	127	545	0	5
0	3	1	0	25	2	2	2	0	0	120	7	0	0	14	13	7	17	822	0	7
1	4	1	0	41	2	2	1	0	0	99	7	0	0	14	13	7	17	801	4	7
0	2	0	0	62	0	0	0	0	0	67	7	0	0	14	13	5	707	903	0	6
0	2	1	0	69	0	0	0	1	0	46	7	0	0	7	13	8	809	586	0	6
0	3	1	0	41	3	3	1	0	0	118	7	0	0	14	13	5	757	473	4	6
0	1	1	0	47	0	0	0	1	1	70	3	1	5	8	0	5	207	602	0	7
0	4	1	0	66	1	1	0	1	0	187	6	0	0	15	13	8	0	0	7	7
0	2	1	0	82	0	0	0	2	0	50	7	0	0	8	13	8	0	0	7	5
0	2	1	1	44	0	0	0	0	0	100	7	1	2	14	3	8	798	933	4	7
0	2	1	0	31	0	0	0	0	1	185	1	0	5	18	13	3	717	1	0	7
0	2	1	0	62	0	0	0	0	1	271	0	0	3	12	13	8	669	903	0	7
3	3	1	0	40	2	2	0	0	0	99	1	0	0	16	13	4	729	53	0	6
0	3	1	0	67	0	0	0	1	0	69	7	0	0	10	13	8	178	514	0	7
0	4	1	1	39	2	2	0	1	1	102	7	1	2	13	2	1	927	332	2	6
0	2	1	1	70	0	0	0	1	1	109	7	1	2	9	5	8	0	0	7	7
0	2	1	0	49	0	0	0	0	0	129	2	0	0	12	13	4	78	586	0	7
0	3	1	0	47	2	2	1	1	1	500	7	0	0	13	13	6	779	245	5	7
3	2	1	1	36	1	1	0	0	0	112	2	0	0	18	13	4	857	142	3	7
0	2	1	0	39	0	0	0	0	0	106	4	0	0	20	13	4	709	271	0	7
1	4	1	0	38	2	2	0	0	0	33	7	0	0	12	13	7	17	801	4	2
0	2	1	0	66	0	0	0	1	0	174	1	0	0	10	13	4	178	903	0	7
0	3	1	0	46	3	3	0	0	0	166	7	0	0	10	13	5	557	705	0	7
0	3	1	0	64	0	0	0	0	0	89	7	0	0	10	13	4	108	705	0	7
0	3	1	1	28	1	1	1	0	0	136	7	0	0	16	13	4	527	282	0	7
0	3	1	1	31	2	2	0	0	0	67	7	0	0	9	13	4	757	473	0	5
0	3	1	0	39	3	3	1	0	0	153	7	0	0	15	13	7	329	233	0	7
3	3	1	0	39	0	0	0	0	0	165	7	0	0	14	13	8	77	245	0	7
0	1	1	0	22	0	0	0	0	0	86	4	0	0	16	13	0	737	933	0	7
0	3	1	0	29	2	2	2	0	0	116	7	0	0	14	13	4	208	13	0	7
0	3	1	0	33	0	0	0	0	0	190	7	0	0	15	13	5	197	225	0	7
3	3	1	0	33	2	2	1	1	1	126	7	0	0	14	13	7	669	230	4	6
4	2	1	1	62	1	1	0	0	0	40	4	0	0	5	13	8	417	753	0	4
0	3	1	0	55	0	0	0	0	0	91	7	1	2	13	3	3	848	372	2	7
0	3	1	0	26	3	3	3	0	0	141	7	0	0	14	13	3	448	12	0	7

ILLINOIS SAMPLE ($n = 100$)—continued

01	02	03	04	05	06	07	08	09	10	11	12	13	14	15	16	17	18	19	20	21
81	1	2	9	0	1	2	2	3	11	60	1	0	1	0	1	0	1	0	5	6
82	1	4	1	0	4	0	2	3	11	112	0	18	0	19	1	0	2	0	5	6
83	5	4	0	0	4	0	0	1	11	0	3	0	3	0	2	0	3	0	5	0
84	2	5	0	0	5	0	2	2	11	168	0	15	0	15	0	35	2	0	3	6
85	1	3	4	0	3	0	2	3	11	97	0	7	0	15	1	0	2	0	5	6
86	1	6	1	0	5	0	2	3	11	130	1	0	1	0	1	0	2	0	2	6
87	1	2	1	0	4	0	2	3	11	142	0	7	0	2	1	0	1	0	5	6
88	2	5	0	0	5	0	0	2	8	0	3	0	3	0	2	0	3	0	2	6
89	2	2	5	0	3	0	2	3	11	136	0	5	0	1	1	0	1	0	5	6
90	2	2	0	0	3	0	0	2	6	0	3	0	3	0	2	0	3	0	5	6
91	1	2	9	0	3	0	2	3	11	324	0	14	2	0	1	0	2	0	3	6
92	1	4	2	0	6	0	0	3	11	0	3	0	3	0	2	0	3	0	5	6
93	1	5	5	0	3	0	2	3	11	152	0	12	0	10	1	0	2	0	5	6
94	2	2	9	0	3	0	2	3	11	175	0	10	1	0	1	0	2	0	1	6
95	1	6	0	0	4	0	0	2	5	0	3	0	3	0	2	0	3	0	5	0
96	1	2	3	0	3	0	0	3	11	0	3	0	3	0	2	0	3	0	5	6
97	1	9	5	0	4	2	2	3	11	117	0	8	0	4	1	0	2	0	5	6
98	1	8	9	0	4	0	2	3	11	144	0	10	1	0	1	0	2	0	2	6
99	1	6	1	0	4	0	0	3	11	0	3	0	3	0	2	0	3	0	5	6
100	2	5	0	0	5	0	0	2	8	0	3	0	3	0	2	0	3	0	3	6

ILLINOIS SAMPLE ($n = 100$)—continued

Variable Number

22	23	24	25	26	27	28	29	30	31	32	33	34	35	36	37	38	39	40	41	42
4	1	1	1	66	0	0	0	2	0	12	5	0	0	9	13	8	407	912	0	1
0	2	1	0	24	2	2	2	0	0	100	1	0	0	13	13	8	408	563	0	6
0	3	1	0	41	2	2	0	0	0	100	7	0	0	13	13	2	297	690	0	6
0	3	1	0	35	3	3	1	0	0	86	3	1	3	12	4	2	628	310	0	5
0	2	1	0	44	1	1	0	0	0	86	1	1	2	10	5	8	0	0	7	7
0	3	1	1	38	5	5	1	0	0	999	6	1	5	12	5	8	287	402	0	0
3	2	1	0	57	0	0	0	0	0	54	4	1	3	14	5	4	669	915		6
0	3	1	0	35	3	3	1	0	0	112	7	0	0	18	13	8	747	715	0	6
0	1	1	0	22	0	0	0	0	0	108	2	0	0	20	13	8	558	303	0	7
0	1	1	0	67	0	0	0	2	0	205	7	0	0	8	13	0	267	696	0	7
3	2	1	0	63	0	0	0	0	0	300	1	0	0	14	13	4	337	245	0	7
0	3	1	1	57	1	1	0	1	1	126	7	0	0	14	13	4	139	715	0	7
0	2	1	1	68	0	0	0	1	0	20	5	0	0	13	13	4	399	726	0	0
0	2	1	0	27	0	0	0	0	0	200	1	0	0	20	13	4	377	21	0	7
0	3	1	1	38	3	3	0	0	0	67	7	1	3	13	4	4	838	75	0	5
4	1	1	0	59	0	0	0	0	0	161	7	0	0	14	13	4	267	586	0	7
4	3	1	1	40	6	6	4	0	0	70	3	0	0	14	13	8	139	796	0	3
0	3	1	1	44	5	5	0	0	0	253	0	0	0	8	13	4	299	643	0	7
3	3	1	0	37	4	4	1	0	0	169	7	0	0	14	13	4	69	415	0	7
0	3	1	0	46	3	3	0	0	0	178	7	0	0	14	13	6	407	395	0	7

OREGON SAMPLE ($n = 30$)

Variable Number

	01	02	03	04	05	06	07	08	09	10	11	12	13	14	15	16	17	18	19	20	21
1	2	3	0	0	3	0	2	2	2	11	140	0	15	0	8	0	90	2	0	5	6
2	2	4	0	0	5	0	2	2	2	11	166	0	9	0	17	0	4	2	0	3	6
3	0	2	9	0	4	0	2	2	3	11	125	0	20	2	0	1	0	2	0	1	6
4	3	4	5	0	3	0	2	2	3	11	102	0	7	2	0	1	0	2	0	2	6
5	3	8	0	0	5	0	2	2	2	11	120	0	25	2	0	0	124	2	0	4	6
6	1	2	9	0	2	0	2	2	3	11	157	0	17	2	0	1	0	2	0	1	6
7	1	4	0	0	5	0	2	2	2	11	150	0	15	2	0	0	32	0	210	3	6
8	1	2	0	0	3	0	2	2	2	11	105	0	10	2	0	0	25	0	100	3	6
9	1	6	0	0	3	0	2	2	2	11	88	0	12	2	0	0	12	0	60	3	6
10	3	6	3	0	4	0	2	2	3	11	97	0	12	2	0	1	0	2	0	3	6
11	0	6	0	0	6	0	2	2	2	11	128	0	10	2	0	0	40	0	300	5	6
12	0	2	0	0	1	0	2	2	2	11	45	1	0	1	0	1	0	2	0	4	6
13	2	2	9	0	3	0	2	2	3	11	132	0	12	2	0	1	0	2	0	2	6
14	0	5	0	0	4	0	2	2	2	11	172	0	25	2	0	0	85	2	0	1	6
15	0	2	0	0	5	0	2	2	2	11	116	0	9	2	0	1	0	0	250	2	6
16	0	2	0	0	5	0	2	2	2	11	69	0	12	2	0	1	0	0	80	4	0
17	0	3	0	0	5	0	2	2	2	11	135	0	18	2	0	0	180	0	140	4	0
18	0	6	0	0	5	0	2	2	2	11	92	0	15	2	0	0	100	0	100	5	6
19	2	2	0	0	4	0	2	2	2	11	211	0	5	0	15	0	60	2	0	1	0
20	2	3	0	0	2	0	2	2	2	11	116	0	10	0	18	0	36	2	0	4	6
21	3	4	0	0	6	0	2	2	2	11	119	0	30	2	0	0	47	2	0	5	6
22	3	3	0	0	3	0	2	2	2	11	101	0	20	2	0	0	75	2	0	3	6
23	2	3	0	0	3	0	2	2	2	11	114	0	9	2	0	0	25	0	150	5	0
24	3	3	9	0	2	0	2	2	3	11	108	0	7	0	16	1	0	2	0	4	6
25	1	2	9	0	3	0	2	2	3	11	72	0	4	0	3	1	0	1	0	5	6
26	3	4	0	0	4	0	2	2	2	11	162	0	14	0	21	1	0	0	25	2	6
27	3	2	0	0	3	0	2	2	2	11	70	0	20	2	0	1	0	2	0	4	6
28	2	3	0	0	2	0	2	2	2	11	94	0	17	2	0	0	24	2	0	4	6
29	2	5	0	0	4	0	2	2	2	11	93	0	11	2	0	0	60	0	150	3	6
30	1	3	0	0	3	0	2	2	2	11	95	1	0	0	20	1	0	2	0	5	6

OREGON SAMPLE ($n = 30$)—continued

Variable Number

22	23	24	25	26	27	28	29	30	31	32	33	34	35	36	37	38	39	40	41	42
0	2	1	0	26	1	1	1	0	0	116	1	0	0	14	13	6	527	690	0	7
0	3	1	0	27	2	2	2	0	0	94	3	0	0	14	13	4	628	245	0	6
2	2	1	0	23	1	1	1	0	0	43	5	1	3	15	1	4	628	310	0	5
2	2	1	0	29	2	2	2	0	0	57	3	0	0	20	13	8	857	141	3	5
2	3	1	0	34	6	6	2	0	0	70	3	0	0	14	13	5	757	473	0	3
2	1	0	0	27	0	0	0	0	0	149	1	0	0	17	13	8	718	216	4	7
3	3	1	0	40	0	0	0	0	0	165	1	0	0	18	13	2	527	282	0	7
3	3	1	0	42	0	0	0	0	0	120	1	0	0	12	13	6	417	715	0	7
3	2	1	0	41	4	4	1	0	0	58	2	0	0	14	13	5	108	690	0	3
2	1	1	0	36	3	3	0	1	1	110	1	0	0	10	13	4	499	976	0	6
3	3	1	0	52	4	4	0	0	0	96	2	0	0	15	13	4	68	375	2	5
1	1	1	0	39	0	0	0	0	0	51	1	0	0	11	13	8	237	415	0	6
2	2	1	0	22	0	0	0	0	0	65	3	0	0	18	13	3	727	245	0	6
2	3	1	0	43	3	3	0	0	0	500	0	0	0	20	13	7	828	65	4	7
3	3	1	0	27	1	1	0	0	0	68	3	1	3	14	1	4	27	220	1	6
3	2	1	0	25	0	0	0	0	0	66	1	0	0	14	13	3	669	912	0	6
3	2	1	0	31	1	1	0	0	0	116	1	0	0	14	13	4	27	25	2	7
5	3	1	0	29	4	4	2	0	0	71	2	0	0	14	13	6	699	726	0	4
0	3	1	0	50	0	0	0	0	0	112	3	0	0	10	13	4	319	374	0	7
3	1	1	0	44	0	0	0	0	0	101	1	1	3	14	2	4	599	396	0	7
2	4	1	0	41	2	2	0	0	0	169	0	0	0	6	13	4	78	586	0	7
2	2	1	0	43	1	1	0	0	1	107	1	0	0	12	13	4	28	752	5	7
3	2	1	8	26	1	1	1	0	0	84	2	0	0	14	13	7	917	961	1	6
0	1	1	0	26	1	1	1	0	0	83	2	0	0	14	13	4	108	753	0	6
3	1	1	0	52	0	0	0	0	0	28	4	1	4	14	3	8	777	901	0	3
0	3	1	0	29	2	2	1	0	0	78	4	0	0	15	13	6	108	753	0	6
2	2	1	0	22	0	0	0	0	0	25	4	0	0	14	13	7	17	822	0	3
2	1	1	0	19	1	1	1	0	0	64	2	0	0	14	13	6	158	441	0	6
3	2	1	0	48	3	3	0	0	0	167	0	0	0	15	13	4	477	485	3	7
0	2	1	0	25	1	1	1	0	0	59	2	0	0	14	13	7	648	473	4	6

MASSACHUSETTS SAMPLE ($n = 40$)

									Variable Number											
01	02	03	04	05	06	07	08	09	10	11	12	13	14	15	16	17	18	19	20	21
1	2	2	0	0	4	0	2	2	11	158	0	10	2	0	1	0	2	0	2	6
2	1	2	0	0	4	0	2	3	11	109	0	7	2	0	1	0	0	150	5	6
3	2	4	0	0	6	0	2	3	11	225	0	15	0	13	1	0	0	450	5	6
4	2	2	1	0	3	0	2	3	11	76	0	6	2	0	1	0	1	0	5	6
5	1	4	1	0	3	0	2	3	11	58	0	5	0	5	1	0	0	100	5	6
6	1	2	3	0	3	0	2	3	11	165	0	10	1	0	1	0	1	0	0	6
7	2	2	0	0	2	0	2	3	11	108	0	7	0	5	1	0	0	250	5	6
8	2	4	1	0	4	0	2	3	11	97	0	10	0	4	1	0	0	396	5	6
9	2	2	0	0	4	0	2	3	11	106	0	6	2	0	1	0	0	240	5	6
10	2	2	1	0	3	0	2	3	11	160	0	15	0	10	1	0	0	125	5	6
11	2	6	1	0	4	0	2	3	11	120	0	15	0	25	1	0	1	0	5	6
12	2	2	0	0	5	0	2	3	11	131	0	8	2	0	1	0	0	275	5	6
13	2	5	0	0	3	0	2	2	11	265	0	15	2	0	1	0	0	300	3	6
14	2	3	3	0	2	0	2	3	11	110	0	5	2	0	1	0	1	0	5	6
15	2	2	2	0	3	0	2	3	11	89	0	9	0	20	1	0	2	0	5	6
16	1	2	2	0	4	0	2	3	11	92	0	10	0	22	1	0	2	0	5	6
17	1	5	2	0	4	0	2	3	11	87	0	7	0	32	1	0	2	0	5	6
18	1	2	1	0	2	0	2	3	11	102	0	7	2	0	1	0	1	0	5	6
19	1	2	5	0	4	0	2	3	11	76	0	6	0	20	1	0	2	0	5	6
20	2	4	0	0	4	0	2	3	11	131	0	10	0	4	1	0	0	200	5	6
21	2	2	4	0	3	0	2	3	11	220	0	10	1	0	1	0	2	0	1	6
22	3	2	2	0	4	0	2	3	11	85	0	6	0	19	1	0	2	0	5	6
23	2	2	0	0	1	0	2	3	11	149	0	14	2	0	1	0	2	0	2	6
24	1	3	5	0	2	0	2	3	11	159	0	20	2	0	1	0	0	170	5	6
25	5	2	0	0	2	0	2	2	11	87	0	7	1	0	1	0	1	0	3	0
26	2	6	1	0	3	0	2	3	11	99	0	15	0	9	1	0	2	0	5	6
27	1	2	9	0	2	0	2	3	11	163	0	15	0	2	1	0	1	0	3	6
28	2	5	0	0	2	0	2	3	11	116	0	6	2	0	1	0	2	0	5	6
29	2	2	2	0	3	0	2	3	11	144	0	15	0	9	1	0	1	0	3	6
30	1	4	0	0	4	0	2	3	11	165	0	20	0	36	1	0	0	450	5	6
31	2	3	5	0	3	0	2	3	11	94	0	9	1	0	1	0	2	0	3	6
32	2	4	2	0	5	0	2	3	11	222	0	7	0	20	1	0	0	660	3	6
33	2	5	2	0	5	0	2	3	11	122	0	7	0	30	1	0	2	0	3	6
34	3	2	1	0	4	0	2	3	11	144	0	10	0	30	1	0	2	0	5	6
35	2	6	1	0	5	0	2	3	11	102	0	10	0	22	1	0	2	0	5	6
36	1	5	4	0	3	0	2	3	11	93	0	25	0	12	1	0	2	0	5	6
37	1	7	2	0	4	0	2	3	11	140	0	10	0	26	0	240	2	0	5	0
38	1	2	3	0	3	0	2	3	11	96	0	8	0	30	1	0	0	220	5	6
39	2	2	0	0	4	0	2	3	11	162	0	7	0	30	1	0	2	0	5	6
40	4	5	0	0	4	0	2	3	11	143	0	10	0	25	1	0	2	0	5	6

MASSACHUSETTS SAMPLE ($n = 40$)—continued

Variable Number

22	23	24	25	26	27	28	29	30	31	32	33	34	35	36	37	38	39	40	41	42
3	3	1	0	42	0	0	0	0	0	44	5	0	0	14	13	0	937	961	3	5
3	2	1	0	24	0	0	0	0	0	65	3	0	0	14	13	3	358	1	0	6
3	4	1	0	54	1	1	0	0	0	213	1	0	0	19	13	5	207	12	0	7
3	1	0	0	62	0	0	0	0	0	183	0	0	0	14	13	4	558	441	0	7
3	2	1	0	33	2	2	1	0	0	70	1	0	0	12	13	4	207	610	0	5
3	2	1	0	28	0	0	0	0	0	96	3	0	0	18	13	5	207	602	4	7
3	1	1	0	56	0	0	0	0	0	123	1	0	0	12	13	4	67	415	0	7
3	3	1	0	36	2	2	0	0	0	92	1	0	0	9	13	6	417	715	0	6
3	2	1	0	59	0	0	0	0	0	112	1	0	0	14	13	6	259	610	0	7
3	2	0	0	50	0	0	0	0	0	40	5	1	2	15	2	8	718	372	0	5
0	2	1	0	26	4	4	4	0	0	90	2	0	0	12	13	4	417	715	0	5
3	3	1	0	59	0	0	0	0	0	108	2	0	0	14	13	4	209	162	0	7
3	2	1	0	29	3	3	3	0	0	120	4	0	0	20	13	1	858	14	0	6
3	2	1	0	72	0	0	0	1	1	58	3	1	2	11	4	8	0	0	7	6
0	1	1	0	53	0	0	0	0	0	72	2	0	0	13	13	5	158	680	0	6
0	1	1	0	22	0	0	0	0	0	108	1	0	0	6	13	8	68	522	0	7
0	3	1	0	42	3	3	0	0	0	123	0	0	0	0	13	4	18	822	0	6
3	1	1	0	23	0	0	0	0	0	74	2	0	0	14	13	4	848	426	0	7
0	2	1	0	55	0	0	0	0	0	109	0	0	0	9	13	2	328	692	0	7
3	2	1	0	26	2	2	2	0	0	71	3	0	0	14	13	4	759	480	0	5
0	2	1	0	72	0	0	0	2	0	32	5	0	0	14	13	8	447	175	0	4
0	3	1	0	42	1	1	0	0	0	63	2	1	3	10	3	4	628	310	0	6
2	1	1	0	60	0	0	0	0	0	38	5	0	0	16	13	8	607	284	4	5
3	2	1	0	76	0	0	0	1	0	130	2	0	2	8	13	8	0	0	7	7
3	1	1	0	20	0	0	0	0	0	8	5	0	0	17	13	0	858	332	2	0
3	2	1	0	44	4	4	1	0	0	82	1	0	0	12	13	4	177	603	0	5
3	1	1	0	62	0	0	0	0	0	152	1	0	0	12	13	0	897	194	4	7
0	1	1	0	39	3	3	1	0	0	97	1	0	0	14	13	4	499	796	3	6
3	2	1	0	54	0	0	0	0	0	51	4	0	0	10	13	8	0	0	7	6
3	2	1	0	27	2	2	2	0	0	70	4	0	0	10	13	5	267	586	0	5
3	2	1	0	46	1	1	0	0	0	94	1	0	0	14	13	5	767	586	0	7
0	3	1	0	26	0	0	0	0	0	122	3	0	0	20	13	6	729	195	0	7
0	3	1	0	54	1	1	0	0	0	122	1	0	0	13	13	4	667	902	0	6
0	2	1	0	23	0	0	0	0	0	89	2	0	0	18	13	4	857	144	3	7
0	4	1	0	44	4	4	0	0	0	123	1	0	0	10	13	7	749	715	0	6
0	2	1	0	49	3	3	0	0	0	19	5	0	0	0	13	4	29	690	0	0
0	3	1	0	51	0	0	0	0	1	136	1	0	0	10	13	8	17	822	0	6
0	2	1	0	29	1	1	0	0	0	0	5	1	4	12	2	8	319	643	0	0
0	2	1	0	25	0	0	0	0	0	171	1	0	0	19	13	6	208	13	0	7
0	3	1	0	29	3	3	1	0	0	91	2	0	0	14	13	5	118	220	0	6

Appendix C

APPENDIX TABLES

TABLE I SQUARES, SQUARE ROOTS, AND RECIPROCALS

n	n^2	$\sqrt{n}$	$\sqrt{10n}$	$1/n$	n	n^2	$\sqrt{n}$	$\sqrt{10n}$	$1/n$
1	1	1.000	3.162	1.00000	51	2,601	7.141	22.583	0.01961
2	4	1.414	4.472	0.50000	52	2,704	7.211	22.804	0.01923
3	9	1.732	5.477	0.33333	53	2,809	7.280	23.022	0.01887
4	16	2.000	6.325	0.25000	54	2,916	7.348	23.238	0.01852
5	25	2.236	7.071	0.20000	55	3,025	7.416	23.452	0.01818
6	36	2.449	7.746	0.16667	56	3,136	7.483	23.664	0.01786
7	49	2.646	8.367	0.14286	57	3,249	7.550	23.875	0.01754
8	64	2.828	8.944	0.12500	58	3,364	7.616	24.083	0.01724
9	81	3.000	9.487	0.11111	59	3,481	7.681	24.290	0.01695
10	100	3.162	10.000	0.10000	60	3,600	7.746	24.495	0.01667
11	121	3.317	10.488	0.09091	61	3,721	7.810	24.698	0.01639
12	144	3.464	10.954	0.08333	62	3,844	7.874	24.900	0.01613
13	169	3.606	11.402	0.07692	63	3,969	7.937	25.100	0.01587
14	196	3.742	11.832	0.07143	64	4,096	8.000	25.298	0.01563
15	225	3.873	12.247	0.06667	65	4,225	8.062	25.495	0.01538
16	256	4.000	12.649	0.06250	66	4,356	8.124	25.690	0.01515
17	289	4.123	13.038	0.05882	67	4,489	8.185	25.884	0.01493
18	324	4.243	13.416	0.05556	68	4,624	8.246	26.077	0.01471
19	361	4.359	13.784	0.05263	69	4,761	8.307	26.268	0.01449
20	400	4.472	14.142	0.05000	70	4,900	8.367	26.458	0.01429
21	441	4.583	14.491	0.04762	71	5,041	8.426	26.646	0.01408
22	484	4.690	14.832	0.04545	72	5,184	8.485	26.833	0.01389
23	529	4.796	15.166	0.04348	73	5,329	8.544	27.019	0.01370
24	576	4.899	15.492	0.04167	74	5,476	8.602	27.203	0.01351
25	625	5.000	15.811	0.04000	75	5,625	8.660	27.386	0.01333
26	676	5.099	16.125	0.03846	76	5,776	8.718	27.568	0.01316
27	729	5.196	16.432	0.03704	77	5,929	8.775	27.749	0.01299
28	784	5.292	16.733	0.03571	78	6,084	8.832	27.928	0.01282
29	841	5.385	17.029	0.03448	79	6,241	8.888	28.107	0.01266
30	900	5.477	17.321	0.03333	80	6,400	8.944	28.284	0.01250
31	961	5.568	17.607	0.03226	81	6,561	9.000	28.460	0.01235
32	1,024	5.657	17.889	0.03125	82	6,724	9.055	28.636	0.01220
33	1,089	5.745	18.166	0.03030	83	6,889	9.110	28.810	0.01205
34	1,156	5.831	18.439	0.02941	84	7,056	9.165	28.983	0.01190
35	1,225	5.916	18.708	0.02857	85	7,225	9.220	29.155	0.01176
36	1,296	6.000	18.974	0.02778	86	7,396	9.274	29.326	0.01163
37	1,369	6.083	19.235	0.02703	87	7,569	9.327	29.496	0.01149
38	1,444	6.164	19.494	0.02632	88	7,744	9.381	29.665	0.01136
39	1,521	6.245	19.748	0.02564	89	7,921	9.434	29.833	0.01124
40	1,600	6.325	20.000	0.02500	90	8,100	9.487	30.000	0.01111

TABLE I SQUARES, SQUARE ROOTS, AND RECIPROCALS—continued

n	n^2	$\sqrt{n}$	$\sqrt{10n}$	$1/n$	n	n^2	$\sqrt{n}$	$\sqrt{10n}$	$1/n$
41	1,681	6.403	20.248	0.02439	91	8,281	9.539	30.166	0.01099
42	1,764	6.481	20.494	0.02381	92	8,464	9.592	30.332	0.01087
43	1,849	6.557	20.736	0.02326	93	8,649	9.644	30.496	0.01075
44	1,936	6.633	20.976	0.02273	94	8,836	9.695	30.659	0.01064
45	2,025	6.708	21.213	0.02222	95	9,025	9.747	30.822	0.01053
46	2,116	6.782	21.448	0.02174	96	9,216	9.798	30.984	0.01042
47	2,209	6.856	21.679	0.02128	97	9,409	9.849	31.145	0.01031
48	2,304	6.928	21.909	0.02083	98	9,604	9.899	31.305	0.01020
49	2,401	7.000	22.136	0.02041	99	9,801	9.950	31.464	0.01010
50	2,500	7.071	22.361	0.02000	100	10,000	10.000	31.623	0.01000

TABLE II 5,000 RANDOM DIGITS

(Read across, down, or diagonally, a sequence of random digits occurs.)

12159	66144	05091	13446	45653	13684	66024	91410	51351	22772
30156	90519	95785	47544	66735	35754	11088	67310	19720	08379
59069	01722	53338	41942	65118	71236	01932	70343	25812	62275
54107	58081	82470	59407	13475	95872	16268	78436	39251	64247
99681	81295	06315	28212	45029	57701	96327	85436	33614	29070
27252	37875	53679	01889	35714	63534	63791	76342	47717	73684
93259	74585	11863	78985	03881	46567	93696	93521	54970	37607
84068	43759	75814	32261	12728	09636	22336	75629	01017	45503
68582	97054	28251	63787	57285	18854	35006	16343	51867	67979
60646	11298	19680	10087	66391	70853	24423	73007	74958	29020
97437	52922	80739	59178	50628	61017	51652	40915	94696	67843
58009	20681	98823	50979	01237	70152	13711	73916	87902	84759
77211	70110	93803	60135	22881	13423	30999	07104	27400	25414
54256	84591	65302	99257	92970	28924	36632	54044	91798	78018
37493	69330	94069	39544	14050	03476	25804	49350	92525	87941
87569	22661	55970	52623	35419	76660	42394	63210	62626	00581
22896	62237	39635	63725	10463	87944	92075	90914	30599	35671
02697	33230	64527	97210	41359	79399	13941	88378	68503	33609
20080	15652	37216	00679	02088	34138	13953	68939	05630	27653
20550	95151	60557	57449	77115	87372	02574	07851	22428	39189
72771	11672	67492	42904	64647	94354	45994	42538	54885	15983
38472	43379	76295	69406	96510	16529	83500	28590	49787	29822
24511	56510	72654	13277	45031	42235	96502	25567	23653	36707
01054	06674	58283	82831	97048	42983	06471	12350	49990	04809
94437	94907	95274	26487	60496	78222	43032	04276	70800	17378
97842	69095	25982	03484	25173	05982	14624	31653	17170	92785
53047	13486	69712	33567	82313	87631	03197	02438	12374	40329
40770	47013	63306	48154	80970	87976	04939	21233	20572	31013
52733	66251	69661	58387	72096	21355	51659	19003	75556	33095
41749	46502	18378	83141	63920	85516	75743	66317	45428	45940
10271	85184	46468	38860	24039	80949	51211	35411	40470	16070
98791	48848	68129	51024	53044	55039	71290	26484	70682	56255
30196	09295	47685	56768	29285	06272	98789	47188	35063	24158
99373	64343	92433	06388	65713	35386	43370	19254	55014	98621
27768	27552	42156	23239	46823	91077	06306	17756	84459	92513
67791	35910	56921	51976	78475	15336	92544	82601	17996	72268
64018	44004	08136	56129	77024	82650	18163	29158	33935	94262
79715	33859	10835	94936	02857	87486	70613	41909	80667	52176
20190	40737	82688	07099	65255	52767	65930	45861	32575	93731
82421	01208	49762	66360	00231	87540	88302	62686	38456	25872
00083	81269	35320	72064	10472	92080	80447	15259	62654	70882
56558	09762	20813	48719	35530	96437	96343	21212	32567	34305
41183	20460	08608	75283	43401	25888	73405	35639	92114	48006
39977	10603	35052	53751	64219	36235	84687	42091	42587	16996
29310	84031	03052	51356	44747	19678	14619	03600	08066	93899
47360	03571	95657	85065	80919	14890	97623	57375	77855	15735
48481	98262	50414	41929	05977	78903	47602	52154	47901	84523
48097	56362	16342	75261	27751	28715	21871	37943	17850	90999
20648	30751	96515	51581	43877	94494	80164	02115	09738	51938
60704	10107	59220	64220	23944	34684	83696	82344	19020	84834

Reproduced with permission from the RAND Corporation, *A Million Random Digits with 100,000 Normal Deviates*, The Free Press, Glencoe, Ill., 1964.

TABLE II   5,000 RANDOM DIGITS—continued

(Read across, down, or diagonally, a sequence of random digits occurs.)

03689	33090	43465	96789	56688	32389	88206	06534	10558	14478
43367	46409	44751	73410	35138	24910	70748	57336	56043	68550
45357	52080	62670	73877	20604	40408	98060	96733	65094	80335
62683	03171	77195	92515	78041	27590	42651	00254	73179	10159
04841	40918	69047	68986	08150	87984	08887	76083	37702	28523
85963	06992	65321	43521	46393	40491	06028	43865	58190	28142
03720	78942	61990	90812	98452	74098	69738	83272	39212	42817
10159	85560	35619	58248	65498	77977	02896	45198	10655	13973
80162	35686	57877	19552	63931	44171	40879	94532	17828	31848
74388	92906	65829	24572	79417	38460	96294	79201	47755	90980
12660	09571	29743	45447	64063	46295	44191	53957	62393	42229
81852	60620	87757	72165	23875	87844	84038	04994	93466	27418
03068	61317	65305	64944	27319	55263	84514	38374	11657	67723
29623	58530	17274	16908	39253	37595	57497	74780	88624	93333
30520	50588	51231	83816	01075	33098	81308	59036	49152	86262
93694	02984	91350	33929	41724	32403	42566	14232	55085	65628
86736	40641	37958	25415	19922	65966	98044	39583	26828	50919
28141	15630	37675	52545	24813	22075	05142	15374	84533	12933
79804	05165	21620	98400	55290	71877	60052	46320	79055	45913
63763	49985	88853	70681	52762	17670	62337	12199	44123	37993
49618	47068	63331	62675	51788	58283	04295	72904	05378	98085
26502	68980	26545	14204	34304	50284	47730	57299	73966	02566
13549	86048	27912	56733	14987	09850	78217	85168	09538	92347
89221	78076	40306	34045	52557	52383	67796	41382	50490	30117
97809	34056	76778	60417	05153	83827	67369	08602	56163	28793
65668	44694	34151	51741	11484	13226	49516	17391	39956	34839
53653	59804	59051	95074	38307	99546	32962	26962	86252	50704
34922	95041	17398	32789	26860	55536	82415	82911	42208	62725
74880	65198	61357	90209	71543	71114	94868	05645	44154	72254
66036	48794	30021	92601	21615	16952	18433	44903	51322	90379
39044	99503	11442	81344	57068	74662	90382	59433	48440	38146
87756	71151	68543	08358	10183	06432	97482	90301	76114	83778
47117	45575	29524	02522	08041	70698	80260	73588	86415	72523
71572	02109	96722	21684	64331	71644	18933	32801	11644	12364
35609	58072	63209	48429	53108	59173	55337	22445	85940	43707
81703	70069	74981	12197	48426	77365	26769	65078	27849	41311
88979	88161	56531	46443	47148	42773	18601	38532	22594	12395
90279	42308	00380	17181	38757	09071	89804	15232	99007	39495
49266	18921	06498	88005	72736	81848	92716	96279	94582	22792
50897	22569	48402	80376	65470	19157	49729	19615	79087	47039
20950	65643	52280	37103	66977	65141	18522	39333	59824	73084
32686	51645	11382	75341	03189	94128	06275	22345	86856	77394
72525	65092	65086	47094	14781	61486	61895	85698	53028	61682
70502	57550	29699	36797	35862	90894	93217	96158	94321	12012
63087	03802	03142	72582	44267	56028	01576	69840	67727	77419
16418	07903	74344	89861	62952	49262	86210	65676	96617	38081
67730	17532	39489	28035	13415	83494	26750	01440	01161	16346
27274	98848	59506	28124	33596	89623	21006	94898	03550	88629
44250	52829	22614	21323	28597	66402	15425	39845	01823	19639
57476	33687	81784	05811	66625	17690	46170	93914	82346	82851

TABLE III INDIVIDUAL BINOMIAL PROBABILITIES

$P(r|n, \pi)$ or the probability of obtaining r successes in n trials where π is the probability of a success on a single trial.

							π					
n	r	1/20 = 0.0500	1/10 = 0.1000	1/9 = 0.1111	1/8 = 0.1250	1/7 = 0.1429	3/20 = 0.1500	1/6 = 0.1667	1/5 = 0.2000	2/9 = 0.2222	1/4 = 0.2500	
1	0	0.9500	0.9000	0.8889	0.8750	0.8571	0.8500	0.8333	0.8000	0.7778	0.7500	1
	1	0.0500	0.1000	0.1111	0.1250	0.1429	0.1500	0.1667	0.2000	0.2222	0.2500	0
2	0	0.9025	0.8100	0.7901	0.7656	0.7347	0.7225	0.6944	0.6400	0.6049	0.5625	2
	1	0.0950	0.1800	0.1975	0.2188	0.2449	0.2550	0.2778	0.3200	0.3457	0.3750	1
	2	0.0025	0.0100	0.0124	0.0156	0.0204	0.0225	0.0278	0.0400	0.0494	0.0625	0
3	0	0.8574	0.7290	0.7023	0.6699	0.6297	0.6141	0.5787	0.5120	0.4705	0.4219	3
	1	0.1354	0.2430	0.2634	0.2871	0.3149	0.3251	0.3472	0.3840	0.4033	0.4219	2
	2	0.0071	0.0270	0.0329	0.0410	0.0525	0.0574	0.0694	0.0960	0.1152	0.1406	1
	3	0.0001	0.0010	0.0014	0.0020	0.0029	0.0034	0.0046	0.0080	0.0120	0.0156	0
4	0	0.8145	0.6561	0.6243	0.5862	0.5398	0.5220	0.4822	0.4096	0.3660	0.3164	4
	1	0.1715	0.2916	0.3122	0.3350	0.3598	0.3685	0.3858	0.4096	0.4182	0.4219	3
	2	0.0135	0.0486	0.0585	0.0718	0.0900	0.0975	0.1157	0.1536	0.1792	0.2109	2
	3	0.0005	0.0036	0.0049	0.0068	0.0100	0.0115	0.0154	0.0256	0.0341	0.0469	1
	4	0.0000	0.0001	0.0002	0.0002	0.0004	0.0005	0.0008	0.0016	0.0024	0.0039	0
5	0	0.7738	0.5905	0.5549	0.5129	0.4627	0.4437	0.4019	0.3277	0.2846	0.2373	5
	1	0.2036	0.3280	0.3468	0.3664	0.3856	0.3915	0.4019	0.4096	0.4066	0.3955	4
	2	0.0214	0.0729	0.0867	0.1047	0.1285	0.1382	0.1608	0.2048	0.2324	0.2637	3
	3	0.0011	0.0081	0.0108	0.0150	0.0214	0.0244	0.0322	0.0512	0.0664	0.0879	2
	4	0.0000	0.0004	0.0007	0.0011	0.0018	0.0022	0.0032	0.0064	0.0095	0.0146	1
	5	0.0000	0.0000	0.0000	0.0000	0.0001	0.0001	0.0001	0.0003	0.0005	0.0010	0
6	0	0.7351	0.5314	0.4933	0.4488	0.3966	0.3772	0.3349	0.2621	0.2214	0.1780	6
	1	0.2321	0.3543	0.3700	0.3847	0.3966	0.3993	0.4019	0.3932	0.3795	0.3560	5
	2	0.0305	0.0984	0.1156	0.1374	0.1652	0.1762	0.2009	0.2458	0.2711	0.2966	4
	3	0.0021	0.0146	0.0193	0.0262	0.0367	0.0414	0.0536	0.0819	0.1031	0.1318	3
	4	0.0001	0.0012	0.0018	0.0028	0.0046	0.0055	0.0080	0.0154	0.0221	0.0330	2
	5	0.0000	0.0000	0.0001	0.0002	0.0003	0.0004	0.0006	0.0015	0.0025	0.0044	1
	6	0.0000	0.0000	0.0000	0.0000	0.0000	0.0000	0.0000	0.0001	0.0001	0.0002	0
7	0	0.6983	0.4783	0.4385	0.3927	0.3399	0.3206	0.2791	0.2097	0.1722	0.1335	7
	1	0.2573	0.3720	0.3836	0.3927	0.3966	0.3960	0.3907	0.3670	0.3444	0.3115	6
	2	0.0406	0.1240	0.1439	0.1683	0.1983	0.2096	0.2344	0.2752	0.2952	0.3115	5
	3	0.0036	0.0230	0.0300	0.0401	0.0551	0.0617	0.0781	0.1147	0.1406	0.1730	4
	4	0.0002	0.0026	0.0038	0.0057	0.0092	0.0109	0.0156	0.0287	0.0402	0.0577	3
	5	0.0000	0.0002	0.0003	0.0005	0.0009	0.0012	0.0019	0.0043	0.0069	0.0115	2
	6	0.0000	0.0000	0.0000	0.0000	0.0000	0.0001	0.0001	0.0004	0.0007	0.0013	1
	7	0.0000	0.0000	0.0000	0.0000	0.0000	0.0000	0.0000	0.0000	0.0000	0.0001	0
		19/20 0.9500	9/10 0.9000	8/9 0.8889	7/8 0.8750	6/7 0.8571	17/20 0.8500	5/6 0.8333	4/5 0.8000	7/9 0.7778	3/4 0.7500	r
							π					

TABLE III INDIVIDUAL BINOMIAL PROBABILITIES—continued

$$P(r \mid n, \pi)$$

						π						
n	r	2/7 = 0.2857	3/10 = 0.3000	1/3 = 0.3333	7/20 = 0.3500	3/8 = 0.3750	2/5 = 0.4000	3/7 = 0.4286	4/9 = 0.4444	9/20 = 0.4500	1/2 = 0.5000	
1	0	0.7143	0.7000	0.6667	0.6500	0.6250	0.6000	0.5714	0.5556	0.5500	0.5000	1
	1	0.2857	0.3000	0.3333	0.3500	0.3750	0.4000	0.4286	0.4444	0.4500	0.5000	0
2	0	0.5102	0.4900	0.4444	0.4225	0.3906	0.3600	0.3265	0.3086	0.3025	0.2500	2
	1	0.4082	0.4200	0.4444	0.4550	0.4688	0.4800	0.4898	0.4938	0.4950	0.5000	1
	2	0.0816	0.0900	0.1111	0.1225	0.1406	0.1600	0.1837	0.1975	0.2025	0.2500	0
3	0	0.3644	0.3430	0.2963	0.2746	0.2441	0.2160	0.1866	0.1715	0.1664	0.1250	3
	1	0.4373	0.4410	0.4444	0.4436	0.4394	0.4320	0.4198	0.4115	0.4084	0.3750	2
	2	0.1749	0.1890	0.2222	0.2389	0.2637	0.2880	0.3149	0.3292	0.3341	0.3750	1
	3	0.0233	0.0270	0.0370	0.0429	0.0527	0.0640	0.0787	0.0878	0.0911	0.1250	0
4	0	0.2603	0.2401	0.1975	0.1785	0.1526	0.1296	0.1066	0.0953	0.0915	0.0625	4
	1	0.4165	0.4116	0.3951	0.3845	0.3662	0.3456	0.3199	0.3048	0.2995	0.2500	3
	2	0.2499	0.2646	0.2963	0.3105	0.3296	0.3456	0.3598	0.3658	0.3675	0.3750	2
	3	0.0666	0.0756	0.0988	0.1115	0.1318	0.1536	0.1799	0.1951	0.2005	0.2500	1
	4	0.0067	0.0081	0.0124	0.0150	0.0198	0.0256	0.0337	0.0390	0.0410	0.0625	0
5	0	0.1859	0.1681	0.1317	0.1160	0.0954	0.0778	0.0609	0.0529	0.0503	0.0312	5
	1	0.3719	0.3602	0.3292	0.3124	0.2861	0.2592	0.2285	0.2117	0.2059	0.1562	4
	2	0.2975	0.3087	0.3292	0.3364	0.3433	0.3456	0.3427	0.3387	0.3369	0.3125	3
	3	0.1190	0.1323	0.1646	0.1812	0.2060	0.2304	0.2570	0.2710	0.2756	0.3125	2
	4	0.0238	0.0284	0.0412	0.0488	0.0618	0.0768	0.0964	0.1084	0.1128	0.1562	1
	5	0.0019	0.0024	0.0041	0.0052	0.0074	0.0102	0.0145	0.0173	0.0184	0.0312	0
6	0	0.1328	0.1176	0.0878	0.0754	0.0596	0.0467	0.0348	0.0294	0.0277	0.0156	6
	1	0.3187	0.3025	0.2634	0.2437	0.2146	0.1866	0.1567	0.1411	0.1359	0.0938	5
	2	0.3187	0.3241	0.3292	0.3280	0.3219	0.3110	0.2938	0.2822	0.2780	0.2344	4
	3	0.1700	0.1852	0.2195	0.2355	0.2575	0.2765	0.2938	0.3011	0.3032	0.3125	3
	4	0.0510	0.0595	0.0823	0.0951	0.1159	0.1382	0.1652	0.1806	0.1861	0.2344	2
	5	0.0082	0.0102	0.0165	0.0205	0.0278	0.0369	0.0496	0.0578	0.0609	0.0938	1
	6	0.0005	0.0007	0.0014	0.0018	0.0028	0.0041	0.0062	0.0077	0.0083	0.0156	0
7	0	0.0949	0.0824	0.0585	0.0490	0.0372	0.0280	0.0199	0.0163	0.0152	0.0078	7
	1	0.2656	0.2471	0.2048	0.1848	0.1565	0.1306	0.1044	0.0915	0.0872	0.0547	6
	2	0.3187	0.3176	0.3073	0.2985	0.2816	0.2613	0.2350	0.2195	0.2140	0.1641	5
	3	0.2125	0.2269	0.2561	0.2679	0.2816	0.2903	0.2938	0.2927	0.2918	0.2734	4
	4	0.0850	0.0972	0.1280	0.1442	0.1690	0.1935	0.2203	0.2342	0.2388	0.2734	3
	5	0.0204	0.0250	0.0384	0.0466	0.0608	0.0774	0.0991	0.1124	0.1172	0.1641	2
	6	0.0027	0.0036	0.0064	0.0084	0.0122	0.0172	0.0248	0.0300	0.0320	0.0547	1
	7	0.0002	0.0002	0.0005	0.0006	0.0010	0.0016	0.0027	0.0034	0.0037	0.0078	0
		5/7 0.7143	7/10 0.7000	2/3 0.6667	13/20 0.6500	5/8 0.6250	3/5 0.6000	4/7 0.5714	5/9 0.5556	11/20 0.5500	1/2 0.5000	r
							π						

TABLE III INDIVIDUAL BINOMIAL PROBABILITIES—continued

$P(r \mid n, \pi)$ or the probability of obtaining r successes in n trials where π is the probability of a success on a single trial.

							π					
		1/20 =	1/10 =	1/9 =	1/8 =	1/7 =	3/20 =	1/6 =	1/5 =	2/9 =	1/4 =	
n	r	0.0500	0.1000	0.1111	0.1250	0.1429	0.1500	0.1667	0.2000	0.2222	0.2500	
8	0	0.6634	0.4305	0.3897	0.3436	0.2914	0.2725	0.2326	0.1687	0.1339	0.1001	8
	1	0.2793	0.3826	0.3897	0.3927	0.3885	0.3847	0.3721	0.3355	0.3061	0.2670	7
	2	0.0515	0.1488	0.1705	0.1964	0.2266	0.2376	0.2605	0.2936	0.3061	0.3115	6
	3	0.0054	0.0331	0.0426	0.0561	0.0755	0.0839	0.1042	0.1468	0.1749	0.2076	5
	4	0.0004	0.0046	0.0067	0.0100	0.0157	0.0185	0.0260	0.0459	0.0625	0.0865	4
	5	0.0000	0.0004	0.0007	0.0011	0.0021	0.0026	0.0042	0.0092	0.0143	0.0231	3
	6	0.0000	0.0000	0.0000	0.0001	0.0002	0.0002	0.0004	0.0012	0.0020	0.0038	2
	7	0.0000	0.0000	0.0000	0.0000	0.0000	0.0000	0.0000	0.0001	0.0002	0.0004	1
	8	0.0000	0.0000	0.0000	0.0000	0.0000	0.0000	0.0000	0.0000	0.0000	0.0000	0
9	0	0.6302	0.3874	0.3464	0.3007	0.2497	0.2316	0.1938	0.1342	0.1042	0.0751	9
	1	0.2985	0.3874	0.3897	0.3866	0.3746	0.3679	0.3488	0.3020	0.2678	0.2252	8
	2	0.0628	0.1722	0.1949	0.2209	0.2497	0.2597	0.2791	0.3020	0.3061	0.3003	7
	3	0.0077	0.0446	0.0568	0.0736	0.0971	0.1069	0.1302	0.1762	0.2041	0.2336	6
	4	0.0006	0.0074	0.0107	0.0158	0.0243	0.0283	0.0391	0.0661	0.0875	0.1168	5
	5	0.0000	0.0008	0.0013	0.0022	0.0040	0.0050	0.0078	0.0165	0.0250	0.0389	4
	6	0.0000	0.0001	0.0001	0.0002	0.0004	0.0006	0.0010	0.0028	0.0048	0.0086	3
	7	0.0000	0.0000	0.0000	0.0000	0.0000	0.0000	0.0001	0.0003	0.0006	0.0012	2
	8	0.0000	0.0000	0.0000	0.0000	0.0000	0.0000	0.0000	0.0000	0.0000	0.0001	1
	9	0.0000	0.0000	0.0000	0.0000	0.0000	0.0000	0.0000	0.0000	0.0000	0.0000	0
10	0	0.5987	0.3487	0.3080	0.2631	0.2141	0.1969	0.1615	0.1074	0.0810	0.0563	10
	1	0.3151	0.3874	0.3849	0.3758	0.3568	0.3474	0.3230	0.2684	0.2315	0.1877	9
	2	0.0746	0.1937	0.2165	0.2416	0.2676	0.2759	0.2907	0.3020	0.2976	0.2816	8
	3	0.0105	0.0574	0.0722	0.0920	0.1189	0.1298	0.1550	0.2013	0.2267	0.2503	7
	4	0.0010	0.0112	0.0158	0.0230	0.0347	0.0401	0.0543	0.0881	0.1134	0.1460	6
	5	0.0001	0.0015	0.0024	0.0039	0.0069	0.0085	0.0130	0.0264	0.0389	0.0584	5
	6	0.0000	0.0001	0.0002	0.0005	0.0010	0.0012	0.0022	0.0055	0.0092	0.0162	4
	7	0.0000	0.0000	0.0000	0.0000	0.0001	0.0001	0.0002	0.0008	0.0015	0.0031	3
	8	0.0000	0.0000	0.0000	0.0000	0.0000	0.0000	0.0000	0.0001	0.0002	0.0004	2
	9	0.0000	0.0000	0.0000	0.0000	0.0000	0.0000	0.0000	0.0000	0.0000	0.0000	1
	10	0.0000	0.0000	0.0000	0.0000	0.0000	0.0000	0.0000	0.0000	0.0000	0.0000	0
11	0	0.5688	0.3138	0.2737	0.2302	0.1835	0.1673	0.1346	0.0859	0.0630	0.0422	11
	1	0.3293	0.3836	0.3764	0.3617	0.3364	0.3248	0.2961	0.2362	0.1980	0.1549	10
	2	0.0867	0.2131	0.2352	0.2584	0.2803	0.2866	0.2961	0.2953	0.2829	0.2581	9
	3	0.0137	0.0710	0.0882	0.1107	0.1402	0.1517	0.1777	0.2215	0.2425	0.2581	8
		19/20	9/10	8/9	7/8	6/7	17/20	5/6	4/5	7/9	3/4	r
		0.9500	0.9000	0.8889	0.8750	0.8571	0.8500	0.8333	0.8000	0.7778	0.7500	
							π					

TABLE III INDIVIDUAL BINOMIAL PROBABILITIES—continued

$$P(r \mid n, \pi)$$

						π						
n	r	2/7 = 0.2857	3/10 = 0.3000	1/3 = 0.3333	7/20 = 0.3500	3/8 = 0.3750	2/5 = 0.4000	3/7 = 0.4286	4/9 = 0.4444	9/20 = 0.4500	1/2 = 0.5000	
8	0	0.0678	0.0576	0.0390	0.0319	0.0233	0.0168	0.0114	0.0091	0.0084	0.0039	8
	1	0.2168	0.1976	0.1561	0.1373	0.1118	0.0896	0.0682	0.0581	0.0548	0.0312	7
	2	0.3036	0.2965	0.2731	0.2587	0.2347	0.2090	0.1790	0.1626	0.1570	0.1094	6
	3	0.2428	0.2541	0.2731	0.2786	0.2816	0.2787	0.2688	0.2602	0.2568	0.2188	5
	4	0.1214	0.1361	0.1707	0.1875	0.2112	0.2322	0.2518	0.2602	0.2627	0.2734	4
	5	0.0389	0.0467	0.0683	0.0808	0.1014	0.1239	0.1511	0.1665	0.1719	0.2188	3
	6	0.0078	0.0100	0.0171	0.0218	0.0304	0.0413	0.0566	0.0666	0.0703	0.1094	2
	7	0.0009	0.0012	0.0024	0.0034	0.0052	0.0079	0.0121	0.0152	0.0164	0.0312	1
	8	0.0000	0.0001	0.0002	0.0002	0.0004	0.0007	0.0011	0.0015	0.0017	0.0039	
9	0	0.0484	0.0404	0.0260	0.0207	0.0146	0.0101	0.0065	0.0050	0.0046	0.0020	9
	1	0.1742	0.1556	0.1171	0.1004	0.0786	0.0605	0.0438	0.0363	0.0339	0.0176	8
	2	0.2788	0.2668	0.2341	0.2162	0.1886	0.1612	0.1316	0.1162	0.1110	0.0703	7
	3	0.2602	0.2668	0.2731	0.2716	0.2640	0.2508	0.2302	0.2168	0.2119	0.1641	6
	4	0.1561	0.1715	0.2048	0.2194	0.2376	0.2508	0.2590	0.2602	0.2600	0.2461	5
	5	0.0624	0.0735	0.1024	0.1181	0.1426	0.1672	0.1942	0.2082	0.2128	0.2461	4
	6	0.0166	0.0210	0.0341	0.0424	0.0570	0.0743	0.0971	0.1110	0.1160	0.1641	3
	7	0.0028	0.0039	0.0073	0.0098	0.0147	0.0212	0.0312	0.0381	0.0407	0.0703	2
	8	0.0003	0.0004	0.0009	0.0013	0.0022	0.0035	0.0058	0.0076	0.0083	0.0176	1
	9	0.0000	0.0000	0.0000	0.0001	0.0002	0.0003	0.0005	0.0007	0.0008	0.0020	0
10	0	0.0346	0.0282	0.0173	0.0135	0.0091	0.0060	0.0037	0.0028	0.0025	0.0010	10
	1	0.1383	0.1211	0.0867	0.0725	0.0546	0.0403	0.0278	0.0224	0.0207	0.0098	9
	2	0.2489	0.2335	0.1951	0.1756	0.1473	0.1209	0.0940	0.0807	0.0763	0.0440	8
	3	0.2655	0.2668	0.2601	0.2522	0.2357	0.2150	0.1879	0.1721	0.1665	0.1172	7
	4	0.1859	0.2001	0.2276	0.2377	0.2475	0.2508	0.2466	0.2409	0.2384	0.2051	6
	5	0.0892	0.1029	0.1366	0.1536	0.1782	0.2007	0.2220	0.2313	0.2340	0.2461	5
	6	0.0297	0.0368	0.0569	0.0689	0.0891	0.1115	0.1387	0.1542	0.1596	0.2051	4
	7	0.0068	0.0090	0.0163	0.0212	0.0306	0.0425	0.0595	0.0705	0.0746	0.1172	3
	8	0.0010	0.0014	0.0030	0.0043	0.0069	0.0106	0.0167	0.0211	0.0229	0.0440	2
	9	0.0001	0.0001	0.0003	0.0005	0.0009	0.0016	0.0028	0.0038	0.0042	0.0098	1
	10	0.0000	0.0000	0.0000	0.0000	0.0000	0.0001	0.0002	0.0003	0.0003	0.0010	0
11	0	0.0247	0.0198	0.0116	0.0088	0.0057	0.0036	0.0021	0.0016	0.0014	0.0005	11
	1	0.1086	0.0932	0.0636	0.0518	0.0375	0.0266	0.0175	0.0137	0.0125	0.0054	10
	2	0.2173	0.1998	0.1590	0.1396	0.1126	0.0887	0.0656	0.0548	0.0513	0.0269	9
	3	0.2608	0.2568	0.2384	0.2254	0.2026	0.1774	0.1477	0.1314	0.1259	0.0806	8
		5/7 0.7143	7/10 0.7000	2/3 0.6667	13/20 0.6500	5/8 0.6250	3/5 0.6000	4/7 0.5714	5/9 0.5556	11/20 0.5500	1/2 0.5000	r
							π						

TABLE III INDIVIDUAL BINOMIAL PROBABILITIES—continued

$P(r|n, \pi)$ or the probability of obtaining r successes in n trials where π is the probability of a success on a single trial.

						π						
n	r	1/20 = 0.0500	1/10 = 0.1000	1/9 = 0.1111	1/8 = 0.1250	1/7 = 0.1429	3/20 = 0.1500	1/6 = 0.1667	1/5 = 0.2000	2/9 = 0.2222	1/4 = 0.2500	
	4	0.0014	0.0158	0.0220	0.0316	0.0467	0.0536	0.0711	0.1107	0.1386	0.1721	7
	5	0.0001	0.0025	0.0039	0.0063	0.0109	0.0132	0.0199	0.0388	0.0554	0.0803	6
	6	0.0000	0.0003	0.0005	0.0009	0.0018	0.0023	0.0040	0.0097	0.0158	0.0268	5
	7	0.0000	0.0000	0.0000	0.0001	0.0002	0.0003	0.0006	0.0017	0.0032	0.0064	4
	8	0.0000	0.0000	0.0000	0.0000	0.0000	0.0000	0.0001	0.0002	0.0005	0.0011	3
	9	0.0000	0.0000	0.0000	0.0000	0.0000	0.0000	0.0000	0.0000	0.0000	0.0001	2
	10	0.0000	0.0000	0.0000	0.0000	0.0000	0.0000	0.0000	0.0000	0.0000	0.0000	1
	11	0.0000	0.0000	0.0000	0.0000	0.0000	0.0000	0.0000	0.0000	0.0000	0.0000	0
12	0	0.5404	0.2824	0.2433	0.2014	0.1573	0.1422	0.1122	0.0687	0.0490	0.0317	12
	1	0.3413	0.3766	0.3650	0.3453	0.3145	0.3012	0.2692	0.2062	0.1680	0.1267	11
	2	0.0988	0.2301	0.2509	0.2713	0.2883	0.2924	0.2961	0.2835	0.2640	0.2323	10
	3	0.0173	0.0852	0.1046	0.1292	0.1602	0.1720	0.1974	0.2362	0.2515	0.2581	9
	4	0.0020	0.0213	0.0294	0.0415	0.0601	0.0683	0.0888	0.1329	0.1617	0.1936	8
	5	0.0002	0.0038	0.0059	0.0095	0.0160	0.0193	0.0284	0.0532	0.0739	0.1032	7
	6	0.0000	0.0005	0.0009	0.0016	0.0031	0.0040	0.0066	0.0155	0.0246	0.0402	6
	7	0.0000	0.0000	0.0001	0.0002	0.0004	0.0006	0.0011	0.0033	0.0060	0.0115	5
	8	0.0000	0.0000	0.0000	0.0000	0.0000	0.0001	0.0001	0.0005	0.0011	0.0024	4
	9	0.0000	0.0000	0.0000	0.0000	0.0000	0.0000	0.0000	0.0001	0.0001	0.0004	3
	10	0.0000	0.0000	0.0000	0.0000	0.0000	0.0000	0.0000	0.0000	0.0000	0.0000	2
	11	0.0000	0.0000	0.0000	0.0000	0.0000	0.0000	0.0000	0.0000	0.0000	0.0000	1
	12	0.0000	0.0000	0.0000	0.0000	0.0000	0.0000	0.0000	0.0000	0.0000	0.0000	0
13	0	0.5133	0.2542	0.2163	0.1762	0.1348	0.1209	0.0935	0.0550	0.0381	0.0238	13
	1	0.3512	0.3672	0.3515	0.3273	0.2921	0.2774	0.2430	0.1787	0.1416	0.1030	12
	2	0.1109	0.2448	0.2636	0.2806	0.2921	0.2937	0.2916	0.2680	0.2427	0.2059	11
	3	0.0214	0.0997	0.1208	0.1470	0.1785	0.1900	0.2138	0.2457	0.2542	0.2516	10
	4	0.0028	0.0277	0.0378	0.0525	0.0744	0.0838	0.1069	0.1536	0.1816	0.2097	9
	5	0.0003	0.0055	0.0085	0.0135	0.0223	0.0266	0.0385	0.0691	0.0934	0.1258	8
	6	0.0000	0.0008	0.0014	0.0026	0.0050	0.0063	0.0103	0.0230	0.0356	0.0559	7
	7	0.0000	0.0001	0.0002	0.0004	0.0008	0.0011	0.0020	0.0058	0.0102	0.0186	6
	8	0.0000	0.0000	0.0000	0.0000	0.0001	0.0002	0.0003	0.0011	0.0022	0.0047	5
	9	0.0000	0.0000	0.0000	0.0000	0.0000	0.0000	0.0000	0.0002	0.0004	0.0009	4
	10	0.0000	0.0000	0.0000	0.0000	0.0000	0.0000	0.0000	0.0000	0.0000	0.0001	3
	11	0.0000	0.0000	0.0000	0.0000	0.0000	0.0000	0.0000	0.0000	0.0000	0.0000	2
	12	0.0000	0.0000	0.0000	0.0000	0.0000	0.0000	0.0000	0.0000	0.0000	0.0000	1
	13	0.0000	0.0000	0.0000	0.0000	0.0000	0.0000	0.0000	0.0000	0.0000	0.0000	0
		19/20 0.9500	9/10 0.9000	8/9 0.8889	7/8 0.8750	6/7 0.8571	17/20 0.8500	5/6 0.8333	4/5 0.8000	7/9 0.7778	3/4 0.7500	r
							π						

TABLE III INDIVIDUAL BINOMIAL PROBABILITIES—continued

$$P(r\,|\,n,\pi)$$

						π						
n	r	2/7 = 0.2857	3/10 = 0.3000	1/3 = 0.3333	7/20 = 0.3500	3/8 = 0.3750	2/5 = 0.4000	3/7 = 0.4286	4/9 = 0.4444	9/20 = 0.4500	1/2 = 0.5000	
	4	0.2086	0.2201	0.2384	0.2428	0.2431	0.2365	0.2215	0.2103	0.2060	0.1611	7
	5	0.1168	0.1321	0.1669	0.1830	0.2042	0.2207	0.2326	0.2356	0.2360	0.2256	6
	6	0.0467	0.0566	0.0835	0.0985	0.1225	0.1472	0.1744	0.1884	0.1931	0.2256	5
	7	0.0134	0.0173	0.0298	0.0379	0.0525	0.0701	0.0934	0.1077	0.1128	0.1611	4
	8	0.0027	0.0037	0.0074	0.0102	0.0158	0.0234	0.0350	0.0431	0.0467	0.0806	3
	9	0.0004	0.0005	0.0012	0.0018	0.0032	0.0052	0.0088	0.0115	0.0126	0.0269	2
	10	0.0000	0.0000	0.0001	0.0002	0.0004	0.0007	0.0013	0.0018	0.0021	0.0054	1
	11	0.0000	0.0000	0.0000	0.0000	0.0000	0.0000	0.0001	0.0001	0.0002	0.0005	0
12	0	0.0176	0.0138	0.0077	0.0057	0.0036	0.0022	0.0012	0.0009	0.0008	0.0002	12
	1	0.0847	0.0712	0.0462	0.0368	0.0256	0.0174	0.0109	0.0083	0.0075	0.0029	11
	2	0.1863	0.1678	0.1272	0.1088	0.0844	0.0638	0.0450	0.0365	0.0338	0.0161	10
	3	0.2484	0.2397	0.2120	0.1954	0.1688	0.1419	0.1125	0.0974	0.0923	0.0537	9
	4	0.2235	0.2311	0.2384	0.2367	0.2279	0.2128	0.1898	0.1753	0.1700	0.1208	8
	5	0.1430	0.1585	0.1908	0.2039	0.2188	0.2270	0.2278	0.2243	0.2225	0.1934	7
	6	0.0668	0.0792	0.1113	0.1281	0.1532	0.1766	0.1993	0.2094	0.2124	0.2256	6
	7	0.0229	0.0291	0.0477	0.0591	0.0788	0.1009	0.1281	0.1436	0.1490	0.1934	5
	8	0.0057	0.0078	0.0149	0.0199	0.0295	0.0420	0.0601	0.0718	0.0762	0.1208	4
	9	0.0010	0.0015	0.0033	0.0048	0.0079	0.0125	0.0200	0.0255	0.0277	0.0537	3
	10	0.0001	0.0002	0.0005	0.0008	0.0014	0.0025	0.0045	0.0061	0.0068	0.0161	2
	11	0.0000	0.0000	0.0000	0.0001	0.0002	0.0003	0.0006	0.0009	0.0010	0.0029	1
	12	0.0000	0.0000	0.0000	0.0000	0.0000	0.0000	0.0000	0.0001	0.0001	0.0002	0
13	0	0.0126	0.0097	0.0051	0.0037	0.0022	0.0013	0.0007	0.0005	0.0004	0.0001	13
	1	0.0655	0.0540	0.0334	0.0259	0.0173	0.0113	0.0068	0.0050	0.0045	0.0016	12
	2	0.1572	0.1388	0.1002	0.0836	0.0624	0.0453	0.0304	0.0240	0.0220	0.0095	11
	3	0.2306	0.2181	0.1837	0.1651	0.1372	0.1107	0.0836	0.0703	0.0660	0.0349	10
	4	0.2306	0.2337	0.2296	0.2222	0.2058	0.1845	0.1567	0.1406	0.1350	0.0873	9
	5	0.1660	0.1803	0.2066	0.2154	0.2222	0.2214	0.2115	0.2025	0.1989	0.1571	8
	6	0.0886	0.1030	0.1378	0.1546	0.1778	0.1968	0.2115	0.2160	0.2169	0.2095	7
	7	0.0354	0.0442	0.0689	0.0833	0.1067	0.1312	0.1586	0.1728	0.1775	0.2095	6
	8	0.0106	0.0142	0.0258	0.0336	0.0480	0.0656	0.0892	0.1037	0.1089	0.1571	5
	9	0.0024	0.0034	0.0072	0.0101	0.0160	0.0243	0.0372	0.0461	0.0495	0.0873	4
	10	0.0004	0.0006	0.0014	0.0022	0.0038	0.0065	0.0112	0.0148	0.0162	0.0349	3
	11	0.0000	0.0001	0.0002	0.0003	0.0006	0.0012	0.0023	0.0032	0.0036	0.0095	2
	12	0.0000	0.0000	0.0000	0.0000	0.0001	0.0001	0.0003	0.0004	0.0005	0.0016	1
	13	0.0000	0.0000	0.0000	0.0000	0.0000	0.0000	0.0000	0.0000	0.0000	0.0001	0
		5/7 0.7143	7/10 0.7000	2/3 0.6667	13/20 0.6500	5/8 0.6250	3/5 0.6000	4/7 0.5714	5/9 0.5556	11/20 0.5500	1/2 0.5000	r

					π					

TABLE III INDIVIDUAL BINOMIAL PROBABILITIES—continued

$P(r|n, \pi)$ or the probability of obtaining r successes in n trials where π is the probability of a success on a single trial.

		π										
n	r	1/20 = 0.0500	1/10 = 0.1000	1/9 = 0.1111	1/8 = 0.1250	1/7 = 0.1429	3/20 = 0.1500	1/6 = 0.1667	1/5 = 0.2000	2/9 = 0.2222	1/4 = 0.2500	
14	0	0.4877	0.2288	0.1922	0.1542	0.1155	0.1028	0.0779	0.0440	0.0296	0.0178	14
	1	0.3593	0.3559	0.3364	0.3084	0.2696	0.2539	0.2181	0.1539	0.1186	0.0832	13
	2	0.1229	0.2570	0.2734	0.2864	0.2921	0.2912	0.2835	0.2501	0.2202	0.1802	12
	3	0.0259	0.1142	0.1367	0.1636	0.1947	0.2056	0.2268	0.2501	0.2517	0.2402	11
	4	0.0038	0.0349	0.0470	0.0643	0.0892	0.0998	0.1247	0.1720	0.1978	0.2202	10
	5	0.0004	0.0078	0.0118	0.0184	0.0298	0.0352	0.0499	0.0860	0.1130	0.1468	9
	6	0.0000	0.0013	0.0022	0.0039	0.0074	0.0093	0.0150	0.0322	0.0484	0.0734	8
	7	0.0000	0.0002	0.0003	0.0006	0.0014	0.0019	0.0034	0.0092	0.0158	0.0280	7
	8	0.0000	0.0000	0.0000	0.0001	0.0002	0.0003	0.0006	0.0020	0.0040	0.0082	6
	9	0.0000	0.0000	0.0000	0.0000	0.0000	0.0000	0.0001	0.0003	0.0008	0.0018	5
	10	0.0000	0.0000	0.0000	0.0000	0.0000	0.0000	0.0000	0.0000	0.0001	0.0003	4
	11	0.0000	0.0000	0.0000	0.0000	0.0000	0.0000	0.0000	0.0000	0.0000	0.0000	3
	12	0.0000	0.0000	0.0000	0.0000	0.0000	0.0000	0.0000	0.0000	0.0000	0.0000	2
	13	0.0000	0.0000	0.0000	0.0000	0.0000	0.0000	0.0000	0.0000	0.0000	0.0000	1
	14	0.0000	0.0000	0.0000	0.0000	0.0000	0.0000	0.0000	0.0000	0.0000	0.0000	0
15	0	0.4633	0.2059	0.1709	0.1349	0.0990	0.0874	0.0649	0.0352	0.0231	0.0134	15
	1	0.3658	0.3432	0.3204	0.2891	0.2476	0.2312	0.1947	0.1319	0.0988	0.0668	14
	2	0.1348	0.2669	0.2804	0.2891	0.2889	0.2856	0.2726	0.2309	0.1976	0.1559	13
	3	0.0307	0.1285	0.1519	0.1790	0.2086	0.2184	0.2363	0.2501	0.2447	0.2252	12
	4	0.0048	0.0428	0.0570	0.0767	0.1043	0.1156	0.1418	0.1876	0.2098	0.2252	11
	5	0.0006	0.0105	0.0157	0.0241	0.0382	0.0449	0.0624	0.1032	0.1318	0.1652	10
	6	0.0000	0.0019	0.0033	0.0057	0.0106	0.0132	0.0208	0.0430	0.0628	0.0918	9
	7	0.0000	0.0003	0.0005	0.0010	0.0023	0.0030	0.0054	0.0138	0.0231	0.0393	8
	8	0.0000	0.0000	0.0001	0.0002	0.0004	0.0005	0.0011	0.0034	0.0066	0.0131	7
	9	0.0000	0.0000	0.0000	0.0000	0.0000	0.0001	0.0002	0.0007	0.0015	0.0034	6
	10	0.0000	0.0000	0.0000	0.0000	0.0000	0.0000	0.0000	0.0001	0.0002	0.0007	5
	11	0.0000	0.0000	0.0000	0.0000	0.0000	0.0000	0.0000	0.0000	0.0000	0.0001	4
	12	0.0000	0.0000	0.0000	0.0000	0.0000	0.0000	0.0000	0.0000	0.0000	0.0000	3
	13	0.0000	0.0000	0.0000	0.0000	0.0000	0.0000	0.0000	0.0000	0.0000	0.0000	2
	14	0.0000	0.0000	0.0000	0.0000	0.0000	0.0000	0.0000	0.0000	0.0000	0.0000	1
	15	0.0000	0.0000	0.0000	0.0000	0.0000	0.0000	0.0000	0.0000	0.0000	0.0000	0
16	0	0.4401	0.1853	0.1519	0.1181	0.0849	0.0742	0.0541	0.0282	0.0179	0.0100	16
	1	0.3706	0.3294	0.3038	0.2699	0.2264	0.2096	0.1731	0.1126	0.0820	0.0534	15
	2	0.1463	0.2745	0.2848	0.2891	0.2830	0.2775	0.2596	0.2111	0.1757	0.1336	14
	3	0.0359	0.1423	0.1661	0.1928	0.2201	0.2285	0.2423	0.2463	0.2342	0.2079	13
	4	0.0062	0.0514	0.0675	0.0895	0.1192	0.1311	0.1575	0.2001	0.2175	0.2252	12
		19/20 0.9500	9/10 0.9000	8/9 0.8889	7/8 0.8750	6/7 0.8571	17/20 0.8500	5/6 0.8333	4/5 0.8000	7/9 0.7778	3/4 0.7500	r
							π					

TABLE III INDIVIDUAL BINOMIAL PROBABILITIES—continued

$$P(r\,|\,n,\,\pi)$$

						π						
n	r	2/7 = 0.2857	3/10 = 0.3000	1/3 = 0.3333	7/20 = 0.3500	3/8 = 0.3750	2/5 = 0.4000	3/7 = 0.4286	4/9 = 0.4444	9/20 = 0.4500	1/2 = 0.5000	
14	0	0.0090	0.0068	0.0034	0.0024	0.0014	0.0008	0.0004	0.0003	0.0002	0.0001	14
	1	0.0504	0.0407	0.0240	0.0181	0.0117	0.0073	0.0042	0.0030	0.0026	0.0008	13
	2	0.1310	0.1134	0.0779	0.0634	0.0455	0.0317	0.0203	0.0155	0.0141	0.0056	12
	3	0.2096	0.1943	0.1559	0.1366	0.1091	0.0845	0.0608	0.0497	0.0462	0.0222	11
	4	0.2306	0.2290	0.2143	0.2022	0.1800	0.1550	0.1254	0.1094	0.1040	0.0611	10
	5	0.1845	0.1963	0.2143	0.2178	0.2160	0.2066	0.1880	0.1750	0.1701	0.1222	9
	6	0.1107	0.1262	0.1607	0.1759	0.1944	0.2066	0.2115	0.2100	0.2088	0.1833	8
	7	0.0506	0.0618	0.0918	0.1082	0.1333	0.1574	0.1813	0.1920	0.1952	0.2095	7
	8	0.0177	0.0232	0.0402	0.0510	0.0700	0.0918	0.1190	0.1344	0.1398	0.1833	6
	9	0.0047	0.0066	0.0134	0.0183	0.0280	0.0408	0.0595	0.0717	0.0762	0.1222	5
	10	0.0009	0.0014	0.0034	0.0049	0.0084	0.0136	0.0223	0.0287	0.0312	0.0611	4
	11	0.0001	0.0002	0.0006	0.0010	0.0018	0.0033	0.0061	0.0083	0.0093	0.0222	3
	12	0.0000	0.0000	0.0001	0.0001	0.0003	0.0006	0.0011	0.0017	0.0019	0.0056	2
	13	0.0000	0.0000	0.0000	0.0000	0.0000	0.0001	0.0001	0.0002	0.0002	0.0008	1
	14	0.0000	0.0000	0.0000	0.0000	0.0000	0.0000	0.0000	0.0000	0.0000	0.0001	0
15	0	0.0064	0.0048	0.0023	0.0016	0.0009	0.0005	0.0002	0.0002	0.0001	0.0000	15
	1	0.0386	0.0305	0.0171	0.0126	0.0078	0.0047	0.0025	0.0018	0.0016	0.0005	14
	2	0.1080	0.0916	0.0600	0.0476	0.0328	0.0219	0.0134	0.0100	0.0090	0.0032	13
	3	0.1872	0.1700	0.1299	0.1110	0.0852	0.0634	0.0434	0.0345	0.0318	0.0139	12
	4	0.2246	0.2186	0.1948	0.1792	0.1534	0.1268	0.0977	0.0829	0.0780	0.0417	11
	5	0.1977	0.2061	0.2143	0.2123	0.2025	0.1859	0.1612	0.1458	0.1404	0.0916	10
	6	0.1318	0.1472	0.1786	0.1906	0.2025	0.2066	0.2015	0.1945	0.1914	0.1527	9
	7	0.0678	0.0811	0.1148	0.1319	0.1562	0.1771	0.1943	0.2000	0.2013	0.1964	8
	8	0.0271	0.0348	0.0574	0.0710	0.0938	0.1181	0.1457	0.1600	0.1647	0.1964	7
	9	0.0084	0.0116	0.0223	0.0298	0.0438	0.0612	0.0850	0.0996	0.1048	0.1527	6
	10	0.0020	0.0030	0.0067	0.0096	0.0158	0.0245	0.0382	0.0478	0.0515	0.0916	5
	11	0.0004	0.0006	0.0015	0.0024	0.0043	0.0074	0.0130	0.0174	0.0191	0.0417	4
	12	0.0000	0.0001	0.0002	0.0004	0.0009	0.0016	0.0033	0.0046	0.0052	0.0139	3
	13	0.0000	0.0000	0.0000	0.0000	0.0001	0.0002	0.0006	0.0009	0.0010	0.0032	2
	14	0.0000	0.0000	0.0000	0.0000	0.0000	0.0000	0.0001	0.0001	0.0001	0.0005	1
	15	0.0000	0.0000	0.0000	0.0000	0.0000	0.0000	0.0000	0.0000	0.0000	0.0000	0
16	0	0.0046	0.0033	0.0015	0.0010	0.0005	0.0003	0.0001	0.0001	0.0001	0.0000	16
	1	0.0294	0.0228	0.0122	0.0088	0.0052	0.0030	0.0016	0.0010	0.0009	0.0002	15
	2	0.0882	0.0732	0.0457	0.0353	0.0234	0.0150	0.0087	0.0063	0.0056	0.0018	14
	3	0.1646	0.1465	0.1066	0.0888	0.0656	0.0468	0.0305	0.0236	0.0215	0.0085	13
	4	0.2139	0.2040	0.1732	0.1554	0.1279	0.1014	0.0744	0.0614	0.0572	0.0278	12
		5/7 0.7143	7/10 0.7000	2/3 0.6667	13/20 0.6500	5/8 0.6250	3/5 0.6000	4/7 0.5714	5/9 0.5556	11/20 0.5500	1/2 0.5000	r
							π						

TABLE III INDIVIDUAL BINOMIAL PROBABILITIES—continued

$P(r|n, \pi)$ or the probability of obtaining r successes in n trials where π is the probability of a success on a single trial.

		π										
n	r	1/20 = 0.0500	1/10 = 0.1000	1/9 = 0.1111	1/8 = 0.1250	1/7 = 0.1429	3/20 = 0.1500	1/6 = 0.1667	1/5 = 0.2000	2/9 = 0.2222	1/4 = 0.2500	
	5	0.0008	0.0137	0.0202	0.0307	0.0477	0.0555	0.0756	0.1201	0.1492	0.1802	11
	6	0.0001	0.0028	0.0046	0.0080	0.0146	0.0180	0.0277	0.0550	0.0781	0.1101	10
	7	0.0000	0.0004	0.0008	0.0016	0.0035	0.0045	0.0079	0.0196	0.0319	0.0524	9
	8	0.0000	0.0001	0.0001	0.0003	0.0006	0.0009	0.0018	0.0055	0.0102	0.0197	8
	9	0.0000	0.0000	0.0000	0.0000	0.0001	0.0001	0.0003	0.0012	0.0026	0.0058	7
	10	0.0000	0.0000	0.0000	0.0000	0.0000	0.0000	0.0000	0.0002	0.0005	0.0014	6
	11	0.0000	0.0000	0.0000	0.0000	0.0000	0.0000	0.0000	0.0000	0.0001	0.0002	5
	12	0.0000	0.0000	0.0000	0.0000	0.0000	0.0000	0.0000	0.0000	0.0000	0.0000	4
	13	0.0000	0.0000	0.0000	0.0000	0.0000	0.0000	0.0000	0.0000	0.0000	0.0000	3
	14	0.0000	0.0000	0.0000	0.0000	0.0000	0.0000	0.0000	0.0000	0.0000	0.0000	2
	15	0.0000	0.0000	0.0000	0.0000	0.0000	0.0000	0.0000	0.0000	0.0000	0.0000	1
	16	0.0000	0.0000	0.0000	0.0000	0.0000	0.0000	0.0000	0.0000	0.0000	0.0000	0
		19/20 0.9500	9/10 0.9000	8/9 0.8889	7/8 0.8750	6/7 0.8571	17/20 0.8500	5/6 0.8333	4/5 0.8000	7/9 0.7778	3/4 0.7500	r
		π										

TABLE III   INDIVIDUAL BINOMIAL PROBABILITIES—continued

$$P(r \mid n, \pi)$$

| | | \multicolumn{10}{c}{π} | |
n	r	2/7 = 0.2857	3/10 = 0.3000	1/3 = 0.3333	7/20 = 0.3500	3/8 = 0.3750	2/5 = 0.4000	3/7 = 0.4286	4/9 = 0.4444	9/20 = 0.4500	1/2 = 0.5000	
	5	0.2054	0.2099	0.2078	0.2008	0.1841	0.1623	0.1340	0.1179	0.1123	0.0666	11
	6	0.1506	0.1649	0.1905	0.1982	0.2025	0.1983	0.1842	0.1729	0.1684	0.1222	10
	7	0.0861	0.1010	0.1361	0.1524	0.1736	0.1889	0.1974	0.1976	0.1969	0.1746	9
	8	0.0387	0.0487	0.0765	0.0924	0.1172	0.1417	0.1665	0.1778	0.1812	0.1964	8
	9	0.0138	0.0185	0.0340	0.0442	0.0625	0.0840	0.1110	0.1264	0.1318	0.1746	7
	10	0.0039	0.0056	0.0119	0.0167	0.0262	0.0392	0.0583	0.0708	0.0755	0.1222	6
	11	0.0008	0.0013	0.0032	0.0049	0.0086	0.0142	0.0238	0.0309	0.0337	0.0666	5
	12	0.0001	0.0002	0.0007	0.0011	0.0022	0.0040	0.0074	0.0103	0.0115	0.0278	4
	13	0.0000	0.0000	0.0001	0.0002	0.0004	0.0008	0.0017	0.0025	0.0029	0.0085	3
	14	0.0000	0.0000	0.0000	0.0000	0.0000	0.0001	0.0003	0.0004	0.0005	0.0018	2
	15	0.0000	0.0000	0.0000	0.0000	0.0000	0.0000	0.0000	0.0000	0.0001	0.0002	1
	16	0.0000	0.0000	0.0000	0.0000	0.0000	0.0000	0.0000	0.0000	0.0000	0.0000	0
		5/7 0.7143	7/10 0.7000	2/3 0.6667	13/20 0.6500	5/8 0.6250	3/5 0.6000	4/7 0.5714	5/9 0.5556	11/20 0.5500	1/2 0.5000	r
		\multicolumn{10}{c}{π}	

TABLE IV(A) CENTRAL AREA UNDER THE STANDARD NORMAL DISTRIBUTION

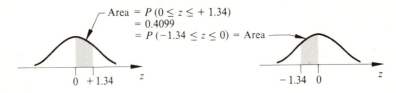

Area $= P(0 \le z \le +1.34)$
$= 0.4099$
$= P(-1.34 \le z \le 0) =$ Area

z To First Decimal	Second Decimal									
	0.00	0.01	0.02	0.03	0.04	0.05	0.06	0.07	0.08	0.09
0.0	0.0000	0.0040	0.0080	0.0120	0.0160	0.0199	0.0239	0.0279	0.0319	0.0359
0.1	0.0398	0.0438	0.0478	0.0517	0.0557	0.0596	0.0636	0.0675	0.0714	0.0753
0.2	0.0793	0.0832	0.0871	0.0910	0.0948	0.0987	0.1026	0.1064	0.1103	0.1141
0.3	0.1179	0.1217	0.1255	0.1293	0.1331	0.1368	0.1406	0.1443	0.1480	0.1517
0.4	0.1554	0.1591	0.1628	0.1664	0.1700	0.1736	0.1772	0.1808	0.1844	0.1879
0.5	0.1915	0.1950	0.1985	0.2019	0.2054	0.2088	0.2123	0.2157	0.2190	0.2224
0.6	0.2257	0.2291	0.2324	0.2357	0.2389	0.2422	0.2454	0.2486	0.2517	0.2549
0.7	0.2580	0.2611	0.2642	0.2673	0.2703	0.2734	0.2764	0.2794	0.2823	0.2852
0.8	0.2881	0.2910	0.2939	0.2967	0.2995	0.3023	0.3051	0.3078	0.3106	0.3133
0.9	0.3159	0.3186	0.3212	0.3238	0.3264	0.3289	0.3315	0.3340	0.3365	0.3389
1.0	0.3413	0.3438	0.3461	0.3485	0.3508	0.3531	0.3554	0.3577	0.3599	0.3621
1.1	0.3643	0.3665	0.3686	0.3708	0.3729	0.3749	0.3770	0.3790	0.3810	0.3830
1.2	0.3849	0.3869	0.3888	0.3907	0.3925	0.3944	0.3962	0.3980	0.3997	0.4015
1.3	0.4032	0.4049	0.4066	0.4082	0.4099	0.4115	0.4131	0.4147	0.4162	0.4177
1.4	0.4192	0.4207	0.4222	0.4236	0.4251	0.4265	0.4279	0.4292	0.4306	0.4319
1.5	0.4332	0.4345	0.4357	0.4370	0.4382	0.4394	0.4406	0.4418	0.4429	0.4441
1.6	0.4452	0.4463	0.4474	0.4484	0.4495	0.4505	0.4515	0.4525	0.4535	0.4545
1.7	0.4554	0.4564	0.4573	0.4582	0.4591	0.4599	0.4608	0.4616	0.4625	0.4633
1.8	0.4641	0.4649	0.4656	0.4664	0.4671	0.4678	0.4686	0.4693	0.4699	0.4706
1.9	0.4713	0.4719	0.4726	0.4732	0.4738	0.4744	0.4750	0.4756	0.4761	0.4767
2.0	0.4772	0.4778	0.4783	0.4788	0.4793	0.4798	0.4803	0.4808	0.4812	0.4817
2.1	0.4821	0.4826	0.4830	0.4834	0.4838	0.4842	0.4846	0.4850	0.4854	0.4857
2.2	0.4861	0.4864	0.4868	0.4871	0.4875	0.4878	0.4881	0.4884	0.4887	0.4890
2.3	0.4893	0.4896	0.4898	0.4901	0.4904	0.4906	0.4909	0.4911	0.4913	0.4916
2.4	0.4918	0.4920	0.4922	0.4925	0.4927	0.4929	0.4931	0.4932	0.4934	0.4936
2.5	0.4938	0.4940	0.4941	0.4943	0.4945	0.4946	0.4948	0.4949	0.4951	0.4952
2.6	0.4953	0.4955	0.4956	0.4957	0.4959	0.4960	0.4961	0.4962	0.4963	0.4964
2.7	0.4965	0.4966	0.4967	0.4968	0.4969	0.4970	0.4971	0.4972	0.4973	0.4974
2.8	0.4974	0.4975	0.4976	0.4977	0.4977	0.4978	0.4979	0.4979	0.4980	0.4981
2.9	0.4981	0.4982	0.4982	0.4983	0.4984	0.4984	0.4985	0.4985	0.4986	0.4986
3.0	0.4987	0.4987	0.4987	0.4988	0.4988	0.4989	0.4989	0.4989	0.4990	0.4990

Tables IV(A) and IV(B) are adapted from Jerome Bracken and Arthur Schleifer, Jr., *Tables for Normal Sampling with Unknown Variance*, Table G, p. 102, with permission of Division of Research, Harvard Business School, Boston, Mass.

TABLE IV(B)   TAIL AREA UNDER THE STANDARD NORMAL DISTRIBUTION

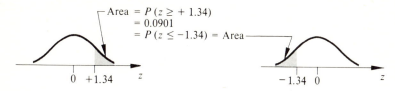

$$\text{Area} = P(z \geq +1.34)$$
$$= 0.0901$$
$$= P(z \leq -1.34) = \text{Area}$$

z To First Decimal	Second Decimal									
	0.00	0.01	0.02	0.03	0.04	0.05	0.06	0.07	0.08	0.09
0.0	0.5000	0.4960	0.4920	0.4880	0.4840	0.4801	0.4761	0.4721	0.4681	0.4641
0.1	0.4602	0.4562	0.4522	0.4483	0.4443	0.4404	0.4364	0.4325	0.4286	0.4247
0.2	0.4207	0.4168	0.4129	0.4090	0.4052	0.4013	0.3974	0.3936	0.3897	0.3859
0.3	0.3821	0.3783	0.3745	0.3707	0.3669	0.3632	0.3594	0.3557	0.3520	0.3483
0.4	0.3446	0.3409	0.3372	0.3336	0.3300	0.3264	0.3228	0.3192	0.3156	0.3121
0.5	0.3085	0.3050	0.3015	0.2981	0.2946	0.2912	0.2877	0.2843	0.2810	0.2776
0.6	0.2743	0.2709	0.2676	0.2643	0.2611	0.2578	0.2546	0.2514	0.2483	0.2451
0.7	0.2420	0.2389	0.2358	0.2327	0.2296	0.2266	0.2236	0.2206	0.2177	0.2148
0.8	0.2119	0.2090	0.2061	0.2033	0.2005	0.1977	0.1949	0.1922	0.1894	0.1867
0.9	0.1841	0.1814	0.1788	0.1762	0.1736	0.1711	0.1685	0.1660	0.1635	0.1611
1.0	0.1587	0.1562	0.1539	0.1515	0.1492	0.1469	0.1446	0.1423	0.1401	0.1379
1.1	0.1357	0.1335	0.1314	0.1292	0.1271	0.1251	0.1230	0.1210	0.1190	0.1170
1.2	0.1151	0.1131	0.1112	0.1093	0.1075	0.1056	0.1038	0.1020	0.1003	0.0985
1.3	0.0968	0.0951	0.0934	0.0918	0.0901	0.0885	0.0869	0.0853	0.0838	0.0823
1.4	0.0808	0.0793	0.0778	0.0764	0.0749	0.0735	0.0721	0.0708	0.0694	0.0681
1.5	0.0668	0.0655	0.0643	0.0630	0.0618	0.0606	0.0594	0.0582	0.0571	0.0559
1.6	0.0548	0.0537	0.0526	0.0516	0.0505	0.0495	0.0485	0.0475	0.0465	0.0455
1.7	0.0446	0.0436	0.0427	0.0418	0.0409	0.0401	0.0392	0.0384	0.0375	0.0367
1.8	0.0359	0.0351	0.0344	0.0336	0.0329	0.0322	0.0314	0.0307	0.0301	0.0294
1.9	0.0287	0.0281	0.0274	0.0268	0.0262	0.0256	0.0250	0.0244	0.0239	0.0233
2.0	0.0228	0.0222	0.0217	0.0212	0.0207	0.0202	0.0197	0.0192	0.0188	0.0183
2.1	0.0179	0.0174	0.0170	0.0166	0.0162	0.0158	0.0154	0.0150	0.0146	0.0143
2.2	0.0139	0.0136	0.0132	0.0129	0.0125	0.0122	0.0119	0.0116	0.0113	0.0110
2.3	0.0107	0.0104	0.0102	0.0099	0.0096	0.0094	0.0091	0.0089	0.0087	0.0084
2.4	0.0082	0.0080	0.0078	0.0075	0.0073	0.0071	0.0069	0.0068	0.0066	0.0064
2.5	0.0062	0.0060	0.0059	0.0057	0.0055	0.0054	0.0052	0.0051	0.0049	0.0048
2.6	0.0047	0.0045	0.0044	0.0043	0.0041	0.0040	0.0039	0.0038	0.0037	0.0036
2.7	0.0035	0.0034	0.0033	0.0032	0.0031	0.0030	0.0029	0.0028	0.0027	0.0026
2.8	0.0026	0.0025	0.0024	0.0023	0.0023	0.0022	0.0021	0.0021	0.0020	0.0019
2.9	0.0019	0.0018	0.0018	0.0017	0.0016	0.0016	0.0015	0.0015	0.0014	0.0014
3.0	0.0013	0.0013	0.0013	0.0012	0.0012	0.0011	0.0011	0.0011	0.0010	0.0010

TABLE V SELECTED VALUES OF THE t DISTRIBUTION

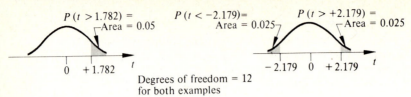

$P(t > 1.782) =$ Area $= 0.05$

$P(t < -2.179) =$ Area $= 0.025$

$P(t > +2.179) =$ Area $= 0.025$

Degrees of freedom $= 12$
for both examples

Degrees of Freedom	Area in One Tail					
	0.10	0.05	0.025	0.01	0.005	0.0005
	Area in Both Tails					
	0.20	0.10	0.05	0.02	0.01	0.001
1	3.078	6.314	12.706	31.821	63.657	636.619
2	1.886	2.920	4.303	6.975	9.925	31.598
3	1.638	2.353	3.182	4.541	5.841	12.941
4	1.533	2.132	2.776	3.747	4.604	8.610
5	1.476	2.015	2.571	3.365	4.032	6.859
6	1.440	1.943	2.447	3.143	3.707	5.959
7	1.415	1.895	2.365	2.998	3.499	5.405
8	1.397	1.860	2.306	2.896	3.355	5.041
9	1.383	1.833	2.262	2.821	3.250	4.781
10	1.372	1.812	2.228	2.764	3.169	4.587
11	1.363	1.796	2.201	2.718	3.106	4.437
12	1.356	1.782	2.179	2.681	3.055	4.318
13	1.350	1.771	2.160	2.650	3.012	4.221
14	1.345	1.761	2.145	2.624	2.977	4.140
15	1.341	1.753	2.131	2.602	2.947	4.073
16	1.337	1.746	2.120	2.583	2.921	4.015
17	1.333	1.740	2.110	2.567	2.898	3.965
18	1.330	1.734	2.101	2.552	2.878	3.922
19	1.328	1.729	2.093	2.539	2.861	3.883
20	1.325	1.725	2.086	2.528	2.845	3.850
21	1.323	1.721	2.080	2.518	2.831	3.819
22	1.321	1.717	2.074	2.508	2.819	3.792
23	1.319	1.714	2.069	2.500	2.807	3.767
24	1.318	1.711	2.064	2.492	2.797	3.745
25	1.316	1.708	2.060	2.485	2.787	3.725
26	1.315	1.706	2.056	2.479	2.779	3.707
27	1.314	1.703	2.052	2.473	2.771	3.690
28	1.313	1.701	2.048	2.467	2.763	3.674
29	1.311	1.699	2.045	2.462	2.756	3.659
30	1.310	1.697	2.042	2.457	2.750	3.646
40	1.303	1.684	2.021	2.423	2.704	3.551
60	1.296	1.671	2.000	2.390	2.660	3.460
120	1.289	1.658	1.980	2.358	2.617	3.373
∞	1.282	1.645	1.960	2.326	2.576	3.291

Adapted from E. S. Pearson and H. O. Hartley (eds.), *Biometrika Tables for Statisticians*, vol. I, 3d ed., 1966, with permission of Biometrika Trustees.

TABLE VI 95 PERCENT CONFIDENCE LIMITS FOR THE POPULATION PROPORTION

Example: In a random sample of 20 observations, if there are 6 successes, then $n = 20$ and $\bar{p} = \frac{6}{20} = 0.30$. Reading up from $\bar{p} = 0.30$, for $n = 20$, we find that $P(0.11 < \pi < 0.55) = 0.95$.

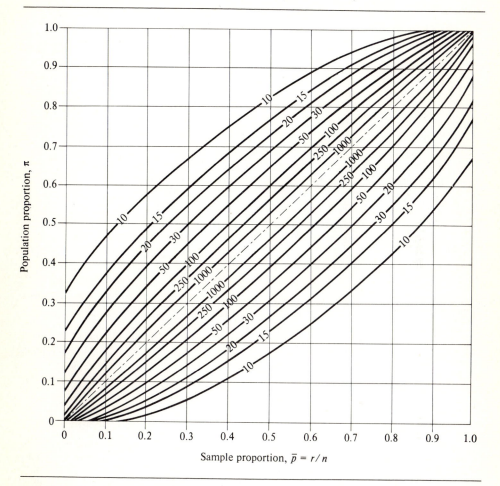

TABLE VII SELECTED VALUES FOR THE CHI-SQUARE DISTRIBUTION

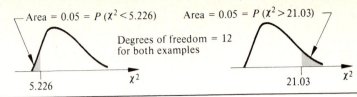

Area = 0.05 = $P(\chi^2 < 5.226)$ Area = 0.05 = $P(\chi^2 > 21.03)$

Degrees of freedom = 12 for both examples

5.226 χ^2 21.03 χ^2

Degrees of Freedom	Area or probability to the left of χ^2 value							
	0.010	0.025	0.050	0.100	0.900	0.950	0.975	0.990
1	0.000157	0.000982	0.00393	0.0158	2.706	3.841	5.024	6.635
2	0.0201	0.0506	0.1026	0.2107	4.605	5.991	7.378	9.210
3	0.1148	0.2158	0.3518	0.5844	6.251	7.815	9.348	11.34
4	0.2971	0.4844	0.7107	1.065	7.779	9.488	11.14	13.28
5	0.5543	0.8312	1.145	1.161	9.236	11.07	12.83	15.09
6	0.8721	1.237	1.635	2.204	10.64	12.59	14.45	16.81
7	1.239	1.690	2.167	2.833	12.02	14.07	16.01	18.48
8	1.646	2.180	2.733	3.490	13.36	15.51	17.53	20.09
9	2.088	2.700	3.325	4.168	14.68	16.92	19.02	21.67
10	2.558	3.247	3.940	4.865	15.99	18.31	20.48	23.21
11	3.053	3.816	4.575	5.578	17.28	19.68	21.92	24.72
12	3.571	4.404	5.226	6.304	18.55	21.03	23.34	26.22
13	4.107	5.009	5.892	7.042	19.81	22.36	24.74	27.69
14	4.660	5.629	6.571	7.790	21.06	23.68	26.12	29.14
15	5.229	6.262	7.261	8.547	22.31	25.00	27.49	30.58
16	5.812	6.910	7.962	9.312	23.54	26.30	28.84	32.00
17	6.408	7.564	8.672	10.08	24.77	27.59	30.19	33.41
18	7.015	8.231	9.390	10.86	25.99	28.87	31.53	34.80
19	7.633	8.906	10.12	11.65	27.20	30.14	32.85	36.19
20	8.260	9.591	10.85	12.44	28.41	31.41	34.17	37.57
21	8.897	10.28	11.59	13.24	29.62	32.67	35.48	38.93
22	9.542	10.98	12.34	14.04	30.81	33.92	36.78	40.29
23	10.20	11.69	13.03	14.85	32.01	35.17	38.08	41.64
24	10.86	12.40	13.85	15.66	33.20	36.42	39.36	42.98
25	11.52	13.12	14.61	16.47	34.38	37.65	40.65	44.31
26	12.20	13.84	15.38	17.29	35.56	38.88	41.92	45.64
27	12.88	14.57	16.15	18.11	36.74	40.11	43.19	46.96
28	13.56	15.31	16.93	18.94	37.92	41.34	44.46	48.28
29	14.26	16.05	17.71	19.77	39.09	42.56	45.72	49.59
30	14.95	16.79	18.49	20.60	40.26	43.77	46.98	50.89
40	22.16	24.43	26.51	29.05	51.80	55.76	59.34	63.69
60	37.48	40.48	43.19	46.46	74.40	79.08	83.30	88.38
80	53.54	57.15	60.39	64.28	96.58	101.9	106.6	112.3
100	70.06	74.22	77.93	82.36	118.5	124.3	129.6	135.8
120	96.92	91.57	95.70	100.6	140.2	146.6	152.2	159.0
	0.010	0.025	0.050	0.100	0.100	0.050	0.025	0.010
		Area in left tail				Area in right tail		

H. Leon Harter, *New Tables of the Incomplete Gamma-Function Ratio and of Percentage Points of the Chi-Square and Beta Distributions*, Aerospace Research Laboratories, Office of Aerospace Research, USAF, 1964.

TABLE VIII   SELECTED VALUES OF THE F-DISTRIBUTION

The boldface table values are for $P(F > F_0) = 0.05$; the other values are for $P(F > F_0) = 0.01$, given f_1 degrees of freedom in the numerator and f_2 degrees of freedom in the denominator of the F ratio. For example, if $f_1 = 5$ and $f_2 = 8$, then $P(F > 3.69) = 0.05$ and $P(F > 6.63) = 0.01$.

f_1 Degrees of Freedom (for Numerator)

f_2	1	2	3	4	5	6	8	10	12	16	20	30	40	50	100	f_2
1	**161** 4,052	**200** 4,999	**216** 5,403	**225** 5,625	**230** 5,764	**234** 5,859	**239** 5,981	**242** 6,056	**244** 6,106	**246** 6,169	**248** 6,208	**250** 6,258	**251** 6,286	**252** 6,302	**253** 6,334	1
2	**18.51** 98.49	**19.00** 99.00	**19.16** 99.17	**19.25** 99.25	**19.30** 99.30	**19.33** 99.33	**19.37** 99.36	**19.39** 99.40	**19.41** 99.42	**19.43** 99.44	**19.44** 99.45	**19.46** 99.47	**19.47** 99.48	**19.47** 99.48	**19.49** 99.49	2
3	**10.13** 34.12	**9.55** 30.82	**9.28** 29.46	**9.12** 28.71	**9.01** 28.24	**8.94** 27.91	**8.84** 27.49	**8.78** 27.23	**8.74** 27.05	**8.69** 26.83	**8.66** 26.69	**8.62** 26.50	**8.60** 26.41	**8.58** 26.35	**8.56** 26.23	3
4	**7.71** 24.20	**6.94** 18.00	**6.59** 16.69	**6.39** 15.98	**6.26** 15.52	**6.16** 15.21	**6.04** 14.80	**5.96** 14.54	**5.91** 14.37	**5.84** 14.15	**5.80** 14.02	**5.74** 13.83	**5.71** 13.74	**5.70** 13.69	**5.66** 13.57	4
5	**6.61** 16.26	**5.79** 13.27	**5.41** 12.06	**5.19** 11.39	**5.05** 10.97	**4.95** 10.67	**4.82** 10.27	**4.74** 10.05	**4.68** 9.89	**4.60** 9.68	**4.56** 9.55	**4.50** 9.38	**4.46** 9.29	**4.44** 9.24	**4.40** 9.13	5
6	**5.99** 13.74	**5.14** 10.92	**4.76** 9.78	**4.53** 9.15	**4.39** 8.75	**4.28** 8.47	**4.15** 8.10	**4.06** 7.87	**4.00** 7.72	**3.92** 7.52	**3.87** 7.39	**3.81** 7.23	**3.77** 7.14	**3.75** 7.09	**3.71** 6.99	6
7	**5.59** 12.25	**4.74** 9.55	**4.35** 8.45	**4.12** 7.85	**3.97** 7.46	**3.87** 7.19	**3.73** 6.84	**3.63** 6.62	**3.57** 6.47	**3.49** 6.27	**3.44** 6.15	**3.38** 5.98	**3.34** 5.85	**3.32** 5.85	**3.28** 5.75	7
8	**5.32** 11.26	**4.46** 8.65	**4.07** 7.59	**3.84** 7.01	**3.69** 6.63	**3.58** 6.37	**3.44** 6.03	**3.34** 5.82	**3.28** 5.67	**3.20** 5.48	**3.15** 5.36	**3.08** 5.20	**3.05** 5.11	**3.03** 5.06	**2.98** 4.96	8
9	**5.12** 10.56	**4.26** 8.02	**3.86** 6.99	**3.63** 6.42	**3.48** 6.06	**3.37** 5.80	**3.23** 5.47	**3.13** 5.26	**3.07** 5.11	**2.98** 4.92	**2.93** 4.80	**2.86** 4.64	**2.82** 4.56	**2.80** 4.51	**2.76** 4.41	9
10	**4.96** 10.04	**4.10** 7.56	**3.71** 6.55	**3.48** 5.99	**3.33** 5.64	**3.22** 5.39	**3.07** 5.06	**2.97** 4.85	**2.91** 4.71	**2.82** 4.52	**2.77** 4.41	**2.70** 4.25	**2.67** 4.17	**2.64** 4.12	**2.59** 4.01	10
11	**4.84** 9.65	**3.98** 7.20	**3.59** 6.22	**3.36** 5.67	**3.20** 5.32	**3.09** 5.07	**2.95** 4.74	**2.86** 4.54	**2.79** 4.40	**2.70** 4.21	**2.65** 4.10	**2.57** 3.94	**2.53** 3.86	**2.50** 3.80	**2.45** 3.70	11

TABLE VIII SELECTED VALUES OF THE F-DISTRIBUTION—continued

The boldface table values are for $P(F > F_0) = 0.05$; the other values are for $P(F > F_0) = 0.01$, given f_1 degrees of freedom in the numerator and f_2 degrees of freedom in the denominator of the F ratio. For example, if $f_1 = 5$ and $f_2 = 8$, then $P(F > 3.69) = 0.05$ and $P(F > 6.63) = 0.01$.

| f_2 | \multicolumn{15}{c}{f_1 Degrees of Freedom (for Numerator)} | f_2 |

f_2	1	2	3	4	5	6	8	10	12	16	20	30	40	50	100	f_2
12	**4.75** 9.33	**3.88** 6.93	**3.49** 5.95	**3.26** 5.41	**3.11** 5.06	**3.00** 4.82	**2.85** 4.50	**2.76** 4.30	**2.69** 4.16	**2.60** 3.98	**2.54** 3.86	**2.46** 3.70	**2.42** 3.61	**2.40** 3.56	**2.35** 3.46	12
13	**4.67** 9.07	**3.80** 6.70	**3.41** 5.74	**3.18** 5.20	**3.02** 4.86	**2.92** 4.62	**2.77** 4.30	**2.67** 4.10	**2.60** 3.96	**2.51** 3.78	**2.46** 3.67	**2.38** 3.51	**2.34** 3.42	**2.32** 3.37	**2.26** 3.27	13
14	**4.60** 8.86	**3.74** 6.51	**3.34** 5.56	**3.11** 5.03	**2.96** 4.69	**2.85** 4.46	**2.70** 4.14	**2.60** 3.94	**2.53** 3.80	**2.44** 3.62	**2.39** 3.51	**2.31** 3.34	**2.27** 3.26	**2.24** 3.21	**2.19** 3.11	14
15	**4.54** 8.68	**3.68** 6.36	**3.29** 5.42	**3.06** 4.89	**2.90** 4.56	**2.79** 4.32	**2.64** 4.00	**2.55** 3.80	**2.48** 3.67	**2.39** 3.48	**2.33** 3.36	**2.25** 3.20	**2.21** 3.12	**2.18** 3.07	**2.12** 2.97	15
16	**4.49** 8.53	**3.63** 6.23	**3.24** 5.29	**3.01** 4.77	**2.85** 4.44	**2.74** 4.20	**2.59** 3.89	**2.49** 3.69	**2.42** 3.55	**2.33** 3.37	**2.28** 3.25	**2.20** 3.10	**2.16** 3.01	**2.13** 2.96	**2.07** 2.86	16
17	**4.45** 8.40	**3.59** 6.11	**3.20** 5.18	**2.96** 4.67	**2.81** 4.34	**2.70** 4.10	**2.55** 3.79	**2.45** 3.59	**2.38** 3.45	**2.29** 3.27	**2.23** 3.16	**2.15** 3.00	**2.11** 2.92	**2.08** 2.86	**2.02** 2.76	17
18	**4.41** 8.28	**3.55** 6.01	**3.16** 5.09	**2.93** 4.58	**2.77** 4.25	**2.66** 4.01	**2.51** 3.71	**2.41** 3.51	**2.34** 3.37	**2.25** 3.19	**2.19** 3.07	**2.11** 2.91	**2.07** 2.83	**2.04** 2.78	**1.98** 2.68	18
19	**4.38** 8.18	**3.52** 5.93	**3.13** 5.01	**2.90** 4.50	**2.74** 4.17	**2.63** 3.94	**2.48** 3.63	**2.38** 3.43	**2.31** 3.30	**2.21** 3.12	**2.15** 3.00	**2.07** 2.84	**2.02** 2.76	**2.00** 2.70	**1.94** 2.60	19
20	**4.35** 8.10	**3.49** 5.85	**3.10** 4.94	**2.87** 4.49	**2.71** 4.10	**2.60** 3.87	**2.45** 3.56	**2.35** 3.37	**2.28** 3.23	**2.18** 3.05	**2.12** 2.94	**2.04** 2.77	**1.99** 2.69	**1.96** 2.63	**1.90** 2.53	20
25	**4.24** 7.77	**3.38** 5.57	**2.99** 4.68	**2.76** 4.18	**2.60** 3.86	**2.49** 3.63	**2.34** 3.32	**2.24** 3.13	**2.16** 2.99	**2.06** 2.81	**2.00** 2.70	**1.92** 2.54	**1.87** 2.45	**1.84** 2.40	**1.77** 2.29	25
30	**4.17** 7.56	**3.32** 5.39	**2.92** 4.51	**2.69** 4.02	**2.53** 3.70	**2.42** 3.47	**2.27** 3.17	**2.16** 2.98	**2.09** 2.84	**1.99** 2.66	**1.93** 2.55	**1.84** 2.38	**1.79** 2.29	**1.76** 2.24	**1.69** 2.13	30

TABLE VIII SELECTED VALUES OF THE F-DISTRIBUTION—continued

The boldface table values are for $P(F > F_0) = 0.05$; the other values are for $P(F > F_0) = 0.01$, given f_1 degrees of freedom in the numerator and f_2 degrees of freedom in the denominator of the F ratio. For example, if $f_1 = 5$ and $f_2 = 8$, then $P(F > 3.69) = 0.05$ and $P(F > 6.63) = 0.01$.

f_2	f_1 Degrees of Freedom (for Numerator)															f_2
	1	2	3	4	5	6	8	10	12	16	20	30	40	50	100	
40	**4.08**	**3.23**	**2.84**	**2.61**	**2.45**	**2.34**	**2.18**	**2.07**	**2.00**	**1.90**	**1.84**	**1.74**	**1.69**	**1.66**	**1.59**	40
	7.31	5.18	4.31	3.83	3.51	3.29	2.99	2.80	2.66	2.49	2.37	2.20	2.11	2.05	1.94	
50	**4.03**	**3.18**	**2.79**	**2.56**	**2.40**	**2.29**	**2.13**	**2.02**	**1.95**	**1.85**	**1.78**	**1.69**	**1.63**	**1.60**	**1.52**	50
	7.17	5.06	4.20	3.72	3.41	3.18	2.88	2.70	2.56	2.39	2.26	2.10	2.00	1.94	1.82	
60	**4.00**	**3.15**	**2.76**	**2.52**	**2.37**	**2.25**	**2.10**	**1.99**	**1.92**	**1.81**	**1.75**	**1.65**	**1.59**	**1.56**	**1.48**	60
	7.08	4.98	4.13	3.65	3.34	3.12	2.82	2.63	2.50	2.32	2.20	2.03	1.93	1.87	1.74	
80	**3.96**	**3.11**	**2.72**	**2.48**	**2.33**	**2.21**	**2.05**	**1.95**	**1.88**	**1.77**	**1.70**	**1.60**	**1.54**	**1.51**	**1.42**	80
	6.96	4.88	4.04	3.56	3.25	3.04	2.74	2.55	2.41	2.24	2.11	1.94	1.84	1.78	1.65	
100	**3.94**	**3.09**	**2.70**	**2.46**	**2.30**	**2.19**	**2.03**	**1.92**	**1.85**	**1.75**	**1.68**	**1.57**	**1.51**	**1.48**	**1.39**	100
	6.90	4.82	3.98	3.51	3.20	2.99	2.69	2.51	2.36	2.19	2.06	1.89	1.79	1.73	1.59	
150	**3.91**	**3.06**	**2.67**	**2.43**	**2.27**	**2.16**	**2.00**	**1.89**	**1.82**	**1.71**	**1.64**	**1.54**	**1.47**	**1.44**	**1.34**	150
	6.81	4.75	3.91	3.44	3.14	2.82	2.62	2.44	2.30	2.12	2.00	1.83	1.72	1.66	1.51	
200	**3.89**	**3.04**	**2.65**	**2.41**	**2.26**	**2.14**	**1.98**	**1.87**	**1.80**	**1.69**	**1.62**	**1.52**	**1.45**	**1.42**	**1.32**	200
	6.76	4.71	3.88	3.41	3.11	2.90	2.60	2.41	2.28	2.09	1.97	1.79	1.69	1.62	1.48	
400	**3.86**	**3.02**	**2.62**	**2.39**	**2.23**	**2.12**	**1.96**	**1.85**	**1.78**	**1.67**	**1.60**	**1.49**	**1.40**	**1.38**	**1.28**	400
	6.70	4.66	3.83	3.36	3.06	2.85	2.55	2.37	2.23	2.04	1.92	1.74	1.64	1.57	1.42	
1000	**3.85**	**3.00**	**2.61**	**2.38**	**2.22**	**2.10**	**1.95**	**1.84**	**1.76**	**1.65**	**1.58**	**1.47**	**1.41**	**1.36**	**1.26**	1000
	6.66	4.62	3.80	3.34	3.04	2.82	2.53	2.34	2.20	2.01	1.89	1.71	1.61	1.54	1.38	
∞	**3.84**	**2.99**	**2.60**	**2.37**	**2.21**	**2.09**	**1.94**	**1.83**	**1.75**	**1.64**	**1.57**	**1.46**	**1.40**	**1.35**	**1.24**	∞
	6.64	4.60	3.78	3.32	3.02	2.80	2.51	2.32	2.18	1.99	1.87	1.69	1.59	1.52	1.36	

Adapted from E. S. Pearson and H. O. Hartley (eds.), *Biometrika Tables for Statisticians*, vol. I, 3d ed., 1966, with permission of Biometrika Trustees.

TABLE IX CRITICAL VALUES OF R IN ONE-SAMPLE RUNS TEST

Table IX(L)   Lower Rejection Limits (R_L)

$P(R \le R_L \,|\, n_A, n_B) = 0.05$

n_B	\multicolumn{19}{c}{n_A}

n_B	2	3	4	5	6	7	8	9	10	11	12	13	14	15	16	17	18	19	20
2											2	2	2	2	2	2	2	2	2
3			2	2	2	2	2	2	2	2	2	2	2	3	3	3	3	3	3
4			2	2	2	3	3	3	3	3	3	3	3	3	4	4	4	4	4
5		2	2	3	3	3	3	3	4	4	4	4	4	4	4	4	5	5	5
6		2	2	3	3	3	3	4	4	4	4	5	5	5	5	5	5	6	6
7		2	2	3	3	3	4	4	5	5	5	5	5	6	6	6	6	6	6
8		2	3	3	3	4	4	5	5	5	6	6	6	6	6	7	7	7	7
9		2	3	3	4	4	5	5	5	6	6	6	7	7	7	7	8	8	8
10		2	3	3	4	5	5	5	6	6	7	7	7	7	8	8	8	8	9
11		2	3	4	4	5	5	6	6	7	7	7	8	8	8	9	9	9	9
12	2	2	3	4	4	5	6	6	7	7	7	8	8	8	9	9	9	10	10
13	2	2	3	4	5	5	6	6	7	7	8	8	9	9	9	10	10	10	10
14	2	2	3	4	5	5	6	7	7	8	8	9	9	9	10	10	10	11	11
15	2	3	3	4	5	6	6	7	7	8	8	9	9	10	10	11	11	11	12
16	2	3	4	4	5	6	6	7	8	8	9	9	10	10	11	11	11	12	12
17	2	3	4	4	5	6	7	7	8	9	9	10	10	11	11	11	12	12	13
18	2	3	4	5	5	6	7	8	8	9	9	10	10	11	11	12	12	13	13
19	2	3	4	5	6	6	7	8	8	9	10	10	11	11	12	12	13	13	13
20	2	3	4	5	6	6	7	8	9	9	10	10	11	12	12	13	13	13	14

Adapted from Frieda S. Swed and C. Eisenhart, "Tables for Testing Randomness of Grouping in a Sequence of Alternatives," *The Annals of Mathematical Statistics*, vol. 14, 1943, pp. 83–86, by permission of The Institute of Mathematical Statistics.

TABLE IX CRITICAL VALUES OF R IN ONE-SAMPLE RUNS TEST—continued

Table IX(U) Upper Rejection Limits (R_U)

$P(R \geq R_U \,|\, n_A, n_B) = 0.05$

n_B	2	3	4	5	6	7	8	9	10	11	12	13	14	15	16	17	18	19	20
2																			
3																			
4				9	9														
5			9	10	10	11	11												
6			9	10	11	12	12	13	13	13	13								
7				11	12	13	13	14	14	14	14	15	15	15					
8				11	12	13	14	14	15	15	16	16	16	16	17	17	17	17	17
9					13	14	14	15	16	16	16	17	17	18	18	18	18	18	18
10					13	14	15	16	16	17	17	18	18	18	19	19	19	20	20
11					13	14	15	16	17	17	18	19	19	19	20	20	20	21	21
12					13	14	16	16	17	18	19	19	20	20	21	21	21	22	22
13						15	16	17	18	19	19	20	20	21	21	22	22	23	23
14						15	16	17	18	19	20	20	21	22	22	23	23	23	24
15						15	16	18	18	19	20	21	22	22	23	23	24	24	25
16							17	18	19	20	21	21	22	23	23	24	25	25	25
17							17	18	19	20	21	22	23	23	24	25	25	26	26
18							17	18	19	20	21	22	23	24	25	25	26	26	27
19							17	18	20	21	22	23	23	24	25	26	26	27	27
20							17	18	20	21	22	23	24	25	25	26	27	27	28

TABLE X SELECTED VALUES FOR THE VON NEUMANN RATIO

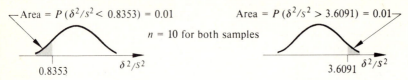

Area = $P(\delta^2/s^2 < 0.8353) = 0.01$ Area = $P(\delta^2/s^2 > 3.6091) = 0.01$

$n = 10$ for both samples

0.8353 δ^2/s^2 3.6091 δ^2/s^2

Number of Observations	Left Tail			Right Tail		
	0.05	0.01	0.001	0.05	0.01	0.001
4	1.0406	0.8341	0.7864	4.2927	4.4992	4.5469
5	1.0255	0.6724	0.5201	3.9745	4.3276	4.4799
6	1.0682	0.6738	0.4361	3.7318	4.1262	4.3639
7	1.0919	0.7163	0.4311	3.5748	3.9504	4.2356
8	1.1228	0.7575	0.4612	3.4486	3.8139	4.1102
9	1.1524	0.7974	0.4973	3.3476	3.7025	4.0027
10	1.1803	0.8353	0.5351	3.2642	3.6091	3.9093
11	1.2962	0.8706	0.5717	3.1938	3.5294	3.8283
12	1.2301	0.9033	0.6062	3.1335	3.4603	3.7574
13	1.2521	0.9336	0.6390	3.0812	3.3996	3.6944
14	1.2725	0.9618	0.6702	3.0352	3.3458	3.6375
15	1.2914	0.9880	0.6999	2.9943	3.2977	3.5858
16	1.3090	1.0124	0.7281	2.9577	3.2543	3.5386
17	1.3253	1.0352	0.7548	2.9247	3.2148	3.4952
18	1.3405	1.0566	0.7801	2.8948	3.1787	3.4552
19	1.3547	1.0766	0.8040	2.8675	3.1456	3.4182
20	1.3680	1.0954	0.8265	2.8425	3.1151	3.3840
21	1.3805	1.1131	0.8477	2.8195	3.0869	3.3523
22	1.3923	1.1298	0.8677	2.7982	3.0607	3.3228
23	1.4035	1.1456	0.8866	2.7784	3.0362	3.2953
24	1.4141	1.1606	0.9045	2.7599	3.0133	3.2695
25	1.4241	1.1748	0.9215	2.7426	2.9919	3.2452
26	1.4336	1.1883	0.9378	2.7264	2.9718	3.2222
27	1.4426	1.2012	0.9535	2.7112	2.9528	3.2003
28	1.4512	1.2135	0.9687	2.6969	2.9348	3.1794
29	1.4594	1.2252	0.9835	2.6834	2.9177	3.1594
30	1.4672	1.2363	0.9978	2.6707	2.9016	3.1402
31	1.4746	1.2469	1.0115	2.6587	2.8864	3.1219
32	1.4817	1.2570	1.0245	2.6473	2.8720	3.1046
33	1.4885	1.2667	1.0369	2.6365	2.8583	3.0882
34	1.4951	1.2761	1.0488	2.6262	2.8451	3.0725
35	1.5014	1.2852	1.0603	2.6163	2.8324	3.0574
36	1.5075	1.2940	1.0714	2.6068	2.8202	3.0429
37	1.5135	1.3025	1.0822	2.5977	2.8085	3.0289
38	1.5193	1.3108	1.0927	2.5889	2.7973	3.0154
39	1.5249	1.3188	1.1029	2.5804	2.7865	3.0024
40	1.5304	1.3266	1.1128	2.5722	2.7760	2.9898

TABLE X SELECTED VALUES FOR THE VON NEUMANN RATIO—continued

Number of Observations	Left Tail			Right Tail		
	0.05	0.01	0.001	0.05	0.01	0.001
41	1.5357	1.3342	1.1224	2.5643	2.7658	2.9776
42	1.5408	1.3415	1.1317	2.5567	2.7560	2.9658
43	1.5458	1.3486	1.1407	2.5494	2.7466	2.9545
44	1.5506	1.3554	1.1494	2.5424	2.7376	2.9436
45	1.5552	1.3620	1.1577	2.5357	2.7289	2.9332
46	1.5596	1.3684	1.1657	2.5293	2.7205	2.9232
47	1.5638	1.3745	1.1734	2.5232	2.7125	2.9136
48	1.5678	1.3802	1.1807	2.5173	2.7049	2.9044
49	1.5716	1.3856	1.1877	2.5117	2.6977	2.8956
50	1.5752	1.3907	1.1944	2.5064	2.6908	2.8872
51	1.5787	1.3957	1.2010	2.5013	2.6842	2.8790
52	1.5822	1.4007	1.2075	2.4963	2.6777	2.8709
53	1.5856	1.4057	1.2139	2.4914	2.6712	2.8630
54	1.5890	1.4107	1.2202	2.4866	2.6648	2.8553
55	1.5923	1.4156	1.2264	2.4819	2.6585	2.8477
56	1.5955	1.4203	1.2324	2.4773	2.6524	2.8403
57	1.5987	1.4249	1.2383	2.4728	2.6465	2.8331
58	1.6019	1.4294	1.2442	2.4684	2.6407	2.8260
59	1.6051	1.4339	1.2500	2.4640	2.6350	2.8190
60	1.6082	1.4384	1.2558	2.4596	2.6294	2.8120

Adapted from B. I. Hart, "Significance Levels for the Ratio of the Mean Square Successive Difference to the Variance," *The Annals of Mathematical Statistics*, vol. 13, 1942, p. 446, by permission of the Institute of Mathematical Statistics.

BIBLIOGRAPHY

CHAPTER 1

Measurement Scales

*Pfanzagl, J.: *Theory of Measurement*, John Wiley & Sons, Inc., New York, 1968.
*Roberts, Blaine, and David L. Schulze: *Modern Mathematics and Economic Analysis*, W. W. Norton & Company, New York, 1973, chapter 3.

CHAPTERS 2 AND 3

Chou, Ya-Lun: *Statistical Analysis*, Holt, Rinehart and Winston, New York, 1975, chapters 4 to 6.
Mendenhall, William: *Introduction to Probability and Statistics*, 2d ed., Wadsworth Publishing Company, Inc., Belmont, Calif., 1967.
Shook, Robert C., and Harold J. Highland: *Probability Models with Business Applications*, Richard D. Irwin, Inc., Homewood, Ill., 1969.
Tanur, Judith M. (ed.): *Statistics: A Guide to the Unknown*, Holden-Day, Inc., Publisher, San Francisco, Calif., 1972. *A fascinating book of readings!*

CHAPTER 4

Descriptive Statistics

Croxton, Frederick E., Dudley J. Cowden, and Sidney Klein: *Applied General Statistics*, Prentice-Hall, Inc., Englewood Cliffs, N.J., 1967.
Huff, Darrell: *How to Lie with Statistics*, W. W. Norton & Company, Inc., New York, 1954.

Index Numbers

Mudgett, Bruce D.: *Index Numbers*, John Wiley & Sons, Inc., New York, 1951.
Wonnacott, Thomas H., and Ronald J. Wonnacott: *Introductory Statistics for Business and Economics*, John Wiley & Sons, Inc., New York, 1972, chapter 23.

---

* Indicates a more advanced mathematical treatment.

CHAPTER 5

Daniel, Wayne W., and James C. Terrell: *Business Statistics*, Houghton Mifflin Company, Boston, 1975, chapter 5.
*Hoel, Paul G.: *Introduction to Mathematical Statistics*, 3d ed., John Wiley & Sons, Inc., New York, 1962, chapter 6.

CHAPTER 6

Harnett, Donald L., and James L. Murphy: *Introductory Statistical Analysis*, Addison-Wesley Publishing Company, Inc., Reading, Mass., 1975, chapter 7.
Huntsburger, David V., Patrick Billingsley, and D. James Croft: *Statistical Inference for Management and Economics*, Allyn and Bacon, Inc., Boston, 1975, chapters 7 and 9.
Summers, George W., and William S. Peters: *Basic Statistics for Business and Economics*, Wadsworth Publishing Company, Inc., Belmont, Calif., 1973, chapter 9.

CHAPTER 7

Hamburg, Morris: *Statistical Analysis for Decision Making*, Harcourt, Brace, & World, Inc., New York, 1970, chapter 7.
Hays, William L., and Robert L. Winkler, *Statistics*, vol. 1, Holt, Rinehart and Winston, Inc., New York, 1970, chapter 7.

CHAPTER 8

Hoel, Paul G., and Raymond J. Jessen: *Statistics for Business and Economics*, John Wiley & Sons, Inc., New York, 1971, chapters 10 and 13.
Yamane, Taro: *Statistics*, 3d ed., Harper & Row, Publishers, Incorporated, New York, 1973, chapters 21 to 22.

CHAPTER 9

*Siegel, Sidney: *Nonparametric Statistics*, McGraw-Hill Book Company, New York, 1956.

CHAPTER 10

*Beals, Ralph E.: *Statistics for Economists*, Rand McNally & Company, Chicago, 1972, chapters 9 and 10.

Boot, John C. G., and Edwin B. Cox: *Statistical Analysis for Managerial Decisions*, 2d ed., McGraw-Hill Book Company, New York, 1974, chapter 13.

Merrill, William C., and Karl A. Fox: *Introduction to Economic Statistics*, John Wiley & Sons, Inc., New York, 1970, chapter 9.

CHAPTER 11

*Kane, Edward J.: *Economic Statistics and Econometrics*, Harper & Row, Publishers, Incorporated, New York, 1968, chapters 11 through 14.

Lapin, Lawrence L.: *Statistics for Modern Business Decisions*, Harcourt Brace Jovanovich, New York, 1973, chapter 14.

*Neter, John, and William Wasserman: *Applied Linear Statistical Models*, Richard D. Irwin, Inc., Homewood, Ill., 1974, chapters 7 to 8.

CHAPTER 12

Clark, Charles T., and Lawrence L. Schkade: *Statistical Analysis for Administrative Decisions*, 2d ed., South-Western Publishing Company, Incorporated, Cincinnati, 1974, chapters 18 to 19.

Neter, John, William Wasserman, and G. A. Whitmore: *Fundamental Statistics for Business and Economics*, 4th ed., Allyn and Bacon, Inc., Boston, 1973, Unit IX.

CHAPTER 13

Brennan, Michael J.: *Preface to Econometrics*, 3d ed., South-Western Publishing Company, Incorporated, Cincinnati, 1973, chapter 23.

*Kelejian, Harry H., and Wallace E. Oates: *Introduction to Econometrics*, Harper & Row, Publishers, Incorporated, New York, 1974, chapter 7.

*Theil, Henri: *Applied Economic Forecasting*, North-Holland Publishing Company, Amsterdam, 1966.

CHAPTER 14

Jedamus, Paul, and Robert Frame: *Business Decision Theory*, McGraw-Hill Book Company, New York, 1969.

Thompson, Gerald E.: *Statistics for Decisions*, Little, Brown and Company, Boston, 1972.

INDEX